THE HANDY SCIENCE ANSWER BOOK

About the Authors

The Carnegie Library of Pittsburgh has answered hundreds of thousands of questions in the fields of science and technology. Established in 1902, The Carnegie Library of Pittsburgh was the first major library to devote a separate department to science and technology. Since then, it has patiently answered reference questions from patrons at a rate of more than 80,000 per year. The answers to many of those questions have been collected and compiled for *The Handy Science Answer Book*.

James Bobick helped compile *The Handy Science Answer Book* during his sixteen years as Head of the Science and Technology Department at the Carnegie Library of Pittsburgh, and he has been instrumental in every revision since it first published over twenty-five years ago. He previously taught at the University of Pittsburgh's School of Information Sciences, and with G. Lynn Berard of Carnegie Mellon University he wrote *Science and Technology Resources: A Guide for Information Professionals and Researchers*. He has master's degrees in both biology and library science. He lives in Pittsburgh.

Naomi E. Balaban, a reference librarian for more than twenty-five years at the Carnegie Library of Pittsburgh, has extensive experience in the areas of science and consumer health. She edited, with James Bobick, all the previous editions of *The Handy Science Answer Book* as well as *The Handy Anatomy Answer Book*, *The Handy Biology Answer Book*, and *The Handy Technology Answer Book*. She has a background in linguistics and a master's degree in library science. She resides in Pittsburgh.

THE HANDY SCIENCE ANSWER BOOK

FIFTH EDITION

Compiled by the Carnegie Library of Pittsburgh

Edited by James E. Bobick and Naomi E. Balaban

Detroit

Also from Visible Ink Press

The Handy Accounting Answer Book
by Amber K. Gray
ISBN: 978-1-57859-675-1

The Handy African American History Answer Book
by Jessie Carnie Smith
ISBN: 978-1-57859-452-8

The Handy American Government Answer Book: How Washington, Politics, and Elections Work
by Gina Misiroglu
ISBN: 978-1-57859-639-3

The Handy American History Answer Book
by David L. Hudson Jr.
ISBN: 978-1-57859-471-9

The Handy Anatomy Answer Book, 2nd edition
by Patricia Barnes-Svarney and Thomas E. Svarney
ISBN: 978-1-57859-542-6

The Handy Answer Book for Kids (and Parents), 2nd edition
by Gina Misiroglu
ISBN: 978-1-57859-219-7

The Handy Art History Answer Book
by Madelynn Dickerson
ISBN: 978-1-57859-417-7

The Handy Astronomy Answer Book, 3rd edition
by Charles Liu
ISBN: 978-1-57859-419-1

The Handy Bible Answer Book
by Jennifer R. Prince
ISBN: 978-1-57859-478-8

The Handy Biology Answer Book, 2nd edition
by Patricia Barnes Svarney and Thomas E. Svarney
ISBN: 978-1-57859-490-0

The Handy Boston Answer Book
by Samuel Willard Crompton
ISBN: 978-1-57859-593-8

The Handy California Answer Book
by Kevin Hile
ISBN: 978-1-57859-591-4

The Handy Chemistry Answer Book
by Ian C. Stewart and Justin P. Lamont
ISBN: 978-1-57859-374-3

The Handy Christianity Answer Book
by Steve Werner
ISBN: 978-1-57859-686-7

The Handy Civil War Answer Book
by Samuel Willard Crompton
ISBN: 978-1-57859-476-4

The Handy Communication Answer Book
By Lauren Sergy
ISBN: 978-1-57859-587-7

The Handy Diabetes Answer Book
by Patricia Barnes-Svarney and Thomas E. Svarney
ISBN: 978-1-57859-597-6

The Handy Dinosaur Answer Book, 2nd edition
by Patricia Barnes-Svarney and Thomas E. Svarney
ISBN: 978-1-57859-218-0

The Handy English Grammar Answer Book
by Christine A. Hult, Ph.D.
ISBN: 978-1-57859-520-4

The Handy Forensic Science Answer Book: Reading Clues at the Crime Scene, Crime Lab, and in Court
by Patricia Barnes-Svarney and Thomas E. Svarney
ISBN: 978-1-57859-621-8

The Handy Geography Answer Book, 3rd edition
by Paul A. Tucci
ISBN: 978-1-57859-576-1

The Handy Geology Answer Book
by Patricia Barnes-Svarney and Thomas E. Svarney
ISBN: 978-1-57859-156-5

The Handy History Answer Book, 3rd edition
by David L. Hudson, Jr., J.D.
ISBN: 978-1-57859-372-9

The Handy Hockey Answer Book
by Stan Fischler
ISBN: 978-1-57859-513-6

The Handy Investing Answer Book
by Paul A. Tucci
ISBN: 978-1-57859-486-3

The Handy Islam Answer Book
by John Renard, Ph.D.
ISBN: 978-1-57859-510-5

The Handy Law Answer Book
by David L. Hudson, Jr., J.D.
ISBN: 978-1-57859-217-3

The Handy Literature Answer Book: An Engaging Guide to Unraveling Symbols, Signs, and Meanings in Great Works
By Daniel S. Burt, Ph.D., and Deborah G. Felder
ISBN: 978-1-57859-635-5

The Handy Math Answer Book, 2nd edition
by Patricia Barnes-Svarney and Thomas E. Svarney
ISBN: 978-1-57859-373-6

The Handy Military History Answer Book
by Samuel Willard Crompton
ISBN: 978-1-57859-509-9

The Handy Mythology Answer Book
by David A. Leeming, Ph.D.
ISBN: 978-1-57859-475-7

The Handy New York City Answer Book
by Chris Barsanti
ISBN: 978-1-57859-586-0

The Handy Nutrition Answer Book
by Patricia Barnes-Svarney and Thomas E. Svarney
ISBN: 978-1-57859-484-9

The Handy Ocean Answer Book
by Patricia Barnes-Svarney and Thomas E. Svarney
ISBN: 978-1-57859-063-6

The Handy Personal Finance Answer Book
by Paul A. Tucci
ISBN: 978-1-57859-322-4

The Handy Philosophy Answer Book
by Naomi Zack, Ph.D.
ISBN: 978-1-57859-226-5

The Handy Physics Answer Book, 2nd edition
By Paul W. Zitzewitz, Ph.D.
ISBN: 978-1-57859-305-7

The Handy Presidents Answer Book, 2nd edition
by David L. Hudson
ISB N: 978-1-57859-317-0

The Handy Psychology Answer Book, 2nd edition
by Lisa J. Cohen, Ph.D.
ISBN: 978-1-57859-508-2

The Handy Religion Answer Book, 2nd edition
by John Renard, Ph.D.
ISBN: 978-1-57859-379-8

The Handy Science Answer Book, 5th edition
by The Carnegie Library of Pittsburgh
ISBN: 978-1-57859-691-1

The Handy State-by-State Answer Book: Faces, Places, and Famous Dates for All Fifty States
by Samuel Willard Crompton
ISBN: 978-1-57859-565-5

The Handy Supreme Court Answer Book
by David L Hudson, Jr.
ISBN: 978-1-57859-196-1

The Handy Technology Answer Book
by Naomi E. Balaban and James Bobick
ISBN: 978-1-57859-563-1

The Handy Texas Answer Book
by James L. Haley
ISBN: 978-1-57859-634-8

The Handy Weather Answer Book, 2nd edition
by Kevin S. Hile
ISBN: 978-1-57859-221-0

The Handy Wisconsin Answer Book
by Terri Schlichenmeyer and Mark Meier
ISBN: 978-1-57859-661-4

PLEASE VISIT THE "HANDY ANSWERS" SERIES WEBSITE AT WWW.HANDYANSWERS.COM.

THE HANDY SCIENCE ANSWER BOOK

Visible Ink Press®
43311 Joy Rd., #414
Canton, MI 48187-2075
Visible Ink Press is a registered trademark of Visible Ink Press LLC.

Most Visible Ink Press books are available at special quantity discounts when purchased in bulk by corporations, organizations, or groups. Customized printings, special imprints, messages, and excerpts can be produced to meet your needs. For more information, contact Special Markets Director, Visible Ink Press, www.visibleink.com, or 734-667-3211.

Managing Editor: Kevin S. Hile
Art Director: Mary Claire Krzewinski
Typesetting: Marco Divita
Proofreaders: Larry Baker and Shoshana Hurwitz
Indexer: Shoshana Hurwitz
Cover images: Shutterstock.

Library of Congress Cataloging-in-Publication Data

Names: Bobick, James E. | Balaban, Naomi E. | Carnegie Library of Pittsburgh.

Title : The handy science answer book / by The Carnegie Library of Pittsburgh ; [edited by James Bobick, Naomi E. Balaban].

Description : 5th edition. | Canton, MI : Visible Ink Press, [2020] | Identifiers: LCCN 2019014765 (print) | LCCN 2019017609 (ebook) | ISBN 9781578597024 (epub) | ISBN 9781578596911 (pbk. : alk. paper)

Subjects : LCSH: Science–Miscellanea. Classification: LCC Q173 (ebook) | LCC Q173 .H24 2020 (print) | DDC 500–dc23

LC record available at https://lccn.loc.gov/2019014765

10 9 8 7 6 5 4 3 2 1

Printed in the United States of America

Table of Contents

Acknowledgments

Jim and Naomi dedicate this edition to Sandi and Carey: "We owe you a lot!" In addition, the authors thank their families for the ongoing interest, encouragement, support, and especially their understanding while this edition was being revised.

Photo Sources

Salix Alba: p. 27.
Andrevruas (Wikicommons): p. 214.
Toni Barros: p. 320.
J. Brew: p. 22.
Carafe (Wikicommons): p. 532.
Carny (Wikicommons): p. 328.
Daderot (Wikicommons): p. 105.
Valentin de Bruyn / Coton: p. 235.
Stephen C. Dickson: p. 87.
Efbrazil (Wikicommons): p. 210.
Electrical Review: p. 79.
ESO/L. Calçada and Nick Risinger (skysurvey.org): p. 139.
Executive Office of the President of the United States: p. 466.
Gawthorpe Hall: p. 91.
Georesearch Volcanedo Germany: p. 227.
German Federal Archives: p. 56.
Mike Goren: p. 365.
Hannes Grobe/AWI: p. 175.
Alex Handy: p. 536.
Fred Hartsook: p. 333.
Kevin Hile: p. 40 (left).
Houghton Library: p. 4.
Kjoonlee (Wikicommons): p. 28.
Kowloonese (Wikicommons): p. 198.
Robert Krewaldt: p. 71.
Library of Congress: pp. 90, 555.
Life magazine: p. 19.
Russ London: p. 315.
Walter Marius: p. 319.
Mercury13 (Wikicommons): p. 537.
Paul Nadar: p. 72.
NASA: pp. 158, 160 (right), 163 (left), 178, 231.
NASA Glenn Research Center (NASA-GRC): p. 160 (left).
NASA/JPL-Caltech: p. 148.
NASA/JPL/DLR: p. 146.
National Institutes of Health: p. 260 (left).
National Maritime Museum: p. 118 (left).
National Museum of Fine Arts, Sweden: p. 263.
National Oceanic and Atmospheric Administration: pp. 248, 375, 499.
National Science Foundation: p. 233.
New York Public Library: p. 260 (right).
Old Farmer's Almanac: p. 223.
Sergey Prokudin-Gorsky: p. 150.
Dustin M. Ramsey: p. 482.
Shutterstock: pp. 2, 6, 9, 34, 36, 42, 43, 45, 47, 48, 58, 61, 63, 66, 69, 76, 82, 89, 93, 101, 107, 121, 124, 126, 128, 132, 134, 137, 143, 153, 155, 163 (right), 166, 179, 184, 187, 192, 193, 196, 201, 202, 205, 208, 219, 220, 229, 238, 242, 244, 247, 251, 254, 255, 267, 270, 274, 276, 279, 283, 285, 289, 291, 293, 301, 308, 309, 313, 337, 339, 341, 344, 345, 349, 351, 354, 358, 359, 367, 369, 371, 378, 381, 384, 387, 389, 393, 394, 396, 399, 401, 404, 406, 409, 411, 413, 419, 422, 424, 428, 432, 435, 438, 442, 445, 446, 448, 452, 454, 457, 459, 469, 471, 472, 476, 481, 484, 487, 492, 494, 497, 504, 506, 509, 511, 513, 515, 517, 521, 523, 540, 542, 545, 547, 551, 553, 556, 561, 563.
Smithsonian Institution Archives Collection: p. 171.
Speeding Cars (Wikicommons): p. 19 (top).
Amber Stuver: p. 74.
Tamiko Thiel: p. 80.
Yunuskhuja Tuygunkhujaev: p. 38.
Tyomitch (Wikicommons): p. 538.
U.S. Air Force: p. 68.
U.S. Department of the Treasury: p. 218.
U.S. Fish and Wildlife Service: p. 490.
U.S. Geological Service: p. 182.
U.S. Navy: p. 500.
Stefan Zachow: p. 12.
Public domain: pp. 16, 31, 40 (right), 49, 78, 111, 117, 118 (right), 124 (inset), 190, 299, 305, 321, 324, 326, 335, 416, 464, 530 (top and bottom), 558.

Introduction

In the twenty-five years since the first edition of *The Handy Science Answer Book* was published in 1994, innumerable discoveries and advancements have been made in all fields of the biological and physical sciences. These accomplishments range from the microscopic to the global—from gene sequencing and CRISPR technology to advances in particle and quantum physics to the discovery of exoplanets. As a society, we have increased our awareness of the environment and the sustainability of our resources.

This newly updated fifth edition continues to be an educational resource that is both informative and enjoyable. The questions are interesting, unusual, frequently asked, or difficult to answer. Statistical data have been updated for this new edition. Both of us are pleased and excited about the various changes, including additions and improvements in this new edition, which continues to add to and enhance the original publication presented by the Science and Technology Department of the Carnegie Library of Pittsburgh.

INTRODUCTION AND HISTORICAL BACKGROUND

What is science?

The field of science involves the observation, description, and experimentation of the natural world in an attempt to explain the "whys" and "hows" of our world. It is a way of thinking and an ongoing method of looking at the world. Science is a way of discovering how the world works by using a set of rules devised by scientists.

Who were the earliest scientists?

Inquisitive individuals have always attempted to explain the physical world. The early Babylonians and Egyptians were aware of natural phenomena and events. Many times, they tried to explain these events in terms of their gods. The early Greeks were among the first people to look for explanations of natural phenomena based on discovery and knowledge. Greek philosopher Thales (c. 624–c. 547 B.C.E.) is often credited as being the first to look for an answer to the question, "What is the world made of?" Although most of the writings of Thales have been lost, we know he proposed water as the single substance from which everything in the world was made.

Who is considered the most influential scientist in the history and development of Western science?

Most experts seem to agree that Isaac Newton (1642–1727) is the most influential figure in the history of Western science. He was considered a great intellectual in his lifetime, and the admiration within the scientific community continues today after some three hundred years.

Why is Isaac Newton also considered the father of modern science?

Isaac Newton earned his place as the father of modern science by totally changing the way science was viewed in the evolution of human understanding of the universe, especially regarding his concepts and theories of motion, gravity, and mechanics.

What other scientists have been particularly influential in the history and development of science through the ages?

Many scholars regard the following list as individuals who have played a major role in the history and development of science:

1. Isaac Newton (1642–1727) and the Newtonian Revolution
2. Albert Einstein (1879–1955) and Twentieth-Century Physics including the Theory of Relativity
3. Galileo Galilei (1564–1642) and the New Science
4. Johannes Kepler (1571–1630) and Motion of the Planets
5. Nicolaus Copernicus (1473–1543) and the Heliocentric Universe
6. Niels Bohr (1885–1962) and the Atom
7. Antoine-Laurent Lavoisier (1743–1794) and the Revolution in Chemistry
8. Charles Darwin (1809–1882) and Evolution
9. Louis Pasteur (1822–1895) and the Germ Theory of Disease
10. Sigmund Freud (1856–1939) and Psychology of the Unconscious

What are some of the historical time periods of science?

The historical periods in the development of science include:

1. Antiquity: The period of time in which practical goals, such as establishing a reliable calendar or determining how to cure a variety of illnesses, existed simultaneously with abstract investigations known as natural philosophy.
2. Medieval Science: Few major contributions in fields of science occurred during the Medieval years. Exceptions were the emergence of science in the first established universities and the formulation of the Scientific Method.

Sir Isaac Newton was the central figure in illuminating to the world how key physical laws of the universe operate.

3. Renaissance and Early Modern Science: During this time period, Nicolaus Copernicus (1473–1543) formulated a heliocentric model of the solar system unlike the geocentric model of Claudius Ptolemy (c. 100–c. 170). Johannes Kepler (1571–1630), through his laws of planetary motion, improved upon Copernicus's heliocentric model. A major technological development was the invention of the printing press.
4. Age of Enlightenment: Science during the Enlightenment was dominated by scientific societies and academies, which had largely replaced universities as centers of scientific research and development. Science became increasingly popular among the educated and literate population. This time period saw advances in mathematics and physics; the development of biological taxonomy; a new understanding of gases as well as magnetism and electricity; and the maturation of chemistry as a discipline.
5. Nineteenth Century: The discoveries and achievements of the nineteenth century brought a close to the era of classical science and set the stage for the development of science as we know it today. During this period, new discoveries occurred about electricity and magnetism, genetics and evolution, the age of Earth, the stars and the planets, and the nature of infection and disease. These discoveries revolutionized the way people lived and perceived the world around them.
6. Twentieth Century: The twentieth century was a time of extraordinary scientific activity. In the life sciences, scientists discovered the structure and function of DNA and uncovered the process by which genetic traits are passed from one generation to the next. New drugs conquered formerly fatal diseases. In the physical sciences, radioactivity and X-rays were discovered as well as the development of the atomic bomb. The move toward increased specialization in all fields of science occurred during this time period. The use of computers in scientific research became common.
7. Twenty-First Century: The expanding horizons of science from the study of subatomic particles to missions deeper into outer space continue to reveal new and exciting, cutting-edge information during the early years of the twenty-first century.

What is the scientific method?

The scientific method is the basis of scientific investigation. A scientist will pose a question and formulate a hypothesis as a potential explanation or answer to the question. The hypothesis will be tested through a series of experiments. The results of the experiments will either prove or disprove the hypothesis. Hypotheses that are consistent with available data are conditionally accepted.

What are the steps of the scientific method?

Research scientists follow these steps:

1. State a hypothesis.
2. Design an experiment to "prove" the hypothesis.

3. Assemble the materials and set up the experiment.
4. Do the experiment and collect data.
5. Analyze the data using quantitative methods.
6. Draw conclusions.
7. Write up and publish the results.

Who is one of the first individuals associated with the scientific method?

Abu Ali al-Hasan ibn al-Haytham (c. 966–1039), whose name is usually Latinized to Alhazen or Alhacen, is known as the "father" of the science of optics and was also one of the earliest experimental scientists. Between the tenth and fourteenth centuries, Muslim scholars were responsible for the development of the scientific method. These individuals were the first to use experiments and observation as the basis of science, and many historians regard science as starting during this period. Alhazen is regarded as the architect of the scientific method. His scientific method involved the following steps:

1. Observe the natural world
2. State a definite problem
3. Formulate a hypothesis
4. Test the hypothesis through experimentation
5. Assess and analyze the results
6. Interpret the data and draw conclusions
7. Publish the findings

Alhazen (left) and Galileo (right) are represented as symbolizing reason and the senses in this engraving for the title page of astronomer Johannes Hevelius's *Selenographia, sive, Lunae descriptio* (1647).

In addition to Alhazen, which other scientists are associated with the development of the scientific method?

Roger Bacon (1214–1292) is an important individual who belongs with Aristotle (384–322 B.C.E.), Avicenna (980–1037), Galileo Galilei (1564–1642), and Newton as one of the great minds behind the formation of the scientific method. Bacon took the work of these individuals together with Robert Grosseteste (1175–1253) and used it to propose the idea of induction as the cornerstone of empiricism. He described the method of observation, prediction (hypothesis), and experimentation, also adding that results should be independently verified, documenting his results in great detail so that others might repeat the experiment.

How does deductive reasoning differ from inductive reasoning?

Deductive reasoning, often used in mathematics and philosophy, uses general principles to examine specific cases. Inductive reasoning is the method of discovering general principles by close examination of specific cases. Inductive reasoning first became important to science in the 1600s, when Francis Bacon (1561–1626), Isaac Newton, and their contemporaries began to use the results of specific experiments to infer general scientific principles.

What is a variable?

A variable is something that is changed or altered in an experiment. For example, to determine the effect of light on plant growth, growing one plant in a sunny window and one in a dark closet will provide evidence as to the effect of light on plant growth. The variable is light.

How does an independent variable differ from a dependent variable?

An independent variable is manipulated and controlled by the researcher. A dependent variable is the variable that the researcher watches and/or measures. It is called a dependent variable because it depends upon and is affected by the independent variable. For example, a researcher may investigate the effect of sunlight on plant growth by exposing some plants to eight hours of sunlight per day and others to only four hours of sunlight per day. The plant growth rate is dependent upon the amount of sunlight, which is controlled by the researcher.

What is a control group?

A control group is the experimental group tested without changing the variable. For example, to determine the effect of temperature on seed germination, one group of seeds may be heated to a certain temperature. The percentage of seeds in this group that germinate and the time it takes them to germinate is then compared to another group of

How do scientific laws differ from theories?

A scientific law is a statement of how something in nature behaves that has proven to be true every time it is tested. Scientific laws are statements of fact that describe what happens in nature. Unlike the general usage of the term "theory," which often means an educated guess, a scientific theory explains a phenomenon that is based on observation, experimentation, and reasoning. A scientific theory may explain a law, but theories do not become laws.

seeds (the control group) that have not been heated. All other variables, such as light and water, will remain the same for each group.

What is a double-blind study?

In a double-blind study, neither the subjects of the experiment nor the persons administering the experiment know the critical aspects of the experiment. This method is used to guard against both experimenter bias and placebo effects.

SOCIETIES, PUBLICATIONS, AND AWARDS

What was the first important scientific society in the United States?

The first significant scientific society in the United States was the American Philosophical Society, organized in 1743 in Philadelphia, Pennsylvania, by Benjamin Franklin (1706–1790). During colonial times, the quest to understand nature and seek information about the natural world was called natural philosophy.

What was the first national scientific society organized in the United States?

The first national scientific society organized in the United States was the American Association for the Advancement of Science (AAAS). It was established on September 20,

A statue of Albert Einstein by Robert Berks sits outside the Washington, D.C., National Academy of Sciences building. The NAS, established by President Abraham Lincoln, also has facilities in Irvine, California, and Woods Hole, Massachusetts.

1848, in Philadelphia, Pennsylvania, for the purpose of "advancing science in every way." The first president of the AAAS was William Charles Redfield (1789–1857).

What was the first national science institute?

On March 3, 1863, President Abraham Lincoln (1809–1865) signed a congressional charter creating the National Academy of Sciences, which stipulated that "the Academy shall, whenever called upon by any department of the government, investigate, examine, experiment, and report upon any subject of science of art, the actual expense of such investigations, examinations, experiments, and reports to be paid from appropriations which may be made for the purpose, but the Academy shall receive no compensation whatever for any services to the Government of the United States." Today, the Academy and its sister organizations—the National Academy of Engineering, established in 1964, and the Institute of Medicine, established in 1970—serve as the country's preeminent sources of advice on science and technology and their bearing on the nation's welfare.

The National Research Council was established in 1916 by the National Academy of Sciences at the request of President Woodrow Wilson (1856–1924) "to bring into cooperation existing governmental, educational, industrial and other research organizations, with the object of encouraging the investigation of natural phenomena, the increased use of scientific research in the development of American industries, the employment of scientific methods in strengthening the national defense, and such other applications of science as will promote the national security and welfare."

The National Academy of Sciences, the National Academy of Engineering, and the Institute of Medicine work through the National Research Council of the United States, one of the world's most important advisory bodies. More than six thousand scientists, engineers, industrialists, and health and other professionals participating in numerous committees comprise the National Research Council.

Who was the first president of the National Academy of Sciences?

The first president of the National Academy of Sciences was Alexander Dallas Bache (1806–1867). Bache served as president from 1863 until his death in 1867. He was the great-grandson of Benjamin Franklin.

What was the first national physics society organized in the United States?

The first national physics society in the United States was the American Physical Society, organized on May 20, 1899, at Columbia University in New York City. The first president was physicist Henry Augustus Rowland (1848–1901).

What was the first national chemical society organized in the United States?

The first national chemical society in the United States was the American Chemical Society, organized in New York City on April 20, 1876. The first president was John William Draper (1811–1882).

What was the first mathematical society organized in the United States?

The first mathematical society in the United States was the American Mathematical Society, founded in 1888 to further the interests of mathematics research and scholarship. The first president was John Howard Van Amringe (1835–1915).

What book is considered the most important and most influential scientific work?

That would be Isaac Newton's 1687 book *Philosophiae Naturalis Principia Mathematica* (known most commonly as the abbreviated *Principia*). Newton wrote *Principia* in eighteen months, summarizing his work and covering almost every aspect of modern science. Newton introduced gravity as a universal force, explaining that the motion of a planet responds to gravitational forces in inverse proportion to the planet's mass. Newton was able to explain tides and the motion of planets, moons, and comets using gravity. He also showed that spinning bodies, such as Earth, are flattened at the poles.

What was the first scientific journal?

The first scientific journal was *Journal des Sçavans*, published and edited by Denys de Sallo (1626–1669). The first issue appeared on January 5, 1665. It contained reviews of books, obituaries of famous men, experimental findings in chemistry and physics, and other general-interest information. Publication was suspended following the thirteenth issue in March 1665. Although the official reason for the suspension of the publication was that Sallo was not submitting his proofs for official approval prior to publication, some speculate that the real reason for the suspension in publication was his criticism of the work of important people, papal policy, and the old orthodox views of science. It was reinstated in January 1666 and continued as a weekly publication until 1724. The journal was then published on a monthly basis until the French Revolution in 1792. It was published briefly in 1797 under the title *Journal des Savants*. It began regular publication again in 1816 under the auspices of the Institut de France, evolving as a general-interest publication.

What is the oldest continuously published scientific journal?

The *Philosophical Transactions* of the Royal Society of London, first published a few months after the first issue of the *Journal des Sçavans* on March 6, 1665, is the oldest, continuously published scientific journal.

Who financed the initial printing of *Principia*?

The first printing of *Principia* produced only five hundred copies. It was published originally at the expense of Newton's friend Edmond Halley (1656–1742)—of Halley's Comet fame. Halley financed the project because the Royal Society of London, which had intended to publish *Principia*, encountered financial difficulties. The Royal Society had spent its entire budget on producing a history of fish.

What scientific article has the most authors?

The article "Combined Measurement of the Higgs Boson Mass in *pp* Collisions at $\sqrt{s}$ = 7 and 8 TeV with the ATLAS and CMS Experiments," published in *Physical Review Letters*, volume 114, issue 19 (May 15, 2015), page 191803, has 5,154 authors. It is a thirty-three-page article that lists the authors and their institution affiliations on twenty-four of the thirty-three pages. The paper has more authors than words in the text! The article is a collaboration between two teams of scientists working at the Large Hadron Collider (LHC) at CERN, the European Organization for Nuclear Research. It provides the most precise estimate of the mass of the Higgs boson known to date.

What is the most frequently cited scientific journal article?

Web of Science compiled a list of the one hundred most highly cited papers. According to the list, published in 2014, the most frequently cited scientific article is "Protein Measurement with the Folin Phenol Reagent" by Oliver Howe Lowry (1910–1996) and coworkers, published in 1951 in the *Journal of Biological Chemistry*, volume 193, issue 1, pages 265–275. The article had 305,148 citations since it was first published in 1951. The second most frequently cited article is "Cleavage of Structural Proteins during the Assembly of the Head of Bacteriophage T4" by Ulrich K. Laemmli (1940–), published in 1970 in the journal *Nature*, volume 227, pages 680–685, with 213,005 citations. Google Scholar published a similar list of top one hundred cited articles. According to the Google Scholar list, the number one and number two articles were reversed. The Laemmli article had 223,131 citations, and the Lowry article had 192,710 citations. Google Scholar based its list of searched-for references on a much greater literature base, including books.

What is the journal impact factor?

The journal impact factor (IF) is a tool used to measure the importance or rank of a journal in a field. It does not measure the importance or impact of a particular article or author published in the journal. Impact factor is calculated by dividing the number of times articles in a journal were cited in a two-year period by the total number of articles published in the journal during the same two-year period. Journals with a high impact factor are perceived to be more prestigious, although many important and significant research findings are reported in journals with low impact factors.

One of the most distinguished awards in the world is the Nobel Prize. In addition to prizes in literature, peace, and economics, the Nobel is awarded to worthy scientists in the categories of physiology or medicine, chemistry, and physics.

When was the Nobel Prize first awarded?

The Nobel Prize was established by Alfred Nobel (1833–1896) to recognize individuals whose achievements during the preceding year had conferred the greatest benefit to mankind. Five prizes were to be conferred each year in the areas of physics, chemistry, physiology or medicine, literature, and peace. Although Nobel passed away in 1896, the first prizes were not awarded until 1901. The Nobel Prize in Economic Sciences was established in 1968. It was first awarded in 1969. The recipients, called laureates, receive a diploma, a gold medal, and a cash prize that generally exceeds $1 million. All prizes in the fields of physics, chemistry, and physiology or medicine are awarded to individuals. The rules stipulate that each prize can be shared by no more than three individuals.

How many Nobel prizes have been awarded in the three fields of science?

Between 1901 and 2018, 331 have been awarded in the fields of chemistry, physics, and physiology or medicine.

Nobel Prize	Number of Prizes	Total Number of Laureates	Number of Prizes Awarded to One Laureate
Chemistry	110	181	63
Physics	112	210	47
Physiology or Medicine	109	216	39

Who are the youngest and oldest Nobel laureates in the areas of physics, chemistry, and physiology or medicine?

Youngest Nobel Laureates

Category	Nobel Laureate	Age	Year
Chemistry	Frédéric Joliet (1900–1958)	35	1935
Physics	William Lawrence Bragg (1890–1971)	25	1915
Physiology or Medicine	Frederick Banting (1891–1941)	32	1923

Oldest Nobel Laureates

Category	Nobel Laureate	Age	Year
Chemistry	John B. Fenn (1917–2010)	85	2002
Physics	Arthur Ashkin (1922–)	96	2018
Physiology or Medicine	Peyton Rous (1879–1970)	87	1966

Have any Nobel Prize winners won multiple times?

Four individuals have received multiple Nobel prizes. They are Marie Curie (1867–1934), physics in 1903 and chemistry in 1911; John Bardeen (1908–1991), physics in 1956 and

What is the average age of a Nobel laureate?

The average age of the Nobel laureates—in the year they were awarded the prize—is fifty-eight in chemistry, fifty-six in physics, and fifty-eight in physiology or medicine.

1972; Linus Pauling (1901–1994), chemistry in 1954 and peace in 1962; and Frederick Sanger (1918–2013), chemistry in 1958 and 1980.

Who was the first woman to receive the Nobel Prize?

Marie Curie was the first woman to receive the Nobel Prize. She received the Nobel Prize in Physics in 1903 for her work on radioactivity in collaboration with her husband, Pierre Curie (1859–1906) and Antoine Henri Becquerel (1852–1908). The 1903 prize in physics was shared by all three individuals. Marie Curie was also the first person to be awarded two Nobel prizes and is one of only two individuals who have been awarded a Nobel Prize in two different fields.

How many women have been awarded the Nobel Prize in Chemistry, Physics, and Physiology or Medicine?

Since 1901, the Nobel Prize in Chemistry, Physics, and Physiology or Medicine has been awarded to women twenty times to nineteen different women. Marie Curie (1867–1934) was the only woman and one of the few individuals to receive the Nobel Prize twice. The first time women were awarded the Nobel Prize in the fields of both chemistry and physics was 2018.

Women Recipients of the Nobel Prize

Category	Number of Women Recipients
Chemistry	5
Physics	3
Physiology or Medicine	12

Year	Nobel Laureate	Category
1903	Marie Curie (1867–1934)	Physics
1911	Marie Curie (1867–1934)	Chemistry
1935	Irène Joliot-Curie (1897–1956)	Chemistry
1947	Gerty Theresa Cori (1896–1957)	Physiology or Medicine
1963	Maria Goeppert-Mayer (1906–1972)	Physics
1964	Dorothy Crowfoot Hodgkin (1910–1994)	Chemistry
1977	Rosalyn Yalow (1921–2011)	Physiology or Medicine
1983	Barbara McClintock (1902–1992)	Physiology or Medicine
1986	Rita Levi-Montalcini (1909–2012)	Physiology or Medicine

Year	Nobel Laureate	Category
1988	Gertrude B. Elion (1918–1999)	Physiology or Medicine
1995	Christianne Nüsslein-Volhard (1942–)	Physiology or Medicine
2004	Linda B. Buck (1947–)	Physiology or Medicine
2008	Françoise Barré-Sinoussi (1947–)	Physiology or Medicine
2009	Ada E. Yonath (1939–)	Chemistry
2009	Carol W. Greider (1961–)	Physiology or Medicine
2009	Elizabeth H. Blackburn (1948–)	Physiology or Medicine
2014	May-Britt Moser (1963–)	Physiology or Medicine
2015	Youyou Tu (1930–)	Physiology or Medicine
2018	Frances Arnold (1956–)	Chemistry
2018	Donna Strickland (1959–)	Physics

When was the first time two women shared the Nobel Prize in the same field?

It was not until 2009 that two women shared the Nobel Prize in the same field. Carol W. Greider (1961–) and Elizabeth H. Blackburn (1948–) shared the prize in physiology or medicine, along with Jack W. Szostak (1952–), for their discovery of how chromosomes are protected by telomeres and the enzyme telomerase.

Does a Nobel Prize in mathematics exist?

We do not know for certain why Alfred Nobel did not establish a prize in mathematics. Several theories revolve around his relationship and dislike for Gosta Mittag-Leffler (1846–1927), the leading Swedish mathematician in Nobel's time. Most likely, it never occurred to Nobel, or he decided against another prize. The Fields Medal in mathematics is generally considered as prestigious as the Nobel Prize. It is awarded every four years to individuals who are younger than forty to recognize outstanding mathematical achievement for existing work and for the promise of future achievement. The Fields Medal was first awarded in 1936. Since 1950, it has been awarded every four years. In 1966, the number of awardees in a given year was increased from two to four. The 2018 awardees were Caucher Birkar (1978–), Alessio Figalli (1984–), Peter Scholze (1987–), and Akshay Venkatesh (1981–).

With no Nobel category for mathematics, the Fields Medal, established in 1936, has filled the role of most prestigious award in math.

Who was the first woman to receive the Fields Medal?

Maryam Mirzakhani (1977–2017) was the first woman to receive the Fields Medal for her work in theoretical mathematics in 2014.

What is the National Medal of Science?

The National Medal of Science was established by the U.S. Congress in 1959. It is a Presidential Award bestowed upon individuals who have made important contributions for the advancement of knowledge in the physical, biological, mathematical, or engineering sciences. In 1980, the U.S. Congress decided to include the fields of social and behavioral sciences. The National Medal of Science has been awarded to 506 scientists and engineers.

What is the Copley Medal?

The Copley Medal, awarded by the Royal Society of London, is the world's oldest scientific prize. It was first awarded in 1731 for the most important scientific discovery or the greatest contribution by an experiment. It is awarded annually for outstanding achievements in research in any branch of science (odd years for physical sciences and even years for biological sciences).

LABORATORY TOOLS AND TECHNIQUES

What is the SI system of measurement?

French scientists as far back as the seventeenth and eighteenth centuries questioned the hodgepodge of the many illogical and imprecise standards used for measurement, so they began a crusade to make a comprehensive, logical, precise, and universal measurement system called Système Internationale d'Unités, or SI for short. It uses the metric system as its base. Since all the units are in multiples of 10, calculations are simplified. Today, all countries except the United States, Myanmar (formerly Burma), and Liberia use this system. However, some elements within American society do use SI—scientists, exporting/importing industries, and federal agencies.

The SI or metric system has seven fundamental standards: the meter (for length), the kilogram (for mass), the second (for time), the ampere (for electric current), the kelvin (for temperature), the candela (for luminous intensity), and the mole (for amount of substance). In addition, two supplementary units, the radian (plane angle) and steradian (solid angle), and a large number of derived units compose the current system, which is still evolving. Some derived units, which use special names, are the hertz, newton, pascal, joule, watt, coulomb, volt, farad, ohm, siemens, weber, tesla, henry, lumen, lux, becquerel, gray, and sievert. Its unit of volume or capacity is the cubic decimeter, but many still use "liter" in its place. Very large or very small dimensions are expressed through a series of prefixes, which increase or decrease in multiples of ten. For example, a decimeter is 1/10 of a meter, a centimeter is 1/100 of a meter, and a millimeter is 1/1000 of a meter. A dekameter is 10 meters, a hectometer is 100 meters, and a kilometer is 1,000 meters. The use of these prefixes enables the system to express these units in an orderly way and avoid inventing new names and new relationships.

How was the length of a meter originally determined?

It was originally intended that the meter should represent one ten-millionth of the distance along the meridian running from the North Pole to the equator through Dunkirk, France, and Barcelona, Spain. French scientists determined this distance, working nearly six years to complete the task in November 1798. They decided to use a platinum–iridium bar as the physical replica of the meter. Although the surveyors made an error of about two miles, the error was not discovered until much later. Rather than change the length of the meter to conform to the actual distance, scientists in 1889 chose the platinum–iridium bar as the international prototype. It was used until 1960. Numerous copies of it are in other parts of the world, including the U.S. National Bureau of Standards.

How is the length of a meter presently determined?

The meter is equal to 39.37 inches. It is presently defined as the distance traveled by light in a vacuum in 1/299,792,458 of a second. From 1960 to 1983, the length of a meter had been defined as 1,650,763.73 times the wavelength of the orange light emitted when a gas consisting of the pure krypton isotope of mass number 86 is excited in an electrical discharge.

Will the definition of other SI base units be changed?

Four other base units of the SI—the ampere, kilogram, mole, and kelvin—were revised at the 2018 meeting of the General Conference on Weights and Measures. Each of these units are now based on a physical constant. They will no longer need to be modified to accommodate future improvements in the technologies used to realize them.

- The *kilogram* is defined in terms of the Planck constant
- The *ampere* is defined in terms of the elementary charge
- The *kelvin* is defined in terms of the Boltzmann constant
- The *mole* is defined in terms of the Avogadro constant

How are names for large and small quantities constructed in the metric system?

Each prefix listed below can be used in the metric system and with some customary units. For example, centi + meter = centimeter, meaning one-hundredth of a meter.

Metric Prefixes

Prefix	Power	Numerals
Exa-	10^{18}	1,000,000,000,000,000,000
Peta-	10^{15}	1,000,000,000,000,000
Tera-	10^{12}	1,000,000,000,000
Giga-	10^{9}	1,000,000,000
Mega-	10^{6}	1,000,000
Myria-	10^{5}	100,000

Prefix	Power	Numerals
Kilo-	10^3	1,000
Hecto-	10^2	100
Deca-	10^1	10
Deci-	10^{-1}	0.1
Centi-	10^{-2}	0.01
Milli-	10^{-3}	0.001
Micro-	10^{-6}	0.000001
Nano-	10^{-9}	0.000000001
Pico-	10^{-12}	0.000000000001
Femto-	10^{-15}	0.000000000000001
Atto-	10^{-18}	0.000000000000000001

Why do scientists express numbers in scientific notation?

Scientific notation allows scientists to easily manipulate very large or very small numbers. It is based on the fact that all numbers can be expressed as the product of two numbers, one of which is the power of the number 10 (written as the small superscript next to the number 10 and called the exponent). Positive exponents indicate how many times the number must be multiplied by 10, while negative exponents indicate how many times a number must be divided by 10.

Numbers Greater Than 1*	Numbers Less Than 1**
$1{,}000{,}000{,}000 = 1 \times 10^9$	$0.000000001 = 1 \times 10^{-9}$
$100{,}000{,}000 = 1 \times 10^8$	$0.00000001 = 1 \times 10^{-8}$
$10{,}000{,}000 = 1 \times 10^7$	$0.0000001 = 1 \times 10^{-7}$
$1{,}000{,}000 = 1 \times 10^6$	$0.000001 = 1 \times 10^{-6}$
$100{,}000 = 1 \times 10^5$	$0.00001 = 1 \times 10^{-5}$
$10{,}000 = 1 \times 10^4$	$0.0001 = 1 \times 10^{-4}$
$1{,}000 = 1 \times 10^3$	$0.001 = 1 \times 10^{-3}$
$100 = 1 \times 10^2$	$0.01 = 1 \times 10^{-2}$
$10 = 1 \times 10^1$	$0.1 = 1 \times 10^{-1}$
$1 = 1 \times 10^0$	$1 = 1 \times 10^0$

*The exponent to which the power of 10 is raised is equal to the number of zeros to the right of 1.

**The negative exponent to which the power of 10 is raised is equal to the number of zeros to the left of 1, minus one zero.

Who invented the thermometer?

The Greeks of Alexandria knew that air expanded as it was heated. Heron (Hero) of Alexandria (first century C.E.) and Philo of Byzantium (280–220 B.C.E.) made simple "thermoscopes," but they were not real thermometers. In 1592, Galileo Galilei made a kind of thermometer that also functioned as a barometer, and in 1612, his friend Santorio Santorio (1561–1636) adapted the air thermometer (a device in which a colored liquid was driven down by the expansion of air) to measure the body's temperature change during illness and recovery. Still, it was not until 1713 that Daniel Fahrenheit (1686–1736) began developing a thermometer with a fixed scale. He worked out his scale from two "fixed" points: the melting point of ice and the heat of the healthy human body. He realized that the melting point of ice was a constant temperature, whereas the freezing point of water varied. Fahrenheit put his thermometer into a mixture of ice, water, and salt (which he marked off as 0°) and, using this as a starting point, marked off melting ice at 32° and blood heat at 96°. In 1835, it was discovered that normal blood measured 98.6°F. Sometimes, Fahrenheit used spirit of wine as the liquid in the thermometer tube, but more often, he used specially purified mercury. Later, the boiling point of water (212°F) became the upper fixed point.

The creator of the scale that bears his name, Daniel Fahrenheit made important advancements in thermometrics. He also improved the hygrometer, which measures humidity in the air.

What was unusual about the original Celsius temperature scale?

In 1742, Swedish astronomer Anders Celsius (1701–1744) set the freezing point of water at 100°C and the boiling point of water at 0°C. It was Carolus Linnaeus (1707–1778) who reversed the scale, but a later textbook attributed the modified scale to Celsius, and the name has remained.

When did the name of the temperature scale change from Centigrade to Celsius?

The temperature scale that ranges from 0 (freezing point) to 100 (boiling point) had once been named degrees Centigrade. In 1948, the General Conference on Weights and Measures officially changed the name to degrees Celsius. Over the next several decades, the term "Centigrade" has disappeared from usage in textbooks, scholarly papers, and general usage.

William Thomson, Lord Kelvin, was a Scots-Irish engineer and physicist born in Belfast. He studied electricity and contributed to the formation of the first and second laws of thermodynamics. The Kelvin temperature scale is named after him.

What is the Kelvin temperature scale?

Temperature is the level of heat in a gas, liquid, or solid. The freezing and boiling points of water are used as standard reference levels in both the metric (Celsius) and the British system (Fahrenheit). In the metric system, the difference between freezing and boiling is divided into 100 equal intervals called degrees Celsius (°C). In the British system, the intervals are divided into 180 units, with one unit called degree Fahrenheit (°F). However, temperature can be measured from absolute zero (no heat, no motion); this principle defines thermodynamic temperature and establishes a method to measure it upward. This scale of temperature is called the Kelvin temperature scale after its inventor, William Thomson, Lord Kelvin (1824–1907), who devised it in 1848. The Kelvin (K) has the same magnitude as the degree Celsius (the difference between freezing and boiling water is 100 degrees), but the two temperatures differ by 273.15 degrees (0 K equals −273.15°C). Below is a comparison of the three temperatures:

Characteristic	K	°C	°F
Absolute zero	0	–273.15	–459.67
Freezing point of water	273.15	0	32
Normal human body temperature	310.15	37	98.6
100°F	290.95	17.8	100
Boiling point of water (at one atmosphere pressure)	373.15	100	212
Point of equality	233.15	–40.0	–40.0

To convert Celsius to Kelvin: Add 273.15 to the temperature ($K = C + 273.15$).

To convert Fahrenheit to Celsius: Subtract 32 from the temperature and multiply the difference by 5, then divide the product by 9 ($C = 5/9[F - 32]$).

To convert Celsius to Fahrenheit: Multiply the temperature by 1.8, then add 32 ($F = 9/5C + 32$).

How is "absolute zero" defined?

Absolute zero is the theoretical temperature at which all substances have zero thermal energy. Originally conceived as the temperature at which an ideal gas at constant pressure would contract to zero volume, absolute zero is of great significance in thermodynamics and is used as the fixed point for absolute temperature scales. Absolute zero is equivalent to 0 K, –459.67°F, or –273.15°C.

The velocity of a substance's molecules determines its temperature; the faster the molecules move, the more volume they require and the higher the temperature becomes. The lowest actual temperature ever reached was two-billionth of a degree above absolute zero (2×10^{-9}K) by a team at the Low Temperature Laboratory in the Helsinki University of Technology, Finland, in October 1989.

What is the pH scale?

The pH scale is the measurement of the H^+ concentration (hydrogen ions) in a solution. It is used to measure the acidity or alkalinity of a solution. The pH scale ranges from 0 to 14. A neutral solution has a pH of 7; one with a pH greater than 7 is basic, or alkaline; and one with a pH less than 7 is acidic. The lower the pH below 7, the more acidic the solution. Each whole-number drop in pH represents a tenfold increase in acidity.

pH Value	Examples of Solutions
0	Hydrochloric acid (HCl), battery acid
1	Stomach acid (1.0–3.0)
2	Lemon juice (2.3)
3	Vinegar, wine, soft drinks, beer, orange juice, some acid rain
4	Tomatoes, grapes, banana (4.6)
5	Black coffee, most shaving lotions, bread, normal rainwater
6	Urine (5–7), milk (6.6), saliva (6.2–7.4)
7	Pure water, blood (7.3–7.5)
8	Egg white (8.0), seawater (7.8–8.3)
9	Baking soda, Clorox, Tums
10	Soap solutions, milk of magnesia
11	Household ammonia (10.5–11.9), nonphosphate detergents
12	Washing soda (sodium carbonate)
13	Hair remover, oven cleaner
14	Sodium hydroxide (NaOH)

Who developed the pH scale?

The potential of hydrogen, or pH scale, was developed by Danish biochemist Søren Peter Lauritz Sørensen (1868–1939). He introduced this scale in 1909 to measure the acidity and alkalinity of substances.

What is spectroscopy?

Spectroscopy includes a range of techniques to study the composition, structure, and bonding of elements and compounds. The different methods of spectroscopy use differ-

How is red cabbage used as a pH indicator?

Red cabbage contains a pigment called flavin (an anthocyanin). This water-soluble pigment is also found in apple skin, plums, poppies, cornflowers, and grapes. To prepare a solution of red cabbage juice indicator, chop some red cabbage into small pieces and cover them with boiling water. Allow the mixture to sit for approximately 10 minutes. The indicator may now be used to test various solutions as to their acidity. Add a few drops of the cooled red cabbage mixture to a solution. Very acidic solutions will turn anthocyanin a red color. Neutral solutions result in a purplish color. Basic solutions appear greenish-yellow. Therefore, it is possible to determine the pH of a solution based on the color it turns the anthocyanin pigments in red cabbage juice.

ent wavelengths of the electromagnetic spectrum to study atoms, molecules, ions, and the bonding between them.

Type of Spectroscopy	Wavelength Used
Nuclear magnetic resonance spectroscopy	Radio waves
Infrared spectroscopy	Infrared radiation
Atomic absorption spectroscopy, atomic emission spectroscopy, and ultraviolet spectroscopy	Visible and UV radiation
X-ray spectroscopy	X-rays

What is nuclear magnetic resonance?

Nuclear magnetic resonance (NMR) is a process in which the nuclei of certain atoms absorb energy from an external magnetic field. Analytical chemists use NMR spectroscopy to identify unknown compounds, check for impurities, and study the shapes of molecules. They use the knowledge that different atoms will absorb electromagnetic energy at slightly different frequencies.

A nuclear magnetic resonance spectrometer is shown here at the Canadian National Ultrahigh-field NMR Facility for Solids in Ottawa, Canada. These devices use magnetic fields to analyze the atoms within molecules.

When was nuclear magnetic resonance discovered?

Nuclear magnetic resonance was discovered by Felix Bloch (1905–1983), a physicist at Stanford University, and Edward M. Purcell (1912–1997), a physicist at Harvard University, in 1945. Bloch demonstrated NMR in liquid water, while Purcell demonstrated it in solid paraffin. They shared the 1952 Nobel Prize in Physics for their discovery.

A Prussian mechanical engineer and physicist, William Roentgen discovered the radiation now known as X-rays. He earned a Nobel Prize in Physics for his work.

What are X-rays?

X-rays are electromagnetic radiation with short wavelengths (10^{-3} nanometers) and a great amount of energy. They were discovered in 1898 by William Conrad Roentgen (1845–1923). X-rays are frequently used in medicine because they are able to pass through opaque, dense structures, such as bone, and form an image on a photographic plate. They are especially helpful in assessing damage to bones, identifying certain tumors, and examining the chest—heart and lungs—and abdomen.

Who invented chromatography?

Chromatography was invented by Russian botanist Mikhail Tswett (1872–1919) in the early 1900s. He presented a lecture titled "On a New Category of Adsorption Phenomena and Their Application to Biochemical Analysis" to the Biological Section of the Warsaw Society of Natural Sciences in March 1903. He used the technique to separate different pigments found in plants, thereby identifying versions of chlorophyll. The term comes from the Greek words *chroma*, meaning "color," and *graphein*, meaning "writing or drawing."

What are the most common chromatographic techniques?

The most common chromatographic techniques are paper chromatography, gas–liquid chromatography (also called gas chromatography), thin-layer chromatography, and high-pressure (or high-performance) liquid chromatography (HPLC).

How is chromatography used to identify individual compounds?

Chromatography techniques are useful to 1) separate and identify the chemicals in a mixture; 2) check the purity of a chemical product; 3) identify impurities in a product; and 4) purify a chemical product. All methods of chromatography share common characteristics. The process is based on the principle that different chemical compounds will stick to a solid surface, or dissolve in a film of liquid, to different degrees. Chromatography involves a sample (or sample extract) being dissolved in a mobile phase (which may be a gas, a liquid, or a supercritical fluid). The mobile phase is then forced through an immobile, immiscible stationary phase. The phases are chosen such that components of the sample have differing solubilities in each phase. The least soluble component is separated first, and as the separation process continues, the components are separated by increasing solubility.

What are some of the areas of science that use chromatography?

Many different areas of science utilize chromatography techniques. It is used to separate and identify amino acids, carbohydrates, fatty acids, and other natural substances. The food industry uses chromatography to detect contaminants such as aflatoxin. Environmental testing laboratories use chromatography to identify trace quantities of contaminants such as PCBs in waste oil and pesticides such as DDT in groundwater. It is also used to test drinking water and test air quality. Pharmaceutical companies use chromatography to prepare quantities of extremely pure materials. It is also used in forensics and crime scene investigations to determine whether alcohol or drugs or poisons were present at the time of death.

What is electrophoresis?

Electrophoresis is a technique used to separate biological molecules, such as nucleic acids, carbohydrates, and amino acids, based on their movement due to the influence of a direct electric current in a buffered solution. Positively charged molecules move toward the negative electrode, while negatively charged molecules move toward the positive electrode.

What are some of the applications of centrifugation?

Centrifugation is the separation of immiscible liquids or solids from liquids by applying centrifugal force. Since the centrifugal force can be very great, it speeds the process of separating these liquids instead of relying on gravity. Biologists primarily use centrifugation to isolate and determine the biological properties and functions of subcellular organelles and large molecules. They study the effects of centrifugal forces on cells, developing embryos, and protozoa. These techniques have allowed scientists to determine certain properties about cells, including surface tension, relative viscosity of the cytoplasm, and the spatial and functional interrelationship of cell organelles when redistributed in intact cells.

What distinguishes the different types of microscopes?

Microscopes allow scientists to observe cells and cell structures that are not visible to the human eye. All microscopes require a source of illumination and a system of lenses to focus the illumination on the object or specimen being observed to form an image. The two basic types of microscopes are light microscopes and electron microscopes. Light microscopes and electron microscopes differ in their source of illumination and the construction of the lenses. Light microscopes utilize visible light as the source of illumination and a series of glass lenses. Electron microscopes utilize a beam of electrons emitted by a heated tungsten filament as the source of illumination. The lens system consists of a series of electromagnets.

Recent advances using optical techniques have led to the development of specialized light microscopes, including fluorescence microscopy, phase-contrast microscopy, and differential interference contrast microscopy. In fluorescence microscopy, a fluorescent dye is introduced to specific molecules. Both phase-contrast microscopy and differential interference contrast microscopy utilize techniques that enhance and amplify slight changes in the phase of transmitted light as it passes through a structure that has a different refractive index than the surrounding medium.

What is the difference between magnification and resolution?

Magnification—making smaller objects seem larger—is the measure of how much an object is enlarged. Resolution is the minimum distance that two points can be separated and still be seen as two distinct points.

Who invented the compound microscope?

The principle of the compound microscope, in which two or more lenses are arranged to form an enlarged image of an object, occurred independently, at about the same time, to more than one person. Certainly, many opticians were active in the construction of telescopes at the end of the sixteenth century, especially in Holland, so it is likely that the idea of the microscope may have occurred to several of them independently. In all probability, the date may be placed within the period 1590–1609, and the credit should go to three spectacle makers in Holland. Hans Janssen, his son Zacharias Janssen (1580–1638), and Hans Lippershey (1570–1619) have all been cited at various times as deserving chief credit.

What is the resolving power of various lens systems?

Lens System	Resolving Power
Human eye	0.1 mm = 100 µm
Light microscope	0.4 µm
Oil immersion light microscope	0.2 µm
Ultraviolet microscope	0.1 µm
Scanning electron microscope	10 nm = 0.010 µm
Transmission electron microscope	0.2 nm = 0.0002 µm

What was the first publication devoted to microscopic observations?

British scientist Robert Hooke (1635–1703) improved the resolution of a compound microscope to make observations of common items, such as the point of a needle, plants, insects, molds, bird feathers, and other objects. His book *Micrographia*, published in 1665, contains some of the most beautiful drawings of microscopic observations ever made.

This is a replica of a 1933 electron microscope built by Ernst Ruska. It is on display at the Deutsches Museum in Munich, Germany.

Who invented the electron microscope?

The theoretical and practical limits to the use of the optical microscope were set by the wavelength of light. When the oscilloscope was developed, it was realized that cathode-ray beams could be used to resolve much finer detail because their wavelength was so much shorter than that of light. In 1928, Ernst Ruska (1906–1988) and Max Knoll (1897–1969), using magnetic fields to "focus" electrons in a cathode-ray beam, produced a crude instrument that gave a magnification of 17, and by 1932, they had developed an electron microscope having a magnification of 400. By 1937, James Hillier (1915–2007) had advanced this magnification to 7,000. The 1939 instrument Vladimir Zworykin (1889–1982) developed gave fifty times more detail than any optical microscope ever could, with a magnification up to two million. The electron microscope revolutionized biological research; for the first time, scientists could see the molecules of cell structures, proteins, and viruses.

MATHEMATICS

INTRODUCTION AND HISTORICAL BACKGROUND

When and where did the concept of "numbers" and counting first develop?

The human adult (including some of the higher animals) can discern the numbers one through four without any training. After that, people must learn to count. To count requires a system of number manipulation skills, a scheme to name the numbers, and some way to record the numbers. Early people began with fingers and toes and progressed to shells and pebbles. In the fourth millennium B.C.E. in Elam (near what is today Iran along the Persian Gulf), accountants began using unbaked clay tokens instead of pebbles. Each represented one order in a numbering system: a stick shape for the number one, a pellet for ten, a ball for one hundred, and so on. During the same period, another clay-based civilization in Sumer, lower Mesopotamia, invented the same system.

What is the most enduring mathematical work of all time?

The *Elements of Euclid* (c. 300 B.C.E.) has been the most enduring and influential mathematical work of all time. In it, the ancient Greek mathematician Euclid (c. 325–c. 270 B.C.E.) presented the work of earlier mathematicians and included many of his own innovations. He established the system of postulants (statements that are true without proof) and proofs that are still in use today in geometry. The *Elements* is divided into thirteen books: the first six cover plane geometry; seven to nine address arithmetic and number theory; ten treats irrational numbers; and eleven to thirteen discuss solid geometry. Euclid's thirteen-volume treatise remains the definitive work on geometry. The geometry that many students learn in middle and high school today is largely based on Euclid's original ideas on the subject. In presenting his theorems, Euclid used the synthetic approach, in which one proceeds from the known to the unknown by logical steps. This method became the standard procedure for scientific investigation for many centuries, and the *Elements* probably had a greater influence on scientific thinking than any other work.

Who was the first prominent female mathematician?

The first known prominent female mathematician was Hypatia of Alexandria (370–415). In addition to being a mathematician, she was a philosopher and an astronomer. In classical antiquity, astronomy was seen as being essentially mathematics. Hypatia is credited with editing the surviving texts of Euclid's *Elements*. In addition, she wrote a thirteen-volume commentary on Diophantus's (210–290 C.E.) *Arithmetica* and an eight-volume popularization of Apollonius of Perga's (262–190 B.C.E.) treatise on conic sections, including the circle, ellipse, parabola, and hyperbola.

Are any problems in mathematics unsolved?

The earliest challenges and contests to solve important problems in mathematics date back to the sixteenth and seventeenth centuries. Some of these problems have continued to challenge mathematicians until modern times. For example, Pierre de Fermat (1601–1665) issued a set of mathematical challenges in 1657, many on prime numbers and divisibility. The solution to what is now known as Fermat's Last Theorem was not established until the late 1990s by Andrew Wiles (1953–). David Hilbert (1862–1943), a German mathematician, identified twenty-three unsolved problems in 1900 with the hope that these problems would be solved in the twenty-first century. Although some of the problems were solved, others remain unsolved to this day.

Which of the seven Millennium Prize Problems has been solved?

In 2000, the Clay Mathematics Institute named seven mathematical problems that had not been solved with the hope that they could be solved in the twenty-first century. A $1 million prize will be awarded for solving each of these seven problems. The seven Millennium Prize Problems named by the Clay Mathematics Institute were:

1. Birch and Swinnerton-Dyer Conjecture
2. Hodge Conjecture
3. Poincaré Conjecture
4. Riemann Hypothesis
5. Solution of the Navier-Stokes equations
6. Formulation of the Yang-Mills theory
7. P Versus NP

The first Millennium Prize Problem solved was the Poincaré Conjecture. It was solved by a Russian mathematician, Grigoriy Perelman (1966–), in 2010. Perelman declined to accept the cash award.

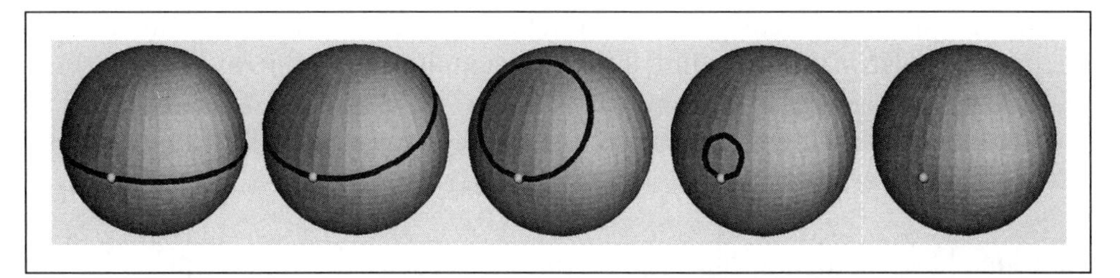

In the above illustration, a loop residing on the surface of a sphere (or 2-sphere because its surface is two-dimensional) can be tightened to a point no matter where it is on the sphere. The Poincaré Conjecture states that the same is true for a 3-sphere (a sphere in four-dimensional space), but this was not proved mathematically until Grigoriy Perelman did so in 2010.

NUMBERS

When was a symbol for the concept of zero first used?

Surprisingly, the symbol for zero emerged later than the concept for the other numbers. Although the Babylonians (600 B.C.E. and earlier) had a symbol for zero, it was merely a placeholder and not used for computational purposes. The ancient Greeks, for instance, conceived of logic and geometry, concepts providing the foundation for all mathematics, yet they never had a symbol for zero. The Maya peoples also had a symbol for zero as a placeholder in the fourth century, but they also did not use zero in computations. Hindu mathematicians are usually given credit for developing a symbol for the concept of "zero." They recognized zero as representing the absence of quantity and developed its use in mathematical calculations. It appears in an inscription at Gwalior dated 870 C.E. However, it is found even earlier than that in inscriptions dating from the seventh century in Cambodia, Sumatra, and Bangka Island (off Sumatra). While no documented evidence exists in printed material for the zero in China before 1247, some historians maintain that a blank space existed on the Chinese counting board, representing zero, as early as the fourth century B.C.E.

What are the seven basic Roman numerals?

Roman numerals are symbols that stand for numbers. They are written using seven basic symbols: I (1), V (5), X (10), L (50), C (100), D (500), and M (1,000). Sometimes, a bar is placed over a numeral to multiply it by 1,000. A smaller numeral appearing before a larger numeral indicates that the smaller numeral is subtracted from the larger one. This notation is generally used for 4s and 9s; for example, 4 is written IV, 9 is IX, 40 is XL, and 90 is XC.

What is pi?

Pi (π) represents the ratio of the circumference of a circle to its diameter, used in calculating the area of a circle (πr^2) and the volume of a cylinder ($\pi r^2 h$) or cone. It is a "transcendental number," an irrational number with an exact value that can be measured to any degree of accuracy but that can't be expressed as the ratio of two integers. In theory, the decimal extends into infinity, though it is generally rounded to 3.1416. Welsh-born mathematician William Jones (1675–1749) selected the Greek symbol π for pi. Rounded to thirty digits past the decimal point, it equals 3.1415926535897932384 62643383279.

How long have mathematicians been trying to calculate the value of pi?

Mathematicians, scientists, and curious individuals have been calculating the value of pi for centuries. Archimedes (c. 287–212 B.C.E.) drew polygons both around the outside and within the interior of circles. By measuring the perimeters of the circles, he determined the range of the upper and lower bounds of the range containing pi. Chinese mathematician Liu Hui (c. 225–295) independently created a polygon-based iterative algorithm at about the same time. Christoph Grienberger (1561–1636), an Austrian astronomer, calculated pi to thirty-eight digits using polygonal algorithms. This is considered the most accurate approximation of the value of pi calculated manually.

In 1665, Isaac Newton used calculus to compute pi to fifteen digits. Other mathematicians continued to calculate pi to more and more digits. They attained seventy-one digits in 1699, one hundred digits in 1706, and 620 digits in 1956—without using a calculator or computer.

The invention of electronic computers has allowed calculations of pi to more and more digits. In 1946, ENIAC, the first computer, calculated pi to 2,037 digits in 70 hours. In 1989, Gregory (1952–) and David Chudnovsky (1947–) at Columbia University in New York City calculated the value of pi to 1,011,961,691 decimal places. They performed the calculation twice on an IBM 3090 mainframe and on a CRAY–2 supercomputer with matching results. In 1991, they calculated pi to 2,260,321,336 decimal places. The Chudnovsky algorithm is used frequently to calculate pi.

Who holds the world record for calculating the number of digits of π (pi)?

Calculating the value of pi (π) to more and more decimal points has become a challenge for many individuals. It is often a test of computing systems. In November 2016, a new world record for pi was set by Peter Trueb. He succeeded in calculating pi to 22,459,157,718,361 (22.4 trillion digits). The calculation took 105 days. The computer file of the 22.4 trillion digits is nearly nine terabytes in size. If printed, it would

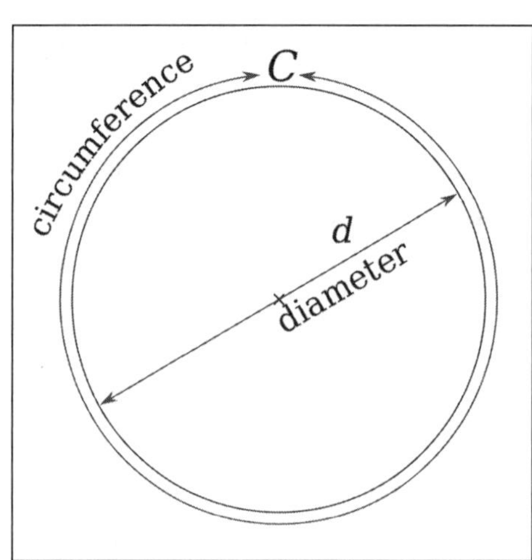

Pi is the value one gets when the circumferance of a circle is divided by the length of its own diameter.

fill several million books each with one thousand pages. The calculation was performed on a computer Trueb built with twenty-four hard drives, each containing six terabytes of memory. He used the computer program y-cruncher that was developed by Alexander Yee and uses the Chudnovsky algorithm. The program is available online for free for anyone to download.

What is another name for natural numbers?

Natural numbers are also known as "counting numbers." They are any positive, whole number. For example, 5 is a natural number, but –5 and 0.5 are not natural numbers.

What is the difference between an irrational number and a rational number?

Numbers that cannot be expressed as an exact ratio are called irrational numbers; numbers that can be expressed as an exact ratio are called rational numbers. For instance, 1/2 (one half, or 50 percent of something) is rational. Examples of irrational numbers are 1.61803 (ϕ), 3.14159 (π), and 1.41421 ($\sqrt{2}$). Irrational numbers are nonrepeating, nonterminating decimal numbers. History claims that Pythagoras (c. 580–c. 500 B.C.E.) in the sixth century B.C.E. first used the term when he discovered that the square root of 2 could not be expressed as a fraction.

What are real numbers?

Real numbers are a combination of rational numbers and irrational numbers. Real numbers can be thought of as points on an infinitely long number line. In other words, most of the numbers that we use on a daily basis are real numbers.

What is a perfect number?

A perfect number is a number equal to the sum of all its proper divisors (divisors smaller than the number), including 1. The number 6 is the smallest perfect number; the sum of its divisors 1, 2, and 3 equals 6. The next three perfect numbers are 28, 496, and 8,126. To date, all known perfect numbers end in either 6 or 8. No odd perfect numbers are known. The largest-known perfect number is:

$$2^{77,232,917} \times (2^{77,232,917} - 1)$$

This number is over forty-five million digits. It was discovered in 2017.

What are imaginary numbers?

Imaginary numbers are the square roots of negative numbers. Since the square is the product of two equal numbers with like signs, it is always positive. Therefore, no number multiplied by itself can give a negative real number. The symbol *i* is used to indicate an imaginary number.

How do imperfect numbers differ from perfect numbers?

Imperfect numbers are numbers whose factors (excluding the number itself), when added together, equal a number that is either larger or smaller than the number itself. The sum of the factors of 4—1 and 2—equals 3. The sum of the factors of 12—1, 2, 3, 4, and 6—equals 16. Imperfect numbers whose sum is less than the number itself are known as deficient numbers. Imperfect numbers whose sum is greater than the number itself are known as abundant numbers.

What are Fibonacci numbers?

Fibonacci numbers are a series of numbers where each, after the second term, is the sum of the two preceding numbers—for example, 1, 1, 2, 3, 5, 8, 13, 21, and so on. They were first described by Leonardo Fibonacci (c. 1180–c. 1250), also known as Leonardo of Pisa, as part of a thesis on series in his most famous book *Liber abaci* (*The Book of the Calculator*), published in 1202 and later revised by him.

What are prime numbers?

A prime number is one that is evenly divisible only by itself and 1. The integers 1, 2, 3, 5, 7, 11, 13, 17, and 19 are prime numbers. Euclid (c. 325–270 B.C.E.) proved that no "largest prime number" exists because any attempt to define the largest results in a paradox. If a largest prime number (P) existed, adding 1 to the product of all primes up to and including P, 1 1 (1 3 2 3 3 3 5 3 … 3 P) yields a number that is itself a prime number because it cannot be divided evenly by any of the known primes.

Who discovered the largest prime number presently known?

The largest-known prime—and fiftieth known Mersenne prime—was discovered on December 26, 2017, by Jonathan Pace (1964–). The number is $2^{77,232,917}$–1 and has 23,249,425 digits. The newest Mersenne prime number is also known as M77232917. Four different individuals verified M77232917 using four different programs on four different computers. Writing out the number with over twenty-three million digits could fill an entire bookshelf with nine thousand pages. If someone would write, by hand, five digits to an inch every second, it would take fifty-four days to write out the number and would extend over seventy-three miles (118 kilometers)!

Who was Marin Mersenne?

Marin Mersenne (1588–1648) was a French monk who did the first work studying prime numbers in the seventeenth century. Mersenne numbers are 1 less than a power of 2: 2^{n-1}.

What is the Great Internet Mersenne Prime Search (GIMPS)?

The Great Internet Mersenne Prime Search (GIMPS) was formed in January 1996 by George Woltman (1957–). The purpose of GIMPS is to discover new world-record-sized prime numbers. GIMPS relies on the computing efforts of thousands of small, personal computers around the world. Interested participants can become involved in the search for primes by going to *https://www.mersenne.org/gettingstarted/*. Since 1996, volunteers involved in GIMPS have discovered sixteen new Mersenne primes.

Marin Mersenne was a French polymath who was famous for coming up with Mersenne prime numbers. He is also famous for *Harmonie universelle* (1636), a music theory book.

What is the Sieve of Eratosthenes?

Eratosthenes (c. 276–c. 194 B.C.E.) was a Greek mathematician and philosopher who devised a method to identify (or "sift" out) prime numbers from a list of natural numbers arranged in order. It is a simple method, although it becomes tedious to identify large prime numbers. The steps of the sieve are:

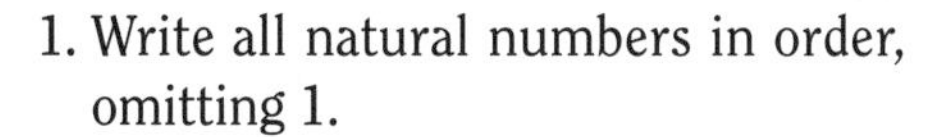

1. Write all natural numbers in order, omitting 1.
2. Circle the number 2 and then cross out every other number. Every second number will be a multiple of 2 and, hence, is not a prime number.
3. Circle the number 3 and then cross out every third number, which will be a multiple of 3 and, therefore, not a prime number.
4. The numbers that are circled are prime, and those that are crossed out are composite numbers.

Why is the number 10 considered important?

One reason is that the metric system is based on the number 10. The metric system emerged in the late eighteenth century out of a need to bring standardization to measurement, which had up until then been fickle, depending upon the preference of the ruler of the day, but 10 was important well before the metric system. Nicomachus of Gerasa (c. 60–c. 120), a second-century neo-Pythagorean from Judea, considered 10 a "perfect" number, the figure of divinity present in creation with mankind's fingers and toes. Pythagoreans believed 10 to be "the first-born of the numbers, the mother of them all, the one that never wavers and gives the key to all things." Also, shepherds of West Africa counted sheep in their flocks by colored shells based on 10, and 10 had evolved as a "base" of most numbering schemes. Some scholars believe the reason 10 developed as a base number had more to do with ease: 10 is easily counted on fingers, and the rules of addition, subtraction, multiplication, and division for the number 10 are easily memorized.

What prefixes define the powers of 10?

Various prefixes are used to represent powers of 10. These prefixes are usually derived from Greek (*kilo*, "thousand") or Latin (*mille*, "thousand").

Prefix	Value	Common Name
tera	10^{12}	Trillion
giga	10^{9}	Billion
mega	10^{6}	Million
kilo	10^{3}	Thousand
hecto	10^{2}	Hundred
deka	10^{1}	Ten
deci	10^{-1}	Tenth
centi	10^{-2}	Hundredth
milli	10^{-3}	Thousandth
micro	10^{-6}	Millionth
nano	10^{-9}	Billionth
pico	10^{-12}	Trillionth

How does the decimal system differ from the binary system?

Different number systems are based on different numbers for counting and measuring. The most common numeration system in the world is the decimal or base 10 system. The decimal system has digits for 0 through 9, with larger numbers identified with values in the place-value columns. For example, the number 100 is written "100": 0 in the ones column, 0 in the tens column, and 1 in the hundreds column.

The binary system, also known as base 2, includes the digits 0 and 1. Larger numbers are expressed as multiples of 2. The binary system is used for programming computers. The following chart expresses the numbers 1 through 10 in both the decimal and binary systems.

Decimal (Base 10)	Binary (Base 2)	Expansion
0	0	0 ones
1	1	1 one
2	10	1 two and 0 ones
3	11	1 two and 1 one
4	100	1 four, 0 twos, and 0 ones
5	101	1 four, 0 twos, and 1 one
6	110	1 four, 1 two, and 0 ones
7	111	1 four, 1 two, and 1 one
8	1000	1 eight, 0 fours, 0 twos, and 0 ones
9	1001	1 eight, 0 fours, 0 twos, and 1 ones
10	1010	1 eight, 0 fours, 1 two, and 0 ones

What are some very large numbers?

Value in Name	Number Powers of 10	Number of Groups of 0s	Number of Three 0s after 1,000
Billion	10^{9}	9	2
Trillion	10^{12}	12	3
Quadrillion	10^{15}	15	4
Quintillion	10^{18}	18	5
Sextillion	10^{21}	21	6
Septillion	10^{24}	24	7
Octillion	10^{27}	27	8
Nonillion	10^{30}	30	9
Decillion	10^{33}	33	10
Undecillion	10^{36}	36	11
Duodecillion	10^{39}	39	12
Tredecillion	10^{42}	42	13
Quattuor-decillion	10^{45}	45	14
Quindecillion	10^{48}	48	15
Sexdecillion	10^{51}	51	16
Septen-decillion	10^{54}	54	17
Octodecillion	10^{57}	57	18
Novemdecillion	10^{60}	60	19
Vigintillion	10^{63}	63	20
Centillion	10^{303}	303	100

The British, French, and Germans use a different system for naming denominations above one million. The googol and googolplex are rarely used outside the United States.

How large is a googol?

A googol is 10^{100} (the number 1 followed by 100 zeros). Unlike most other names for numbers, it does not relate to any other numbering scale. American mathematician Edward Kasner (1878–1955) first used the term in 1938; when searching for a term for this large number, Kasner asked his nephew, Milton Sirotta (1911–1981), then about nine years old, to suggest a name. The googolplex is 10 followed by a googol of zeros, represented as 10^{googol}. The popular Web search engine *Google.com* is named after the concept of a googol.

Is it possible to count to infinity?

No. Very large finite numbers are not the same as infinite numbers. Infinite numbers are defined as being unbounded, or without limit. Any number that can be reached by counting or by representation of a number followed by billions of zeros is a finite number.

How long has the abacus been used?

The abacus grew out of early counting boards, with hollows in a board holding pebbles or beads used to calculate. It has been documented in Mesopotamia back to around 3500 B.C.E. The current form, with beads sliding on rods, dates back at least to fifteenth-century China. Prior to the use of decimal number systems, which allowed the familiar paper-and-pencil methods of calculation, the abacus was essential for almost all multiplication and division. Unlike the modern calculator, the abacus does not perform any mathematical computations. The person using the abacus performs calculations in their head, relying on the abacus as a physical aid to keep track of the sums. It has become a valuable tool for teaching arithmetic to blind students.

Can a person using an abacus calculate more rapidly than someone using a calculator?

In 1946, the Tokyo staff of *Stars and Stripes* sponsored a contest between a Japanese abacus expert and an American accountant using the best electric adding machine then available. The abacus operator proved faster in all calculations except the multiplication of very large numbers. While today's electronic calculators are much faster and easier to use than the adding machines used in 1946, undocumented tests still show that an expert can add and subtract faster on an abacus than someone using an electronic calculator. It also allows long division and multiplication problems with more digits than a hand calculator can accommodate.

A teacher shows her students how to use an abacus, an ancient device that can be used for several types of calculations such as addition, multiplication, division, and even square roots.

What are Napier's bones?

In the sixteenth century, Scottish mathematician John Napier (1550–1617), Baron of Merchiston, developed a method of simplifying the processes of multiplication and division using exponents of 10, which Napier called logarithms (commonly abbreviated as logs). Using this system, multiplication is reduced to addition and division to subtraction. For example, the log of 100 (10^2) is 2; the log of 1,000 (10^3) is 3; the multiplication of 100 by 1000, $100 \times 1000 = 100{,}000$, can be accomplished by adding their logs: $\log[(100)(1000)] = \log(100) + \log(1000) = 2 + 3 = 5 = \log(100{,}000)$. Napier published his methodology in *A Description of the Admirable Table of Logarithms* in 1614. In 1617, he published a method of using a device, made up of a series of rods in a frame marked with the digits 1 through 9, to multiply and divide using the principles of logarithms. This device was commonly called "Napier's bones" or "Napier's rods."

What is the history of the slide rule?

Most engineering and design calculations for buildings, bridges, automobiles, airplanes, and roads were done on a slide rule until the mid-1970s. A slide rule is an apparatus with movable scales based on logarithms, which were invented by John Napier, Baron of Merchiston. The slide rule can, among other things, quickly multiply, divide, square root, or find the logarithm of a number. In 1620, Edmund Gunter (1581–1626) of Gresham College, London, England, described an immediate forerunner of the slide rule as his "logarithmic line of numbers." William Oughtred (1574–1660), rector of Aldbury, England, made the first rectilinear slide rule in 1621. This slide rule consisted of two logarithmic scales that could be manipulated together for calculation. His former pupil, Richard Delamain (1600–1644), published a description of a circular slide rule in 1630 (and received a patent at about that time for it), three years before Oughtred published a description of his invention (at least one source says that Delamain published it in 1620). Oughtred accused Delamain of stealing his idea, but evidence indicates that the inventions were probably arrived at independently.

The earliest existing straight slide rule using the modern design of a slider moving in a fixed stock dates from 1654. A wide variety of specialized slide rules were developed by the end of the seventeenth century for trades such as masonry, carpentry, and excise tax collecting. Peter Mark Roget (1779–1869), best known for his *Thesaurus of English Words and Phrases*, invented a log-log slide rule for calculating the roots and powers of numbers in 1814.

In 1967, Hewlett-Packard produced the first pocket calculators. Within a decade, slide rules became the subject of science trivia and collector's books. Interestingly, slide rules were carried on five of the Apollo space missions, including a trip to the Moon. They were known to be accurate and efficient in the event of a computer malfunction.

What are Cuisenaire rods?

The Cuisenaire method is a teaching system used to help young students independently discover basic mathematical principles. Developed by Emile-Georges Cuisenaire (1891–

1976), a Belgian schoolteacher, the method uses rods of ten different colors and lengths that are easy to handle. The rods help students understand mathematical principles rather than merely memorizing them. They are also used to teach elementary arithmetic properties such as associative, commutative, and distributive properties.

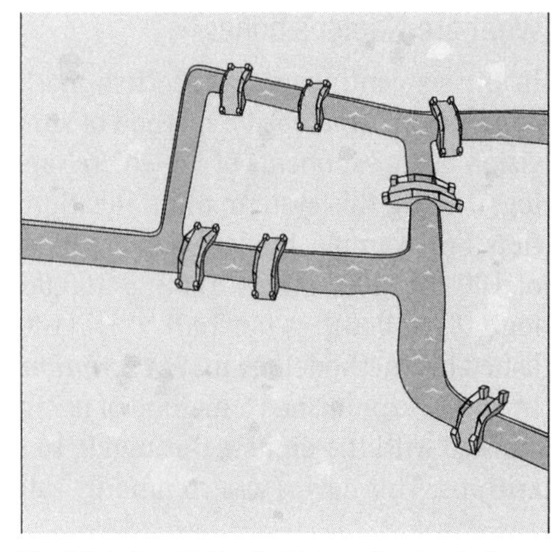

The Königsberg Bridge Problem invites the challenge of crossing all seven bridges in the pattern shown here only once (and without swimming a river haha). Leonhard Euler proved it was not possible.

What is the Königsberg Bridge Problem?

The city of Königsberg was located in Prussia on the Pregel River. Two islands in the river were connected by seven bridges. By the eighteenth century, it had become a tradition for the citizens of Königsberg to go for a walk through the town trying to cross each bridge only once. No one was able to succeed, and the question was asked whether it was possible to do so. In 1736, Leonhard Euler (1707–1783) proved that it was not possible to cross the Königsberg bridges only once. Euler's solution led to the development of two new areas of mathematics: graph theory, which deals with questions about networks of points that are connected by lines; and topology, which is the study of those aspects of the shape of an object that do not depend on length measurements.

What is the four-color map theorem?

The four-color map theorem was first posed by Francis Guthrie (1831–1899) in 1852. While coloring a map of the British counties, Guthrie discovered that he could do it with only four colors, and no two adjacent counties would be the same color. He extrapolated the question to whether every map, no matter how complicated and how many countries are on the map, could be colored using only four colors with no two adjacent countries being the same color. The theorem was not proved until 1976, 124 years after the question had been raised, by Kenneth Appel (1932–2013) and Wolfgang Haken (1928–). Appel and Haken's solution required approximately 1,200 hours of computer time to examine around 1,500 configurations. Their proof is considered correct, although it relies on computers for the calculations. No known simple way exists to check the proof by hand.

What is chaos?

Chaos, or chaotic behavior, is the behavior of a system whose final state depends very sensitively on the initial conditions. The behavior is unpredictable and cannot be distinguished from a random process, even though it is strictly determinate in a mathematical sense. Chaos studies the complex and irregular behavior of many systems in nature, such as changing weather patterns, flow of turbulent fluids, and swinging pendulums. Scientists once thought they could make exact predictions about such systems but found that a tiny difference in starting conditions can lead to greatly different results. Chaotic systems do obey certain rules, which mathematicians have described with equations, but the science of chaos demonstrates the difficulty of predicting the long-range behavior of chaotic systems.

Who pioneered the theory of chaos in the twentieth century?

Edward Lorenz (1917–2008) was an early pioneer in modern chaos theory. His interest in chaos came about accidentally while completing research in the field of weather prediction, which was his main area of research. His discovery came in 1961, when a computer model he had been running was actually saved using three-digit numbers rather than the six digits he had been using for calculations. This small rounding error produced very different results. He discovered that small changes in initial conditions can produce large changes in the long-term outcome. This behavior is known as deterministic chaos, or simply chaos. The theory of chaos was summarized by Lorenz as: When the present determines the future but the approximate present does not approximately determine the future.

Chaos theory was born from observing weather patterns, but it has become applicable to a variety of other situations in many areas of the physical sciences, the natural sciences, engineering, economics, finance, philosophy, anthropology, politics, psychology, and robotics.

What is Zeno's paradox?

Zeno of Elea (c. 490–c. 425 B.C.E.), a Greek philosopher and mathematician, is famous for his paradoxes, which deal with the continuity of motion. One form of the paradox is: If an object moves with constant speed along a straight line from point 0 to point 1, the object must first cover half the distance (1/2), then half the remaining distance (1/4), then half the remaining distance (1/8), and so on without end. The conclusion is that the object never reaches point 1. Because some distance is always left to be covered, motion is impossible. In another approach to this paradox, Zeno used an allegory telling of a race between a tortoise and Achilles (who could run 100 times as fast), where the tortoise started running 10 rods (165 feet) in front of Achilles. Because the tortoise always advanced 1/100 of the distance that Achilles advanced in the same time period, it was theoretically impossible for Achilles to pass him. British mathematician and writer Charles Dodgson (1832–1898), better known as Lewis Carroll, used the characters of Achilles and the tortoise to illustrate his paradox of infinity.

ALGEBRA

What is algebra?

Algebra is a branch of mathematics that uses variables and symbols to solve equations. It combines basic arithmetic, including addition, subtraction, multiplication, and division, with mathematical rules to express quantitative concepts. The term "algebra" originates from the title of the book *al-Kitab al-mukhtasar fi hisab al-jabr wa'l-muqabala* by Arab mathematician Muhammad ibn Mūsā al-Khwārizmī (c. 780–c. 850).

Which book is credited with the origin of the term "algebra"?

Arab mathematician Muḥammad ibn Mūsā al-Khwārizmī wrote *Al-kitāb al-mukhta Ṣar fī Ḥisāb al-Ğabr wa'l-muqābala*. It is believed that he wrote this treatise during the years 825–830 C.E. The translation of the title of the work is *The Compendious Book on Calculation by Completion and Balancing*. The term *al-jabr* referred to "restoring" or moving known or unknown quantities of the same power to one side of the equation. The term *wa'l-muqabala* referred to "balancing" the equation. This involved reducing or subtracting positive quantities of the same power on both sides of the equation. *Al-jabr* became the name of the branch of mathematics we know as "algebra" today.

A statue of Muḥammad ibn Mūsā al-Khwārizmī was erected in his home of Khiva, Uzbekistan.

Who are some of the mathematicians who played a significant role in the development of algebra?

The development of algebra dates back to Babylonian times. Babylonian clay tablets (1800–1600 B.C.E.) exist documenting a procedure to solve the first quadratic equation. Evidence exists from writings on papyrus (1650 B.C.E.) that the Egyptians solved problems in algebra, including fractions. Greek mathematician Diophantus (210–290) is considered by many to be the "father of algebra." He wrote the book *Arithmetica*, which included many solutions to indeterminate equations. He introduced the concept of using symbols instead of words to write out problems and solutions. He accepted any positive number, including fractions, when solving equations. His work later influenced Muhammad ibn Mūsā al-Khwārizmī

What is an equation?

An equation is a statement that says that two things are equal. The equal sign "=" will appear in the statement. For example, $x + 3 = 15$ is an equation. It means that what is on the left of the equal sign (=) is the same as on the right of the equal sign (=).

How does an equation differ from a mathematical expression?

A mathematical expression is a finite combination of symbols that may include numbers, variables, or operations. Examples of mathematical expressions are:

x

$0 + 7$

$8x - 5$

An equation has two expressions that are equal to each other and have an equal sign ("=") between the two expressions, such as:

$x - 4 = 6 + y$

When does $0 \times 0 = 1$?

Factorials are the product of a given number and all the factors less than that number. The notation n! is used to express this idea. For example, 5! (five factorial) is $5 \times 4 \times 3 \times 2 \times 1 = 120$. For completeness, 0! is assigned the value 1, so $0 \times 0 = 1$.

When did the concept of square root originate?

A square root of a number is a number that, when multiplied by itself, equals the given number. For instance, the square root of 25 is 5 ($5 \times 5 = 25$). The concept of the square root has been in existence for many thousands of years. Exactly how it was discovered is not known, but several different methods of exacting square roots were used by early mathematicians. Babylonian clay tablets between 1900 and 1600 B.C.E. contain the squares and cubes of integers 1 through 30. The early Egyptians used square roots around 1700 B.C.E., and during the Greek Classical period (600–300 B.C.E.), better arithmetic methods improved square root operations. In the sixteenth century, French mathematician René Descartes (1596–1650 C.E.) was the first to use the square root symbol, called "the radical sign," $\sqrt{}$.

What is abstract algebra?

Abstract algebra is the set of advanced topics of algebra that deal with abstract algebraic structures rather than the usual number systems. Algebraic structures include groups, rings, fields, lattices, and other concepts.

What is Boolean algebra?

Boolean algebra is an abstract mathematical system used to express the relationship between sets (groups of objects or concepts). British mathematician George Boole (1815–1864) was the first to develop this type of logic by demonstrating the algebraic manipulation of logical statements, showing whether or not a statement is true, and showing how a statement can be made into a simpler, more convenient form without changing its overall meaning. Boolean algebra is important in the study of information theory, the theory of probability, and the geometry of sets. The use of Boolean notation in electrical networks aided the development of switching theory and the eventual design of computers.

What are Venn diagrams?

Venn diagrams are graphical representations of set theory, which use circles to show the logical relationships of the elements of different sets using the logical operators (also called in computer parlance "Boolean Operators") and, or, and not. John Venn

English mathematician and logician John Venn created the Venn diagram, a helpful, visual way to study sets of things and how they relate to each other.

(1834–1923) first used them in his 1881 *Symbolic Logic*, in which he interpreted and corrected the work of George Boole and Augustus de Morgan (1806–1871). While Venn's attempts to clarify perceived inconsistencies and ambiguities in Boole's work are not widely accepted, his new method of diagramming was considered to be an improvement. Venn used shading to better illustrate inclusion and exclusion. Charles Dodgson, better known by his pseudonym Lewis Carroll, refined Venn's system, in particular by enclosing the diagram to represent the universal set.

GEOMETRY

What is geometry?

Geometry is a branch of mathematics that studies the sizes, shapes, positions, and dimensions of things. Squares, circles, and triangles are some of the simplest shapes in plane geometry, while cubes, cylinders, cones, and spheres are simple shapes in solid geometry.

What are the divisions of geometry?

Geometry can be divided into four major divisions, including:

1. Plane geometry, which studies lines, squares, circles, triangles, and polygons
2. Solid geometry, which studies three-dimensional shapes such as cubes, cylinders, cones, spheres, and polyhedrons
3. Spherical geometry, which deals with objects such as spherical triangles and spherical polygons
4. Analytic geometry, which deals with figures in terms of their positions, configurations, separations, and other coordinates

Many other specialized areas of geometry exist, such as absolute geometry, combinatorial geometry, enumerative geometry, inversive geometry, non-Euclidian geometry, ordered geometry, projective geometry, and stochastic geometry.

What is an angle?

An angle is a figure formed by two rays (or lines), called the sides of the angle, that share a common end point, the vertex of the angle.

What are the various types of angles?

The four basic types of angles are:

1. Straight angle, which is equal to ½ turn or 180 degrees
2. Right angle, which is equal to ¼ turn or 90 degrees
3. Acute angle, which is smaller than a right angle (less than 90 degrees)

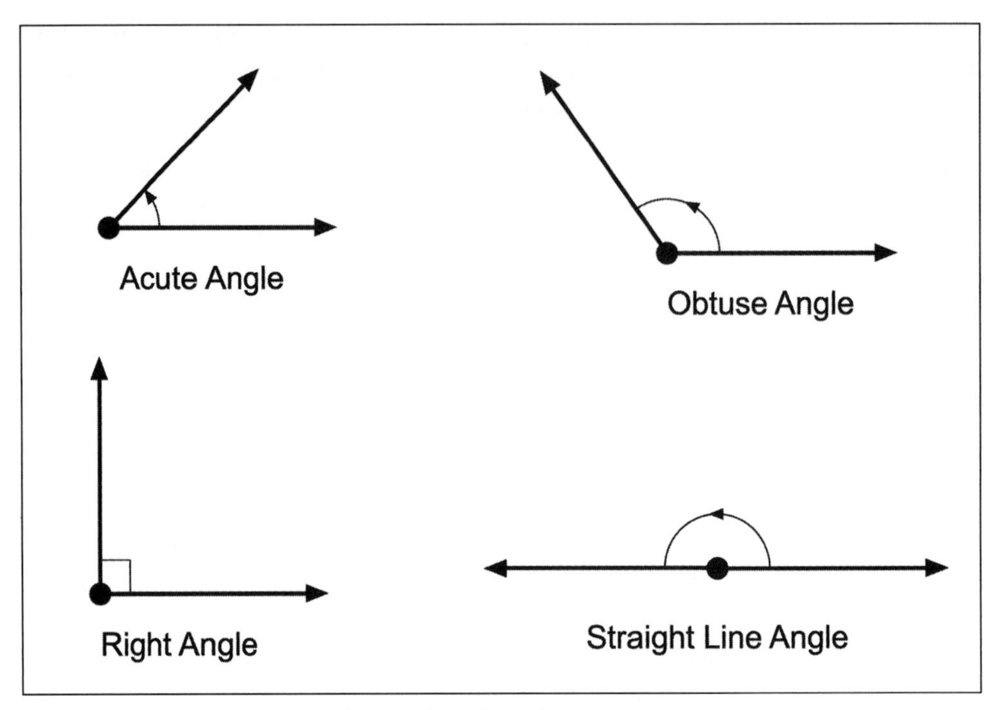

The four basic types of angles: acute, obtuse, right, and straight.

4. Obtuse angle, which is larger than a right angle and smaller than a straight angle (between 90 and 180 degrees).

What are the most common mathematical formulas for area?

Area of a rectangle:
 length times width — $l \times w$

Area of a circle:
 pi times the radius squared — πr^2 or $A = \frac{1}{4}\pi d^2$

Area of a triangle:
 one half the altitude times the base — $\frac{1}{2}ab$

Area of the surface of a sphere:
 four times pi times the radius squared — $4\pi r^2$ or $A = \pi d^2$

Area of a square:
 length times width, or length of one side squared — lw or $A = s^2$

Area of a cube:
 square of the length of one side times 6 — $6s^2$

Area of an ellipse:
 long diameter times short diameter times 0.7854.

What are the various types of triangles?

Triangles are polygons, or figures with three sides. The three interior angles of a triangle are always equal to 180 degrees. The six basic types of triangles are:

1. Acute triangle, in which all three angles are less than 90 degrees or the triangle has three acute angles
2. Obtuse triangle, in which one angle is greater than 90 degrees or the triangle has one obtuse angle
3. Right triangle, in which one angle is 90 degrees

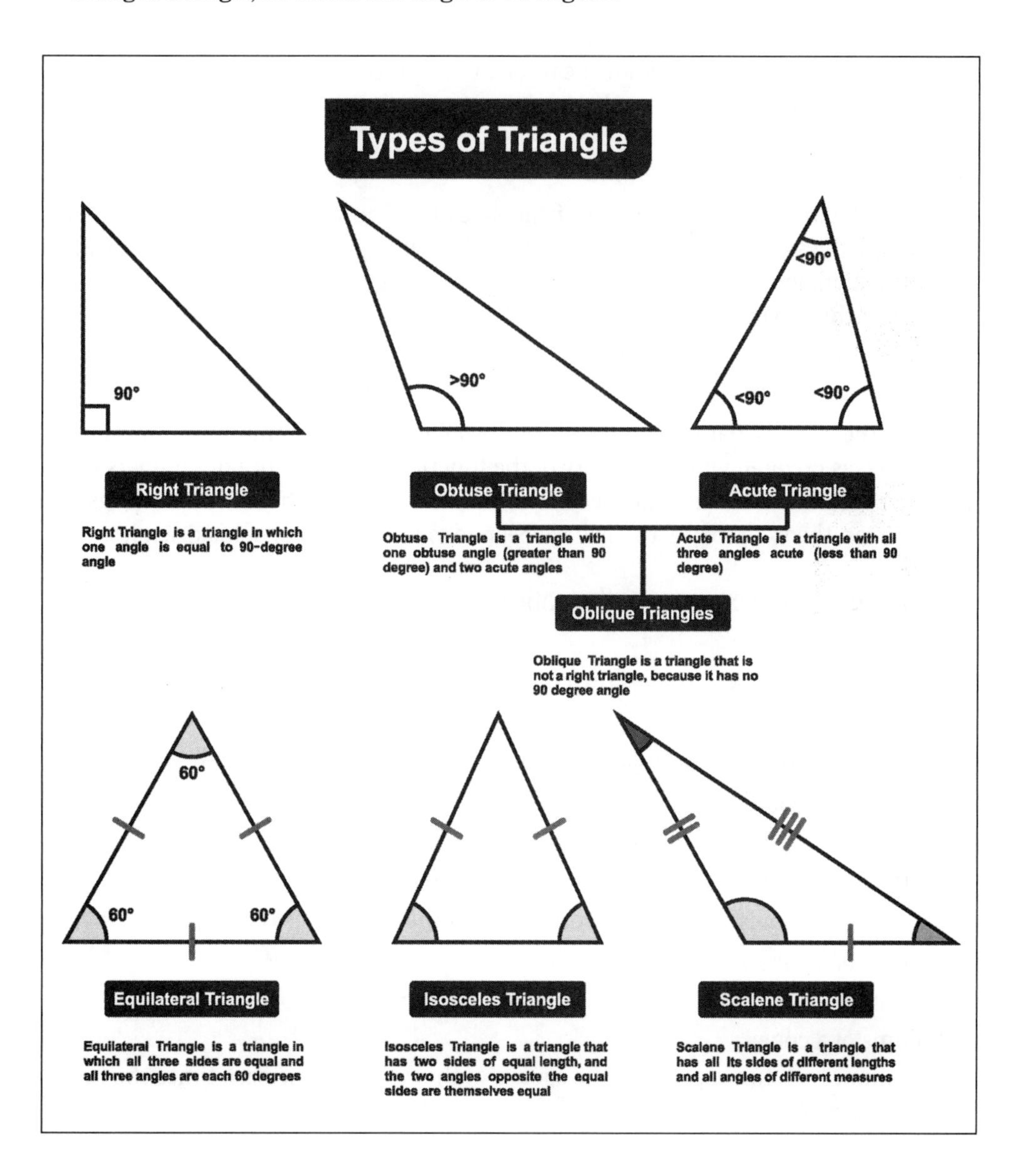

4. Scalene triangle, in which no sides and, therefore, no angles are equal
5. Isosceles triangle, in which two of the sides are equal and, therefore, the base angles are equal
6. Equilateral triangle, which is an acute triangle in which all three angles are equal

Who discovered the formula for the area of a triangle?

Heron (Hero) of Alexandria (first century C.E.) is best known in the history of mathematics for the formula that bears his name. This formula calculates the area of a triangle with sides a, b, and c with s = half the perimeter: $A = \sqrt{[(2 - a)(s - b)(s - c)]}$. The Arab mathematicians who preserved and transmitted the mathematics of the Greeks reported that this formula was known earlier to Archimedes, but the earliest proof now known is that appearing in Heron's *Metrica*.

What is the Pythagorean theorem?

In a right triangle (one where two of the sides meet in a 90-degree angle), the hypotenuse is the side opposite the right angle. The Pythagorean theorem, also known as the rule of Pythagoras, states that the square of the length of the hypotenuse is equal to the sum of the squares of the other two sides ($h^2 = a^2 + b^2$). If the lengths of the sides are h = 5 inches, a = 4 inches, and b = 3 inches, then:

$$h = \sqrt{(a^2 + b^2)} = \sqrt{(4^2 + 3^2)} = \sqrt{(16 + 9)} = \sqrt{25} = 5$$

The theorem is named for Greek philosopher and mathematician Pythagoras. Pythagoras is credited with the theory of the functional significance of numbers in the objective world and numerical theories of musical pitch. As he left no writings, the Pythagorean theorem may actually have been formulated by one of his disciples.

What are the most common mathematical formulas for volume?

Volume of a sphere:

Volume = ⅘ times pi times the cube of the radius – $V = \frac{4}{5} \times \pi r^3$

Volume of a pyramid:

Volume = ⅓ times the area of the base times the height – $V = \frac{1}{3}bh$

Volume of a cylinder:

Volume = area of the base times the height – $V = bh$

Volume of a circular cylinder (with circular base):

Volume = pi times the square of the radius of the base times the height – $V = \pi r^2 h$

Volume of a cube:

Volume = the length of one side cubed – $V = S^3$

Volume of a cone:

Volume = ⅓ times pi times the square of the radius of the base times the height $V = \frac{1}{3}\pi r^2 h$

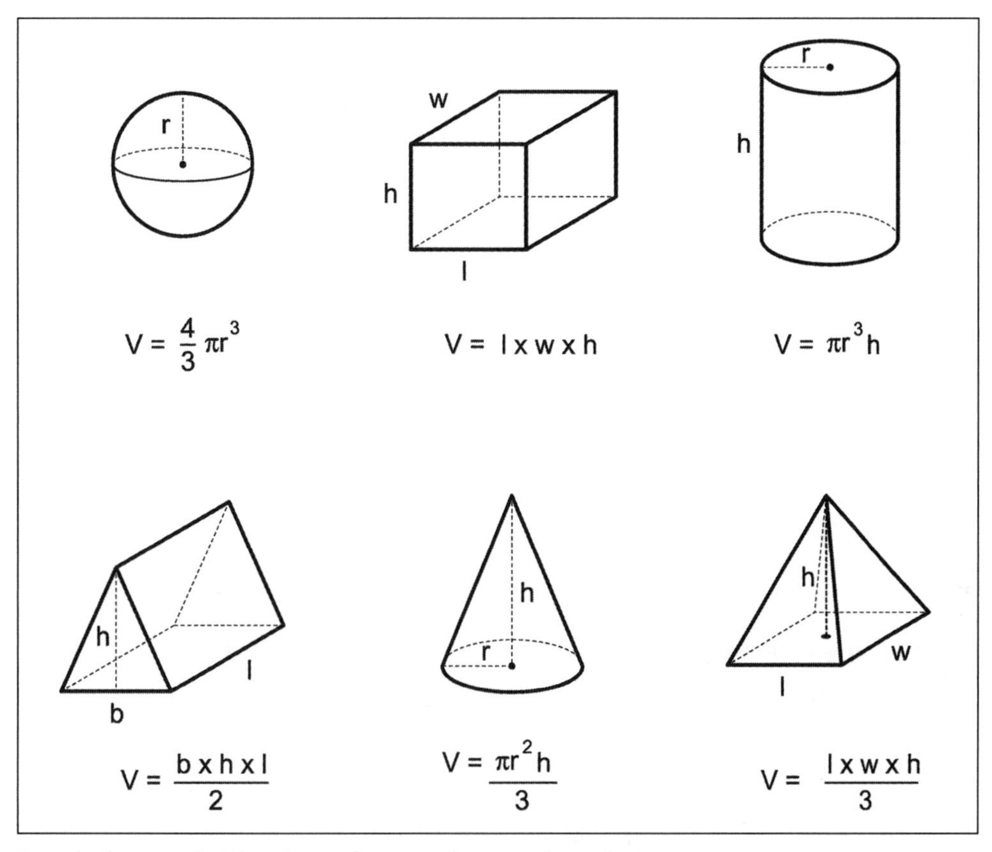

Formulas for computing the volumes of common shapes are shown above.

Volume of a rectangular solid:
Volume = length times width times height – $V = lwh$

What are axioms, theorems, and proofs?

An axiom is a simple idea that is thought to be true. An example of an axiom is that a straight line is the shortest distance between two points. A theorem is a statement demonstrated to be true by accepted mathematical operations and arguments. A theorem is usually based on some general principle that makes it part of a larger theory. It differs from an axiom in that a proof is required for its acceptance. A proof is simply the process of showing a theorem to be correct.

What are the names for some regular polygons?

Polygons are named for the number of sides that are in the figure. Not all polygons have names. A polygon with n sides is called an n-gon. It is also possible to substitute "n-gon" when the name of a polygon is not known, such as one with twenty-three sides. It would be referred to as a 23-gon.

Names for Some Regular Polygons

Sides	Name
3	Triangle
4	Quadrilateral or Tetragon
5	Pentagon
6	Hexagon
7	Heptagon
8	Octagon
9	Nonagon
10	Decagon
12	Dodecagon
15	Pentadecagon
20	Isosagon
50	Pentacontagon
100	Hectogon
1,000	Chiliagon
10,000	Myriagon

What are the Platonic solids?

The Platonic solids are the five regular polyhedra: the four-sided tetrahedron, the six-sided cube or hexahedron, the eight-sided octahedron, the twelve-sided dodecahedron, and the twenty-sided icosahedron. While they had been studied as long ago as the time of Pythagoras, they are called the Platonic solids because they were first described in detail by Plato (427–347 B.C.E.) around 400 B.C.E. The ancient Greeks gave mystical significance to the Platonic solids: the tetrahedron represented fire, the icosahedron represented water, the stable cube represented Earth, and the octahedron represented the air. The twelve faces of the dodecahedron corresponded to the twelve signs of the zodiac, and this figure represented the entire universe.

What is the ancient Greek problem of squaring the circle?

This problem was to construct, with a straight-edge and compass, a square having the same area as a given circle. The Greeks were unable to solve the problem because the task is impossible, as was shown by German mathematician Ferdinand von Lindemann (1852–1939) in 1882.

What is a golden section?

A golden section, also called the divine proportion, is the division of a line segment so that the ratio of the whole segment to the larger part is equal to the ratio of the larger part to the smaller part. The ratio is approximately 1.61803 to 1. The number 1.61803 is called the golden number (also called Phi [with a capital P]). The golden number is the limit of the ratios of consecutive Fibonacci numbers such as, for instance, 21/13 and 34/21. A golden rectangle is one whose length and width correspond to this ratio.

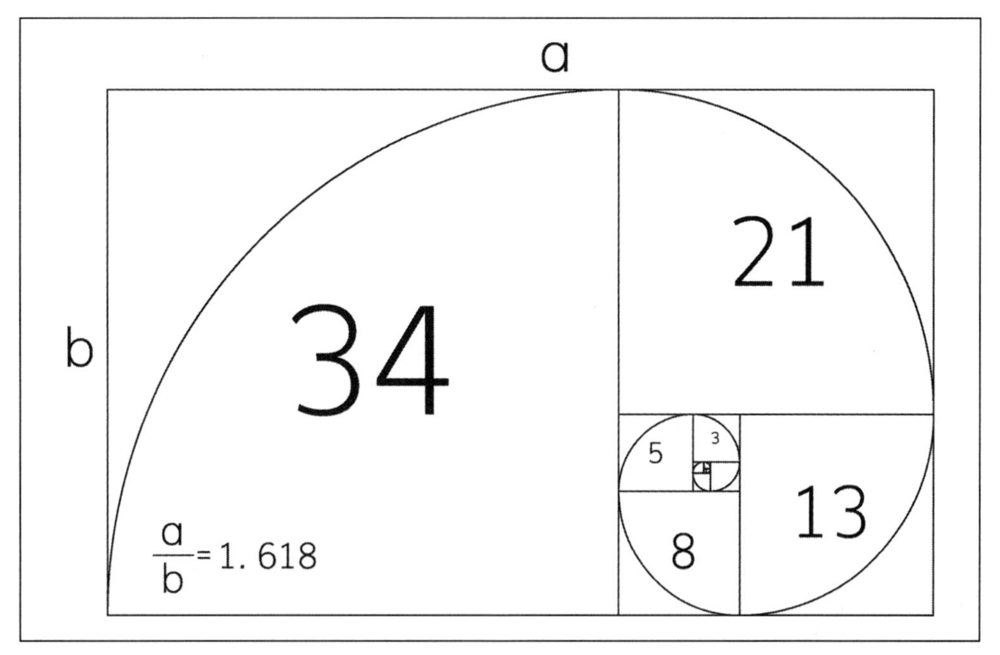

The divine proportions of the golden section can be found often in nature, such as in the shell of a chambered nautilus, the seeds in a sunflower, or the structure of a pine cone, and more.

The ancient Greeks thought this shape had the most pleasing proportions. Many famous painters have used the golden rectangle in their paintings, and architects have used it in their design of buildings, the most famous example being the Greek Parthenon.

What is Pascal's triangle?

Pascal's triangle is an array of numbers arranged so that every number is equal to the sum of the two numbers above it on either side. It can be represented in several slightly different triangles, but this is the most common form:

```
              1
            1   1
          1   2   1
        1   3   3   1
      1   4   6   4   1
    1   5  10  10   5   1
```

How is Pascal's triangle used?

The triangle is used to determine the numerical coefficients resulting from the computation of higher powers of a binomial (two numbers added together). When a binomial is raised to a higher power, the result is expanded using the numbers in that row of the triangle. For example, $(a + b)1 = a1 + b1$ using the coefficients in the second line of the triangle. $(a + b)^2 = a^2 + 2ab + b^2$ using the coefficients in the next line of the triangle. (The first line of the triangle correlates to [a = b]0.) While the calculation of coefficients is fairly straightforward, the triangle is useful in calculating them for the higher powers without needing to multiply them out. Binomial coefficients are useful in calculating probabilities; Blaise Pascal (1623–1662) was one of the pioneers in developing laws of probability.

As with many other mathematical developments, some evidence exists of a previous appearance of the triangle in China. Around 1100 C.E., Chinese mathematician Chia Hsien (1010–1070) wrote about "the tabulation system for unlocking binomial coefficients"; the first publication of the triangle was probably in a book called *Piling-Up Powers and Unlocking Coefficients* by Liu Ju-Hsieh.

What is a Möbius strip?

A Möbius strip is a surface with only one side, usually made by connecting the two ends of a rectangular strip of paper after putting a half twist (180 degrees relative to the opposite side) in the strip. Cutting a Möbius strip in half down the center of the length of the strip results in a single band with four half twists. Devised by German mathematician August Ferdinand Möbius (1790–1868) to illustrate the properties of one-sided surfaces, it was presented in a paper that was not discovered or published until after his death. Another nineteenth-century German mathematician, Johann Benedict Listing (1808–1882), developed the idea independently at the same time.

What are fractals?

A fractal is a set of points that are too irregular to be described by traditional geometric terms but that often have some degree of self-similarity; that is, they are made of parts that resemble the whole. They are used in image processing to compress data and to depict apparently chaotic objects in nature such as mountains or coastlines. Scientists also use fractals to better comprehend rainfall trends, patterns formed by clouds and waves, and the distribution of vegetation. Fractals are also used to create computer-generated art.

Fractal patterns, which can be seen in nature such as in this romanesco broccoli, result when a set of points add up to resemble the whole of something.

CALCULUS

What is calculus?

Calculus is a mathematical discipline that deals with instantaneous rates of change of quantities (differentiation) and with the accumulation of quantities (integration). It grew out of a desire to understand such physical phenomena as the orbits of planets and the effects of gravity. The immediate success of calculus in formulating physical laws and predicting their consequences led to a new division in mathematics called analysis, of which calculus is an important component. Today, calculus is the essential language of science and engineering, providing the means by which physical laws are expressed in mathematical terms.

Who invented calculus?

The invention of calculus marked the beginning of higher mathematics. Gottfried Wilhelm Leibniz (1646–1716) and Isaac Newton are the two scientists generally recognized as having laid the foundations for calculus. The question of who invented calculus was debated throughout their lives, and most scientists on the continent of Europe gave credit to Leibniz, while those in England gave credit to Newton. History suggests that both Leibniz and Newton independently discovered the fundamental theorem of calculus, which describes the relationship between derivatives and integrals. Leibniz is credited for creating a notation for the integral. The integral symbol looks like an elongated "S" ($\int$). Leibniz is also credited with developing a notation for finding a derivative. Both of these symbols are still used in calculus today. Newton was interested in the study of "fluxions," or methods that are used to describe how things change over time. The motion of an object often changes over time and can be described using derivatives. Newton is also credited with finding many different applications of calculus in the physical world. As to who was first, Leibniz began investigating calculus ten years after Newton and may not have been aware of Newton's efforts, yet Leibniz published his results twenty years before Newton published his. Newton was typically very late in publishing his work, so although Newton discovered many of the concepts in calculus eight to ten years before Leibniz, Leibniz was the first to make his work public by publishing the

Gottfried Wilhelm Leibniz is credited with inventing calculus at the same time Isaac Newton was developing the same new math.

first paper on calculus, *Nova methodus pro maximis et minimis, itemque tangentibus, ..., & singulare pro illis calculi genus* (translated into English: *New method for the maximum and minimum, and also tangents, ..., and a singular [remarkable] type of calculus for them*), in 1684.

What are the two major fields of calculus?

The two major fields of calculus are differential calculus and integral calculus. Differential calculus uses the concept of function derivatives to study the behavior and rate of how different quantities change. Using the process of differentiation, the graph of a function can actually be computed, analyzed, and predicted. The second subfield is called integral calculus. Integration is actually the reverse process of differentiation.

What is a function?

In the simplest terms, a function shows the relationship between input and output.

What is a derivative?

The concept of derivative is at the core of calculus. The definition of the derivative can be approached in two different ways. One is geometrical (as a slope of a curve), and the other is physical (as a rate of change). In either case, the emphasis is on the use of the derivative as a tool.

Who uses calculus?

Calculus is the language of engineers, scientists, and economists. The work of these professionals has a major impact on our daily lives. Calculus gives a researcher the ability to find the effects of changing conditions on a system. A researcher can then learn how to control a system to make it behave in a desired way. With the ability to model and control systems, calculus provides extraordinary power over the material world. Here are some specific examples:

- Credit card companies use calculus to set the minimum payments due on credit card statements at the exact time the statement is processed by considering multiple variables such as changing interest rates and a fluctuating available balance.
- Biologists use differential calculus to determine the exact rate of growth in a bacterial culture when different variables, such as temperature and nutrition source, are changed. This research can help increase the rate of growth of beneficial bacteria or decrease the rate of growth for harmful and potentially threatening bacteria.
- An electrical engineer uses integration to determine the exact length of power cable needed to connect two substations that are miles apart. Because the cable is hung from poles, it is constantly curving. Calculus allows a precise figure to be determined.
- A physicist uses calculus to find the center of mass of a sports-utility vehicle to design appropriate safety features that must adhere to federal specifications on different road surfaces and at different speeds.

- An architect will use integration to determine the amount of materials necessary to construct a curved dome over a new sports arena as well as calculate the weight of that dome and determine the type of support structure required.

STATISTICS AND PROBABILITY

What is statistics?

Statistics is a branch of mathematics that deals with the collection, organization, presentation, analysis, and interpretation of data. Statistics is an integral component of studies in such fields as the biological, medical, and physical sciences, engineering, and the social and behavioral sciences as well as business, economics, and finance.

What is probability?

Probability is a branch of mathematics that considers and measures the likelihood that an event will occur. The higher the probability of an event, the more likely it is that the event will occur. Events can be as simple as a coin toss—what is the probability that it will land "heads"—to more complex ones such as the probability of a major earthquake in San Francisco next year or the probability that an individual will be diagnosed with pancreatic cancer after surviving Stage 3 Hodgkin's disease.

What is a sample?

In statistics and quantitative research studies, a sample is a set of data collected and/or selected from a statistical population by a defined procedure. The elements of a sample are known as sample points or sampling units. In many statistical studies, the population is very large, making a complete study or enumeration impractical or impossible. The sample usually represents a subset of manageable size that can provide statistical information that can be used to make inferences or extrapolations from the sample to the larger population.

What is the difference between the median, the mean, and the mode in statistics?

In mathematics, mean has several definitions, but most commonly for a data set, it refers to the arithmetic mean, which is the sum of values of a data set divided by the number

What is chance?

Chance is mathematically related to probability. Chance is a measure of how likely an event will occur. Such events relating to chance range from purchasing a winning lottery ticket to meeting someone at a concert or sporting event that you have not seen in ten years.

of values. Mean is frequently referred to as the average. In this example: (1 + 2 + 2 + 3 + 4 + 7 + 0) / 7, the mean is 4. Median is the middle value separating the greater and lesser halves of a data set. In this example: 1, 2, 2, 3, 4, 7, 9 the median is 3. Mode is the most frequent value in a data set. In this example: 1, 2, 2, 3, 4, 7, 9 the mode is 2.

What is a normal distribution?

The normal distribution (also called the Gaussian or Gauss or Laplace-Gauss) distribution in probability theory is a very common, continuous probability distribution. The normal distribution is sometimes informally called the bell curve. In the normal distribution, or bell curve, more than two-thirds of the sampling or measurements fall in the central region of the graph, and about one-sixth of them are found on either side.

What are ratios and proportions?

A ratio is a relationship between two numbers or two objects that defines the quantity of the first in comparison to the second. For example, in canvassing a local neighborhood, it was determined that four bicycles were present for every tricycle. The ratio is 4:1 (bicycles to tricycles). Ratios can also be written in the fractional form, so comparing three boys with five girls may be written as 3:5 or 3/5. We often think of fractions in the form of part to whole, but a ratio is not necessarily comparing a part to a whole. In the above example of boys and girls, the denominator, 5, represents girls. If we wanted to compare boys to the total number of people, the ratio would be 3:8 = 3/8.

A proportion is a mathematical comparison between two numbers. Frequently, these numbers can represent a comparison between people or objects. For example, if a fruit display contains six limes and eight lemons, the mathematical proportions may be written in two ways. The first way is to use the colon symbol, :. The second way is to write the proportion in the form of equivalent fractions: i.e., 5:10 or 5/10 = 1/2.

What is percent?

A percent (using the symbol %) is the ratio of one number to another. Percents are quantitative terms in which *n* percent of a number is *n* one-hundredths of the number. They are usually expressed as the equivalent ratio of some number to the number 100. Mathematical operations can be conducted with percents when they are translated into ratios and fractions, such as 25 percent is equal to 0.25 or 1/4.

How is percentage of increase calculated?

To find the percentage of increase, divide the amount of increase by the base amount. Multiply the result by 100 percent. For example, a raise in salary from $10,000 to $12,000 would have a percentage of increase = (2,000/10,000) × 100% = 20%.

What is the difference between simple interest and compound interest?

Simple interest is calculated on the amount of principal only. Compound interest is calculated on the amount of principal plus any previous interest already earned. For ex-

How many ways can a deck of cards be shuffled?

It is possible to shuffle a deck of cards 80,658,175,170,943,878,571,660,636,856, 403,766,975,289,505,440,883,277,824,000,000,000,000 ways.

ample, $100 invested at a rate of 5 percent for one year will earn $5.00 after one year earning simple interest. The same $100 will earn $5.12 if compounded monthly.

How many different bridge games are possible?

Roughly fifty-four octillion different bridge games are possible.

What is the frequency of multiple births in the United States?

The twin birth rate in the United States in 2015 (data released in 2017) was 33.5 twins per 1,000 live births, or 3.35 percent of live births. The birth rate for triplets or higher-order births was 103.6 per 100,000 live births.

If thirty people are chosen at random, what is the probability that at least two of them have their birthday on the same day?

The probability that at least two people in a group of thirty share the same birthday is about 70 percent.

What is the probability of a successful triple play occurring in a single baseball game?

The odds against a triple play in a game of baseball are 1,400 to 1.

INTRODUCTION AND HISTORICAL BACKGROUND

What is classical physics?

Physics investigates the matter and energy in the universe and the interactions and relations between them. It is concerned with the forces that exist between objects. Classical physics focuses on large-scale phenomena on the macroscopic level. The fields of light, heat, sound, electricity, magnetism, and mechanics are studied in classical physics.

When did the era of modern physics begin?

The era of modern physics began in the early twentieth century. The concepts of quantum theory and relativity established modern physics. The three most influential physicists in the development of modern physics were Max Planck (1858–1947), Albert Einstein (1879–1955), and Werner Karl Heisenberg (1901–1976). Modern physics focuses on the nature and behavior of particles on the submicroscopic level. The fields of atomic physics, particle physics, and high-energy physics are investigated in modern physics.

What was Max Planck's contribution to the development of modern physics?

Max Planck proposed that atoms could only absorb or emit energy at certain frequencies. These frequencies were only whole numbers that were multiples of a base frequency h. The h became known as the Planck constant. He introduced the concept of the "quanta" of energy—small packets of energy. Planck was awarded the Nobel Prize in Physics in 1918 for his work in quantum physics.

Who is generally regarded as the founder of quantum mechanics?

German theoretical physicist and Nobel Prize winner Werner Heisenberg remains one of the giants of his field for his work in quantum mechanics.

German mathematical physicist Werner Karl Heisenberg is regarded as the father of quantum mechanics (the theory of small-scale physical phenomena). His theory of uncertainty in 1927 overturned traditional classical mechanics and electromagnetic theory regarding energy and motions when applied to subatomic particles such as electrons and parts of atomic nuclei. The theory states that while it is impossible to specify precisely both the position and the simultaneous momentum (mass × velocity) of a particle, they can only be predicted. This means that the result of an action can be expressed only in terms of probability that a certain effect will occur. Heisenberg was awarded the Nobel Prize in Physics in 1932.

Who is considered the founder of modern theoretical physics?

Albert Einstein (1879–1955) was the principal founder of modern theoretical physics; his theory of relativity (speed of light is a constant and not relative to the observer or source of light) and the relationship of mass and energy ($E = mc^2$) fundamentally changed human understanding of the physical world.

Einstein produced three landmark papers in 1905. These papers dealt with the nature of particle movement known as Brownian motion, the quantum nature of electromagnetic radiation as demonstrated by the photoelectric effect, and the special theory of relativity. Although Einstein is probably best known for the last of these works, it was for his quantum explanation of the photoelectric effect that he was awarded the 1921 Nobel Prize in Physics.

His stature as a scientist, together with his strong humanitarian stance on major political and social issues, made him one of the outstanding men of the twentieth century.

ENERGY, MOTION, AND FORCE

What is energy?

Physicists define energy as the capacity to do work. Work is defined as the force required to move an object some distance. Examples of the different kinds of energy are thermal (heat) energy, light (radiant) energy, mechanical energy, electrical energy, and nuclear energy. The law of the conservation of energy states that within an isolated system, energy may be transformed from one form to another, but it cannot be created nor can it be destroyed.

What are the two forms of energy?

The two forms of energy are kinetic (working) energy and potential (stored) energy. Kinetic energy is the energy possessed by an object as a result of its motion, while potential energy is the energy possessed (stored) by an object as a result of its position. As an example, a ball sitting on top of a fence has potential energy. When the ball falls off the fence, it has kinetic energy. The potential energy is transformed into kinetic energy. Some examples of different forms of energy are:

- Chemical energy is a form of potential energy. It is the energy stored in the bonds of atoms and molecules. Some examples of chemical energy are batteries, biomass, petroleum, natural gas, and coal.
- Mechanical energy is also a form of potential energy. It is stored in objects by tension. Examples of mechanical energy are compressed springs and stretched rubber bands.
- Nuclear energy is energy stored in the nucleus of an atom. Large amounts of energy may be released when the nuclei are combined or split apart. Nuclear energy is a form of potential energy.
- Gravitational energy is energy stored in an object's height. The higher and heavier the object, the more gravitational energy is stored. As an object goes down a hill and picks up speed, the gravitational energy is converted to motion energy. Gravitational energy is also a form of potential energy.
- Radiant energy is electromagnetic energy that travels in transverse waves. Light and sunshine are examples of radiant energy. It is a form of kinetic energy.
- Thermal energy (heat) is the energy that comes from the movement of atoms and molecules in a substance. Thermal energy is a form of kinetic energy.
- Motion energy is energy stored in the movement of objects. The faster an object moves, the more energy is stored. Motion energy is also a form of kinetic energy.

How is thermal energy transferred?

Heat always flows from the hotter (energy sources) to cooler objects (energy receivers). Thermal energy can be transferred in three ways: conduction, convection, and radiation. Conduction occurs when two objects are in contact. An example of conduction is when you place your hand in hot water. The heat of the water warms your hand. Convection is the motion of a fluid, usually air or water. The fluid is heated by the hotter object, then moves until it contacts a colder object, where it heats that object. Radiation is infrared waves that are emitted by hotter objects and absorbed by colder ones. If you bring your hands near a hot electric burner on a stove, you can feel radiation. The Sun heats Earth by radiation.

Which amusement park ride provides an example of the transformation of energy?

A roller-coaster ride is a fun example of the transformation of energy from potential to kinetic. Potential energy builds up at the top of the hill and becomes kinetic energy as the coaster descends.

Roller-coaster rides provide an example of the transformation of energy. When a roller coaster is pulled to the top of the first hill, it has the greatest potential energy—the energy of position, in this case, the height of the first hill. As the roller coaster drops and gains speed, the potential energy is converted to kinetic energy—the energy of motion. When the roller coaster reaches the bottom of the first hill, it has very little potential energy but the greatest amount of kinetic energy. The kinetic energy is used to propel the roller coaster to the top of the next hill, increasing the potential energy. Roller-coaster rides are a continuous exchange between potential energy and kinetic energy.

What is inertia?

Inertia is the tendency of all objects and matter in the universe to stay still or, if moving, to continue moving in the same direction unless acted on by some outside force. This forms the first law of motion formulated by Isaac Newton. To move a body at rest, enough external force must be used to overcome the object's inertia; the larger the object is, the more force is required to move it. In his *Philosophiae Naturalis Principia Mathematica,* published in 1687, Newton sets forth all three laws of motion. Newton's second law is that the force to move a body is equal to its mass times its acceleration (F = MA), and the third law states that every action has an equal and opposite reaction.

Which hill in a roller-coaster ride will always be the highest?

The first hill in a roller-coaster ride will always be the highest. The maximum potential energy is at the top of the highest hill. The greatest amount of kinetic energy (speed) is at the bottom of the highest hill. Since energy can only be transferred and not gained during a ride, a roller-coaster would not be able to reach the top of a hill higher than the first in the middle of the ride.

What is superconductivity?

Superconductivity is a condition in which many metals, alloys, organic compounds, and ceramics conduct electricity without resistance, usually at low temperatures. Heike Kamerlingh Onnes (1853–1926), a Dutch physicist, discovered superconductivity in 1911. He was awarded the Nobel Prize in Physics in 1913 for his low-temperature studies. The modern theory regarding the phenomenon was developed by three American physicists—John Bardeen (1908–1991), Leon N. Cooper (1930–), and John Robert Schrieffer (1931–). Known as the BCS theory after the three scientists, it postulates that superconductivity occurs in certain materials because the electrons in them, rather than remaining free to collide with imperfections and scatter, form pairs that can flow easily around imperfections and do not lose their energy. Bardeen, Cooper, and Schrieffer received the Nobel Prize in Physics for their work in 1972.

A further breakthrough in superconductivity was made in 1986 by J. Georg Bednorz (1950–) and K. Alex Müller (1927–). Bednorz and Müller discovered a ceramic material consisting of lanthanum, barium, copper, and oxygen, which became superconductive at 35 K (–238°C)—much higher than any other material. Bednorz and Müller won the Nobel Prize in Physics in 1987. This was a significant accomplishment since in most situations, the Nobel Prize is awarded for discoveries made as many as twenty to forty years earlier.

What are some practical applications of superconductivity?

A variety of uses have been proposed for superconductivity in fields as diverse as electronics, transportation, and power. Research continues to develop more powerful, more efficient electric motors and devices that measure extremely small magnetic fields for medical diagnosis. The field of electric power transmission has much to gain by developing superconducting materials since 15 percent of the electricity generated must be used to overcome the resistance of traditional copper wire. More powerful electromagnets will be utilized to build high-speed magnetically levitated trains, known as "maglevs."

What is friction?

Friction is defined as the force that resists motion when the surface of one object slides over or comes into contact with the surface of another object. The three laws that govern the friction of an object at rest and the surface with which it is in contact state:

- Friction is proportional to the weight of an object.
- Friction is not determined by the surface area of the object.
- Friction is independent of the speed at which an object is moving along a surface provided the speed is not zero.

Although friction reduces the efficiency of machines and opposes movement, it is an essential force. Without friction, it would be impossible to walk, drive a car, or even strike a match.

Why is friction important when striking a match?

The head at the tip of a strike-anywhere match contains all the chemicals required to create a spark. A strike-anywhere match only needs to be rubbed against a surface with a high coefficient of friction, such as sandpaper, to create enough frictional heat to ignite the match. Safety matches differ from strike-anywhere matches since the chemicals necessary for ignition are divided between the match head and the treated strip found on the matchbox or matchbook. The friction between the match head and the treated strip will ignite the match. Matches fail to ignite when wet because water reduces friction.

Why is a lubricant, such as oil, often used to counter the force of friction?

Lubricants, such as oil, are used to reduce friction. For example, in machines consisting of metal parts, the continuous rubbing of the parts together increases the temperature and creates heat. To prevent serious wear and damage to the machines, grease and oil are applied to reduce the friction.

Who successfully demonstrated that curveballs actually curve?

In 1959, Lyman Briggs (1874–1963) demonstrated that a baseball can curve up to 17.5 inches (44.45 centimeters) over the 60 feet, 6 inches (18.4 meters) it travels between a pitcher and a batter, ending the debate of whether curveballs actually curve or if the apparent change in course was merely an optical illusion. Briggs studied the effect of spin and speed on the trajectory and established the relationship between the amount of curvature and the spin of the ball.

A rapidly spinning baseball experiences two lift forces that cause it to curve in flight. One is the Magnus force, named after H. G. Magnus (1802–1870), the German physicist who discovered it, and the other is the wake deflection force. The Magnus force causes the curveball to move sideways because the pressure forces on the ball's sides do not balance each other. The stitches on a baseball cause the pressure on one side of the ball to be less than on its opposite side. This forces the ball to move faster on one side than the

Why do golf balls have dimples?

The dimples minimize the drag (a force that makes a body lose energy as it moves through a fluid or gas), allowing the ball to travel farther than a smooth ball would travel. The air, as it passes over a dimpled ball, tends to cling to the ball longer, reducing the eddies or wake effects that drain the ball's energy. A dimpled ball can travel up to 300 yards (275 meters), but a smooth ball only goes 70 yards (65 meters). A ball can have three hundred to five hundred dimples that can be 0.01 inches (0.25 millimeters) deep. Another effect to get distance is to give the ball a backspin. A backspin creates less air pressure on the top of the ball, so the ball stays aloft longer (much like an airplane).

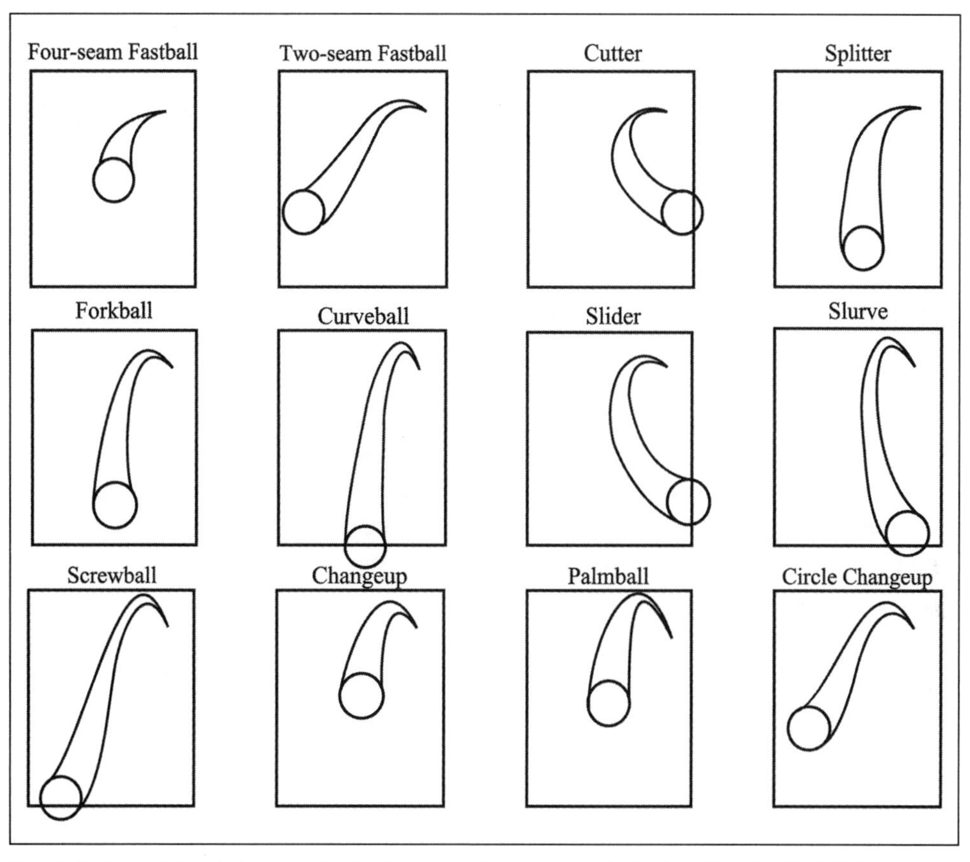

Curveball pitches in baseball provide fascinating, real-life examples of physics in action. A skilled pitcher can hold a ball a certain way and approach the throw from different angles to make the ball swerve, drop, and spin in ways that can annoy a batter.

other and forces the ball to "curve." The wake deflection force also causes the ball to curve to one side. It occurs because the air flowing around the ball in the direction of its rotation remains attached to the ball longer, and the ball's wake is deflected.

Why does a boomerang return to its thrower?

Two well-known scientific principles dictate the characteristic flight of a boomerang: (1) the force of lift on a curved surface caused by air flowing over it and (2) the unwillingness of a spinning gyroscope to move from its position.

When a person throws a boomerang properly, he or she causes it to spin vertically. As a result, the boomerang will generate lift, but it will be to one side rather than upward. As the boomerang spins vertically and moves forward, air flows faster over the top arm at a particular moment than over the bottom arm. Accordingly, the top arm produces more lift than the bottom arm and the boomerang tries to twist itself, but because it is spinning fast, it acts like a gyroscope and turns to the side in an arc. If the

What is Maxwell's demon?

An imaginary creature that, by opening and shutting a tiny door between two volumes of gases, could, in principle, concentrate slower molecules in one (making it colder) and faster molecules in the other (making it hotter), thus breaking the second law of thermodynamics. Essentially, this law states that heat does not naturally flow from a colder body to a hotter body; work must be expended to make it do so. This hypothesis was formulated in 1871 by James Clerk Maxwell (1831–1879), who is considered by many to be the greatest theoretical physicist of the nineteenth century. The demon would bring about an effective flow of molecular kinetic energy. This excess energy would be useful to perform work, and the system would be a perpetual motion machine. In about 1950, French physicist Léon Brillouin (1889–1969) disproved Maxwell's hypothesis by demonstrating that the decrease in entropy resulting from the demon's actions would be exceeded by the increase in entropy in choosing between the fast and slow molecules.

boomerang stays in the air long enough, it will turn a full circle and return to the thrower. Every boomerang has a built-in orbit diameter, which is not affected by a person throwing the boomerang harder or spinning it faster.

Does the Coriolis effect have an impact on water in a bathtub?

Many people believe that water draining from a bathtub, sink, or toilet bowl in the Northern Hemisphere swirls counterclockwise, while in the Southern Hemisphere, the water drains clockwise due to the Coriolis effect. First described by French mathematician and engineer Gaspard Gustav de Coriolis (1792–1843), the Coriolis effect is the apparent deflection of air masses and fluids caused by Earth's rotation. Although it does have an effect on fluids over great distances or long lengths of time, such as hurricanes, it is too weak to control fluids on a small scale, such as bathtubs, sinks, or toilet bowls. These can drain in either direction in both hemispheres. The direction is determined by numerous factors, including the shape of the container, the shape of the drain, the initial water velocity, and the tilt of the sink.

LIGHT, SOUND, AND OTHER WAVES

What are the primary colors in light?

Color is determined by the wavelength of visible light (the distance between one crest of the light wave and the next). Those colors that blend to form "white light" are, from shortest wavelength to longest: red, orange, yellow, green, blue, indigo, and violet. All these monochromatic colors, except indigo, occupy large areas of the spectrum (the entire range of wavelengths produced when a beam of electromagnetic radiation is broken up). These colors can be seen when a light beam is refracted through a prism. Some consider the primary colors to be six monochromatic colors that occupy large areas of the spectrum: red, orange, yellow, green, blue, and violet. Many physicists recognize three primary colors: red, yellow, and blue. All other colors can be made from these by adding two primary colors in various proportions. Within the spectrum, scientists have discovered fifty-five distinct hues. Infrared and ultraviolet rays at each end of the spectrum are invisible to the human eye.

What is the speed of light?

Light travels at 186,282 miles (299,792 kilometers) per second or 12 million miles (19.3 million kilometers) per minute.

How do polarized sunglasses reduce glare?

Sunlight reflected from the horizontal surface of water, glass, and snow is partially polarized, with the direction of polarization chiefly in the horizontal plane. Such reflected light may be so intense as to cause glare. Polarized sunglasses contain filters that block (absorb) light that is polarized in a direction perpendicular to the transmission axes. The transmission axes of the lenses of polarized sunglasses are oriented vertically.

Why does the color of clothing appear different in sunlight than it does in a store under fluorescent light?

White light is a blend of all the colors, and each color has a different wavelength. Although sunlight and fluorescent light both appear as "white light," they each contain

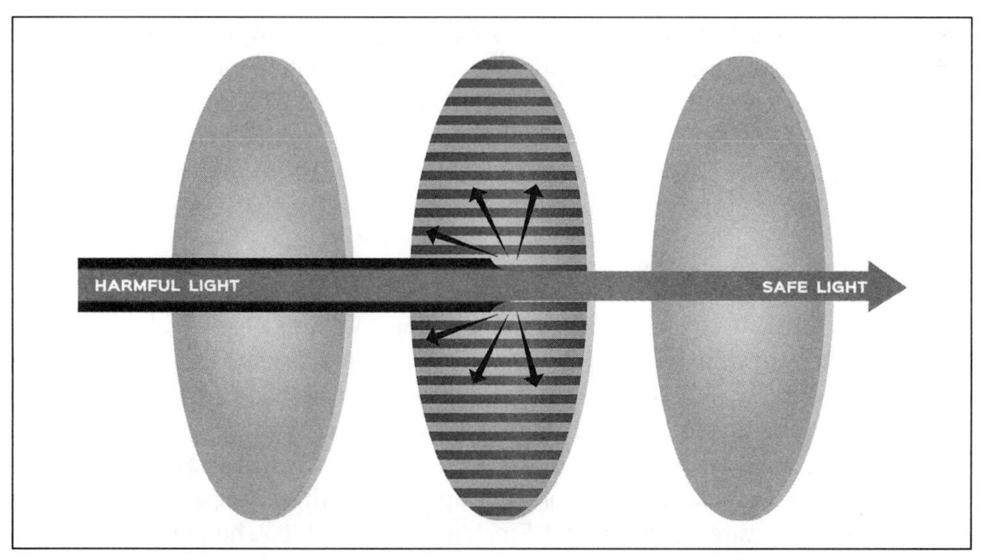

Polarized sunglasses work by blocking horizontal light waves that cause glare while allowing vertical light rays to pass through the filters.

slightly different mixtures of these varying wavelengths. When sunlight and fluorescent light (white light) are absorbed by a piece of clothing, only some of the wavelengths (composing white light) reflect from the clothing. When the retina of the eye perceives the "color" of the clothing, it is really perceiving these reflected wavelengths. The mixture of wavelengths determines the color perceived. This is why an article of clothing sometimes appears to be a different color in the store than it does on the street.

When were liquid crystals discovered?

Liquid crystals were observed by Austrian botanist Friedrich Reinitzer (1857–1927) in 1888. He noticed that the solid organic compound cholesteryl benzoate became a cloudy liquid at 293°F (145°C) and a clear liquid at 354°F (179°C). The following year, German physicist Otto Lehmann (1855–1922) used a microscope with a heating stage to determine that some molecules do not melt directly but first pass through a phase when they flow like a liquid but maintain the molecular structure and properties of a solid. He coined the phrase "liquid crystal" to describe this substance. Further experimentation showed that if an electrical charge is passed through a liquid crystal material, the liquid will line up according to the direction of the electrical field.

What are some uses of liquid crystal displays?

Liquid crystal displays (LCDs) are used for electronic panel displays, including televisions, laptop computer screens, virtual reality helmet displays, and watch faces.

What were Anders Ångström's contributions to the development of spectroscopy?

Swedish physicist and astronomer Anders Jonas Ångström (1814–1874) was one of the founders of spectroscopy. His early work provided the foundation for spectrum analysis (analysis of the ranges of electromagnetic radiation emitted or absorbed). He investigated the Sun spectra as well as that of the aurora borealis. In 1868, he established measurements for wavelengths of greater than 100 Frauenhofer. In 1907, the angstrom (Å, equal to 10^{-10} m), a unit of wavelength measurement, was officially adopted.

Is light a wave or a particle?

Scientists have debated for centuries whether light is a wave or a particle. Isaac Newton was one of the early proponents of the particulate (or corpuscular) theory of light. According to this theory, light travels as a stream of particles that come from a source, such as the Sun, travel to an object, and are then reflected to an observer. One of the early proponents of the wave theory of light was Dutch physicist Christiaan Huygens (1629–1695). According to the wave theory of light, light travels through space in the form of a wave, similar to water waves. Albert Einstein's work in 1905 showed that light is a bundle of tiny particles, called photons. Scientists now believe that light has properties of both waves and particles, explained as wave–particle duality.

Why was the Michelson–Morley experiment important?

This experiment on light waves, first carried out in 1886 by physicists Albert A. Michelson (1852–1931) and Edward W. Morley (1838–1923) at Western Reserve University in Cleveland, Ohio, is one of the most historically significant experiments in physics and led to the development of Einstein's theory of relativity. The original experiment, using the Michelson interferometer, attempted to detect the velocity of Earth with respect to the hypothetical "luminiferous ether," a medium in space proposed to carry light waves. The procedure measured the speed of light in the direction of Earth and the speed of light at right angles to Earth's motion. No difference was found. This result discredited the ether theory and ultimately led to the proposal by Albert Einstein that the speed of light is a universal constant.

What are optical fibers?

An optical fiber consists of a core, cladding, and a protective coating. The core is a very thin—approximately .0004 inches (10μm) in diameter—strand of coated glass with a high refractive index. The cladding is a thicker material with a lower refractive index. The decreased index of refraction allows light to bounce off the interface. A plastic coating surrounds the core, and cladding protects them from damage. The core, cladding, and protective coating has a diameter of approximately .0005 inches (125μm)—slightly larger than a strand of human hair, which measures approximately 100μm. A fiber optic cable may contain hundreds or thousands of optical fibers bound together.

How do optical fibers transmit light?

The principle of total internal reflection is essential for light to travel through optical fibers. Total internal reflection occurs when a ray of light in a medium with a higher index of refraction strikes the interface between that medium (the core) and one with a lower index of refraction (the cladding). When a source of light is sent down an optical fiber, the light strikes the interface between the core and the cladding and is reflected back into the cable. The light continues to move down the core of the fiber.

What are some uses of fiber optics?

Fiber optics are used for telecommunications, including telephone, computer, and television signals. Endoscopes in the medical field use fiber optics to allow physicians to see internal organs without invasive procedures. Fiber optic sensors are used to monitor industrial processes.

How does the light from a laser differ from other light?

The word "laser" is an acronym for light amplification by stimulated emission of radiation. If the electrons in special atoms in glasses, crystals, or gases are energized into excited atomic states, they will emit light photons in response to a weak laser pulse. Several differences exist between the light from lasers and other light. Ordinary visible light travels in different wavelengths (different colors of light have different wavelengths), with the peaks and troughs in different cycles. Laser light travels in one wavelength (one specific color) with the peaks and troughs in phase, meaning that the peaks and troughs are coherent with each other. Another difference is that ordinary light, e.g., the light from a flashlight, is diffused and spreads out the farther it is from the source. Laser light is focused and does not spread very much, allowing it to travel great distances. The beams of laser light are narrow, very bright, may be focused on a very small spot, may travel very long distances, and are able to concentrate a lot of energy on a very small area.

When was the first laser demonstrated?

Several scientists had been exploring the amplification of microwaves (masers) during the 1940s and 1950s. Theodore Maiman (1927–2007) was the first person to demonstrate an amplification at light wavelengths (laser) on May 16, 1960, while working as a researcher at Hughes Aircraft Company. The first laser was produced by using pulses of light from a flash lamp to excite the atoms in artificial ruby. It was a very short burst of light, but it formed a highly directed, coherent beam of light with very high intensity. Maiman published his paper, "Stimulated Optical Radiation in Ruby," in the August 6, 1960, issue of *Nature* magazine (volume 187, pages 495–494). Maiman's first working laser is stored in a safe deposit box in a bank in Vancouver, British Columbia. Maiman lived in Vancouver from 1999 until the time of his death in 2007.

What are some uses of lasers?

Lasers are found in many products and technologies. Lasers are used in tools to cut through diamonds or thick metal. They are used to record and retrieve information, including CD and DVD players, barcode scanners, and laser printers. They are in measuring systems and have been used to measure the distance from Earth to the Moon. Lasers are also used for surgical procedures.

There are many practical applications for lasers, such as in industry. Here we see a laser spot welding automobile parts; they can also be used for precision cuts in sheet metal.

What is the difference between special and general relativity?

Albert Einstein developed the theory of relativity in the early twentieth century. He published the theory of special relativity in 1905 and the theory of general relativity in 1916. Special relativity deals only with nonaccelerating (inertial) reference frames, while general relativity deals with accelerating (noninertial) reference frames. Simply stated, according to the theory of special relativity, the laws of nature are the same for all observers whose frames of reference are moving with constant velocity with respect to each other. Published as an addendum to the special theory of relativity was the famous equation $E = mc^2$, representing that mass and energy can be transformed into each other. In contrast, general relativity states that the laws of nature are the same for all observers, even if they are accelerating with respect to each other.

How did a total solar eclipse confirm Einstein's theory of general relativity?

When formulating his theory of general relativity, Albert Einstein proposed that the curvature of space near a massive object like the Sun would bend light that passed close by. For example, a star seen near the edge of the Sun during an eclipse would appear to have shifted by 1.75 arc seconds from its usual place. British astronomer Arthur Eddington (1882–1944) confirmed Einstein's hypothesis during an eclipse on May 29, 1919. The subsequent attention given to Eddington's findings helped establish Einstein's reputation as one of science's greatest figures.

How do sound waves differ from light waves?

Waves consist of a series of motions in regular succession carrying energy from one place to another without moving any matter. Periodic waves include ocean waves, sound waves, and electromagnetic waves. Visible light and radio waves are electromagnetic waves. Mechanical waves, such as ocean waves and sound waves, involve matter, but it is important to remember that matter is not transported. The water in an ocean wave does not move from one location to another; merely the energy of the wave is transported. Light waves involve only energy without matter.

Does the speed of sound remain the same when it travels through water and air?

The speed of sound is not a constant; it varies depending on the medium in which it travels. The measurement of sound velocity in the medium of air must take into account many factors, including air temperature, pressure, and purity. At sea level and 32°F (0°C), scientists do not agree on a standard figure; estimates range between 740 and 741.5 miles (1,191.6–1,193.2 kilometers) per hour. As air temperature rises, sound velocity increases. Sound travels faster in water than in air and even faster in iron and steel. Sounds traveling 1 mile in air for 5 seconds will travel the same distance in 1 second underwater and one-third of a second in steel.

Who was the first person to break the sound barrier?

On October 14, 1947, Charles E. (Chuck) Yeager (1923–) was the first pilot to break the sound barrier. He flew a Bell X-1, attaining a speed of 750 miles (1,207 kilometers) per hour (Mach 1.06) and an altitude of 70,140 feet (21,379 meters) over the town of Victorville, California. The first woman to break the sound barrier was Jacqueline Cochran

Chuck Yeager named the Bell X-1 he flew to break the sound barrier *Glamorous Glennis* after his wife.

(c. 1906–1980). On May 18, 1953, she flew a North American F-86 Saber over Edwards Air Force Base in California, attaining the speed of 760 miles (1,223 kilometers) per hour.

When is a sonic boom heard?

As long as an airborne object, such as a plane, is moving below the speed of sound (called Mach 1), the disturbed air remains well in front of the craft, but as the craft passes Mach 1 and is flying at supersonic speeds, a sharp air pressure rise occurs in front of the craft. In a sense, the air molecules are crowded together and collectively impact. What is heard is a claplike thunder called a sonic boom or a supersonic bang. Many shocks come from a supersonic aircraft, but these shocks usually combine to form two main shocks, one coming from the nose and one from the aft end of the aircraft. Each of the shocks moves at a different velocity. If the time difference between the two shock waves is greater than 0.10 seconds apart, two sonic booms will be heard. This usually occurs when an aircraft ascends or descends quickly. If the aircraft moves more slowly, the two booms will sound like only one boom to the listener.

What is the Doppler effect?

In 1842, Austrian physicist Christian Doppler (1803–1853) explained the phenomenon of the apparent change in wavelength of radiation—such as sound or light—emitted either by a moving body (source) or by the moving receiver. The frequency of the wavelengths increases, and the wavelength becomes shorter as the moving source approaches, producing high-pitched sounds and bluish light (called blue shift). Likewise,

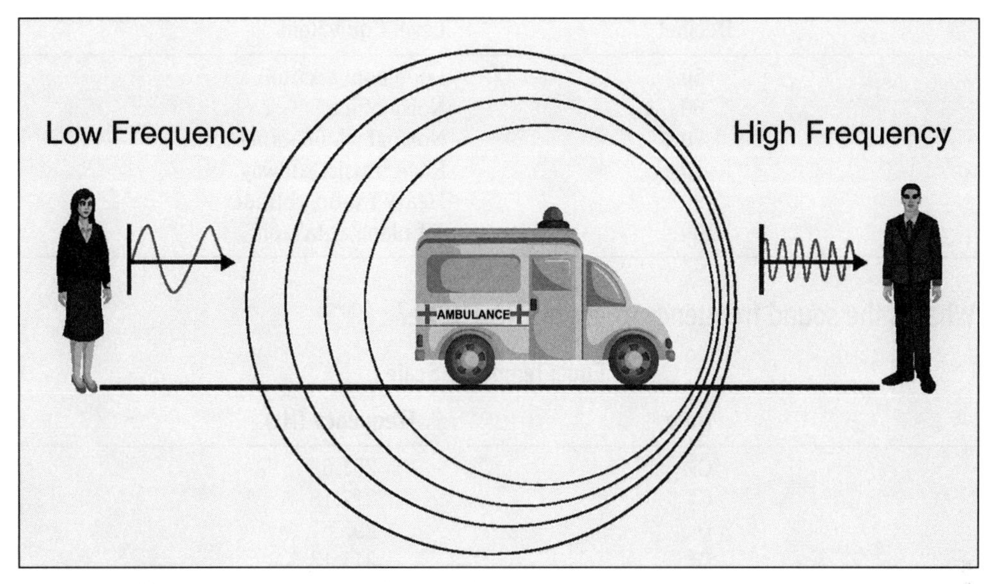

The Doppler effect explains how waves (such as sound or light waves) are distorted when objects move toward or away from one another.

as the source recedes from the receiver, the frequency of the wavelengths decreases, the sound is pitched lower, and the light appears reddish (called red shift). This Doppler effect is commonly demonstrated by the whistle of an approaching train or the roar of a jet aircraft.

Acoustical (sound) and optical (light) Doppler effects have three differences: The optical frequency change is not dependent on what is moving—the source or observer—nor is it affected by the medium through which the waves are moving, but acoustical frequency is affected by such conditions. Optical frequency changes are affected if the source or observer moves at right angles to the line connecting the source and observer. Observed acoustical changes are not affected in such a situation. Applications of the Doppler phenomenon include the Doppler radar and the measurement by astronomers of the motion and direction of celestial bodies.

What is a decibel?

A decibel is a measure of the relative loudness or intensity of sound. A 20-decibel sound is ten times louder than a 10-decibel sound; 30 decibels is one hundred times louder, etc. One decibel is the smallest difference between sounds detectable by the human ear.

Decibel	Level Equivalent
10	Light whisper
20	Quiet conversation
30	Normal conversation
40	Light traffic

Decibel	Level Equivalent
50	Loud conversation
60	Noisy office
70	Normal traffic, quiet train
80	Rock music, subway
90	Heavy traffic, thunder
100	Jet plane at takeoff

What is the sound frequency of the musical scale?

Equal Tempered Scale

Note	Frequency (Hz)
C♭	261.63
C♯	277.18
D	293.67
D♯	311.13
E	329.63
F	349.23
F♯	369.99
G	392.00
G♯	415.31
A	440.00
A♯	466.16
B	493.88
Cn	523.25

Note: ♭ indicates flat; ♯ indicates sharp; n indicates return to natural; Hz is Hertz.

The lowest frequency distinguishable as a note is about 20 hertz. The highest audible frequency is about 20,000 hertz. A hertz (Hz) is a unit of frequency that measures the number of wave cycles per second frequency of a periodic phenomenon whose periodic time is 1 second (cycles per second).

Who predicted electromagnetic waves?

James Clerk Maxwell demonstrated the mathematical relationship between oscillating electric and magnetic fields. In 1873, Maxwell published his work in *A Treatise on Electricity and Magnetism*, thus unifying the concepts of electricity and magnetism. The first of four equations in his publication states that the electric field passes through a particular surface area to a charge within that surface. This is also known as Gauss's law. The second equation states that unlike electrical charges, magnetic charges do not exist. Therefore, magnetic lines of forces must always form closed loops. The third equation, also known as Ampère's law, states that a magnetic field can be induced as an electric current or by a changing electric flux. The fourth equation, also known as Faraday's law, states that voltage is generated in a conductor as it passes through a magnetic field.

The propagation of electromagnetic waves is possible since the change in the electric field gives rise to a magnetic field and a changing magnetic field gives rise to an electric field.

German physicist Heinrich Hertz proved the existence of electromagnetic waves that were first theorized by James Clerk Maxwell.

What was Heinrich Hertz's contribution to the field of electromagnetic waves?

Heinrich Hertz (1857–1894) was a German physicist. He was the first person to demonstrate that electromagnetic waves existed in 1883. He designed a transmitter and receiver that produced waves with a 4-meter wavelength. He used standing waves to measure their wavelength. He showed that they could be reflected, refracted, polarized, and could produce interference. It was Hertz's experimentation that eventually led to the development of radio. The unit of measure, "hertz," is named in his honor. One hertz is equal to the number of electromagnetic waves or cycles in a signal, which is one cycle per second.

What are the seven regions of the electromagnetic spectrum?

The electromagnetic spectrum consists of the wide range of electromagnetic waves. Electromagnetic waves consist of two transverse waves: one an oscillating electric field, the other a corresponding magnetic field perpendicular to it. The electromagnetic spectrum (EM) is divided into seven regions in order of decreasing wavelength and increasing energy and frequency. Wavelength is measured as the difference between two consecutive peaks of a wave. Frequency is the number of waves in a given length of time. All electromagnetic waves travel at the speed of light when they are in a vacuum.

Regions on the Electromagnetic Spectrum

Name	Wavelength	Frequency
Radio waves	> 0.4 inches (approximately 10 millimeters)	Up to about 30 billion hertz or 30 gigahertz (GHz)
Microwaves	0.4 inches (10 millimeters)to 0.004 inches (100 micrometers [μm])	3 gigahertz (GHz) to approximately 30 trillion hertz or 30 terahertz (THz)
Infrared	0.004 inches (100 μm) to 0.00003 inches (740 nanometers [nm])	30 terahertz (THz) to approximately 400 THz
Visible light	0.00003 inches (704 nm) to 0.00015 inches (380 nm)	400 terahertz (THz) to 800 Thz

Name	Wavelength	Frequency
Ultraviolet	0.00015 inches (380 nm) to approximately 0.0000004 inches (10 nm)	8×10^{14} to 3×10^{16} hertz (Hz)
X-rays	4×10^{-7} inches (10 nm) to approximately 4×10^{-8} inches (100 picometers [pm])	3×10^{16} Hz to approximately 10^{18} Hz
Gamma rays	$< 4 \times 10^{-9}$ inches (< 100 pm)	$> 10^{18}$ Hz

How do microwave ovens cook food?

Microwave ovens are a common gadget in many kitchens. A microwave oven generates 2.4 GHz waves and scatters them throughout the oven. The microwaves excite water and fat molecules into resonance and cause them to rotate, increasing their thermal energy. Different kinds of molecules absorb the energy at different rates, so some foods are heated more than others. Microwave-safe containers are made of materials that do not absorb microwave energy, remaining cool during the heating and cooking process.

Which regions of the electromagnetic spectrum are used for lidar?

Lidar, light detection and ranging, is similar to radar but uses the infrared, visible, or ultraviolet range of the electromagnetic spectrum to transmit pulses of light. Radar transmission is in the radio or microwave regions of the electromagnetic spectrum. Lidar is a remote sensing system that consists of a laser transmitter and an optical receiver.

What are the characteristics of alpha, beta, and gamma radiation?

Radiation is a term that describes all the ways energy is emitted by the atom as X-rays, gamma rays, neutrons, or charged particles. Most atoms, being stable, are nonradioactive; others are unstable and give off either particles or gamma radiation. Substances bombarded by radioactive particles can become radioactive and yield alpha particles, beta particles, and gamma rays.

Alpha particles, first identified by Antoine Henri Becquerel, have a positive electrical charge and consist of two protons and two neutrons. Because of their great mass, alpha particles can travel only a short distance, around 2 inches (5 centimeters) in air, and can be stopped by a sheet of paper.

French physicist and Nobel laureate Antoine Henri Becquerel discovered the first evidence for radioactivity. The becquerel (bq), the SI united for measuring radioactivity, is named after him.

Beta particles, identified by Ernest Rutherford (1871–1937), are extremely high-speed electrons that move at the speed of light. They can travel far in air and can pass through solid matter several millimeters thick.

Gamma rays, identified by Marie and Pierre Curie, are similar to X-rays, but they usually have a shorter wavelength. These rays, which are bursts of photons, or very short-wave electromagnetic radiation, travel at the speed of light. They are much more penetrating than either the alpha or beta particles and can go through 7 inches (18 centimeters) of lead.

What is the difference between nuclear fission and nuclear fusion?

Nuclear fission is the splitting of an atomic nucleus into at least two fragments. Nuclear fusion is a nuclear reaction in which the nuclei of atoms of low atomic numbers, such as hydrogen and helium, fuse to form a heavier nucleus. Although in both nuclear fission and nuclear fusion substantial amounts of energy are produced, the amount of energy produced in fusion is far greater than the amount of energy produced in fission.

Who discovered nuclear fission?

Three scientists, Otto Hahn (1879–1968), Lise Meitner (1878–1968), and Fritz Strassman (1902–1980), are credited with the discovery of nuclear fission. The groundwork for their discovery was established in 1934 when Enrico Fermi (1901–1954) bombarded uranium with neutrons. Fermi thought he had produced the first elements that were heavier than uranium. However, in fact, the uranium had broken into smaller elements. Hahn and Strassman were chemists and performed the chemical analysis, while Meitner, a physicist, explained the nuclear processes when the uranium was bombarded with neutrons. Meitner also collaborated with Otto Frisch (1904–1979). Meitner and Frisch submitted their findings to the journal *Nature*. Hahn and Strassman submitted their own paper. Hahn was awarded the Nobel Prize in Chemistry in 1944 for his contributions to the discovery of nuclear fission.

Who coined the term "nuclear fission"?

Otto Frisch and Lise Meitner coined the term "nuclear fission" to describe the process since it was similar to the fission observed in cell division in biology.

What are gravitational waves?

Gravitational waves are ripples in space-time caused by major astronomical events, such as the collision or merger of two black holes or neutron stars. This is comparable to dropping two stones into a pool of water. Each stone sends out ripples in concentric circles that eventually intersect. Once they meet, the shape of the waves changes—becoming smaller or larger or cancelling out each other. The new wave pattern is an interference pattern. Gravitational waves travel through the universe at the speed of light, carrying information about their origins as well as clues about the nature of gravity itself.

When were gravitational waves first detected?

The existence of gravitational waves was predicted by Albert Einstein in 1916 in his general theory of relativity. He predicted that massive accelerating objects, such as neutron stars or black holes, orbiting around each other would disrupt space-time in such a way that ripples or waves would radiate from the source. The proof of the existence of gravitational waves was demonstrated in 1974 by two astronomers working in Puerto Rico. They discovered a binary pulsar—two extremely dense and heavy stars in orbit around each other. They began to monitor the two stars and determined that they were getting closer to each other at precisely the rate predicted by general relativity if they were emitting gravitational waves. Observations of this system for more than forty years indicate that it is emitting gravitational waves.

Gravitational waves were first physically detected on September 14, 2015, at 5:51 A.M. Eastern Daylight Time. The twin Laser Interferometer Gravitational-Wave Observatory (LIGO) detectors physically sensed the distortion in space-time caused by passing gravitational waves produced during the final fraction of a second by the merger of two colliding black holes nearly 1.3 billion light years away. They formed a massive, spinning black hole.

In 2018, a gravitational wave from two binary stars colliding with each other rippled across the fabric of space-time. Calculated to have occurred 130 million years ago, the gravitational wave was detected in August 2018. The event was also captured in various types of light: ultraviolet, radio, infrared, optical, gamma rays, and X-rays.

Are LIGO's interferometers the same as the Michelson interferometer?

The Laser Interferometer Gravitational-Wave Observatory (LIGO) interferometers are based on the same principles as the interferometer used in the Michelson–Morley ex-

The LIGO control room is located in Livingston, Louisiana. Together with the other facility in Hanford, Washington, the scientists here and at the Virgo Collaboration in Italy announced the detection of gravitational waves emanating from black holes in 2015.

Why are scientists interested in gravitational waves?

Scientists will continue to study gravitational waves to learn more about neutron stars and black holes. Studying gravitational waves will help provide information as to what happens during the most violent explosions in the universe—possibly even during the earliest moments of the universe.

periment in 1887. The interferometer used in the Michelson–Morley experiment and LIGO are both L-shaped. Both have mirrors at the ends of the arms to reflect light in order to combine light beams and create an interference pattern. Both measure patterns and intensity of a resulting light beam after two beams have been superimposed. Differences also exist between the two interferometers, with LIGO being much larger and more sensitive than the interferometer used in the Michelson–Morley experiment.

Where is the LIGO interferometer located?

The Laser Interferometer Gravitational-Wave Observatory (LIGO) consists of two identical interferometers located thousands of miles apart from each other—one in Hanford, Washington, and the other in Livingston, Louisiana. Each interferometer consists of two L-shaped arms that are 2.5 miles (4 kilometers) long. Each arm has a Fabry-Perot "cavity" created by adding mirrors to the beam splitter. These mirrors reflect parts of the laser beam back and forth within the 4-kilometer-long arms, effectively making the arms 696 miles (1,120 kilometers) long. More mirrors are placed between the laser source and the beam splitter, boosting the power of the laser to 750 kilowatts. The result is a detector that is so sensitive it can detect the tiniest vibrations on Earth, including those close to the detector and from sources thousands of miles away from the interferometer. The reason the two detectors are located so far apart is that while each detector will feel local vibrations—for example, from local traffic—they will both feel a gravitational wave vibration at virtually the same time. Analyzing the data from the two sites, scientists ignore vibrations that differ between the sites and focus on identical signals that occur at the same time at both locations.

ELECTRICITY AND MAGNETISM

Who is the founder of the science of magnetism?

British scientist William Gilbert (1544–1603) regarded Earth as a giant magnet and investigated its magnetic field in terms of dip and variation. He explored many other magnetic and electrostatic phenomena. The Gilbert (Gb), a unit of magnetism, is named for him.

John H. Van Vleck (1899–1980), an American physicist, made significant contributions to modern magnetic theory. He explained the magnetic, electrical, and optical properties of many elements and compounds with the ligand field theory, demonstrated the effect of temperature on paramagnetic materials (called Van Vleck paramagnetism), and developed a theory on the magnetic properties of atoms and their components.

What is a Leyden jar?

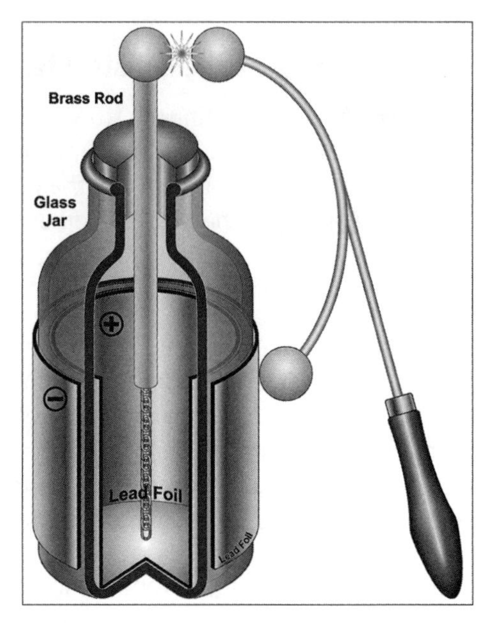

A schematic of a Leyden jar.

A Leyden jar, the earliest form of capacitor, is a device for storing an electrical charge. First described in 1745 by E. Georg van Kleist (c. 1700–1748), it was also used by Pieter van Musschenbroek (1692–1761), a professor of physics at the University of Leyden. The device came to be known as a Leyden jar and was the first device that could store large amounts of electric charge. The jars contained an inner wire electrode in contact with water, mercury, or wire. The outer electrode was a human hand holding the jar. An improved version coated the jar inside and outside with separate metal foils, with the inner foil connected to a conducting rod and terminated in a conducting sphere. This eliminated the need for the liquid electrolyte. In use, the jar was normally charged from an electrostatic generator. The Leyden jar—which makes hair stand up—is still used for classroom demonstrations of static electricity.

If a magnetic bar is cut in two, will it yield a north pole and a south pole magnet?

Every magnet contains both a north pole and a south pole. No magnets are purely north or south. Cutting a magnetic bar in half will produce two magnets each with a north pole and a south pole.

Who was the first to observe a connection between electricity and magnetism?

Danish physicist Hans Christian Oersted (1777–1851) was the first to observe a connection between electricity and magnetism. While preparing a lecture and demonstration in 1820, he found that a current in a wire caused a nearby compass needle to rotate, thereby establishing that an electric current always produces a magnetic field around itself. When an electric current flows through a wire wrapped around a piece of iron, it creates a magnetic field in the iron. Electromagnets are used in a wide variety of applications, including doorbells, switches, and valves in heating and cooling equipment to telephones, large machinery used to lift heavy loads of scrap metal, and particle accelerators.

How do permanent magnets differ from temporary magnets?

Permanent magnets, also referred to as naturally occurring magnets, remain magnetized until they are demagnetized. Naturally occurring magnets are found in minerals such as magnetite and lodestone. These magnets are known as ferromagnets.

What materials are used to make the most powerful permanent magnets?

The most powerful permanent magnets are made from alloys of iron, boron, and neodymium. Magnetic strength is measured in units called tesla and gauss. One tesla equals 10,000 gauss. Most of the magnets used to decorate refrigerators are 10 gauss. The most powerful permanent magnets produce magnetic fields of approximately 1.5 tesla.

What is static electricity?

Static electricity results from an imbalance between negative and positive charges in an object. Most of the time, the positive and negatives charges in an object are balanced, meaning the object is neutral; it is neither positively nor negatively charged. When the charges build up on an object, they must find a way to be released or transferred to restore the balance between negative and positive charges. During the winter, the air has very little water vapor in it and is dry. In the summer, the air contains more water vapor; the humidity is greater. Water is an electrical conductor, allowing electrons to move from one object to another more easily. The humidity in the summer air allows extra electrons on charged objects to leak off into the air and attach to objects that have too few electrons. The dry air in the winter makes it more difficult for the extra electrons to leak off an object, so static electricity, with its characteristic shock or spark, is more common.

How do dryer sheets and hair conditioner reduce or prevent static electricity?

Since materials that are electrical conductors allow free movements of charges, the goal of dryer sheets and hair conditioner is to turn electrical insulators, such as fabric and hair, into electrical conductors. When wet, fabric and hair are usually negatively charged. Applying a cleansing agent that contains a positively charged detergent molecule to these wet fibers, such as dryer sheets and hair conditioner, clings to the wet fibers and remains in place, giving fabrics and hair the soft, silky, nonclinging, static-free feeling.

How did the electrical term "ampere" originate?

It was named for André-Marie Ampère (1775–1836), the physicist who formulated the basic laws of the science of electrodynamics. The ampere (A), often abbreviated as "amp," is the unit of electric current, defined as the constant current that, maintained in two

The electrical volt is named after Italian chemist and physicist Allesandro Volta (left); the unit of electrical current called the amp is named after French mathematician and physicist André-Marie Ampère (right).

straight parallel infinite conductors placed one meter apart in a vacuum, would produce a force between the conductors of 2×10^{-7}.

How did the electrical unit volt originate?

The unit of voltage is the volt, named after Alessandro Volta (1745–1827), the Italian scientist who built the first modern battery. (A battery, operating with a lead rod and vinegar, was also manufactured in ancient Egypt.) Voltage measures the force or "oomph" with which electrical charges are pushed through a material. Some common voltages are 1.5 volts for a flashlight battery; 12 volts for a car battery; 115 volts for ordinary household receptacles; and 230 volts for a heavy-duty household receptacle.

What is the electrical unit watt?

Named for Scottish engineer and inventor James Watt (1736–1819), the watt is used to measure electric power. An electrical device uses 1 watt when 1 volt of electric current drives 1 ampere of current through it.

What materials are the best and worst conductors of electricity?

Electrical conductivity is the ability of a material to transmit current or the movement of charged particles, most often protons. Materials that carry the flow of electrical current are called conductors. Metals, such as silver and aluminum, are some of the best conductors of electricity. Other good conductors of electricity are copper and gold. Materials that do not permit the flow of electrical current are called nonconductors or insulators. Wood, paper, and most plastics are examples of insulators. Resistance is defined as the extent to which a material prevents the flow of electricity. Materials with a low resistance have a high conductivity, while those with a high resistance have low conductivity. German physicist Georg Simon Ohm (1789–1854) was the first to describe the laws of electrical conductivity and resistance.

The brilliant, visionary inventor Nikola Tesla is shown here in front of a metal coil used for his experiments in wireless power.

How do lead-acid batteries work?

Lead-acid batteries consist of positive and negative lead plates suspended in a diluted sulfuric acid solution called an electrolyte. Everything is contained in a chemically and electrically inert case. As the cell discharges, sulfur molecules from the electrolyte bond with the lead plates, releasing excess electrons. The flow of electrons is called electricity.

Who was Nikola Tesla?

Nikola Tesla (1856–1943) was a leading innovator in the field of electricity. Tesla held over one hundred patents, among which are patents for alternating current and the seminal patents for radio. Tesla's work for Westinghouse in the late 1880s led to the commercial production of electricity, including the Niagara Falls Power Project in 1895. After a bitter and prolonged public feud, Tesla's alternating current system was proven superior to Thomas Edison's (1847–1931) direct current system. Tesla was responsible for many other innovations, including the Tesla coil, radio-controlled boats, and neon and fluorescent lighting.

PARTICLE PHYSICS

When was the cyclotron invented?

The idea for the cyclotron originated in 1929 after Ernest O. Lawrence (1901–1958) read a paper by Norwegian engineer Rolf Wideroe (1902–1996). Lawrence was a member in the physics department at the University of California, Berkeley. Lawrence began experimenting and building models in his laboratory that would accelerate ions in a spiral path between two D-shaped electrodes. The device could accelerate nuclear particles to high velocities without the use of high voltages. A prototype was demonstrated in 1932. Lawrence received a U.S. patent for the cyclotron in 1934. He was awarded the 1939 Nobel Prize in Physics for the invention and development of the cyclotron.

What is string theory?

A relatively recent theory in particle physics, string theory conceives elementary particles not as points but as lines or loops. The idea of these "strings" is purely theoretical since no string has ever been detected experimentally. The ultimate expression of string theory may potentially require a new kind of geometry—perhaps one involving an infinite number of dimensions.

How did the quark get its name?

This theoretical particle, considered to be the fundamental unit of matter, was named by Murray Gell-Mann (1929–), an American theoretical physicist who was awarded the 1969 Nobel Prize in Physics. Its name was initially a playful tag that Gell-Mann invented, sounding something like "kwork." Later, Gell-Mann came across the line "three quarks for Master Marks" in James Joyce's (1882–1941) *Finnegan's Wake*, and the tag became known as a quark.

What was Richard Feynman's contribution to physics?

Richard Feynman (1918–1988) developed a theory of quantum electrodynamics that described the interaction of electrons, positrons, and photons, providing physicists a new way to work with electrons. He reconstructed quantum mechanics and electrodynamics in his own terms, formulating a matrix of measurable quantities visually represented by a series of graphs known as the Feynman diagrams. Feynman was awarded the Nobel Prize in Physics in 1965.

American physicist and Nobel Prize winner Richard Feynman was a giant in his field, making contributions in particle physics, superfluidity, quantum mechanics, and quantum electrodynamics.

What are subatomic particles?

Subatomic particles are particles that are smaller than atoms. Historically, subatomic particles were considered to be electrons, protons, and neutrons. However, the definition of subatomic particles has now been expanded to include elementary particles, which are so small that they do not appear to be made of anything more minute. The physical study of such particles became possible only during the twentieth century with the development of

Why does the Standard Model not provide a complete description of the universe?

Although the Standard Model currently provides the best description of the subatomic universe, it is not perfect. It describes known particles and accurately predicted unknown particles, including the Higgs boson. However, it cannot describe the fourth fundamental force, gravity, dark matter, or dark energy.

increasingly sophisticated apparatus. Many new particles were discovered in the last half of the twentieth century.

What is the Standard Model?

The Standard Model was a theory developed in the 1970s to explain how fundamental particles interact and the forces between them. It consists of two basic types of fundamental matter particles and four fundamental forces. According to the Standard Model, everything in the universe is made from fundamental particles and governed by fundamental forces.

What are the two types of particles in the Standard Model?

The two types of particles in the Standard Model are quarks and leptons. It contains six quarks and six leptons. The quarks are up, down, strange, charm, top, and bottom (or beauty). Like all known particles, a quark has its antimatter opposite, known as an antiquark (having the same mass but opposite charge). Quarks never occur alone in nature. They always combine to form particles called hadrons. According to the Standard Model, all other subatomic particles consist of some combination of quarks and their antiparticles. The leptons are electron, electron neutrino, muon, muon neurtrino, tau, and tau neutrino.

How many fundamental forces are described in the Standard Model?

Four fundamental forces are described in the Standard Model: the strong force, the weak force, the electromagnetic force, and gravity. The forces are carried by particles called bosons—gluons, photons, and W and Z bosons. The gluons carry the strong force, photons carry the electromagnetic force, and the W and Z bosons carry the weak force.

Where is the largest particle accelerator located?

The largest and most powerful particle accelerator is the Large Hadron Collider at CERN. It is 16.8 miles (27 kilometers) in circumference. It has a series of superconducting electromagnets and a number of accelerating structures. Two high-energy particle beams traveling in opposite directions at close to the speed of light collide. Scientists and researchers study results of the collisions to gain knowledge about the universe.

Why is existence of the Higgs boson important?

The existence of the Higgs boson as part of the Standard Model was predicted in the 1960s but was not observed until July 2012 using the Large Hadron Collider at CERN. The Higgs boson particle interacts with particles and gives each particle mass. It was named for Peter Higgs (1929–). Higgs shared the 2013 Nobel Prize in Physics with François Englert (1932–) for their contributions to the understanding of the origin of mass for subatomic particles.

When was CERN founded?

The European Organization for Nuclear Research, known in French as *Conseil Européen pour la Recherche Nucléaire* (CERN), was founded in 1954 by twelve member countries. The goal was to establish a world-class fundamental physics research organization in Europe. The CERN laboratory is located near Geneva, Switzerland. The United States has the status of an observer state. Thousands of scientists visit CERN to do research in particle physics.

What is entanglement?

Entanglement states that two particles, even if they are separated by a long distance, may be part of the same system. Their actions are able to influence each other even across

Particle accelerators at the CERN laboratory are shown here. The European organization maintains the Large Hadron Collider in Switzerland.

the vast distances. Consider what happens when a positron and electron annihilate. Two gammas (high-energy photons) are produced that go off in opposite directions 180° apart. They can be detected many meters from their source and have opposite spins, but which gamma has which spin is a random choice; that is, each gamma has a 50/50 chance that it will be in a particular direction. Suppose you find that the spin of one gamma is pointing up, but the moment you detect the spin of that gamma, the spin of the other gamma must be pointing down. The result of one measurement determines the results of the other measurement. The two detectors would measure the spins of their gammas at the same time, so it is impossible that one gamma could communicate with the other gamma.

Physicists say that the spins of the two photons are entangled and that the spin state of each photon is the superposition of the two possible spin directions. When the spin is measured, the wave function "collapses" and gives a definitive result. Albert Einstein called results a "spooky action at a distance." Others have called it quantum weirdness.

Similar results can be obtained with atoms or ions, in which case the photons are light quanta and may be transported through space (of the air) or by optical fibers. For example, if an atom is excited by the absorption of a photon, it can emit two photons that are entangled the same way the gammas are in the above example.

What was one of the earliest presentations of superposition?

One of the earliest and most famous presentations of superposition in quantum physics was Schrödinger's cat. Erwin Schrödinger (1887–1961) stated that objects could exist in two states, in this case a cat being dead and alive at the same time, in what is known as superposition. While superposition has not been observed in the macroscopic world (cats are either dead or alive), the behavior is confined to quantum physics.

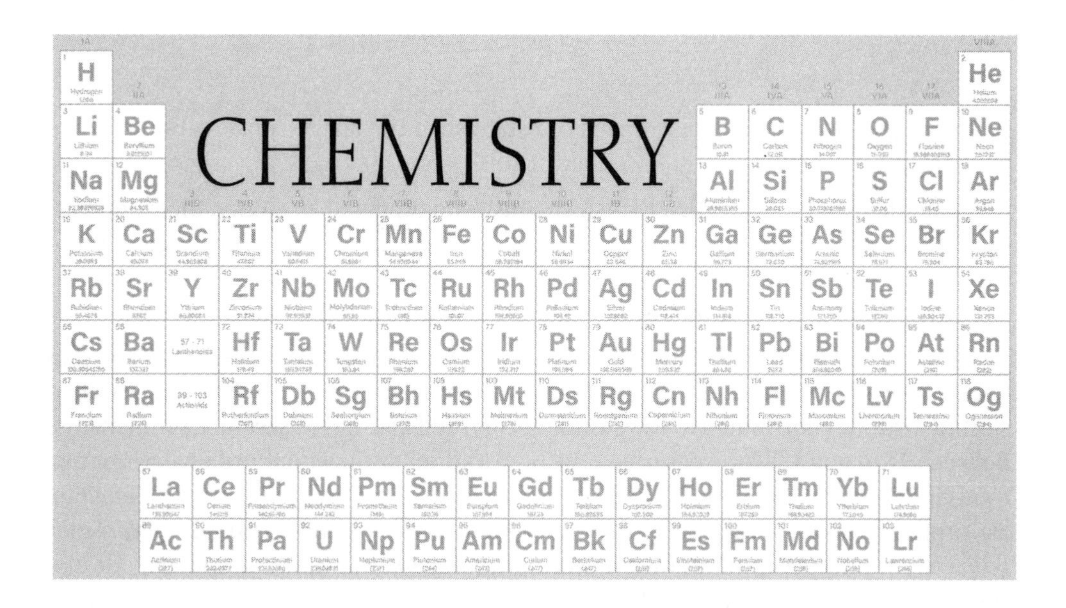

INTRODUCTION AND HISTORICAL BACKGROUND

What are the four major divisions of chemistry?

Chemistry has traditionally been divided into organic, inorganic, analytical, and physical chemistry. Organic chemistry is the study of compounds that contain carbon. More than 90 percent of all known chemicals are organic. Inorganic chemistry is the study of compounds of all elements except carbon. Analytical chemists determine the structure and composition of compounds and mixtures. They also develop and operate instruments and techniques for carrying out the analyses. Physical chemists use the principles of physics to understand chemical phenomena.

Who are some of the founders of modern chemistry?

The history of chemistry in its modern form is often considered to have begun with British scientist Robert Boyle (1627–1691), although its roots can be traced back to the earliest recorded history. Best known for his discovery of Boyle's law (volume of a gas is inversely proportional to its pressure at constant temperature), he was a pioneer in the use of experiments and the scientific method. A founder of the Royal Society, he worked to remove the mystique of alchemy from chemistry to make it a pure science.

French chemist Antoine-Laurent de Lavoisier (1743–1794) is regarded as another important founder of modern chemistry. Indeed, he is considered "the father of modern chemistry." His wide-ranging contributions include the discrediting of the phlogiston theory of combustion, which long had been a stumbling block to a true understanding of chemistry. He established modern terminology for chemical substances and did the first experiments in quantitative organic analysis. He is sometimes credited with having discovered or established the law of conservation of mass in chemical reactions.

John Dalton (1766–1844), a British chemist who proposed an atomic theory of matter that became a basic theory of modern chemistry, is also an important figure in the development of the field. His theory, first proposed in 1803, states that each chemical element is composed of its own kind of atoms, all with the same relative weight.

Another important individual in the development of modern chemistry was Swedish chemist Jöns Jakob Berzelius (1779–1848). He devised chemical symbols, determined atomic weights, contributed to the atomic theory, and discovered several new elements. Between 1810 and 1816, he described the preparation, purification, and analysis of two thousand chemical compounds. Then he determined atomic weights for forty elements. He simplified chemical symbols, introducing a notation—letters with numbers—that replaced the pictorial symbols his predecessors used, and that is still used today. He discovered cerium (in 1803, with Wilhelm Hisinger [1766–1852]), selenium (1818), silicon (1824), and thorium (1829).

What were the two main aims of alchemy?

The main aim of alchemy, the early study of chemical reactions, was the transmutation (or transformation) of common elements into gold. Needless to say, all attempts to change a substance into gold were unsuccessful. A second aim of alchemy was to discover an elixir or universal remedy that would promote everlasting life. Again, this pursuit was unsuccessful.

Who proposed the phlogiston theory?

Phlogiston was a name used in the eighteenth century to identify a supposed substance given off during the process of combustion. The phlogiston theory was developed in the early 1700s by German chemist and physicist Georg Ernst Stahl (1660–1734).

In essence, Stahl held that combustible material, such as coal or wood, was rich in a material substance called "phlogiston." What remained after combustion was without phlogiston and could no longer burn. The rusting of metals also involved a transfer of phlogiston. This accepted theory explained a great deal previously unknown to chemists.

What is a philosopher's stone?

A philosopher's stone is the name of a substance believed by medieval alchemists to have the power to change baser metals into gold or silver. It had, according to some, the power of prolonging life and of curing all injuries and diseases. The pursuit of it by alchemists led to the discovery of several chemical substances; however, the magical philosopher's stone has since proved fictitious.

For instance, metal smelting was consistent with the phlogiston theory, as was the fact that charcoal lost weight when burned. Thus, the loss of phlogiston either decreased or increased weight.

French chemist Antoine-Laurent de Lavoisier (Antoine Lavoisier) demonstrated that the gain of weight when a metal turned to a calx was just equal to the loss of weight of the air in the vessel. Lavoisier also showed that part of the air (oxygen) was indispensable to combustion and that no material would burn in the absence of oxygen. The transition from Stahl's phlogiston theory to Lavoisier's oxygen theory marks the birth of modern chemistry at the end of the eighteenth century.

Antoine-Laurent de Lavoisier convinced the world that the idea of "phlogiston" was incorrect and that oxygen was the key to combustion.

When was spontaneous combustion first recognized?

Spontaneous combustion is the ignition of materials stored in bulk. This is due to internal heat buildup caused by oxidation (generally a reaction in which electrons are lost, specifically when oxygen is combined with a substance or when hydrogen is removed from a compound). Because this oxidation heat cannot be dissipated into the surrounding air, the temperature of the material rises until the material reaches its ignition point and bursts into flame.

A Chinese text written before 290 C.E. recognized this phenomenon in a description of the ignition of stored oilcloth. The first Western acknowledgment of spontaneous combustion was by J. P. F. Duhamel (1730–1816) in 1757 when he discussed the gigantic conflagration of a stack of oil-soaked canvas sails drying in the July sun. Before spontaneous combustion was recognized, such events were usually blamed on arsonists.

Who made the first organic compound to be synthesized from inorganic ingredients?

In 1828, Friedrich Wöhler (1800–1882) synthesized urea from ammonia and cyanic acid. This synthesis dealt a deathblow to the vital-force theory, which held that definite and fundamental differences existed between organic and inorganic compounds. Swedish chemist Jöns Jakob Berzelius had proposed that the two classes of compounds were produced from their elements by entirely different laws. Organic compounds were produced under the influence of a vital force and therefore were incapable of being prepared artificially. This distinction ended with Wöhler's synthesis.

When was the International Union of Pure and Applied Chemistry formed?

The International Union of Pure and Applied Chemistry (IUPAC) was formed in 1919 by a group of chemists from both academia and industry. The goal of the organization was to bring international standardization to the field of chemistry. The IUPAC is responsible for nomenclature and terminology, including names for new elements on the periodic table, standardized methods for measurements, and atomic weights.

MATTER

What is matter?

The most common definition of matter is anything that has mass and occupies space. Matter is composed of very tiny particles. Modern physicists and chemists understand that these tiny particles contain a great amount of energy. Matter has both physical and chemical properties. Examples of physical properties are state, i.e., solid, liquid, or gas; hardness; texture; color; density; odor; and taste. The physical properties make it possible to identify unknown kinds of matter. Chemical properties allow a substance to change its chemical makeup or react with other substances.

What is an atom?

An atom is the smallest unit of an element, containing the unique chemical properties of that element. Atoms are very small—several million atoms could fit in the period at the end of this sentence.

Parts of an Atom

Subatomic Particle	Charge	Location
Proton	Positive	Nucleus
Neutron	Neutral	Nucleus
Electron	Negative	Orbits nucleus

Who is regarded as the founder of modern atomic physics?

In 1897, British physicist Joseph John Thomson (1856–1940) researched electrical conduction in gases, which led to the important discovery that cathode rays consist of negatively charged particles called electrons. The discovery of the electron inaugurated the electrical theory of the atom, and this, along with other work, entitled Thomson to be regarded as the founder of modern atomic physics.

Ernest Rutherford discovered the proton in 1919. He also predicted the existence of the neutron, later discovered by his colleague James Chadwick (1891–1974). Chadwick was awarded the 1935 Nobel Prize in Physics for the discovery of the neutron. Rutherford was awarded the Nobel Prize in Chemistry in 1908 for his work on the chemistry of radioactive substances.

What is a chemical bond?

A chemical bond is an attraction between the electrons present in the outermost energy level or shell of a particular atom. This outermost energy level is known as the valence shell. Atoms with an unfilled outer shell are less stable and tend to share, accept, or donate electrons. When this happens, a chemical bond is formed.

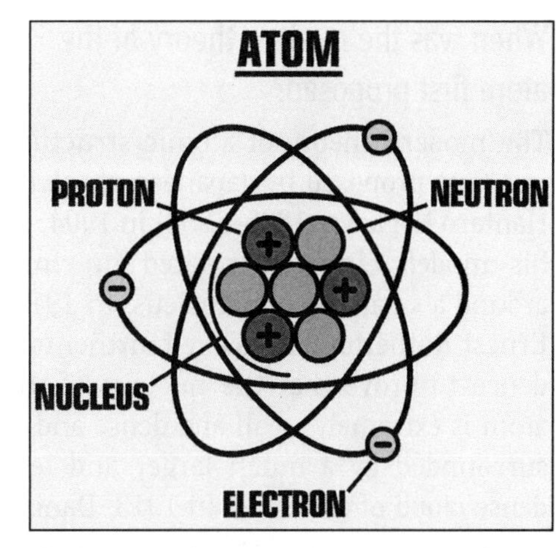

The basic particles in an atom are protons and neutrons in the nucleus and electrons that orbit the nucleus.

What are the major types of bonds?

Chemical bonds are categorized into three types: covalent, ionic, and hydrogen. The type of bond that is established is determined by the electron structure. Ionic bonds are formed when electrons are exchanged between two atoms and the resulting bond is relatively weak. Covalent bonds, the strongest type of bond, occur when electrons are shared between atoms. Hydrogen bonds are temporary, but they are important because they have the ability to be rapidly formed and reformed. The following chart explains the types of bonds and their characteristics:

Three Types of Chemical Bonds

Type	Strength	Examples
Covalent	Strong	Bonds between hydrogen and oxygen in a molecule of water
Ionic	Moderate	Bond between Na^+ and Cl^- in salt
Hydrogen	Weak	Bonds between molecules of water

Who was the first person to propose the atomic theory?

John Dalton (1766–1844), a British natural philosopher, chemist, physicist, and teacher, developed the concepts of an atomic theory in the early nineteenth century. He published *A New System of Chemical Philosophy* in 1808. His main concepts of atomic theory may be summarized as:

- All matter—solid, liquid, and gas—consists of tiny, indivisible particles called atoms.
- All atoms of a given element have the same mass and are identical, but they are different from the atoms of different elements.
- Chemical reactions involve the rearrangement of combinations of those atoms, not the destruction of atoms.
- When elements react to form compounds, they combine in simple, whole-number ratios.

When was the modern theory of the atom first proposed?

British chemist, physicist, and philosopher John Dalton is credited with developing atomic theory.

The modern theory of atomic structure was first proposed by Japanese physicist Hantaro Nagaoka (1865–1950) in 1904. In his model, electrons rotated in rings around a small, central nucleus. In 1911, Ernest Rutherford discovered further evidence to prove that the nucleus of the atom is extremely small and dense and is surrounded by a much larger and less dense cloud of electrons. In 1913, Danish physicist Niels Bohr (1885–1962) proposed a model that is known as the Bohr atom. It suggested that electrons orbit the nucleus in concentric quantum shells at certain well-specified distances from the nucleus corresponding to the electron's energy levels. These orbits are known as Bohr orbits. Several years later, Erwin Schrödinger proposed the Schrödinger wave equation, which provided a firm theoretical basis for the Bohr orbits.

What is the fourth state of matter?

Plasma, a mixture of free electrons and ions or atomic nuclei, is sometimes referred to as a "fourth state of matter." Plasmas occur in thermonuclear reactions as in the Sun, in fluorescent lights, and in stars. When the temperature of gas is raised high enough, the collision of atoms becomes so violent that electrons are knocked loose from their nuclei. The result of a gas having loose, negatively charged electrons and heavier, positively charged nuclei is called a plasma.

All matter is made up of atoms. Animals and plants are organic matter; minerals and water are inorganic matter. Solid, liquid, and gas are the first three states of matter. Whether matter appears as a solid, liquid, or gas depends on how the molecules are held together in their chemical bonds. Solids have a rigid structure in the atoms of the molecules; in liquids, the molecules are close together but not packed; in a gas, the molecules are widely spaced and move around, occasionally colliding but usually not interacting.

What is antimatter?

Antimatter is the exact opposite of normal matter. Antimatter was predicted in a series of equations derived by Paul Dirac (1902–1984). He was attempting to combine the theory of relativity with equations governing the behavior of electrons. In order to make his equations work, he had to predict the existence of a particle that would be similar to the electron but opposite in charge. This particle, discovered in 1932, was the antimatter equivalent of the electron, called the positron (electrons with a positive charge). Other antimatter particles would not be discovered until 1955, when particle accelerators were finally able to confirm the existence of the antineutron and antiproton (protons with a negative charge). Antiatoms (pairings of positrons and antiprotons) are other examples of antimatter.

What is density?

Density (the mass per unit volume or mass/volume) refers to how compact or crowded a substance is. For instance, the density of water is 1 g/cm^3 (gram per cubic centimeter) or 1 kg/l (kilogram per liter); the density of a rock is 3.3 g/cm^3; pure iron is 7.9 g/cm^3; and earth (as a whole) is 5.5 g/cm^3 (average).

Why is liquid water more dense than ice?

Pure liquid water is most dense at 39.2°F (3.98°C) and decreases in density as it freezes. The water molecules in ice are held in a relatively rigid geometric pattern by their hydrogen bonds, producing an open, porous structure. Liquid water has fewer bonds; therefore, more molecules can occupy the same space, making liquid water more dense than ice. Water as a solid (i.e., ice) floats.

How was the principle of buoyancy discovered?

Buoyancy was first discovered by Greek mathematician Archimedes. The famous story recounts how the king of Syracuse, Hieron II (c. 306–c. 215 B.C.E.), asked Archimedes to verify that his crown was made of pure gold without destroying the crown. When Archimedes entered his bath, he noticed that the water overflowed the tub. He realized that the volume water that flowed out of the bath had to be equal to the volume of his own body that was immersed in the bath. Shouting "Eureka," he ran through the streets of Syracuse announcing that he had found a method to determine whether the king's crown was made of pure gold. He could measure the amount of water that was displaced by a block of pure gold of the same weight as the crown. If the crown was made of pure gold, it would displace the same amount

Anglo-Irish natural philosopher and physicist Robert Boyle is considered to be the first modern chemist in history. His Boyle's law describes how pressure, volume, and temperature of gases are related.

of water as the block of gold. The principle of buoyancy, also known as Archimedes's principle, states that the buoyant force acting on an object placed in a fluid is equal to the weight of the fluid displaced by the object.

What are gas laws?

Gas laws are physical laws concerning the behavior of gases. They include Boyle's law, which states that the volume of a given mass of gas at a constant temperature is inversely proportional to its pressure, and Charles's law, which states that the volume of a given mass of gas at a constant pressure is directly proportional to its absolute temperature. These two laws can be combined to give the general or universal gas law, which may be expressed as:

$$(\text{pressure} \times \text{volume})/\text{temperature} = \text{constant}$$

Avogadro's law states that equal volumes of all gases contain the same number of particles if they all have the same pressure and temperature.

The laws are not obeyed exactly by any real gas, but many common gases obey them under certain conditions, particularly at high temperatures and low pressures.

What is STP?

The abbreviation STP is often used for standard temperature and pressure. As a matter of convenience, scientists have chosen a specific temperature and pressure as standards for comparing gas volumes. The standard temperature is 0°C (273°K), and the standard pressure is 760 torr (one atmosphere).

CHEMICAL ELEMENTS

What is an element?

An element is the most basic form of a chemical substance. Elements may not be broken down to a simpler form by any ordinary chemical or physical means. The combination of different elements, as a result of chemical reactions and bonding, produces all of the substances and objects in our world.

Who developed the periodic table of elements?

Dmitry Ivanovich Mendeleyev (1834–1907) was a Russian chemist whose name will always be linked with the development of the periodic table. He was the first chemist to really understand that all elements are related members of a single ordered system. He changed what had been a highly fragmented and speculative branch of chemistry into a true, logical science. His nomination for the 1906 Nobel Prize in Chemistry failed by one vote, but his name became recorded in perpetuity fifty years later when element 101 was called mendelevium.

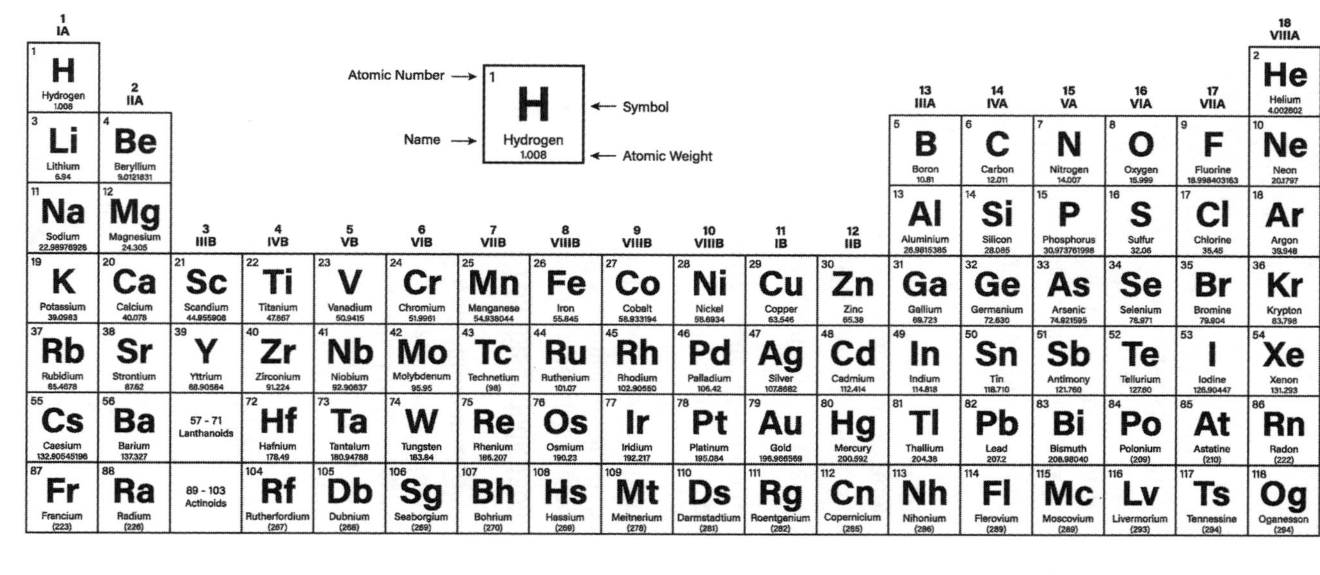
Periodic Table of the Elements
Atomic Number
Symbol
Name
Atomic Weight
1 H Hydrogen 1.008
1 IA
2 IIA
3 IIIB
4 IVB
5 VB
6 VIB
7 VIIB
8 VIIIB
9 VIIIB
10 VIIIB
11 IB
12 IIB
13 IIIA
14 IVA
15 VA
16 VIA
17 VIIA
18 VIIIA
1 H Hydrogen
2 He Helium
3 Li Lithium
4 Be Beryllium
5 B Boron
6 C Carbon
7 N Nitrogen
8 O Oxygen
9 F Fluorine
10 Ne Neon
11 Na Sodium
12 Mg Magnesium
13 Al Aluminium
14 Si Silicon
15 P Phosphorus
16 S Sulfur
17 Cl Chlorine
18 Ar Argon
19 K Potassium
20 Ca Calcium
21 Sc Scandium
22 Ti Titanium
23 V Vanadium
24 Cr Chromium
25 Mn Manganese
26 Fe Iron
27 Co Cobalt
28 Ni Nickel
29 Cu Copper
30 Zn Zinc
31 Ga Gallium
32 Ge Germanium
33 As Arsenic
34 Se Selenium
35 Br Bromine
36 Kr Krypton
37 Rb Rubidium
38 Sr Strontium
39 Y Yttrium
40 Zr Zirconium
41 Nb Niobium
42 Mo Molybdenum
43 Tc Technetium
44 Ru Ruthenium
45 Rh Rhodium
46 Pd Palladium
47 Ag Silver
48 Cd Cadmium
49 In Indium
50 Sn Tin
51 Sb Antimony
52 Te Tellurium
53 I Iodine
54 Xe Xenon
55 Cs Caesium
56 Ba Barium
57 - 71 Lanthanoids
72 Hf Hafnium
73 Ta Tantalum
74 W Tungsten
75 Re Rhenium
76 Os Osmium
77 Ir Iridium
78 Pt Platinum
79 Au Gold
80 Hg Mercury
81 Tl Thallium
82 Pb Lead
83 Bi Bismuth
84 Po Polonium
85 At Astatine
86 Rn Radon
87 Fr Francium
88 Ra Radium
89 - 103 Actinoids
104 Rf Rutherfordium
105 Db Dubnium
106 Sg Seaborgium
107 Bh Bohrium
108 Hs Hassium
109 Mt Meitnerium
110 Ds Darmstadtium
111 Rg Roentgenium
112 Cn Copernicium
113 Nh Nihonium
114 Fl Flerovium
115 Mc Moscovium
116 Lv Livermorium
117 Ts Tennessine
118 Og Oganesson
57 La Lanthanum
58 Ce Cerium
59 Pr Praseodymium
60 Nd Neodymium
61 Pm Promethium
62 Sm Samarium
63 Eu Europium
64 Gd Gadolinium
65 Tb Terbium
66 Dy Dysprosium
67 Ho Holmium
68 Er Erbium
69 Tm Thulium
70 Yb Ytterbium
71 Lu Lutetium
89 Ac Actinium
90 Th Thorium
91 Pa Protactinium
92 U Uranium
93 Np Neptunium
94 Pu Plutonium
95 Am Americium
96 Cm Curium
97 Bk Berkelium
98 Cf Californium
99 Es Einsteinium
100 Fm Fermium
101 Md Mendelevium
102 No Nobelium
103 Lr Lawrencium

What is pitchblende?

Pitchblende is a massive variety of uraninite, or uranium oxide, found in metallic veins. It is a radioactive material and the most important ore of uranium. In 1898, Marie and Pierre Curie discovered that pitchblende contained radium, a rare element that has since been used in medicine and the sciences.

According to Mendeleyev, the properties of the elements, as well as those of their compounds, are periodic functions of their atomic weights (in the 1920s, it was discovered that atomic number was the key rather than weight). Mendeleyev compiled the first true periodic table in 1869 listing all the sixty-three (then-known) elements. In order to make the table work, Mendeleyev had to leave gaps, and he predicted that further elements would eventually be discovered to fill them. Three were discovered in Mendeleyev's lifetime: gallium in 1875, scandium in 1879, and germanium in 1896.

How many of the elements included on the periodic table occur in nature?

The periodic table contains 118 elements. Scientists are not certain of the exact number of elements that occur in nature. For many years, it was believed that elements 1 (hydrogen) through 92 (uranium), with the exception of technetium (number 43), were found in nature. Elements 93–118 were artificially synthesized. More recently, trace amounts of elements 93–98, as well as technetium, have been found in uranium-rich pitchblende. They are produced as a byproduct of radioactive decay.

How many chemical substances are recognized in the CAS Registry?

The CAS (Chemical Abstracts Service) Registry, the most comprehensive database for information about chemical substances, contains information for more than 138 million unique chemical substances. Some types of substances included are organic compounds, inorganic compounds, alloys, metals, minerals, elements, isotopes, and polymers.

What were Lothar Meyer's contributions to the periodic table?

Lothar Meyer (1830–1895), a German chemist, prepared a periodic table that resembled closely Mendeleyev's periodic table. He did not publish his periodic table until after Mendeleyev's paper on the periodic table was published in 1869. It is believed that his work was influential in causing some of the revisions Mendeleyev made in the second version of his periodic table, published in 1870. Specifically, Meyer focused on the periodicity of the physical properties of the elements, while Mendeleyev's focus was the chemical consequences of the periodic law.

What was the first element to be discovered?

Phosphorus was first discovered by German chemist Hennig Brand (c. 1630–c. 1710) in 1669 when he extracted a waxy, white substance from urine that glowed in the dark.

Brand did not publish his findings, though. In 1680, phosphorus was rediscovered by British chemist Robert Boyle.

What are the alkali metals?

These are the elements at the left of the periodic table: lithium (Li, element 3), potassium (K, element 19), rubidium (Rb, element 37), cesium (Cs, element 55), francium (Fr, element 87), and sodium (Na, element 11). The alkali metals are sometimes called the sodium family of elements, or Group I elements. Because of their great chemical reactivity (they easily form positive ions), none exist in nature in the elemental state.

What are the alkaline earth metals?

The alkaline earth metals are beryllium (Be, element 4), magnesium (Mg, element 12), calcium (Ca, element 20), strontium (Sr, element 38), barium (Ba, element 56), and radium (Ra, element 88). They are also called Group II elements. Like the alkali metals, they are never found as free elements in nature and are moderately reactive metals. Harder and less volatile than the alkali metals, these elements all burn in air.

What are the transition elements?

The transition elements are the ten subgroups of elements between Group II and Group XIII, starting with period 4. They include gold (Au, element 79), silver (Ag, element 47), platinum (Pt, element 78), iron (Fe, element 26), copper (Cu, element 29), and other metals. All transition elements are metals. Compared to alkali and alkaline earth metals, they are usually harder and more brittle and have higher melting points. Transition metals are also good conductors of heat and electricity. They have variable valences, and compounds of transition elements are often colored. Transition elements are so named because they comprise a gradual shift from the strongly electropositive elements of Groups I and II to the electronegative elements of Groups VI and VII.

What are the transuranic chemical elements and the names for elements 93 to 118?

Transuranic elements are those elements in the periodic system with atomic numbers greater than 92. Many of these elements are ephemeral, do not exist naturally outside the laboratory, and are not stable.

Elements 93 through 118

Element Number	Name	Symbol
93	Neptunium	Np
94	Plutonium	Pu
95	Americum	Am
96	Curium	Cm
97	Berkelium	Bk
98	Californium	Cf
99	Einsteinium	Es

Element Number	Name	Symbol
100	Fermium	Fm
101	Mendelevium	Md
102	Nobelium	No
103	Lawrencium	Lr
104	Rutherfordium	Rf
105	Dubnium	Db
106	Seaborgium	Sg
107	Bohrium	Bh
108	Hassium	Hs
109	Meitnerium	Mt
110	Darmstadium	Ds
111	Roentgenium	Rg
112	Copernicium	Cn
113*	Nihonium	Nh
114	Flerovium	Fl
115*	Moscovium	Mc
116	Livermorium	Lv
117*	Tennessine	Ts
118*	Oganesson	Og

*The names for elements 113, 115, 117, and 118 were officially approved by the International Union of Pure and Applied Chemistry (IUPAC) in 2016.

Which are the only two elements in the periodic table named after women?

Curium, atomic number 96, was named after the pioneers of radioactive research Marie and Pierre Curie. Meitnerium, atomic number 109, was named after Lise Meitner, one of the founders of nuclear fission.

What is Harkin's rule?

Atoms having even atomic numbers are more abundant in the universe than are atoms having odd atomic numbers. Chemical properties of an element are determined by its atomic number, which is the number of protons in the atom's nucleus.

What are some chemical elements whose symbols are not derived from their English names?

Modern Name	Symbol	Older Name
antimony	Sb	stibium
copper	Cu	cuprum
gold	Au	aurum
iron	Fe	ferrum
lead	Pb	plumbum

How are chemical elements named?

Chemical elements are named after a mythological character, celestial object, mineral, place or country, property of the element (e.g., the color), or scientist. Once the discovery of a new element has been accepted by the IUPAC, the discoverers are invited to propose a name for the element. Following a review process, the name is approved by the IUPAC. Some elements named for mythological creatures are neptunium (Neptune), promethium (Prometheus), tantalum (Tantalus), and titanium (Titan). Examples of elements named for a mineral are aluminum (Latin for alum), carbon (Latin for coal or charcoal), lithium (Greek for stone), and nickel (German for St. Nicholas's Copper or kupfernickel). Elements named for a place are often named for the place of discovery—for example, berkelium, californium, polonium, and yttrium. Elements named for one of their properties include iodine (from the Greek word *ioeides*, meaning "violet-colored"), iridium (from the Latin word *iris*, meaning "rainbow"), platinum (from the Spanish word *patina*, meaning "little silver"), and rhodium (from the Greek word *rhodon*, meaning "rose-colored"). Elements that have been named after scientists include fermium (named for Enrico Fermi), lawrencium (named for Ernest O. Lawrence), mendelevium (named for Dmitry Mendeleyev), and nobelium (named for Alfred Nobel).

Modern Name	Symbol	Older Name
mercury	Hg	hydrargyrum
potassium	K	kalium
silver	Ag	argentum
sodium	Na	natrium
tin	Sn	stannum
tungsten	W	wolfram

Which elements are liquid at room temperature?

"Liquid silver," mercury (Hg, element 80) and bromine (Br, element 35) are liquid at room temperature 68°–70°F (20°–21°C). Gallium (Ga, element 31) with a melting point of 85.6°F (29.8°C), cesium (Cs, element 55) with a melting point of 83°F (28.4°C), and francium (Fr, element 87) with a melting point of 80.6°F (27°C) are liquid at slightly above room temperature and pressure.

Which chemical element is the most abundant in the universe?

Hydrogen (H, element 1) makes up about 75 percent of the mass of the universe. It is estimated that more than 90 percent of all atoms in the universe are hydrogen atoms. Most of the rest are helium (He, element 2) atoms.

Which chemical elements are the most abundant on Earth?

Oxygen (O, element 8) is the most abundant element in Earth's crust, waters, and atmosphere. It composes 49.5 percent of the total mass of these compounds. Silicon (Si, element 14) is the second most abundant element. Silicon dioxide and silicates make up about 87 percent of the materials in Earth's crust.

Why are the rare gases and rare earth elements called "rare"?

Rare gases refers to the elements helium, neon, argon, krypton, and xenon. They are rare in that they are gases of very low density ("rarified") at ordinary temperatures and are found only scattered in minute quantities in the atmosphere and in some substances. In addition, rare gases have zero valence and normally will not combine with other elements to make compounds.

Rare earth elements are elements numbered 58 through 71 in the periodic table plus yttrium (Y, element 39) and thorium (Th, element 90). They are called "rare earths" because they are difficult to extract from monazite ore, where they occur. The term has nothing to do with scarcity or rarity in nature.

What is an isotope?

Elements are identified by the number of protons in an atom's nucleus. Atoms of an element that have different numbers of neutrons are isotopes of the same element. Isotopes of an element have the same atomic number but different mass numbers. Although the physical properties of atoms depend on mass, differences in atomic mass (mass numbers) have very little effect on chemical reactions. Common examples of isotopes are carbon-12 and carbon-14. Carbon-12 has 6 protons, 6 electrons, and 6 neutrons, while carbon-14 has 6 protons, 6 electrons, and 8 neutrons.

Which elements have the most isotopes?

The elements with the most isotopes, with thirty-six each, are xenon (Xe) with nine stable isotopes (identified between 1920 and 1922) and twenty-seven radioactive isotopes (identified between 1939 and 1981) and cesium (Cs) with one stable isotope (identified in 1921) and thirty-five radioactive isotopes (identified between 1935 and 1983).

The element with the fewest number of isotopes is hydrogen (H), with three isotopes, including two stable ones—protium (identified in 1920) and deuterium, often called heavy water (identified in 1931), and one radioactive isotope—tritium (first identified in 1934, although not considered a radioactive isotope until 1939).

What is heavy water?

Heavy water, also called deuterium oxide (D_2O), is composed of oxygen and two hydrogen atoms in the form of deuterium, which has about twice the mass of normal hydrogen. As a result, heavy water has a molecular weight of about 20, while ordinary water has a molecular weight of about 18. Approximately one part heavy water can be found in 6,500 parts of ordinary water, and it may be extracted by fractional distillation. It is used in thermonuclear weapons and nuclear reactors and as an isotopic tracer in studies of chemical and biochemical processes. Heavy water was discovered by Harold C. Urey (1893–1981) in 1931. He received the Nobel Prize in Chemistry in 1934 for his discovery of heavy hydrogen.

Do more chemical compounds have even numbers of carbon atoms or odd?

A team of chemists recently noted that the database of the Beilstein Information System, containing over 9.6 million organic compounds, includes significantly more substances with even numbers of carbon atoms than with odd numbers. Statistical analyses of smaller sets of organic compounds, such as the *Cambridge Crystallographic Database* or the *CRC Handbook of Chemistry and Physics*, led to the same results. A possible explanation for the observed asymmetry might be that organic compounds are ultimately derived from biological sources, and nature frequently utilized acetate, a C_2 building block, in its syntheses of organic compounds. It may therefore be that the manufacturers' and synthetic chemists' preferential use of relatively economical starting materials derived from natural sources has left permanent traces in chemical publications and databases.

What are volatile organic compounds?

Volatile organic compounds (VOCs) are carbon-containing compounds that become vapors or gases easily or exist as gases. Some examples of VOCs are benzene, formaldehyde, toluene, and xylene. Most VOCs are hydrocarbons represented by the chemical formula RH. The R represents either an alkyl group or an aryl group. An alkyl group is an alkane that is lacking one hydrogen, while an aryl group is an aromatic hydrocarbon that is lacking one hydrogen. Volatile organic compounds are found in many consumer products, such as paints, varnishes, other solvents, craft supplies, copiers and printers, cleaners and disinfectants, building materials and furnishings, carpet, and air fresheners. They are also present in emission from fuels, vehicle exhaust, and cigarette smoke.

What does "half-life" mean?

Half-life is the time it takes for the number of radioactive nuclei originally present in a sample to decrease to one-half of their original number. Thus, if a sample has a half-life of one year, its radioactivity will be reduced to half its original amount at the end of a year and to one quarter at the end of two years. The half-life of a particular radionuclide is always the same, independent of temperature, chemical combination, or any other condition. Natural radiation was discovered in 1896 by French physicist Antoine Henri Becquerel. His discovery initiated the science of nuclear physics.

Which elements have the highest and lowest boiling points?

Helium has the lowest boiling point of all the elements at –452.074°F (–268.93°C), followed by hydrogen –423.16°F (–252.87°C). According to the *CRC Handbook of Chemistry and Physics,* the highest boiling point for an element is that of rhenium 10,104.8°F (5,596°C), followed by tungsten 10,031°F (5,555°C). Other sources record tungsten with the highest boiling point, followed by rhenium.

Which element has the highest density?

Either osmium or iridium is the element with the highest density; however, scientists have yet to gather enough conclusive data to choose between the two. When traditional methods of measurement are employed, osmium generally appears to be the densest element, yet when calculations are made based upon the space lattice, which may be a more reliable method given the nature of these elements, the density of iridium is 22.65, compared to 22.61 for osmium.

What is the density of air?

The density of dry air is 1.29 grams per liter at 32°F (0°C) at average sea level and a barometric pressure of 29.92 inches of mercury (760 millimeters).

The weight of one cubic foot of dry air at one atmosphere of barometric pressure is:

Temperature (Fahrenheit)	Weight per Cubic Foot (Pounds)
50°	0.07788
60°	0.07640
70°	0.07495

Which elements are the hardest and softest?

Carbon is both the hardest and softest element, occurring in two different forms as graphite and diamond. A single crystal of diamond scores the absolute maximum value on the Knoop hardness scale of 90. Based on the somewhat less informative abrasive hardness scale of Mohs, diamond has a hardness of 10. Graphite is an extremely soft material, with a Mohs hardness of only 0.5 and a Knoop hardness of 0.12.

What are isomers?

Isomers are compounds with the same molecular formula but different structures due to the different arrangement of the atoms within the molecules. Structural isomers have atoms connected in different ways. Geometric isomers differ in their symmetry about a double bond. Optical isomers are mirror images of each other.

What is a Lewis acid?

Named after American chemist Gilbert Newton Lewis (1875–1946), the Lewis theory defines an acid as a species that can accept an electron pair from another atom and a

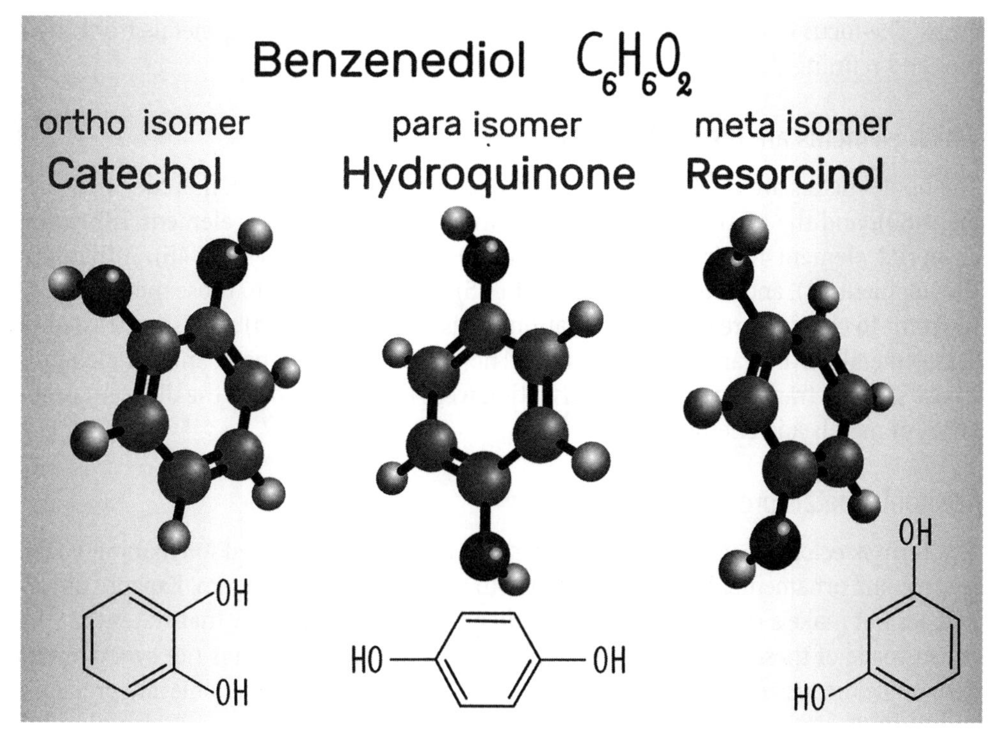

In this example of isomers, the three molecules of Benzenediol all have the same number of carbon, hydrogen, and oxygen atoms, but they are arranged in three different ways.

base as a species that can donate an electron pair to complete the valence shell of another atom. Hydrogen ion (proton) is the simplest substance that will do this, but Lewis acids include many compounds—such as boron trifluoride (BF_3) and aluminum chloride ($AlCl_3$)—that can react with ammonia, for example, to form an addition compound or Lewis salt.

METALS

What are the characteristics of metals?

The three characteristics common to most metals are 1) they are good conductors of electricity; 2) they react, physically or chemically, with other elements to form alloys and compounds; and 3) most are malleable. Malleability means they are flexible, so they can be hammered or rolled into thin sheets or wires.

How many of the chemical elements on the periodic table are metals?

Nearly 75 percent of all the elements on the periodic table are metals. The scientific discipline that studies metals, alloys of metals, and metallic compounds is called metallurgy. The focus of metallurgy is the technical process of extracting metals from their ores and refining them for use.

Which elements are the noble metals?

The noble metals are gold (Au, element 79), silver (Ag, element 47), mercury (Hg, element 80), and the platinum group, which includes platinum (Pt, element 78), palladium (Pd, element 46), iridium (Ir, element 77), rhodium (Rh, element 45), ruthenium (Ru, element 44), and osmium (Os, element 76). The term refers to those metals highly resistant to chemical reaction or oxidation (resistant to corrosion) and is contrasted to "base" metals, which are not so resistant. The term has its origins in ancient alchemy, whose goals of transformation and perfection were pursued through the different properties of metals and chemicals.

Are noble metals precious metals?

The term *precious metals* describes expensive metals that are used for making coins, jewelry, and ornaments. The name is limited to gold, silver, and platinum. Expense or rarity does not make a metal precious, but rather, it is a value set by law that states that the object made of these metals has a certain intrinsic value. The term is not synonymous with noble metals, although a metal (such as platinum) may be both noble and precious.

What distinguishes gold and silver as elements?

Besides their use as precious metals, gold and silver have properties that distinguish them from other chemical elements. Gold is the most ductile and malleable metal—the thinnest gold leaf is 0.0001 millimeters thick. Silver is the most reflective of all metals; thus, it is used in mirrors.

What is 24-karat gold?

The term "karat" refers to the percentage of gold versus the percentage of an alloy in a piece of jewelry or a decorative object. Gold is too soft to be usable in its purest form and has to be mixed with other metals. One karat is equal to 1/24 part fine gold. Thus, 24-

Is white gold really gold?

White gold is the name of a class of jeweler's white alloys used as substitutes for platinum. Different grades vary widely in composition, but usual alloys consist of between 20 percent and 50 percent nickel, with the balance gold. A superior class of white gold is made of 90 percent gold and 10 percent palladium. Other elements used include copper and zinc. The main use of these alloys is to give the gold a white color.

karat gold is 100 percent pure, and 18-karat gold is 18/24 or 75 percent pure. Nickel, copper, and zinc are commonly added to gold for improved strength and hardness.

Karatage	Percentage of Fine Gold
24	100
22	91.75
18	75
14	58.5
12	50.25
10	42
9	37.8
8	33.75

How far can a troy ounce of gold, if formed into a thin wire, be stretched before it breaks?

Ductility is the characteristic of a substance to lend itself to shaping and stretching. Gold is one of the most ductile and malleable substances. A troy ounce of gold (31.1035 grams) can be drawn into a fine wire that is 50 miles (80 kilometers) long.

How thick is gold leaf?

Gold leaf is pure gold that is hammered or rolled into sheets or leaves so extremely thin that it can take three hundred thousand units to make a stack one inch high. The thickness of a single gold leaf is typically 0.0000035 inches (3.5 millionths of an inch), although this may vary widely according to which manufacturer makes it. Also called gold foil, it is used for architectural coverings and for hot-embossed printing on leather.

Which metal was one of the earliest extracted and used by society?

Copper was one of the first metals ever used by humans. It played an important role in the development of civilization. Copper is found in many different mineral deposits. Copper is usually found in association with sulfur. Porphyry copper deposits account for nearly two-thirds of the world's copper. It is often extracted using open-pit mining methods. Copper has high ductility and malleability (properties that allow it to be stretched and shaped), conducts heat and electricity efficiently, and is resistant to corrosion.

In many countries, copper is used in coinage. It is used for wiring and plumbing in appliances, heating and cooling systems in buildings, telecommunications links and motors, wiring, radiators, connectors, brakes, and bearings in cars and trucks. Semiconductor manufacturers use copper for circuitry in silicon chips, producing microprocessors that operate faster and use less energy. A newer use of copper is for surfaces that are frequently touched, such as brass doorknobs. Copper has inherent antimicrobial properties, which reduces the transfer of germs and disease.

How is aluminum foil produced?

Aluminum foil is produced by rolling sheet ingots cast from molten aluminum. The sheets are then rerolled to the desired thickness using heavy rollers. Alternatively, aluminum foil is made using the continuous casting method and then cold rolling. Aluminum foil was first produced in 1903 in France and in 1913 in the United States. The first commercial use of aluminum foil was to package Life Savers.

What is the most common use of zinc?

Zinc is element 30 on the periodic table. More than half of all the zinc that is produced is used for galvinization. Zinc has strong, anticorrosive properties, and it bonds well with other metals. In galvinizing, thin layers of zinc are added to iron or steel to prevent corrosion. The process of galvanizing is named for Italian physicist Luigi Galvani (1737–1798). The earliest method of galvinization was to dip iron into baths of molten zinc. Other uses of zinc include the production of zinc alloys for the die-casting industry, automobile manufacturing, and electrical components. The third significant use of zinc is the production of zinc oxide. Zinc oxide is used to manufacture rubber, batteries, and paints. It is also used as a protective skin ointment.

Why is aluminum an important metal?

Aluminum is the second most abundant metallic element in Earth's crust. It is lightweight, weighing about one-third as much as steel or copper, malleable, ductile, and has excellent corrosion resistance and durability. Many diverse industries and technologies use aluminum, including the automotive and aerospace industries, electronics, building and construction, and the packaging industry.

What is an alloy?

An alloy is a mixture of two or more metals. Alloys may be created by mixing metals while in the molten state or by bonding metal powders. Once two or more metals are mixed to form an alloy, they cannot be separated easily into the individual components of the alloy. The properties and characteristics of an alloy are different from those of the individual components. Alloys are classified as either interstitial or substitutional. In interstitial alloys, smaller elements fill holes that are in the main metallic structure. Steel is an example of an interstitial alloy. In substitutional alloys, some of the atoms of the main metal are substituted with atoms of another metal. Brass is an example of a substitutional alloy.

What is the earliest-known alloy?

The earliest-known alloy is bronze. Bronze dates back to ancient Mesopotamia and was known as early as 3000 B.C.E. The earliest bronze was an alloy of copper and arsenic. Later, tin replaced arsenic to form the alloy bronze. The copper/tin alloy can be traced back to the Sumerians in 2500 B.C.E. Although copper had been used alone for tools and weapons, bronze is much harder than pure copper. Bronze was also used to make coins and for sculptures. The Bronze Age extended until approximately 1000 B.C.E. Bronze is still a popular metal for sculptures.

This Egyptian mirror from the Eighteenth Dynasty more than three thousand years ago is an example of early bronze work.

Which metal is the main component of pewter?

Tin—at least 90 percent. Antimony, copper, and zinc may be added in place of lead to harden and strengthen pewter. Pewter may still contain lead, but high lead content will both tarnish the piece and dissolve into food and drink to become toxic. The alloy used today in fine quality pieces contains 91–95 percent minimum tin, 8 percent maximum antimony, 2.5 percent maximum copper, and 0–5 percent maximum bismuth, as determined by the European Standard for pewter.

Which element is part of the carbide compounds?

Compounds composed of carbon and one of the metallic elements are called carbides. The carbides are very hard, stable materials that are resistant to chemical attack. They have high melting points and are electrically conductive. Most react with water and release acetylene gas. Examples of some carbides are calcium carbide, tungsten carbide, boron carbide, and silicon carbide.

- *Calcium carbide* is also known as acetylenogen. It is commonly used in welding and cutting torches.
- *Tungsten carbide* is used for machining and tools, such as drill bits. Since it is so hard, it improves the tool's ability to hold a cutting edge. It is also used as an abrasive.
- *Boron carbide* is extremely hard. It is used in ceramics and nuclear control rods.
- *Silicon carbide* is used as an abrasive. It is used in industry for grinding wheels and saws to cut rock or concrete but is also used to manufacture emery boards and paper.

How does sterling silver differ from pure silver?

Sterling silver is a high-grade alloy that contains a minimum of 925 parts in 1,000 of silver (92.5 percent silver and 7.5 percent of another metal—usually copper).

Which metals are used to make brass?

Brass is an alloy of copper and zinc. Different types of brasses have various proportions of copper and zinc in the alloy. Brass is used for musical instruments, tools, pipes, fittings, and weapons.

EVERYDAY CHEMISTRY

Where do we find chemistry in our daily lives?

The field of chemistry from the most basic chemical elements and compounds to advanced chemical reactions and processes impact every aspect of our daily lives. Chemistry plays a role in the foods we eat, the clothes we wear, our medicines and cosmetics, our household items, our building materials, and the energy we use. Chemistry is essential to our world and lives.

Which chemical compound is used in greater quantities than any other compound?

Sodium chloride (NaCl), or salt, has over fourteen thousand uses and is probably used in greater quantities and for more applications than any other chemical. Some familiar uses of sodium chloride are flavoring and preserving food, intravenous solutions to alleviate dehydration in patients in the health-care setting, melting ice on roads and sidewalks, and many manufacturing and industrial settings.

What are some familiar chemical compounds that play a role in our lives?

Chemical Compound	Name	Uses
H_2O	Water	Essential for life
$NaCl$	Sodium chloride (salt)	Flavoring, preserving, manufacturing (see above)
$NaHCO_3$	Sodium hydrogen carbonate (baking soda)	Baking, cleaning, whitening
$C_{12}H_{22}O_{11}$	Sucrose (sugar)	Sweetener
$NaClO$	Sodium hypochlorite (bleach)	Stain remover, disinfectant
H_2O_2	Hydrogen peroxide	Bleach alternative, whitener, disinfectant, personal hygiene products

Chemical Compound	Name	Uses
C_3H_6O	Acetone	Solvent, nail polish remover
CH_4	Methane	Natural gas, refine crude oil
CO_2	Carbon dioxide	Carbonation in soft drinks, dry ice
$C_6H_8O_7$	Citric acid	Cooking (found in citrus fruit)
$NaNO_3$	Sodium nitrate	Fertilizer
NH_3	Ammonia	Cleaning products

Who invented bulletproof glass?

Bulletproof glass was invented by French chemist Edouard Benedictus (1878–1930) in 1903. Benedictus accidentally dropped a beaker that had contained cellulose nitrate, a plastic. The cellulose nitrate had dried in the beaker, creating an adhesive film coating on the inside of the beaker. The beaker cracked, but it retained its shape and did not shatter. Bulletproof glass is basically a multiple lamination of glass and plastic layers. It is composed of two sheets of plate glass with a sheet of transparent resin in between, molded together under heat and pressure. When subjected to a severe blow, it will crack without shattering.

What are the major distinguishing characteristics of ceramics?

Ceramics are crystalline compounds of inorganic, nonmetallic elements. They are chemically stable compounds. Ceramics are very hard and among the most rigid of all materials with an almost total absence of ductility. They have the highest-known melting points of materials, with some being as high as 7,000°F (3,870°C) and many that melt at temperatures of 3,500°F (1,927°C). They are good chemical and electrical insulators. Ceramics are oxidation resistant, thus making them the preferred material for use in harsh, corrosive environments.

Traditionally, ceramics include glass, brick, tile, cement and plaster, earthenware pottery, stoneware, and porcelain. Examples of modern or advanced ceramics are spark plugs, tempered glass, synthetic gemstones, quartz watches and clocks, space shuttle tiles, and medical devices.

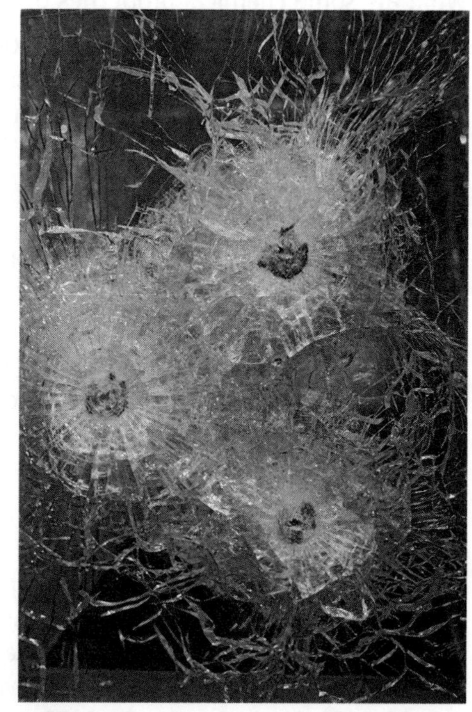

Bulletproof glass was discovered accidentally by French chemist Edouard Benedictus.

Who invented CorningWare®?

Donald Stookey (1915–2014) invented CorningWare®. Originally called Pyroceram®, a glass–ceramic material, Stookey discovered the new product by accident. He had placed a plate of Fotoform glass with added nucleating agents in an oven to heat at what he thought was 1,100°F (593°C). The oven malfunctioned, and the temperature rose to 1,600°F (871°C). Stookey expected to find a molten mess in the oven, but instead, he found an opaque, milky-white plate. As he removed the plate from the oven, it slipped out of the tongs. Instead of shattering and breaking, it bounced. The age of glass ceramics began. Pyroceram®, later called CorningWare®, was first marketed in 1958.

What are polymers?

Polymers are large molecules formed by combining smaller molecules (also called monomers) in a chainlike, repeating pattern. Plastics are synthetic polymers. Polymers are also found in nature. Examples of polymers that occur in nature are DNA and RNA, hair, and spider silk.

What are some chemical properties of plastics?

Plastics are any of a wide range of synthetic or semisynthetic, organic solids that can be shaped or bent without breaking. Plastics are usually organic polymers of high molecular weight, but they often contain other substances. They are usually synthetic, most commonly derived from petrochemicals, but many are partially natural in origin.

Who invented plastic?

In the mid-1850s, Alexander Parkes (1813–1890) experimented with nitrocellulose (also known as guncotton). When he mixed nitrocellulose with camphor, it made a hard but flexible, transparent material. He called the new material "Parkesine." Parkesine was first introduced to the public at the 1862 Great International Exhibition in London. He teamed up with a manufacturer to produce it, but no demand was created for it, and the firm went bankrupt.

What was the first synthetic plastic?

The first synthetic plastic was developed in the 1860s and 1870s. American John Wesley Hyatt (1837–1920) experimented with nitrocellulose and camphor with the goal of producing artificial ivory for billiard balls. The collodion material to coat billiard balls became available in 1868 and was patented in 1869. After making improvements to the formula, John W. Hyatt and his brother Isaiah S. Hyatt received U.S. Patent #105,338 on July 12, 1870, for the process. In 1872, Isaiah Hyatt coined the term "celluloid" for the product that had become successful for the manufacture of billiard balls and many other novelty and fancy goods, including buttons, letter openers, boxes, hatpins, and combs. The material also became the medium for cinematography: celluloid strips coated with a light-sensitive "film" were ideal for shooting and showing moving pictures. Its popularity began to decline in the middle of the twentieth century when other plastics based on synthetic polymers were introduced. John Wesley Hyatt was inducted into the National Inventors Hall of Fame in 2006 for his invention of celluloid. Today, ping-pong balls are almost the only product still made with celluloid.

What was Baekeland's contribution to the development of plastics?

Celluloid was the only plastic material until Belgian-born scientist Leo Hendrik Baekeland (1863–1944) succeeded in producing a synthetic shellac from formaldehyde and phenol. Called Bakelite, it was the first of the thermosetting plastics. These plastics were synthetic materials that, having once been subjected to heat and pressure, became extremely hard and resistant to high temperatures. They continued to maintain their shape even when heated or subjected to various solvents. Baekeland's first patent in the field of plastics was granted in 1906. He announced his invention to the public on February 8, 1909, in a lecture before the New York section of the American Chemical Society. In 1978, he was inducted into the National Inventors Hall of Fame for the invention of Bakelite.

How do thermoset plastics differ from thermoplastics?

The two major types of plastics are 1) thermoset or thermosetting plastics and 2) thermoplastics. Once cooled and hardened, thermosetting plastics retain their shapes and cannot return to their original form. They are hard and durable and can be used for automobile parts and aircraft parts. Examples of thermosetting plastics are polyurethanes,

Who invented Teflon®?

In 1938, American engineer Roy J. Plunkett (1910–1994) at DuPont de Nemours discovered the polymer of tetraluorethylene (PTFE) by accident. This fluorocarbon is marketed under the name of Fluon® in Great Britain and Teflon® in the United States. Plunkett filed for a patent in 1939 and was granted U.S. Patent #2,230,654 in 1941. He was inducted into the National Inventors Hall of Fame in 1985 for the invention of Teflon®. Teflon® was first exploited commercially in 1954. It is resistant to all acids and has exceptional stability and excellent electrical insulating properties. It is used in making piping for corrosive materials, in insulating devices for radio transmitters, in pump gaskets, and in computer microchips. In addition, its nonstick properties make PTFE an ideal material for surface coatings. In 1956, French engineer Marc Gregoire discovered a process whereby he could fix a thin layer of Teflon® on an aluminum surface. He then patented the process of applying it to cookware, and the no-stick frying pan was created.

polyesters, epoxy resins, and phenolic resins. Thermoplastics are less rigid than thermosetting plastics and can soften upon heating and return to their original form. They are easily molded and formed into films, fibers, and packaging. Examples include polyethylene (PE), polypropylene (PP), and polyvinyl chloride (PVC).

What is Styrofoam®, and what are its uses?

Styrofoam® is a type of extruded polystyrene foam currently used for thermal insulation and craft applications. It is a trademarked brand of material that is owned and produced by the Dow Chemical Company. In the United States and Canada, the word "Styrofoam" is frequently and incorrectly used to describe expanded (not extruded) polystyrene beads. This type of expanded polystyrene beads is used to produce disposable coffee cups, coolers, and packaging material, which is usually white in color. Styrofoam® is a good insulator because the material contains millions of trapped gas bubbles, which hinder heat conduction. Styrofoam® materials produced from expanded beads are also excellent insulators. The molecules are so large that they have little movement and hinder heat transfer.

What is the primary use of hydrogen peroxide?

Hydrogen peroxide is a syrupy liquid compound used as a strong bleaching, oxidizing, and disinfecting agent. It is usually made either in anthrahydroquinone autoxidation processes or electrolytically. The primary use of hydrogen peroxide is in bleaching wood pulp. A more familiar use is as a 3 percent solution as an antiseptic and germicide. Undiluted, it can cause burns to human skin and mucous membranes, is a fire and explosion risk, and can be highly toxic.

How is dry ice made?

Dry ice a solid form of carbon dioxide (CO_2) used primarily to refrigerate perishables that are being transported from one location to another. The carbon dioxide, which at normal temperatures is a gas, is stored and shipped as a liquid in tanks that are pressurized at 1,073 pounds per square inch. To make dry ice, the carbon dioxide liquid is withdrawn from the tank and allowed to evaporate at a normal pressure in a porous bag. This rapid evaporation consumes so much heat that part of the liquid CO_2 freezes to a temperature of –109°F (–78°C). The frozen liquid is then compressed by machines into blocks of "dry ice," which will melt into a gas again when set out at room temperature. Although used mostly as a refrigerant or coolant, other uses include medical procedures such as freezing warts, blast cleaning, freeze-branding animals, and creating special effects for live performances and films.

When was dry ice first made commercially?

Dry ice was first made commercially in 1925 by the Prest-Air Devices Company of Long Island City, New York, through the efforts of Thomas Benton Slate (1880–1980). It was used by Schrafft's of New York in July 1925 to keep ice cream from melting. The first large sale of dry ice was made later in that year to Breyer Ice Cream Company of New York.

Who invented dynamite?

Alfred Nobel—yes, the same person whose face is on the prestigious prize—invented dynamite by combining nitroglycerine with stabilizing chemicals.

Dynamite was not an accidental discovery but the result of a methodical search by Swedish technologist Alfred Nobel. Nitroglycerine had been discovered in 1849 by Italian organic chemist Ascanio Sobriero (1812–1888), but it was so sensitive and difficult to control that it was useless. Nobel sought to turn nitroglycerine into a manageable solid by absorbing it into a porous substance. From 1866 to 1867, he tried an unusual mineral, kieselguhr, and created a doughlike explosive that was controllable. He also invented a detonating cap incorporating mercury fulminate, with which nitroglycerine could be detonated at will. The use of dynamite for blasting has played a significant role in the development of mining industries and road building.

Which chemicals are used to produce different colored fireworks?

Fireworks existed in ancient China in the ninth century, where saltpeter (potassium nitrate), sulfur, and charcoal were mixed to produce the dazzling effects. Magnesium burns with a brilliant white light and is widely used in making flares and fireworks. Various other colors can be produced by adding certain substances to the flame. Strontium compounds color the flame scarlet, barium compounds produce yellowish-green, copper produces a blue-green, lithium creates purple, and sodium results in yellow. Iron and aluminum granules give gold and white sparks, respectively.

Who discovered TNT?

TNT is the abbreviation for 2,4,6-trinitrotoluene, a powerful, highly explosive compound widely used in conventional bombs. Discovered by Joseph Wilbrand (1811–1894) in 1863, it is made by treating toluene with nitric acid and sulfuric acid. This yellow crystalline solid with a low melting point has low shock sensitivity and even burns without exploding. This makes it safe to handle and cast, but once detonated, it explodes violently.

How are paper diapers able to absorb moisture?

In 1948, Johnson & Johnson introduced the first mass-marketed disposable diaper in the United States. More recent technical developments have added the water-absorbing chemical sodium polyacrylate to disposable diapers. Sodium polyacrylate forms a gel when mixed with water. This gel has a structure in which water is held the same way that gelatin holds water. Experiments have shown that sodium polyacrylate can absorb as much as several hundred times its weight in tap water.

What are synthetic fibers?

Synthetic fibers are totally made by chemical means or may be fibers of regenerated cellulose. According to U.S. law, the fibers must be labeled in accordance with generic groups as follows:

Synthetic Fiber	Derived From
Acetate	Cellulose acetate
Acrylic	Acrylic resins
Metallic	Any type of fabric made with metallic yarns. Made by twisting thin, metal foil around cotton, silk, linen, or rayon yarns
Modacrylic	Acrylic resins
Nylon	Synthetic polyamides extracted from coal and petroleum
Rayon	Trees, cotton, and woody plants
Saran	Vinylidene chloride
Spandex	Polyurethane
Triacetate	Regenerated cellulose
Vinyl	Polyvinyl chloride

Who developed polyester?

Wallace H. Carothers (1896–1937) was working for DuPont when he discovered that alcohols and carboxyl acids could be successfully combined to form fibers. This was the beginning of polyester, but it was put on the back burner once Carothers discovered nylon. A group of British scientists, J. R. Whinfield (1901–1966), J. T. Dickson, W. K. Birtwhistle, and C. G. Ritchie took up Carothers's work in 1939. In 1941, they created the first polyester fiber caller Terylene. In 1946, DuPont bought all legal rights from the British scientists and developed another polyester fiber that they named Dacron. Polyester was first introduced to the American public in 1951. Carothers was inducted into the National Inventors Hall of Fame in 1984 for his work that led to the invention of nylon, neoprene, and other synthetic fibers.

How are waterproof and water-repellent fabrics manufactured?

Both waterproof and water-repellent fabrics are treated with substances to make them impervious to water. To permanently waterproof a fabric, it is coated with plasticized synthetic resins, then vulcanized or baked. Fabrics treated in this manner become thick and rubberized and may therefore crack. In making a water-repellent fabric, the material is soaked in synthetic resins, metallic compounds, oils, or waxes that allow the fabric to retain its natural characteristics.

What is Kevlar®?

The registered trademark Kevlar® refers to synthetic fiber called liquid crystalline polymers. Discovered by Stephanie Kwolek (1923–2014), Kevlar® is a thin, very strong fiber. It is best known for its use in bulletproof garments.

What are food additives?

Food additives are substances added to food to improve the appearance or texture of food, enhance the flavor or color of a food, enhance the nutritional value of a food, or preserve the freshness and safety of a food item. In ancient times, food additives included salt as a preservative for meats and fish, sugar as a flavor enhancer for fruits, and brine solutions, i.e., vinegar and salt, for pickling. Modern food additives may be from natural products, including plants, animals, and minerals. Artificial food additives are synthesized products.

When were artificial sweeteners discovered?

The first artificial sweetener discovered was saccharin. It was discovered accidentally by Constantin Fahlberg (1850–1910) and Ira Remsen (1846–1927) at Johns Hopkins University in 1879. According to the story, after a day of working in the laboratory, one or both of the men picked up a roll at dinner. The roll tasted sweet. They surmised correctly that something was on their hands from the coal–tar derivative chemicals they had worked with during the day. They tasted the various chemicals in the laboratory and discovered that the compound, benzoic sulfimide, was a sweet compound.

How do artificial sweeteners compare to sugar?

Artificial sweeteners vary in their sweetness relative to table sugar. Since artificial sweeteners are much sweeter than sugar, manufacturers add fillers, such as dextrose and maltodextrin, to add bulk to the small quantity of artificial sweetener used for a teaspoon of the product.

Sweetener (Brand Name)	Relative Sweetness Compared to Sugar
Acesuflame potassium (Ace-K) (Sunett®, Sweet One®)	200 times sweeter than sugar
Aspartame (Equal®, NutraSweet®)	200 times sweeter than sugar
Steviol (Truvia®, PureVia®, Enliten®)	200–400 times sweeter than sugar
Saccharin (Sweet and Low®, Sweet'N Low®)	200–700 times sweeter than sugar
Sucralose (Splenda®)	600 times sweeter than sugar
Neotame (Newtame®)	7,000–13,000 times sweeter than sugar
Advantame	20,000 times sweeter than sugar

INTRODUCTION AND HISTORICAL BACKGROUND

What is astronomy?

Astronomy, from the Greek words *astron*, meaning "star," and *nomos*, meaning "law," is the study of the universe and the celestial bodies in the universe. The celestial bodies include the stars, planets, galaxies, moons, comets, asteroids, and gases in the atmosphere.

Did ancient civilizations study astronomy?

The ancient people and civilizations of the world observed the heavenly bodies. The Mesopotamians established the zodiac and saw the five bright planets: Mercury, Venus, Mars, Jupiter, and Saturn. The ancient Egyptians recognized a 24-hour day and 365-day solar year based on their observations of the Sun and stars. They observed that the appearance of the star Sirius in the eastern sky coincided with the annual flooding of the Nile River. Many astronomical observations were important in the ancient agricultural cultures as indicators for planting and harvesting. The early Chinese astronomers charted the night sky. They recorded solar eclipses, novae, and other celestial phenomena. The Chinese observed what would later be known as Halley's Comet as early as 240 B.C.E. The Mayan civilization learned to predict the movement of the Sun, Moon, and stars. They also devised a calendar.

What is widely considered to be one of the earliest celestial observatories?

Stonehenge, built in England between 2500–1700 B.C.E., is one of the earliest observatories or observatory temples. It is widely believed that its primary function was to observe the midsummer and midwinter solstices.

Who is considered the founder of systematic astronomy?

Greek scientist Hipparchus (c. 190–120 B.C.E.) is considered to be the father of systematic astronomy. He measured as accurately as possible the directions of objects in the sky. He compiled the first catalog of stars, containing about 850 entries, and designated each star's celestial coordinates, indicating its position in the sky. Hipparchus also divided the stars according to their apparent brightness or magnitudes.

What was Ptolemy's contribution to astronomy?

Claudius Ptolemy was an astronomer who lived in Alexandria. He wrote a thirteen-volume set, *Mathematike Syntaxis* in Greek (translated as the *Mathematical Compilation*). When the work was introduced to the Islamic world and translated into Arabic, it became known as simply *The Greatest*. In the middle of the twelfth century, it was translated into Latin and became known as *The Almagest*. The work compiled the knowledge of Greek astronomy based on the works of Aristotle, Hipparchus, Aristarchus (c. 310–230 B.C.E.), and Eratosthenes. It contained information on more than one thousand stars, identified constellations, explained how to calculate latitude and longitude, and predicted solar and lunar eclipses.

How long did the Ptolemaic model of the solar system endure?

The Ptolemaic model of the solar system, introduced by Claudius Ptolemy, is a geocentric model of the universe. In this model, Earth is at the center of the universe, with the Sun and other planets orbiting Earth. Ptolemy added epicycles, or small orbits, to the celestial objects to explain the observed orbits in the context of the geocentric model. The Ptolemaic model endured for more than 1,400 years.

Who proposed a heliocentric model of the solar system?

A heliocentric model of the solar system places the Sun, not Earth, as in Ptolemy's geocentric model, at the center of the solar system, with the planets orbiting the Sun in concentric circles. Although Greek astronomer Aristarchus proposed a heliocentric model of the solar system in the third century B.C.E., his theory was not accepted due to discrepancies between the prediction and observation of planetary motion.

Centuries later, Polish astronomer Nicolaus Copernicus proposed a heliocentric model of the solar system. His first publication detailing the heliocentric model of the solar system was mainly an unpublished, handwritten paper entitled *Commentariolus* ("Little Comment"). In *Commentariolus*, Copernicus outlined that the center of the universe is near the Sun and not Earth. Also, the rotation of Earth accounts for the apparent daily rotation of the stars. Finally, the apparent annual cycles of the movement of the Sun is a result of Earth revolving around the Sun. He detailed his theories and findings in his publication *On the Revolutions of the Heavenly Spheres* (*De Revolutionibus Orbium Coelestium*), published in 1543 shortly before his death. Copernicus is recognized as the father of modern astronomy.

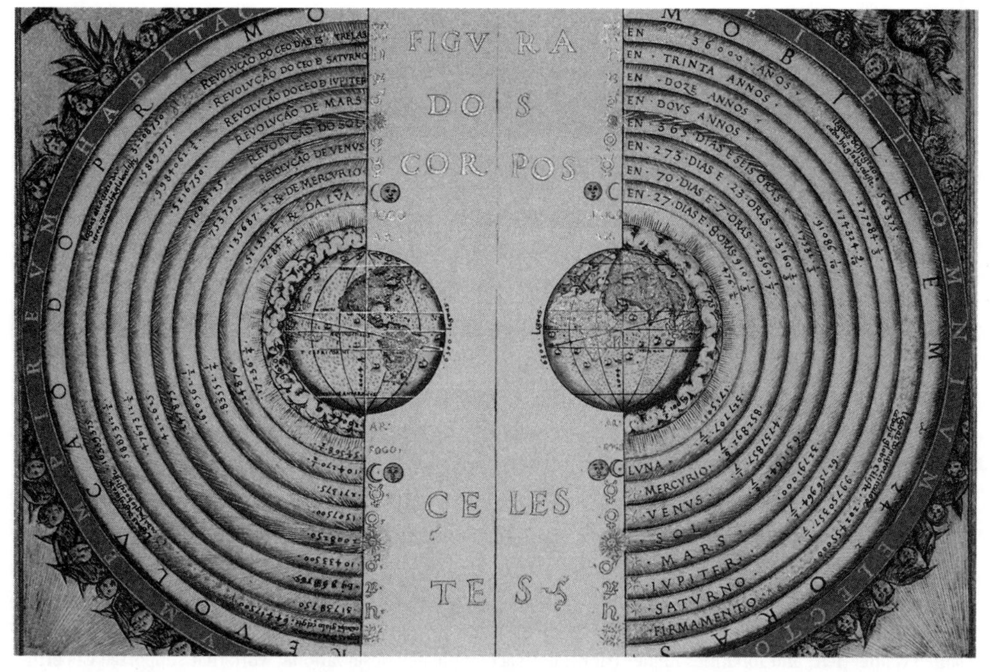

A 1568 illustration by cartographer Bartolomeu Velho explains the Ptolemaic geocentric model of the solar system and universe, a theory that was later supplanted by Copernicus's heliocentric model.

What were Galileo Galilei's contributions to the development of astronomy?

Galileo Galilei was the first scientist to use a telescope to observe the stars, planets, and other celestial objects. The large number of stars visible through a telescope amazed Galileo. Viewing the Moon through the telescope, Galileo was the first to see the rough surface of the Moon, consisting of craters and mountains. Prior to his observations, it was believed that the Moon had a smooth surface. Galileo also discovered sunspots. In 1610, he discovered four moons of Jupiter, known as the Galilean moons or satellites. They are Io, Europa, Callisto, and Ganymede. Galileo also observed the phases of Venus. As a result of Galileo's many observations, he recognized that the heliocentric model of the solar system proposed by Copernicus was correct. He published his ideas in *Dialogue Concerning the Two Chief World Systems Ptolemaic and Copernican* in 1632. These beliefs contradicted the teachings of the Catholic Church. Galileo was forced by the Catholic Church to recant his theories and ideas, condemned by the Inquisition in 1633, and placed under house arrest until his death in 1642. It was not until 1992 that Pope John Paul II (1920–2005) expressed regret for the way Galileo had been treated.

Who presented some of the earliest evidence to support Copernicus's model of the solar system?

Johannes Kepler made many contributions to the study of astronomy. His first work to defend the Copernican hypothesis of a heliocentric solar system was *Mysterium Cos-*

Galileo Galilei (left) and Johannes Kepler (right) both accomplished observational work that supported Copernicus's heliocentric model.

mographicum (*The Mystery of Cosmography*), published in 1596. The publication impressed Tycho Brahe (1546–1601), who invited Kepler to work with him. Brahe had collected data on the orbital motion of Mars, which Kepler analyzed. In 1609, Kepler published *Astronomia Nova* (*New Astronomy*) with evidence supporting Copernicus's theory. *Astronomia Nova* included what has become known as Kepler's first two laws of planetary motion. Kepler's third law of planetary motion was published in 1619 in *Harmonices Mundi* (*Harmonies of the World*).

Kepler made other contributions to astronomy. He observed and studied a supernova in the constellation Ophiuchus, confirmed Galileo's moons of Jupiter, explained how the Moon causes the ocean tides, and published of *Tabulae Rudolphinae* (*The Rudolphine Tables*) in 1627. The *Tabulue Rudolphinae* has tables for calculating planetary positions and tables of refraction and logarithms. It is also a catalog of more than

Who was the first Astronomer Royal?

The first Astronomer Royal was John Flamsteed (1646–1719). He was appointed Astronomer Royal in 1675, when the Royal Greenwich Observatory was founded. Until 1972, the Astronomer Royal also served as the director of the Royal Greenwich Observatory.

What were Katherine Johnson's contributions to NASA?

Katherine Johnson (1918–), an African American woman, began her career with the all-black West Area Computing section at the National Advisory Committee for Aeronautics's (NACA's) Langley Laboratory in 1953. One of her first projects was to analyze data from flight tests in the Maneuver Loads Branch of the Flight Research Division. She provided some of the math for the 1958 document "Notes on Space Technology." When NACA became NASA in 1958, she worked with the Space Task Group. Johnson did the trajectory analysis for America's first human spaceflight, Alan B. Shepard Jr.'s (1923–1998) 1961 mission Freedom 7. In 1962, NASA prepared for John H. Glenn Jr.'s (1921–2016) orbital mission. The calculations had been done on the electronic calculating machines available at the time, but the astronauts were wary of the calculations performed on the machines. Glenn requested that Johnson run the calculations by hand on her mechanical calculating machine to verify the orbital equations that would control the trajectory of Friendship 7.

one thousand stars. Not only did Kepler's work provide the evidence to support Copernicus's heliocentric model of the solar system, it also provided the basis for Isaac Newton to begin his work on gravitation force.

What are Kepler's three laws of planetary motion?

Kepler's first two laws of planetary motion were published in *Astronomia Nova* in 1609. The third law of planetary motion was published in 1619 in *Harmonices Mundi*. Kepler summarized all of his work on planetary motion in a seven-volume work entitled *Epitome Astronomica Copernicanae* ("Epitome of Copernican Astronomy"). The laws state:

1. The orbit of any planet is an ellipse with the Sun at one focus point (the law of ellipses).
2. The law of equal areas states that as each planet orbits the Sun, the line joining the planet and the Sun sweeps out in equal areas in equal times. Basically, this means that when a planet is closer to the Sun, it moves faster, and when it is farther away from the Sun, it moves more slowly.
3. The harmonic law states that the square of the orbital period of any planet is proportional to the cube of its mean distance from the Sun.

When was NASA established?

The National Aeronautics and Space Administration (NASA) was established in 1958. It succeeded the National Advisory Committee for Aeronautics (NACA), which was established in 1915. NASA has programs in aeronautics research, space flight, and exploration of the universe.

UNIVERSE

How old is the universe?

Recent data collected by the Hubble Space Telescope suggests that the universe may only be eight billion years old. This contradicts the previous belief that the universe was somewhere between thirteen and twenty billion years old. The earlier figure was derived from the concept that the universe has been expanding at the same rate since its birth at the Big Bang. The rate of expansion is a ratio known as Hubble's constant. It is calculated by dividing the speed at which the galaxy is moving away from Earth by its distance from Earth. By inverting Hubble's constant—that is, dividing the distance of a galaxy by its recessional speed—the age of the universe can be calculated. The estimates of both the velocity and distance of galaxies from Earth are subject to uncertainties, and not all scientists accept that the universe has always expanded at the same rate. Therefore, many still hold that the age of the universe is open to question.

What was the Big Bang?

The Big Bang theory is the explanation most commonly accepted by astronomers for the origin of the universe. It proposes that the universe began as the result of an explosion of a single particle—the Big Bang—fifteen to twenty billion years ago. Two observations support the basis of this cosmology. First, as Edwin Powell Hubble (1889–1953) demonstrated, the universe is expanding uniformly, with objects at greater distances receding at greater velocities. Secondly, Earth is bathed in a glow of radiation that has the characteristics expected from a remnant of a hot, primeval fireball. This radiation, known as cosmic microwave background radiation, was discovered by Arno A. Penzias (1933–) and Robert W. Wilson (1936–) of Bell Telephone Laboratories. Penzias and Wilson shared the 1978 Nobel Prize in Physics for their discovery. In time, the matter created by the Big Bang came together in huge clumps to form the galaxies. Smaller clumps within the galaxies formed stars. Parts of at least one clump became a group of planets—our solar system.

Who coined the term "big bang"?

The usage of the term "big bang" is attributed to British astronomer Fred Hoyle (1915–2001). Hoyle developed the steady-state theory of the universe with Hermann Bondi (1919–2005) and Thomas Gold (1920–2004) in 1948. According to the steady-state theory, the universe had always existed, it always looked the same, and it would remain the same forever. Hoyle's term "big bang" was intended to ridicule the theory of the universe beginning from a single particle.

Who proposed the Big Bang theory?

Georges Lemaître (1894–1966) is recognized as first formulating and proposing the Big Bang theory. Lemaître published a paper applying Einstein's Theory of Relativity to cosmology to the concept of an expanding universe in 1927. Lemaître was unaware that

Russian mathematician Alexander Friedmann (1888–1925) had proposed a similar mathematical solution in 1922 of an expanding universe in 1922. Lemaître published "The Beginning of the World from the Point of View of Quantum Theory" in *Nature*, volume 127, page 706, on May 9, 1931. His paper stated that at some point in the past, all the matter in the universe was compressed into a single, dense particle. This particle disintegrated in an explosion, giving rise to the universe.

Who was Stephen Hawking?

Stephen Hawking (1943–2018), a British physicist and mathematician, is considered to be the greatest theoretical physicist of the late twentieth century. In spite of being severely handicapped by amyotrophic lateral sclerosis (ALS), he made major contributions to scientific knowledge about black holes and the origin and evolution of the universe though his research into the nature of space-time and its anomalies. For instance, Hawking proposed that a black hole could emit thermal radiation, and he predicted that a black hole would disappear after all its mass had been converted into radiation (called "Hawking's radiation"). Hawking attempted to synthesize quantum mechanics and relativity theory into a theory of quantum gravity. He was the author of several books, including the popular best-selling work *A Brief History of Time*.

What is a black hole?

When a star with a mass greater than about four times that of the Sun collapses, even the neutrons cannot stop the force of gravity. Nothing can stop the contraction, and the star collapses forever. The material is so dense that nothing—not even light—can escape. American physicist John Wheeler (1911–2008) gave this phenomenon the name "black hole" in 1967. Since no light escapes from a black hole, it cannot be observed directly. However, if a black hole existed near another star, it would draw matter from the other star into itself and, in effect, produce X-rays. In the constellation of Cygnus is a strong X-ray source named Cygnus X-1. It is near a star, and the two revolve around each other. The unseen X-ray source has the gravitational pull of at least ten suns and is believed to be a black hole. Another type of black hole, a primordial black hole, may also exist dating from the time of the Big Bang, when regions of gas and dust were highly compressed. Recently, astronomers observed a brief pulse of X-rays from Sagittarius A, a

An artist's depiction of a black hole shows its accretion disc (matter orbiting the black hole that will eventually be sucked inside) and streams of gamma ray radiation shooting out of it.

region near the center of the Milky Way galaxy. The origin of this pulse and its behavior led scientists to conclude that a black hole probably exists in the center of our galaxy.

Four other black holes are possibly in existence: a Schwarzschild black hole has no charge and no angular momentum; a Reissner–Nordstrom black hole has charge but no angular momentum; a Kerr black hole has angular momentum but no charge; and a Kerr–Newman black hole has charge and angular momentum.

How far away is the nearest black hole?

The black hole nearest to Earth is V4641 Sagittarii. It is 1,600 light years (9,399,362,677,500,000 miles) away from Earth.

What is the Big Crunch theory?

According to the Big Crunch theory, at some point in the very distant future, all matter will reverse direction and crunch back into the single point from which it began. Two other theories predict the future of the universe: the Big Bore theory and the Plateau theory. The Big Bore theory, named because it has nothing exciting to describe, claims that all matter will continue to move away from all other matter and the universe will expand forever. According to the Plateau theory, expansion of the universe will slow to the point where it will nearly cease, at which time the universe will reach a plateau and remain essentially the same.

OBSERVATION AND MEASUREMENT

What is a light year?

A light year is a measure of distance, not time. It is the distance that light, which travels in a vacuum at the rate of 186,282 miles (299,792 kilometers) per second, can travel in a year (365.25 days). This is equal to 5.87 trillion miles (9.46 trillion kilometers).

In addition to the light year, what other units are used to measure distances in astronomy?

The astronomical unit (AU) is often used to measure distances within the solar system. One AU is equal to the average distance between Earth and the Sun, or 92,955,630 miles (149,597,870 kilometers). The parsec is equal to 3.26 light years, or about 19.18 trillion miles (30.82 trillion kilometers).

How are new celestial objects named?

Many stars and planets have names that date back to antiquity. The International Astronomical Union (IAU), the professional astronomers' organization, has attempted, in this century, to standardize names given to newly discovered celestial objects and their surface features.

Stars are generally called by their traditional names, most of which are of Greek, Roman, or Arabic origin. They are also identified by the constellation in which they appear, designated in order of brightness by Greek letters. Thus, Sirius is also called Alpha Canis Majoris, which means it is the brightest star in the constellation Canis Major. Other stars are called by catalog numbers, which include the star's coordinates. Several commercial star registries exist, and for a fee, you can submit a star name to them. These names are not officially recognized by the IAU.

The IAU has made some recommendations for naming the surface features of the planets and their satellites. For example, features on Mercury are named for composers, poets, and writers; features of Venus for women; and features on Saturn's moon Mimas for people and places in Arthurian legend.

Comets are named for their discoverers. Newly discovered asteroids are first given a temporary designation consisting of the year of discovery plus two letters. The first letter indicates the half-month of discovery (A = first half of January, B = second half of January, etc.) and the second the order of discovery in that half-month. Thus, asteroid 2002EM was the thirteenth (M) asteroid discovered in the first half of March (E) in 2002. After an asteroid's orbit is determined, it is given a permanent number, and its discoverer is given the honor of naming it. Asteroids have been named after such diverse things as mythological figures (Ceres, Vesta), an airline (Swissair), and the Beatles (Lennon, McCartney, Harrison, Starr).

Who invented the telescope?

Hans Lippershey, a German-Dutch lens grinder and spectacle maker, is generally credited with inventing the telescope in 1608 because he was the first scientist to apply for a patent. Two other inventors, Zacharias Janssen and Jacob Metius (1571–1628), also developed telescopes. Modern historians consider Lippershey and Janssen as the two likely candidates for the title of inventor of the telescope, with Lippershey possessing the

What is an astrolabe?

Invented by the Greeks or Alexandrians around 100 B.C.E., an astrolabe is a two-dimensional working model of the heavens with sights for observations. It consists of two concentric, flat disks: one fixed, representing the observer on Earth; and the other moving, which can be rotated to represent the appearance of the celestial sphere at a given moment. Given latitude, date, and time, the observer can read off the altitude and azimuth of the Sun, the brightest stars, and the planets. By measuring the altitude of a particular body, one can find the time. The astrolabe can also be used to find times of sunrise, sunset, twilight, or the height of a tower or depth of a well. After 1600, it was replaced by the sextant and other more accurate instruments.

strongest claim. Lippershey used his telescope for observing grounded objects from a distance.

In 1609, Galileo also developed his own refractor telescope for astronomical studies. Although small by today's standards, the telescope enabled Galileo to observe the Milky Way and to identify blemishes on the Moon's surface as craters.

What are the differences between reflecting and refracting telescopes?

Reflecting telescopes capture light using a mirror, while refracting telescopes capture light with a lens. The advantages of reflecting telescopes are 1) they collect light with a mirror, so no color fringing occurs, and 2) since a mirror can be supported at the back, size is not limited. In an effort to alleviate the problem of the color fringing that is always associated with lenses, Isaac Newton built a reflecting telescope in 1668 that collected light with mirrors.

When did the Very Large Array (VLA) become operational?

The Very Large Array (VLA) is one of the world's premier astronomical radio observatories. The U.S. Congress approved funding for the VLA in 1972, and construction began in 1973. Since its dedication in 1980, more than three thousand scientists have used it for more than eleven thousand separate observing projects.

The VLA consists of twenty-eight (twenty-seven active plus one spare) antennas arranged in a huge Y pattern up to 22 miles (36 kilometers) across—roughly one and a half times the size of Washington, D.C. Each antenna is 81 feet (25 meters) in diameter; they are combined electronically to give the resolution of an antenna 22 miles (36 kilometers) across with the sensitivity of a dish 422 feet (130 meters) in diameter. Each of the radio telescopes in the VLA is the size of a house and can be moved on train tracks. Some discoveries made by using the VLA are ice on the planet Mercury, radio-bright coronae around ordinary stars, microquasars in our galaxy, gravitationally induced Einstein rings around distant galaxies, and radio counterparts to cosmologically distant gamma-ray bursts. The vast size of the VLA has allowed astronomers to study the details of superfast cosmic jets and even map the center of our galaxy.

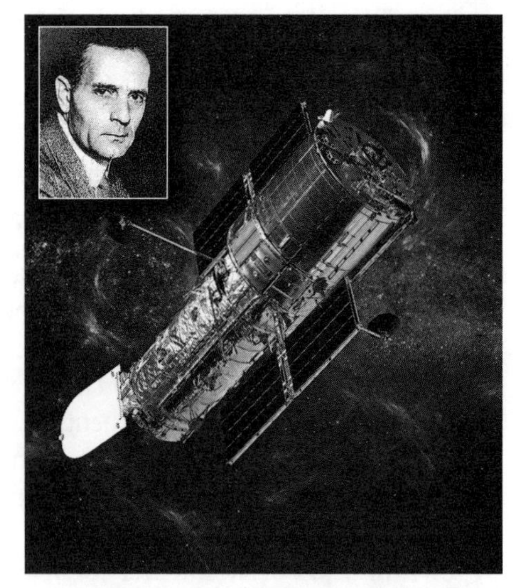

The Hubble Space Telescope is named after American astronomer Edwin Hubble (inset), who was a key figure in the advancement of extragalactic astronomy.

For whom is the Hubble Space Telescope named?

Edwin Powell Hubble was an American astronomer known for his studies of galaxies. His study of nebulae, or clouds—the faint, unresolved luminous patches in the sky—showed that some of them were large groups of many stars. Hubble classified galaxies by their shapes as being spiral, elliptical, or irregular.

Hubble's law establishes a relationship between the velocity of recession of a galaxy and its distance. The speed at which a galaxy is moving away from our solar system (measured by its redshift, the shift of its light to longer wavelengths, presumed to be caused by the Doppler effect) is directly proportional to the galaxy's distance from it.

The Hubble Space Telescope was deployed by the space shuttle Discovery on April 25, 1990. The telescope, which would be free of distortions caused by Earth's atmosphere, was designed to see more deeply into space than any telescope on land. However, on June 27, 1990, the National Aeronautics and Space Administration announced that the telescope had a defect in one of its mirrors that prevented it from properly focusing. Although other instruments, including one designed to make observations in ultraviolet light, were still operating, nearly 40 percent of the telescope's experiments had to be postponed until repairs were made. On December 2, 1993, astronauts were able to make the necessary repairs. Four of Hubble's six gyroscopes were replaced as well as two solar panels. Hubble's primary camera, which had a flawed mirror, was also replaced. Since that mission, four other servicing missions have been conducted, most recently in 2009, dramatically improving the HST's capabilities. Although it will not be repaired or upgraded again, it is expected to operate until 2020 and possibly longer.

What is an observatory?

An observatory is a location where astronomical observations can occur. Observatories may consist of a single telescope or many telescopes. The United States has national observatories, university-run observatories, and private observatories. The National Optical Astronomy Observatory and the National Radio Astronomy Observatory each operate several facilities in both the United States and as far away as Chile. The Kitt Peak Observatory in southern Arizona, part of the National Optical Astronomy Observatory, has one of the largest arrays of optical and radio telescopes in the world. Some well-known, university-owned observatories are the University of California's Lick Observatory and the California Institute of Technology's Palomar Observatory. Many amateur astronomers have made observations on private equipment, which was further investigated by professional researchers.

GALAXIES

What is a galaxy?

A galaxy is a vast collection of stars, gas, dust, other interstellar material, and dark matter that forms a cohesive gravitational unit in the universe. Galaxies range in size from dwarf galaxies containing only one hundred thousand stars to large, massive galaxies containing billions of stars. Galaxies may be spiral, elliptical, or irregular in shape.

What is the Milky Way?

The Milky Way is a hazy band of light that can be seen encircling the night sky. This light comes from the stars that make up the Milky Way galaxy, the galaxy to which our Sun and Earth belong. Astronomers estimate that the Milky Way galaxy contains at least one hundred billion stars and is about one hundred thousand light years in diameter. The galaxy is shaped like a compact disc with a central bulge, or nucleus, and spiral arms curving out from the center.

Which galaxy is closest to us?

The Andromeda galaxy is the galaxy closest to the Milky Way galaxy, where Earth is located. It is estimated to be 2.2 million light years away from Earth. Bigger than the Milky Way, Andromeda is a spiral-shaped galaxy that is also the brightest in Earth's sky.

What are quasars?

The name quasar originated as a contraction of "quasi-stellar" radio source. Quasars appear to be starlike, but they have large redshifts in their spectra, indicating that they are receding from Earth at great speeds, some at up to 90 percent of the speed of light. Their exact nature is still unknown, but many believe quasars to be the cores of distant galaxies, the most distant objects yet seen. Quasars, also called quasi-stellar objects or QSOs, were first identified in 1963 by astronomers at the Palomar Observatory in California.

The Milky Way and Andromeda galaxies are close neighbors that are moving closer and will one day (about 3.75 billion years from now) merge into one supergalaxy.

What are the four types of nebulae?

The four types of nebulae are emission, reflection, dark, and planetary. Primarily the birthplace of stars, nebulae are clouds of gas and dust in space. Emission nebulae and reflection nebulae are bright nebulae. Emission nebulae are colorful and self-luminous. The Orion nebula, visible with the naked eye, is an example of an emission nebula. Reflection nebulae are cool clouds of dust and gas. They are illuminated by the light from nearby stars rather than by their own energy. Dark nebulae, also known as absorption nebulae, are not illuminated and appear as holes in the sky. The Horsehead nebula in the constellation Orion is an example of a dark nebula. Planetary nebulae are the remnants of the death of a star.

STARS

Why do stars twinkle?

Stars actually shine with a more or less constant light. They appear to twinkle to those of us observing them from Earth due mostly to atmospheric interference. Molecules and dust particles float at random in Earth's covering of gases. When such floating particles pass between a star and a person observing it, a brief interruption in the stream of light occurs. Added together, these brief interruptions give rise to twinkling stars.

What does the color of a star indicate?

The color of a star gives an indication of its temperature and age. Stars are classified by their spectral type. From oldest to youngest and hottest to coolest, the types of stars are:

Type	Color	Temp. (°F)	Temp. (°C)
O	Blue	45,000–75,000	25,000–40,000
B	Blue	20,000–45,000	11,000–25,000
A	Blue-White	13,500–20,000	7,500–11,000
F	White	10,800–13,500	6,000–7,500
G	Yellow	9,000–10,800	5,000–6,000
K	Orange	6,300–9,000	3,500–5,000
M	Red	5,400–6,300	3,000–3,500

Each type is further subdivided on a scale of 0 to 9. The Sun is a type G2 star.

What is a binary star?

A binary star is a pair of stars revolving around a common center of gravity. About half of all stars are members of either binary star systems or multiple star systems, which contain more than two stars.

The bright star Sirius, about 8.6 light years away, is composed of two stars: one about 2.3 times the mass of the Sun, the other a white dwarf star about 980 times the mass of Jupiter. Alpha Centauri, the nearest star to Earth after the Sun, is actually three stars: Alpha Centauri A and Alpha Centauri B, two sunlike stars that orbit each other, and Alpha Centauri C, a low-mass red star that orbits around them.

What is a pulsar?

A pulsar is a rotating neutron star that gives off sharp, regular pulses of radio waves at rates ranging from 0.001 to 4 seconds. Stars burn by fusing hydrogen into helium. When they use up their hydrogen, their interiors begin to contract. During this contraction, energy is released, and the outer layers of the star are pushed out. These layers are large and cool; the star is now a red giant. A star with more than twice the mass of the Sun will continue to expand, becoming a supergiant. At that point, it may blow up in an explosion called a supernova. After a supernova, the remaining material of the star's core may be so compressed that the electrons and protons become neutrons. A star 1.4 to 4 times the mass of the Sun can be compressed into a neutron star only about 12 miles (20 kilometers) across. Neutron stars rotate very fast. The neutron star at the center of the Crab Nebula spins thirty times per second.

Some of these neutron stars emit radio signals from their magnetic poles in a direction that reaches Earth. These signals were first detected by Jocelyn Bell (1943–) of Cambridge University in 1967. Because of their regularity, some people speculated that they were extraterrestrial beacons constructed by alien civilizations. This theory was eventually ruled out, and the rotating neutron star came to be accepted as the explanation for these pulsating radio sources.

What is a supernova?

A supernova is the death explosion of a massive star. Immediately after the explosion, the brightness of the star can outshine the entire galaxy, followed by a gradual fading. A supernova is a fairly rare event. The last supernova observed in our galaxy was in 1604. In February 1987, Supernova 1987A appeared in the Large Magellanic Cloud, a nearby galaxy.

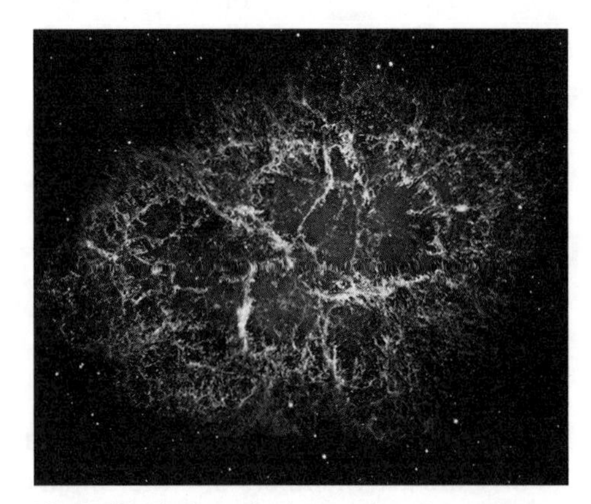

The Crab Nebula marks the remains of a supernova that was observed on Earth in 1054 by Chinese stargazers. At the center of the nebula is a pulsar, the remains of the original star.

What is the most massive star?

The Pistol Star is both the brightest and most massive star known. Located twenty-five thousand light years away in the area of the constellation Sagittarius, this young (one- to three-million-year-old) star is as bright as ten million suns and may have weighed two hundred times the mass of the Sun at one point in its young life.

Which stars are the brightest?

The brightness of a star is called its magnitude. Apparent magnitude is how bright a star appears to the naked eye. The lower the magnitude, the brighter the star. On a clear night, stars of about magnitude 16 can be seen with the naked eye. Large telescopes can detect objects as faint as 127. Very bright objects have negative magnitudes; the Sun is –26.8.

Brightest Stars in Our Night Sky

Star	Constellation	Apparent Magnitude
Sirius	Canis Major	–1.47
Canopus	Carina	–0.72
Arcturus	Boötes	–0.06
Rigil Kentaurus	Centaurus	10.01
Vega	Lyra	10.04
Capella	Auriga	10.05
Rigel	Orion	10.14
Procyon	Canis Minor	10.37
Betelgeuse	Orion	10.41
Achernar	Eridanus	10.51

How many constellations exist, and how were they named?

Constellations are groups of stars that seem to form some particular shape, such as that of a person, animal, or object. They only appear to form this shape and be close to each other from Earth; in actuality, the stars in a constellation are often very distant from each other. Eighty-eight recognized constellations have boundaries that were defined in the 1920s by the International Astronomical Union.

Various cultures in all parts of the world have had their own constellations. However, because modern science is predominantly a product of Western culture, many of the constellations represent characters from Greek and Roman mythology. When Europeans began to explore the Southern Hemisphere in the sixteenth and seventeenth centuries, they derived some of the new star patterns from the technological wonders of their time, such as the microscope.

Names of constellations are usually given in Latin. Individual stars in a constellation are usually designated with Greek letters in the order of brightness; the brightest star is alpha, the second brightest is beta, and so on. The genitive, or possessive, form of the constellation name is used; thus, Alpha Orionis is the brightest star of the constellation Orion.

Which are the ten largest constellations?

The size of a constellation is measured in square degrees. Square degrees is an angular measurement used to describe distances and areas in the sky. Hydra is the largest constellation, 1,303 square degrees, extending from Gemini to the south of Virgo. It has a recognizable long line of stars. The name "hydra" is derived from the water snake monster killed by Hercules in ancient mythology.

Ten Largest Constellations

Constellation	Abbreviation	Meaning	Area (square degrees)
Hydra	Hya	Sea Serpent	1,303
Virgo	Vir	Virgin	1,294
Ursa Major	UMa	Big Bear	1,280
Cetus	Cet	Whale	1,231
Hercules	Her	Hercules	1,225
Eridanus	Eri	River Eridanus	1,138
Pegasus	Peg	Winged Horse	1,121
Draco	Dra	Dragon	1,082
Centaurus	Cen	Centaur	1,060
Aquarius	Aqr	Water Bearer	980

What is the Big Dipper?

The Big Dipper is a group of seven stars that are part of the constellation Ursa Major. They appear to form a sort of bowl, composed of four stars, with a long handle, composed of three stars. The group is known as the Plough in Great Britain. The Big Dipper is almost always visible in the Northern Hemisphere. It serves as a convenient reference point when locating other stars; for example, an imaginary line drawn from the two end stars of the dipper leads to Polaris, the North Star.

Where is the North Star?

If an imaginary line is drawn from the North Pole into space, it will reach a star called Polaris, or the North Star, less than one degree away from the line. As Earth rotates on its axis, Polaris acts as a pivot point around which all the stars visible in the Northern Hemisphere appear to move, while Polaris itself remains motionless. Identifying Polaris was important for navigation since in locating Polaris, it was possible to identify north. In addition, the angle of Polaris above the horizon indicates latitude on Earth.

Was Polaris always the North Star?

Earth has had several North Stars. Earth slowly wobbles on its axis as it spins. This motion is called precession. Earth traces a circle in the sky over a period of twenty-six thou-

Which well-known star is part of the Little Dipper?

The Little Dipper, part of the constellation Ursa Minor, is similar to the Big Dipper. It also has seven bright stars that form the shape of a ladle. Polaris, the North Star, is at the end of the handle of the Little Dipper.

sand years. In Pharonic times, the North Star was Thuban; today, it is Polaris; in around 14,000 C.E., it will be Vega.

Which star is the closest to Earth?

The Sun, at a distance of 92,955,900 miles (149,598,000 kilometers), is the closest star to Earth. After the Sun, the closest stars are the members of the triple-star system known as Alpha Centauri (Alpha Centauri A, Alpha Centauri B, and Alpha Centauri C, sometimes called Proxima Centauri). They are 4.3 light years away.

SUN

How hot is the Sun?

The center of the Sun is about 27,000,000°F (15,000,000°C). The surface, or photosphere, of the Sun is about 10,000°F (5,500°C). Magnetic anomalies in the photosphere cause cooler regions that appear to be darker than the surrounding surface. These sunspots are about 6,700°F (4,000°C). The Sun's layer of lower atmosphere, the chromosphere, is only a few thousand miles thick. At the base, the chromosphere is about 7,800°F (4,300°C), but its temperature rises with altitude to the corona, the Sun's outer layer of atmosphere, which has a temperature of about 1,800,000°F (1,000,000°C).

What is the Sun made of?

The Sun is an incandescent ball of gases. Its mass is 1.8 × 1,027 tons or 1.8 octillion tons (a mass 330,000 times as great as Earth).

Element	Percent of Mass
Hydrogen	73.46
Helium	24.85
Oxygen	0.77
Carbon	0.29
Iron	0.16
Neon	0.12
Nitrogen	0.09
Silicon	0.07
Magnesium	0.05
Sulfur	0.04
Other	0.10

When will the Sun die?

The Sun is approximately 4.5 billion years old. About five billion years from now, the Sun will have burned all of its hydrogen fuel into helium. As this process occurs, the Sun will

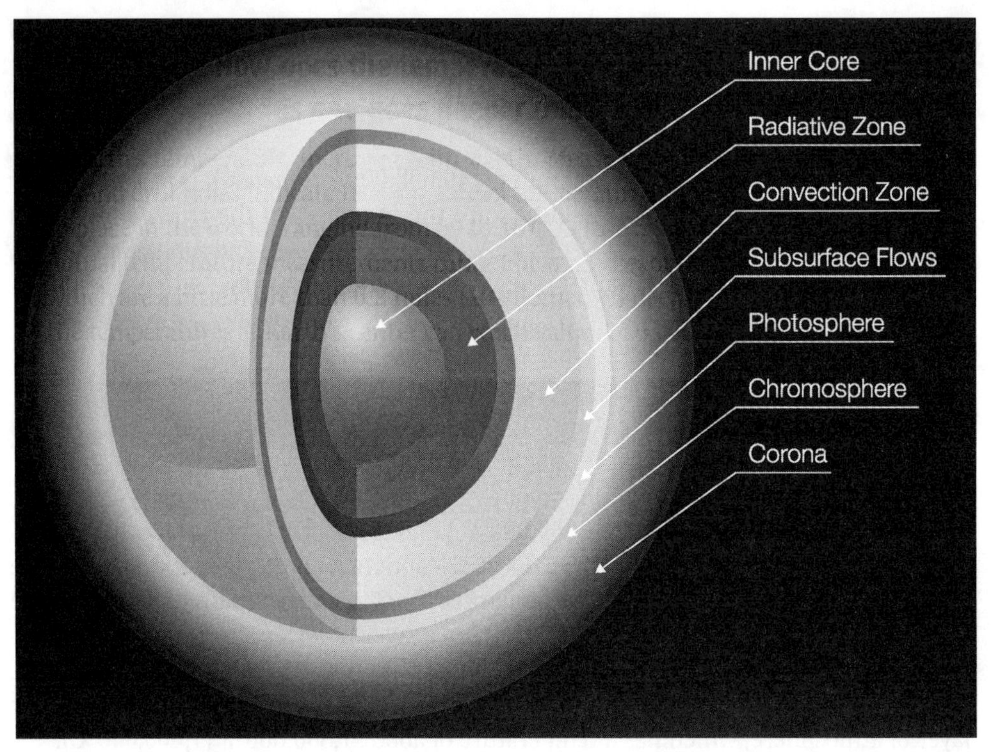

The seven layers of the Sun are shown here. The photosphere is the visible surface of the Sun. The corona can be as hot as 3.5 million degrees Fahrenheit (2 million degrees Celsius).

change from the yellow dwarf as we know it to a red giant. Its diameter will extend well beyond the orbit of Venus and even possibly beyond the orbit of Earth. In either case, Earth will be burned to a cinder.

Why does the color of the Sun vary?

Sunlight contains all the colors of the rainbow, which blend to form white light, making sunlight appear white. At times, some of the color wavelengths, especially blue, become scattered in Earth's atmosphere, and the sunlight appears colored. When the Sun is high in the sky, some of the blue rays are scattered in Earth's atmosphere. At such times, the sky looks blue, and the Sun appears to be yellow. At sunrise or sunset, when the light must follow a longer path through Earth's atmosphere, the Sun looks red (red having the longest wavelengths).

How long does it take light from the Sun to reach Earth?

Sunlight takes about 8 minutes and 20 seconds to reach Earth, traveling at 186,282 miles (299,792 kilometers) per second, although it varies with the position of Earth in its orbit. In January, the light takes about 8 minutes, 15 seconds to reach Earth; in July, the trip takes about 8 minutes, 25 seconds.

How long is a solar cycle?

The solar cycle is the periodic change in the number of sunspots. The cycle is taken as the interval between successive minima and is about 11.1 years long. During an entire cycle, solar flares, sunspots, and other magnetic phenomena move from intense activity to relative calm and back again. The solar cycle is one area of study to be carried out by up to ten ATLAS space missions designed to probe the chemistry and physics of the atmosphere. These studies of the solar cycle will yield a more detailed picture of Earth's atmosphere and its response to changes in the Sun.

What is a solar flare?

A solar flare is a sudden, intense release of energy in the Sun's outer atmosphere. The magnetic energy of a solar flare is released in the form of radiation at speeds of 1 million miles (1.6 million kilometers) per hour. The amount of energy released is equivalent to millions of 100-megaton hydrogen bombs exploding at the same time. The frequency of solar flares is coordinated with the Sun's eleven-year cycle. The number of solar flares is greatest at the peak of the solar cycle. Solar flares may be visible from Earth. The first solar flare observed and recorded was in 1859 by Richard C. Carrington (1826–1875), who saw a sudden, white light while observing sunspots.

What is the sunspot cycle?

It is the fluctuating number of sunspots on the Sun during an eleven-year period. The variation in the number of sunspots seems to correspond with the increase or decrease in the number of solar flares. An increased number of sunspots means an increased number of solar flares.

What is solar wind?

Solar wind is caused by the expansion of gases in the Sun's outermost atmosphere, the corona. Because of the corona's extremely high temperature of 1,800,000°F (1,000,000°C), the gases heat up, and their atoms start to collide. The atoms lose electrons and become electrically charged ions. These ions create the solar wind. Solar wind has a velocity of 310 miles (500 kilometers) per second, and its density is approximately 82 ions per cubic inch (5 ions per cubic centimeter). Because Earth is surrounded by strong magnetic forces, its magnetosphere, it is protected from the solar wind particles. In 1959, the So-

What is a syzygy?

A syzygy (*SIZZ-eh-jee*) is a configuration that occurs when three celestial bodies lie in a straight line, such as the Sun, Earth, and the Moon during a solar or lunar eclipse. The particular syzygy when a planet is on the opposite side of Earth from the Sun is called an opposition.

viet spacecraft Luna 2 acknowledged the existence of solar wind and made the first measurements of its properties.

When do solar eclipses happen?

A solar eclipse occurs when the Moon passes between Earth and the Sun and all three bodies are aligned in the same plane. When the Moon completely blocks Earth's view of the Sun and the umbra, or dark part of the Moon's shadow, reaches Earth, a total eclipse occurs. A total eclipse happens only along a narrow path 100–200 miles (160–320 kilometers) wide called the track of totality. Just before totality, the only parts of the Sun that are visible are a few points of light called Baily's beads shining through valleys on the Moon's surface. Sometimes, a last bright flash of sunlight is seen—the diamond ring ef-

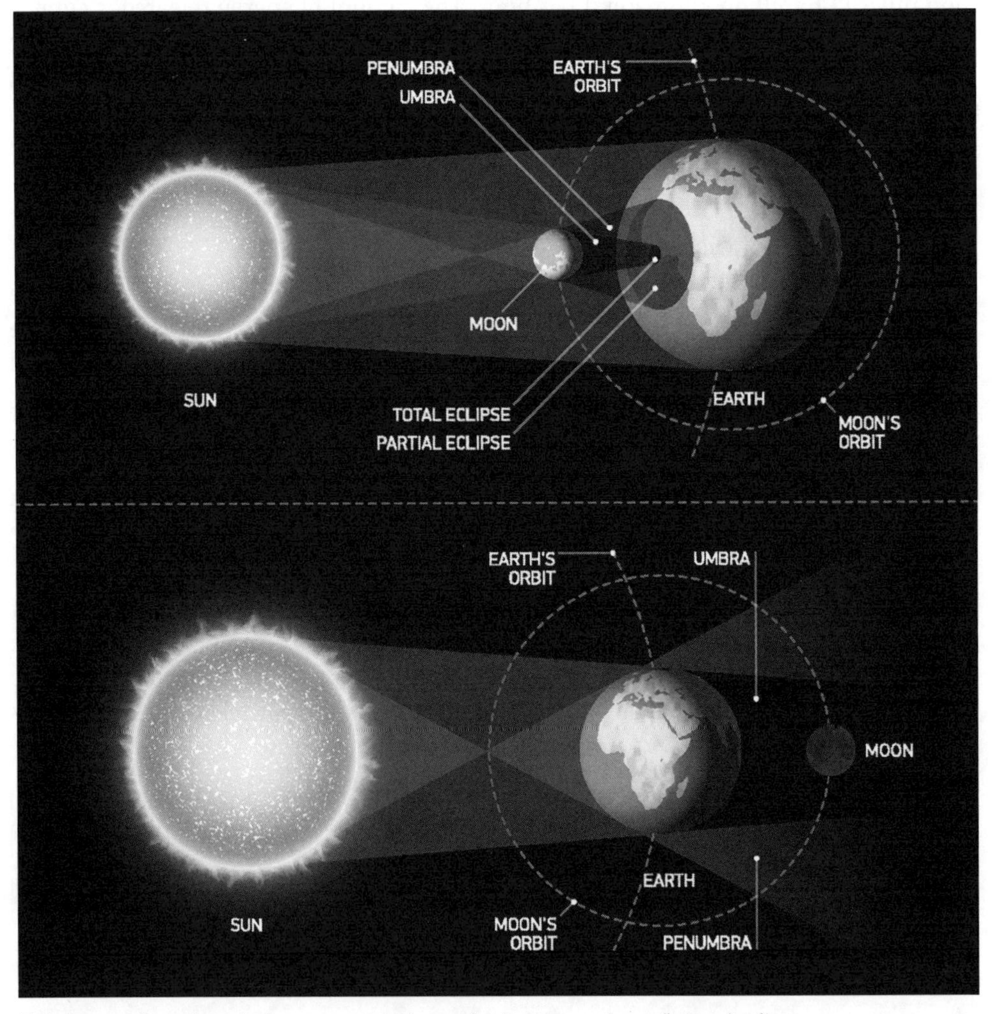

This diagram shows how shadows are cast during a solar (top) versus lunar (bottom) eclipse.

fect. During totality, which averages 2.5 minutes but may last up to 7.5 minutes, the sky is dark, and stars and other planets are easily seen. The corona, the Sun's outer atmosphere, is also visible.

If the Moon does not appear large enough in the sky to completely cover the Sun, it appears silhouetted against the Sun with a ring of sunlight showing around it. This is an annular eclipse. Because the Sun is not completely covered, its corona cannot be seen, and although the sky may darken, it will not be dark enough to see the stars.

During a partial eclipse of the Sun, the penumbra of the Moon's shadow strikes Earth. A partial eclipse can also be seen on either side of the track of totality of an annular or total eclipse. The Moon will cover part of the Sun, and the sky will not darken noticeably during a partial eclipse.

What is the safest way to view a solar eclipse?

Looking directly at the Sun is never safe. The Sun's UV radiation can cause permanent damage to the retina of the eye, including blindness. Special "eclipse eyeglasses" or other solar filters that conform with the international standard ISO 12312-2:2015, "Eye and face protection—Sunglasses and related eyewear—Part 2: Filters for direct observation of the sun," offer protective filters to view the Sun directly. All solar filters and "eclipse eyeglasses" should be inspected to ensure that the filters have no scratches or damage. Viewing a partial or total eclipse with other devices, such as photographic filters, exposed film, smoked glass, camera lenses, telescopes, or binoculars, will cause damage to the eyes.

Alternatively, one may look at a projected image of the eclipsed Sun. The simplest way to project an image of the Sun is with two pieces of stiff, white paper and a pin. Use the pin to punch a hole in the center of one of the sheets of paper. Hold the paper up and aim the hole at the Sun. *Do not look directly at the Sun or through the hole at the Sun.* Adjust the second sheet of paper to view the best image of the Sun.

When and where will the next ten total solar eclipses occur?

A total solar eclipse occurs somewhere on Earth approximately once every eighteen months. However, hundreds of years can elapse for a total solar eclipse to be visible at the same location on Earth. The next total solar eclipse in the United States, April 8, 2024, will be visible from Texas to Ohio, including Arkansas, Missouri, and Indiana. Below is a list of the upcoming solar eclipses across the world.

Upcoming Solar Eclipses

Date	Location of Total Solar Eclipse
December 14, 2020	South Pacific, Chile, Argentina, South Atlantic
December 4, 2021	Antarctica
April 8, 2024	Mexico, central United States, eastern Canada
August 12, 2026	Arctic, Greenland, Iceland, Spain

Date	Location of Total Solar Eclipse
August 2, 2027	Morocco, Spain, Algeria, Libya, Egypt, Saudi Arabia, Yemen, Somalia
July 22, 2028	Australia, New Zealand
November 25, 2030	Botswana, South Africa, Australia
March 30, 2033	Eastern Russia, Alaska
March 20, 2034	Nigeria, Cameroon, Chad, Sudan, Egypt, Saudi Arabia, Iran, Afghanistan, Pakistan, India, China
September 2, 2035	China, Korea, Japan, Pacific

SOLAR SYSTEM

Which celestial objects are part of the solar system?

The solar system consists of the Sun and all the celestial objects that fall within its gravitation's influence. This includes eight planets, dwarf planets, and smaller solar bodies, such as asteroids and comets.

How old is the solar system?

It is currently believed to be 4.5 billion years old. Earth and the rest of the solar system formed from an immense cloud of gas and dust. Gravity and rotational forces caused the cloud to flatten into a disc and much of the cloud's mass to drift into the center. This material became the Sun. The leftover parts of the cloud formed small bodies called planetesimals. These planetesimals collided with each other, gradually forming larger and larger bodies, some of which became the planets. This process is thought to have taken about twenty-five million years.

How large is the solar system?

The size of the solar system can be visualized by imagining the Sun (864,000 miles [1,380,000 kilometers] in diameter) shrunk to a diameter of 1 inch (about the size of a ping-pong ball). Using the same size scale, Earth would be a speck 0.01 inches (0.25 millimeters) in diameter and about 9 feet (2.7 meters) away from the ping-pong-ball-sized Sun. Our Moon would have a diameter of 0.0025 inches (0.06 millimeters; the thickness of a human hair) and be only a little over one-quarter inch (6.3 mm) from Earth. Jupiter, the largest planet in the solar system, appears as the size of a small pea (0.1 inches [2.5 millimeters] in diameter) and 46 feet (14 meters) from the Sun.

How many dwarf planets are in the solar system?

A dwarf planet is an object in orbit around the Sun that has sufficient mass (large enough) to have its own gravity pull itself into a round or near-round shape. Currently, five objects are considered dwarf planets: Ceres, Pluto, Eris, Makemake, and Haumea. Scientists expect to discover additional dwarf planets or reclassify some large asteroids as dwarf planets.

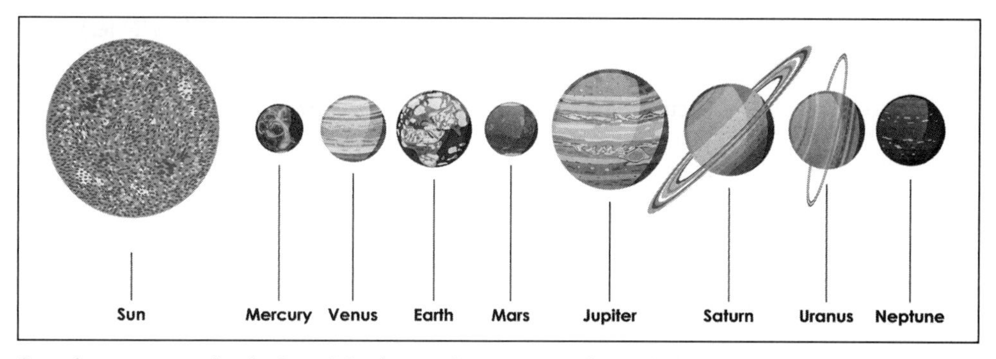

Our solar system contains the Sun, eight planets, plus numerous planetoids, asteroids, and comets (Sun and planets not drawn to scale).

What is a small solar system body?

The International Astronomical Union defines a small solar system body as all objects that orbit the Sun and are too small for their own gravity to pull them into a nearly spherical shape. Simply stated, they are all objects that do not meet the definition of a planet or dwarf planet. Examples of small solar system bodies are asteroids, near-Earth objects, the objects in the Kuiper Belt and Oort Cloud, most Centaurs, most trans-Neptunian objects, and comets.

What are Kuiper Belt Objects?

Kuiper Belt Objects (KBOs) are, as their name implies, objects that originate from or orbit in the Kuiper Belt. Only one KBO was known for more than sixty years: Pluto. Many KBOs have been discovered since 1992, however, and the current estimate is that millions, if not billions, of KBOs exist.

KBOs are basically comets without tails: icy dirtballs that have collected together over billions of years. If they get large enough—like Pluto did—they evolve as other massive, planetlike bodies do, forming dense cores that have a different physical composition than the mantle or crust above it. Most short-period comets—those with relatively short orbital times of a few years to a few centuries—are thought to originate from the Kuiper Belt.

What are the largest Kuiper Belt Objects, and how big are they?

The following table lists the largest KBOs in our solar system that are known of today:

Largest Kuiper Belt Objects

Name	Geometric Mean Diameter (Kilometers/Miles)
Eris	2,600 / 1,616
Pluto	2,390 / 1,485
Sedna	1,500 / 932
Quaoar	1,260 / 783

Name	Geometric Mean Diameter (Kilometers/Miles)
Charon	1,210 / 752
Orcus	940 / 584
Varuna	890 / 553
Ixion	820 / 510
Chaos	560 / 348
Huya	500 / 311

Where are asteroids found?

The asteroids, also called the minor planets, are smaller than any of the eight major planets in the solar system and are not satellites of any major planet. The term asteroid means "starlike" because asteroids appear to be points of light when seen through a telescope. Originally, astronomers thought the asteroids were remnants of a planet that had been destroyed; now they believe asteroids to be material that never became a planet, possibly because it was affected by Jupiter's strong gravity.

Most asteroids are located in the main asteroid belt between Mars and Jupiter, between 2.1 and 3.3 AUs (astronomical units) from the Sun. It is estimated that between 1.1 million and 1.9 million asteroids are larger 0.6 miles (1 kilometer) in diameter in the main asteroid belt. Millions of smaller asteroids are also in the main asteroid belt. The Trojan asteroids move in virtually the same orbit as Jupiter but at points 60° ahead or 60° behind the planet.

The near-Earth asteroids (NEAs) reside in the inner solar system. The NEAs are divided into four groups—Atira, Aten, Apollo, and Amor—based on their perihelion (point where the object is closest to the Sun), the aphelion (point where the object is farthest from the Sun), and their semimajor axes. The Atira have orbits that are contained entirely within the orbit of Earth. The Atens are Earth-crossing NEAs with semimajor axes smaller than Earth's. The Apollos are Earth-crossing NEAs with semimajor axes larger than Earth's. The Amors have orbits that are exterior to Earth's but interior to Mars's. Astronomers have discovered 10,003 NEAs, of which 861 are thought to have a diameter larger than 0.6 miles (1 kilometer).

When was the first object in the asteroid belt discovered?

Ceres, the first member of the asteroid belt to be discovered, was found by Giuseppe Piazzi (1746–1826) on January 1, 1801. Ceres has a radius of 296 miles (476 kilometers).

How many near-Earth asteroids are classified as potentially hazardous?

More than one thousand NEAs are classified as potentially hazardous (PHAs). The PHAs could strike Earth, possibly unleashing cosmic destruction.

Ceres was classified as a dwarf planet in 2006 since it is so much larger and different from its asteroid neighbors.

Since the discovery of Ceres, many, many more asteroids have been discovered. Currently, 775,096 asteroids are known.

Eris (shown here with its moon, Dysnomia, as rendered by an artist) is a trans-Neptunian dwarf planet that is actually larger than Pluto.

When was it decided that Pluto is no a planet?

When Pluto was first discovered in 1930 by American astronomer Clyde Tombaugh (1906–1997), it was considered the ninth planet of our solar system. During the late 1990s and the beginning of the twenty-first century, astronomers began to discover more objects orbiting beyond Neptune in the area known as the trans-Neptunian region. On the night of October 21, 2003, Mike Brown (1965–) from Caltech, Chad Trujillo (1973–) from the Gemini Observatory, and David Rabinowitz (1960–) from Yale University discovered a new object more massive than Pluto with its own satellite (Eris). The International Astronomical Union (IAU) began to debate the question of what constitutes a planet. In 2006, the IAU approved a new definition of a planet. This definition states that a planet is:

1. An object in orbit around the Sun
2. An object with sufficient mass (large enough) to have its self-gravity pull itself into a round (or near-spherical) shape
3. Has cleared the neighborhood around its orbit of other objects.

Pluto no longer meets the definition of a planet because of its size and it is in the trans-Neptunian region, a zone of other similarly sized objects. Instead, Pluto is a dwarf planet.

PLANETS

How far are the planets from the Sun?

The planets revolve around the Sun in elliptical orbits, with the Sun at one focus of the ellipse. Thus, a planet is at times closer to the Sun than at other times. The distances given below are the average distance from the Sun, starting with Mercury, the planet closest to the Sun, and moving outward.

Planet	Average Distance (Miles)	Average Distance (Kilometers)
Mercury	35,983,000	57,909,100
Venus	67,237,700	108,208,600

Planet	Average Distance (Miles)	Average Distance (Kilometers)
Earth	92,955,900	149,598,000
Mars	141,634,800	227,939,200
Jupiter	483,612,200	778,298,400
Saturn	888,184,000	1,427,010,000
Uranus	1,782,000,000	2,869,600,000
Neptune	2,794,000,000	4,496,700,000

Which are the hottest and coldest planets?

The hottest planet is Venus. It has a surface temperature of 900°F (480°C). The carbon dioxide of Venus's atmosphere traps the heat, producing very high surface temperatures. Neptune is the coldest planet, with temperatures of –367°F (–224°C) or 51.7 K.

Which planets have rings?

Jupiter, Saturn, Uranus, and Neptune all have rings. Jupiter's rings were discovered by *Voyager 1* in March 1979. The rings extend 80,240 miles (129,130 kilometers) from the center of the planet. They are about 4,300 miles (7,000 kilometers) in width and less than 20 miles (30 kilometers) thick. A faint inner ring is believed to extend to the edge of Jupiter's atmosphere. Saturn has the largest, most spectacular set of rings in the solar system. Saturn's ring system was first recognized by Dutch astronomer Christiaan Huygens in 1659. Its rings are 169,800 miles (273,200 kilometers) in diameter but less than 10 miles (16 kilometers) thick. Saturn has six different rings, the largest of which appear to be divided into thousands of ringlets. The rings appear to be composed of pieces of water ice ranging in size from tiny grains to blocks several tens of yards in diameter.

In 1977, when Uranus occulted (passed in front of) a star, scientists observed that the light from the star flickered or winked several times before the planet itself covered the star. The same flickering occurred in reverse order after the occultation. The reason for this was determined to be a ring around Uranus. Nine rings were initially identified, and *Voyager 2* observed two more in 1986. The rings are thin, narrow, and very dark.

Voyager 2 also discovered a series of at least four rings around Neptune in 1989. Some of the rings appear to have arcs, areas of a higher density of material than other parts of the ring.

Which planets are visible with the naked eye?

Mercury, Venus, Mars, Jupiter, and Saturn are visible with the naked eye at varying times of the year.

How long do the planets take to orbit the Sun?

Planet	Period of Revolution (Earth Days)	Period of Revolution (Earth Years)
Mercury	88	0.24
Venus	224.7	0.62
Earth	365.26	1.00
Mars	687	1.88
Jupiter	4,332.6	11.86
Saturn	10,759.2	29.46
Uranus	30,685.4	84.01
Neptune	60,189	164.8

What are the diameters of the planets?

Planet	Diameter (Miles)	Diameter (Kilometers)
Mercury	3,031	4,878
Venus	7,520	12,104
Earth	7,926	12,756
Mars	4,221	6,794
Jupiter	88,846	142,984
Saturn	74,898	120,536
Uranus	31,763	51,118
Neptune	31,329	50,530

All diameters are as measured at the planet's equator.

What are the colors of the planets?

Planet	Color
Mercury	Orange
Venus	Yellow
Earth	Blue, brown, green
Mars	Red
Jupiter	Yellow, red, brown, white
Saturn	Yellow
Uranus	Green
Neptune	Blue

What is the gravitational force on each of the planets, the Moon, and the Sun relative to Earth?

If the gravitational force on Earth is taken as 1, the comparative forces are:

Solar System Object	Gravitational Force Compared to Earth
Sun	27.9
Mercury	0.37

What is unique about the rotation of the planet Venus?

Unlike Earth and most of the other planets, Venus rotates in a retrograde, or opposite, direction with relation to its orbital motion about the Sun. It rotates so slowly that only two sunrises and sunsets occur each Venusian year. Uranus's rotation is also retrograde.

Solar System Object	Gravitational Force Compared to Earth
Venus	0.88
Moon	0.16
Mars	0.38
Jupiter	2.64
Saturn	1.15
Uranus	0.93
Neptune	1.22

Weight comparisons can be made by using this table. If a person weighed 100 pounds (45.36 kilograms) on Earth, then the weight of the person on the Moon would be 16 pounds (7.26 kilograms) or 100×0.16.

Which planets are called "inferior" planets and which are "superior" planets?

An inferior planet is one whose orbit is nearer to the Sun than Earth's orbit is. Mercury and Venus are the inferior planets. Superior planets are those whose orbits around the Sun lie beyond that of Earth. Mars, Jupiter, Saturn, Uranus, and Neptune are the superior planets. The terms have nothing to do with the quality of an individual planet.

Is a day the same on all the planets?

No. A day, the period of time it takes for a planet to make one complete turn on its axis, varies from planet to planet. Venus and Uranus display retrograde motion; that is to say, they rotate in the opposite direction from the other planets. The table below lists the length of a day for each planet.

Length of Day

Planet	Earth Days	Hours	Minutes
Mercury	58	15	30
Venus	243		32
Earth		23	56
Mars		24	37
Jupiter		9	50
Saturn		10	39
Uranus		17	14
Neptune		16	3

What are the Jovian and terrestrial planets?

Jupiter, Saturn, Uranus, and Neptune are the Jovian (the adjectival form for the word "Jupiter"), or Jupiter-like, planets. They are gas giant planets, composed primarily of light elements such as hydrogen and helium.

Mercury, Venus, Earth, and Mars are the terrestrial (derived from *terra*, the Latin word for "earth"), or Earth-like, planets. They are small in size, have solid surfaces, and are composed of rocks and iron.

Is it true that the rotational speed of Earth varies?

The rotational speed is at its maximum in late July and early August and at its minimum in April; the difference in the length of the day is about 0.0012 second. Since about 1900, Earth's rotation has been slowing at a rate of approximately 1.7 seconds per year. In the geologic past, Earth's rotational period was much faster; days were shorter, and a year contained more days. About 350 million years ago, the year had 400–410 days; 280 million years ago, a year was 390 days long.

Is Earth closer to the Sun in winter than in summer in the Northern Hemisphere?

Yes. However, Earth's axis, the line around which the planet rotates, is tipped 23.5° with respect to the plane of revolution around the Sun. When Earth is closest to the Sun (its perihelion, about January 3), the Northern Hemisphere is tilted away from the Sun.

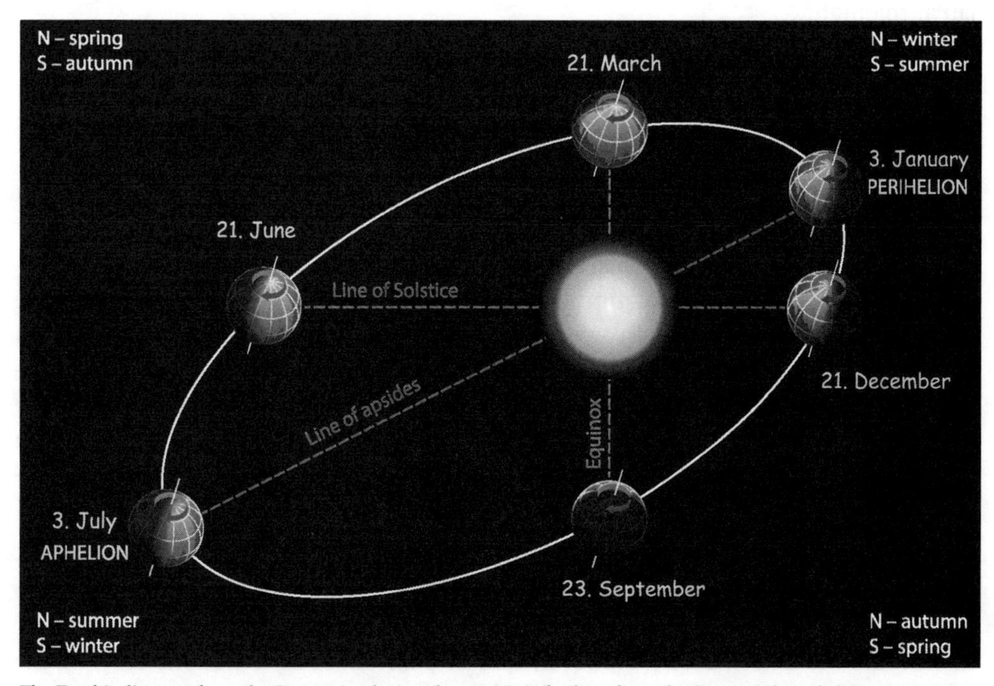

The Earth's distance from the Sun varies during the year. It is farthest from the Sun in July and closest in January.

This causes winter in the Northern Hemisphere while the Southern Hemisphere is having summer. When Earth is farthest from the Sun (its aphelion, around July 4), the situation is reversed, with the Northern Hemisphere tilted toward the Sun. At this time, it is summer in the Northern Hemisphere and winter in the Southern Hemisphere.

What is the circumference of Earth?

Earth is an oblate ellipsoid—a sphere slightly flattened at the poles and bulging at the equator. The distance around Earth at the equator is 24,902 miles (40,075 kilometers). The distance around Earth through the poles is 24,860 miles (40,008 kilometers).

Does life on Mars exist?

The answer to this question still remains inconclusive. Data from the Phoenix mission to Mars confirmed the existence of water ice in 2008. Results of the Viking soil sample data have been disputed. Microfossil-like imprints contained in meteorites that originated from Mars may indicate early forms of life. Exploration of Mars continues with the Mars Exploration rovers Spirit and Opportunity, the Mars Odyssey Orbiter, and the Mars Reconnaissance Orbiter.

What does it mean when a planet is said to be in opposition?

A body in the solar system is in opposition when its longitude differs from the Sun by 180°. In that position, it is exactly opposite the Sun in the sky, and it crosses the meridian at midnight.

EXOPLANETS

What is an exoplanet?

NASA defines an exoplanet as a planet that is outside of our solar system, usually orbiting another star. They are sometimes called "extrasolar" planets since they are outside ("extra") of our solar system. Exoplanets may be large or small; some are larger than Jupiter, while others are small, rocky planets similar in size to Earth or Mars. Their temperature varies from hot enough to boil metal to no warmer than a deep freeze. They are sunless and always in the dark. Some orbit two suns at once. The orbits around their sun may be so "tight" that a year lasts only a few days. NASA continues to explore exoplanets with telescopes on the ground, in space, and in the air.

When was the first exoplanet discovered?

Many scientists recognize the discovery of 51 Pegasi b in 1995 as the first true exoplanet in orbit around a normal star. However, "pulsar planets," rocky remnants orbiting the remnant of a supernova, were observed and confirmed in 1992.

Why is the TRAPPIST-1 solar system special?

TRAPPIST-1 is a star discovered in 1999 and originally named 2MASS J23062928-0502285 because it was spotted with the Two Micron All-Sky Survey (2MASS). Using the Transiting Planets and Planetesimals Small Telescope in Chile, scientists found three planets around the star. Scientists began to refer to the star as TRAPPIST-1 in honor of the telescope in 2016. TRAPPIST-1 appears to be the star in a planetary system. In 2017, scientists described TRAPPIST-1 as a single star with seven Earth-sized, rocky planets orbiting around it. Under the right atmospheric conditions, each of the seven planets could have liquid water—a basic requirement for life as we know it. Three of the seven planets are located in the habitable zone—the area around the parent star most likely to have liquid water. The TRAPPIST-1 planetary system is located 40 light years (235 trillion miles) from Earth.

How big is 51 Pegasi b?

The exoplanet 51 Pegasi b is a "hot Jupiter" planet. It is 47 percent of the mass of Jupiter but 60 percent larger in size. It is fifty light years away from Earth. It orbits its host star or sun every four days. The temperature of 51 Pegasi b is estimated to be 1,800°F (1,000°C).

How many exoplanets have been discovered?

Since the discovery of the first exoplanet twenty years ago, 3,400 exoplanets have been discovered and "confirmed." Another two thousand exoplanet "candidates" require further investigation to confirm that they are in fact exoplanets.

MOONS

How many moons does each planet have?

Planet	Number of Moons	Names of Some of the Moons
Mercury	0	
Venus	0	
Earth	1	The Moon (sometimes called Luna)
Mars	2	Phobos, Deimos
Jupiter	63	Metis, Adrastea, Amalthea, Thebe, Io, Europa, Ganymede, Callisto, Leda, Himalia, Lysithia, Elara, Ananke, Carme, Pasiphae, Sinope, Callirrhoe, Themisto, Megaclite, Taygete, Chaldene, Harpalyke, Kalkye
Saturn	61	Epimetheus, Janus, Mimas, Enceladus, Tethys, Telesto,

		Calypso, Dione, Helene, Rhea, Titan, Hyperion, Iapetus, Phoebe, Atlas, Prometheus, Pandora, Pan, Ymir, Paaliaq
Uranus	27	Cordelia, Ophelia, Bianca, Cressida, Desdemona, Juliet, Portia, Rosalind, Belinda, Puck, Miranda, Ariel, Umbriel, Titania, Oberon, Perdita, Cupid, Francisco, Stephano, Margaret
Neptune	13	Naiad, Thalassa, Despina, Galatea, Larissa, Proteus, Triton, Nereid, Halimede, Psamathe, Sao, Laomedeia, Neso

The four largest moons of Jupiter—Io, Europa, Callisto, and Ganymede—were discovered by Galileo in 1610.

How far is the Moon from Earth?

Since the Moon's orbit is elliptical, its distance varies from about 221,463 miles (356,334 kilometers) at perigee (closest approach to Earth) to 251,968 miles (405,503 kilometers) at apogee (farthest point), with the average distance being 238,857 miles (384,392 kilometers).

Does Earth's Moon have an atmosphere?

The Moon does have an atmosphere; however, it is very slight, having a density of about fifty atoms per cubic centimeter. Recently, scientists have discovered frozen water on the Moon.

What are the diameter and circumference of Earth's Moon?

The Moon's diameter is 2,159 miles (3,475 kilometers), and its circumference is 6,790 miles (10,864 kilometers). The Moon is 27 percent of the size of Earth.

What are the phases of the Moon?

The phases of the Moon are changes in the Moon's appearance during the month, which are caused by the Moon's turning different portions of its illuminated hemisphere toward

The four largest moons of Jupiter are (left to right) Io, Europa, Ganymede, and Callisto. They are known as the Galilean moons because Galileo first observed them.

What are moonquakes?

Similar to earthquakes, moonquakes are a result of the constant shifting of molten or partly molten material in the interior of the Moon. These moonquakes are usually very weak. Other moonquakes may be caused by the impact of meteorites on the Moon's surface. Still others occur at regular intervals during a lunar cycle, suggesting that gravitational forces from Earth have an effect on the Moon similar to ocean tides.

Earth. When the Moon is between Earth and the Sun, its daylight side is turned away from Earth, so it is not seen. This is called the new moon. As the Moon continues its revolution around Earth, more and more of its surface becomes visible. This is called the waxing crescent phase. About a week after the new moon, half the Moon is visible—the first quarter phase. During the next week, more than half of the Moon is seen; this is called the waxing gibbous phase. Finally, about two weeks after the new moon, the Moon and Sun are on opposite sides of Earth.

The side of the Moon facing the Sun is also facing Earth, and all of the Moon's illuminated side is seen as a full moon. In the next two weeks, the Moon goes through the same phases but in reverse, from a waning gibbous to third or last quarter to waning crescent phase. Gradually, less and less of the Moon is visible until a new moon occurs again.

Why does the Moon always keep the same face toward Earth?

Only one side of the Moon is seen because it always rotates in exactly the same length of time that it takes to revolve about Earth. This combination of motions (called "captured rotation") means that it always keeps the same side toward Earth.

What are the names of the full moon during each month?

Month	American Folk Name
January	Wolf Moon
February	Snow Moon
March	Sap Moon
April	Pink Moon
May	Flower Moon
June	Strawberry Moon
July	Buck Moon
August	Sturgeon Moon
September	Harvest Moon
October	Hunter Moon
November	Beaver Moon
December	Cold Moon

Is the Moon really blue during a blue moon?

Although a bluish-looking moon can result from effects of Earth's atmosphere, the term "blue moon" does not refer to the color of the Moon. For example, the phenomenon was widely observed in North America on September 26, 1950, due to Canadian forest fires that had scattered high-altitude dust, which refracted or absorbed certain wavelengths of light.

The popular definition of a "blue moon" is the second full moon in a single calendar month. Based on this definition, a blue moon occurs, on average, every 2.72 years. Since 29.53 days pass between full moons (a synodial month), a blue moon never occurs in February. On rare occasions, a blue moon can be seen twice in one year but only in certain parts of the world.

What is the difference between a hunter's moon and a harvest moon?

The harvest moon is the full moon nearest the autumnal equinox (on or about September 23). It is followed by a period of several successive days when the Moon rises soon after sunset. In the Southern Hemisphere, the harvest moon is the full moon closest to the vernal equinox (on or about March 21). This gives farmers extra hours of light for harvesting crops. The next full moon after the harvest moon is called the hunter's moon.

Who coined the term supermoon?

Richard Nolle (1950–), an astrologer, coined the term "supermoon" in a 1979 article published in *Horoscope* magazine. The term refers to a new or full moon that occurs

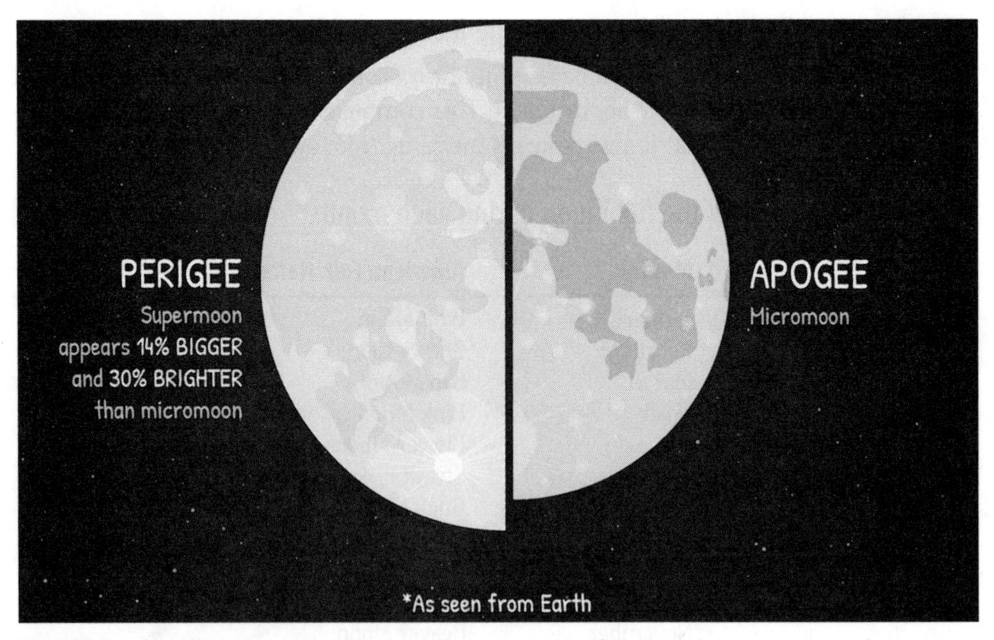

The difference between a supermoon and micromoon are quite significant and obvious, even to an observer on Earth who lacks a telescope or even binoculars.

near or at the time when the Moon is at its closest point (perigee) in its orbit around Earth. A more scientific term describing this phenomenon is a perigean full moon. During a perigean syzygy, Earth, the Moon, and the Sun are all in a line with the Moon in its nearest approach to Earth.

Why do lunar eclipses happen?

A lunar eclipse occurs only during a full moon when the Moon is on one side of Earth, the Sun is on the opposite side, and all three bodies are aligned in the same plane. In this alignment, Earth blocks the Sun's rays to cast a shadow on the Moon. In a total lunar eclipse, the Moon seems to disappear from the sky when the whole Moon passes through the umbra, or total shadow, created by Earth. A total lunar eclipse may last up to 1 hour and 40 minutes. If only part of the Moon enters the umbra, a partial eclipse occurs. A penumbral eclipse takes place if all or part of the Moon passes through the penumbra (partial shadow or "shade") without touching the umbra. It is difficult to detect this type of eclipse from Earth. From the Moon, one could see that Earth blocked only part of the Sun.

When and where will the next ten total lunar eclipses occur?

Upcoming Lunar Eclipses

Date	Location
May 26, 2021	Asia, Australia, Pacific, Americas
May 16, 2022	Americas, Europe, Africa
November 8, 2022	Asia, Australia, Pacific, Americas
March 14, 2025	Pacific, Americas, western Europe, West Africa
September 7, 2025	Europe, Africa, Asia, Australia
March 3, 2026	East Asia, Australia, Pacific, Americas
December 31, 2028	Europe, Africa, Asia, Australia, Pacific
June 26, 2029	Americas, Europe, Africa, Middle East
April 25, 2032	East Africa, Asia, Australia, Pacific
October 18, 2032	Africa, Europe, Asia, Australia

What is the largest crater on the Moon?

The largest crater on the Moon is Bailly. Its diameter is 184 miles (296 kilometers).

What is a blood moon?

During a total lunar eclipse, when the whole moon is in Earth's shadow, some sunlight still reaches the Moon. The sunlight passes through Earth's atmosphere, which filters out most of the Sun's blue light. While the blue light is filtered out, the Moon looks red. The red color has given rise to the nickname "blood moon."

COMETS AND METEORITES

What is a comet?

Comets are basically "snowy dirtballs" or "dirty snowballs"—clumpy collections of rocky material, dust, frozen water, methane, and ammonia that move through the solar system in long, highly elliptical orbits around the Sun. When they are far away from the Sun, comets are simple, solid bodies, but when they get closer to the Sun, they warm up, causing the ice in the comets' outer surface to vaporize. This creates a cloudy "coma" that forms around the solid part of the comet, called the "nucleus." The loosened comet vapor forms long "tails" that can grow to millions of miles in length.

From where do comets originate?

According to a theory developed by Dutch astronomer Jan Oort (1900–1992), a large cloud, now called the Oort Cloud, of gas, dust, and comets is orbiting beyond Pluto out to perhaps 100,000 astronomical units (AU). Occasional stars passing close to this cloud disturb some of the comets from their orbits. Some fall inward toward the Sun. Currently, 3,520 comets are known.

What is a short-period comet?

Most comets have highly elliptical orbits that carry them around the Sun and then fling them back out to the outer reaches of the solar system, never to return. Occasionally, however, a close passage by a comet near one of the planets can alter a comet's orbit, making it stay in the middle or inner solar system. Such a comet is called a short-period comet because it passes close to the Sun at regular intervals. The most famous

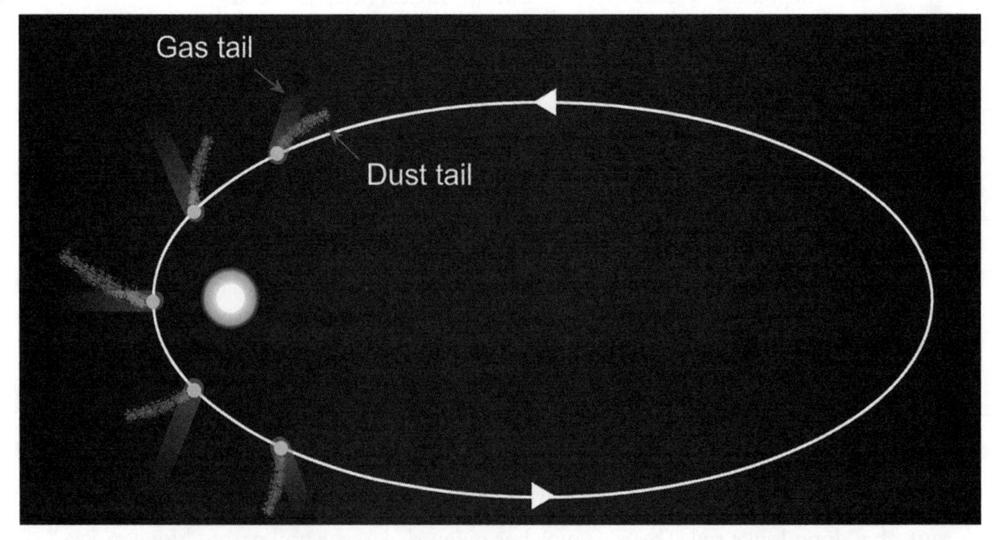

A typical comet orbit is a long elliptical path with one end being close to the Sun and the other somewhere beyond Neptune. As the comet gets close to the Sun, a "coma" of dust and vapors is released, forming a kind of tail.

short-period comet is Halley's Comet, which reaches perihelion (the point in its orbit that is closest to the Sun) about every seventy-six years.

When will Halley's Comet return?

Halley's Comet returns about every seventy-six years. It was most recently seen in 1986 and is predicted to appear again in 2061, then in 2134. Every appearance of what is now known as Comet Halley has been noted by astronomers since the year 239 B.C.E.

The comet is named for Edmond Halley, England's second Astronomer Royal. In 1682, he observed a bright comet and noted that it was moving in an orbit similar to comets seen in 1531 and 1607. He concluded that the three comets were actually one and the same and that the comet had an orbit of seventy-six years. In 1705, Halley published *A Synopsis of the Astronomy of Comets,* in which he predicted that the comet seen in 1531, 1607, and 1682 would return in 1758. On Christmas night of 1758, a German farmer and amateur astronomer named Johann Palitzsch (1723–1788) spotted the comet in just the area of the sky that Halley had foretold.

Prior to Halley, comets appeared at irregular intervals and were often thought to be harbingers of disaster and signs of divine wrath. Halley proved that they are natural objects subject to the laws of gravity.

When was Comet Hale-Bopp first observed?

Comet Hale-Bopp was first observed by Alan Hale (1958–) in New Mexico and Thomas Bopp (1949–2018) in Arizona on July 22, 1995. Their discovery was announced by the International Astronomical Union on July 23, 1995. Comet Hale-Bopp was closest to

How does a meteorite differ from a meteoroid?

A meteorite is a natural object of extraterrestrial origin that survives passage through Earth's atmosphere and hits Earth's surface. A meteorite is often confused with a meteoroid or a meteor. A meteoroid is a small object in outer space, generally less than 30 feet (10 meters) in diameter. A meteor (sometimes called a shooting star) is the flash of light seen when an object passes through Earth's atmosphere and burns as a result of heating caused by friction. A meteoroid becomes a meteor when it enters Earth's atmosphere; if any portion of a meteoroid lands on Earth, it is a meteorite.

Meteorites are divided into three main categories: 1) stone; 2) iron; and 3) stony-iron. The largest group of meteorites are stones. They once formed the outer crust of a planet or asteroid. They often resemble terrestrial rocks, but true stone meteorites are rocks from space. Irons contain 85–95 percent iron; the rest of their mass is mostly nickel. Stony-irons are relatively rare meteorites composed of about 50 percent nickel-iron and 50 percent silicates.

Earth in March 1997, when it was 122 million miles (196 million kilometers) away. It is a very large comet with a nucleus of approximately 25 miles (40 kilometers) in diameter, making it four times as large as Halley's Comet. Although most comets have two tails, Comet Hale-Bopp exhibited three tails. The two tails typical of most comets are the dust tail and the ion tail. The dust tail, consisting of dust and debris from the nucleus, streams behind the comet in its orbit. The ion tail, consisting of the comet's material interacting with the solar wind, faces away from the Sun. Comet Hale-Bopp's third tail was composed of neutral sodium atoms. Comet Hale-Bopp was visible with the naked eye for nearly nineteen months. It is not expected to return for four thousand years.

When do meteor showers occur?

A number of groups of meteoroids orbit the Sun just as Earth is. When Earth's orbit intercepts the path of one of these swarms of meteoroids, some of them enter Earth's atmosphere. When friction with the air causes a meteoroid to burn up, the streak, or shooting star, that is produced is called a meteor. Large numbers of meteors can produce a spectacular shower of light in the night sky. Meteor showers are named for the constellation that occupies the area of the sky from which they originate. Listed below are ten meteor showers and the dates in the year during which they can be seen.

Major Meteor Showers

Name of Shower	Dates
Quadrantids	January 1–6
Lyrids	April 19–24
Eta Aquarids	May 1–8
Perseids	July 25–August 18
Orionids	October 16–26
Taurids	October 20–November 20
Leonids	November 13–17
Phoenicids	December 4–5
Geminids	December 7–15
Ursids	December 17–24

How many meteorites land on Earth in a given year?

Approximately 26,000 meteorites, each weighing over 3.5 ounces (99.2 grams) land on Earth during a given year. Three thousand of these meteorites weigh more than 2.2 pounds (1 kg). This figure is compiled from the number of fireballs visually observed by the Canadian Camera Network. Of that number, only five or six falls are witnessed or cause property damage. The majority fall in the oceans, which cover over 70 percent of Earth's surface.

What are the largest meteorites that have been found in the world?

The famous Willamette (Oregon) iron, displayed at the American Museum of Natural History in New York, is the largest specimen found in the United States. It is 10 feet (3 meters) long and 5 feet (1.5 meters) high.

Largest Meteorites Discovered

Name	Location	Weight (Tons/Tonnes)
Hoba West	Namibia	66.1/60
Ahnighito (the Tent)	Greenland	33.5/30.4
Bacuberito	Mexico	29.8/27
Mbosi	Tanzania	28.7/26
Agpalik	Greenland	22.2/20.1
Armanty	Outer Mongolia	22/20
Willamette	Oregon, USA	15.4/14
Chupaderos	Mexico	15.4/14
Campo del Cielo	Argentina	14.3/13
Mundrabilla	Western Australia	13.2/12
Morito	Mexico	12.1/11

What is an impact crater?

An impact crater is created by large meteorites, comets, and asteroids that hit Earth. A notable impact crater is Meteor Crater near Winslow, Arizona. It is 4,000 feet (1,219 meters) in diameter, 600 feet (183 meters) deep, and is estimated to have been formed about fifty thousand years ago.

Meteor Crater near Winslow, Arizona, was created about 50,000 years ago. While not the largest in the world, at 4,000 feet wide it is definitely impressive.

What are the largest craters on Earth that are observable?

Ten of the largest impact craters known on Earth are listed below, along with their current size and approximate age.

Largest Known Craters on Earth

Name	Location	Diameter (m/km)	Age
Vredefort Crater	South Africa	185/300	2.02 billion years
Sudbury Crater	Ontario, Canada	160/260	1.85 billion years
Chicxulub Crater	Yucatan Peninsula, Mexico	93/150	65 million years*
Manicouagan Crater	Quebec, Canada	62/100	214 million years
Popigai Crater	Siberia, Russia	62/100	35 million years
Acraman Crater	South Australia	56/90	580 million years
Chesapeake Bay Crater	Off the coast of Virginia	53/85	35 million years
Morokweng Crater	Kalahari Desert, South Africa	44/70	145 million years
Kara Crater	Yugorsky Peninsula, Russia	40/65	70.3 million years
Beaverhead Crater	Montana and Idaho	37/60	600 million years

*Scientists theorize the Chicxulub Crater is evidence for a meteor impact that might have contributed to the extinction of dinosaurs. The crater might have originally been as wide as 150 miles (240 kilometers).

SPACE EXPLORATION

Who are the "fathers of space flight"?

In 1903, Konstantin E. Tsiolkovsky (1857–1935), a Russian high school teacher, completed the first scientific paper on the use of rockets for space travel. Several years later, Robert H. Goddard (1882–1945) of the United States and Hermann Oberth (1894–1989) of Germany awakened wider scientific interest in space travel. These three men worked individually on many of the technical problems of rocketry and space travel. They are known, therefore, as the "fathers of space flight."

In 1919, Goddard wrote the paper "A Method of Reaching Extreme Altitudes," which explained how rockets could be used to explore the upper atmosphere and described a way to send a rocket to the Moon. During the 1920s, Tsiolkovsky wrote a series of new studies that included detailed descriptions of multistage rockets. In 1923, Oberth wrote *The Rocket into Interplanetary Space*, which discussed the technical problems of space flight and also described what a spaceship would be like.

What is the difference between zero gravity and microgravity?

Zero gravity is the absence of gravity; the effects of gravity are not felt; total weightlessness is the result. Microgravity is a condition of very low gravity, especially approaching weightlessness. On a spaceship, while in zero- or microgravity, objects would fall freely and float weightlessly. Both terms, however, are technically incorrect. The gravitation in orbit is only slightly less than the gravitation on Earth. A spacecraft and its contents continuously fall toward Earth. It is the spacecraft's immense forward speed that appears to make Earth's surface curve away as the vehicle falls toward it. The continuous falling seems to eliminate the weight of everything inside the spacecraft. For this reason, the condition is sometimes referred to as weightlessness or zero gravity.

Which event is considered the single event that initiated the Space Age?

The Soviet Union launched the world's first satellite, Sputnik 1, into low Earth orbit on October 3, 1957, thus initiating the Space Age. Sputnik 1 was a 184-pound (83.5-kilogram) satellite. It carried instrumentation to study the density and temperature of the upper atmosphere.

Four months later, on January 31, 1958, the United States launched its first satellite, Explorer. This 31-pound (14.06-kilogram) satellite carried instrumentation that led to the discovery of Earth's radiation belts, which would be named after University of Iowa scientist James A. Van Allen (1914–2006).

A replica of Sputnik 1 is shown here at the Moscow's Space Museum. In this display, the satellite is opened up to reveal the equipment used to record temperature and other readings in the upper atmosphere.

Which were some of the important NASA missions?

Dozens of NASA missions have taken place to explore the solar system and beyond. Some of the major missions since the launch of Explorer in 1958 are listed in the chart below.

NASA Missions

Mission: Mercury
Dates: 1958–1963
Goals:
- To orbit a manned spacecraft around Earth
- To investigate man's ability to function in space
- To recover both man and the spacecraft safely
- To test an astronaut's ability to fly long-duration missions

Mission: Gemini (Bridge to the Moon Program)
Dates: 1964–1966
Goals:
- To understand how spacecraft could rendezvous and dock with other vessels in orbit
- To perfect landing and reentry techniques
- To test the effects of longer space flights on astronauts

Mission: Apollo
Dates: 1967–1972
Goals:
- To land on the Moon and carry out scientific exploration of the Moon
- To develop man's capability to work in an environment other than Earth

Mission: Pioneer 10
Dates: Launched in 1972; final communication in 2003
Goals:
- Flyby of Jupiter to take the first close-up pictures of the planet

Mission: Pioneer 11
Dates: Launched in 1973; final communication in 1995
Goals:
- Flyby of Jupiter to gather more pictures
- To obtain the first close-up images of Saturn and its rings

Mission: Skylab (America's first space station)
Dates: 1973–1979
Goals:
- To prove that humans could live and work in space for extended periods of time
- To expand our knowledge of solar astronomy and Earth

Mission: Viking
Dates: 1975–1982 (1982 final communication from the Viking 1 Lander)

When and what was the first animal sent into orbit?

A small, female dog named Laika, aboard the Soviet Sputnik 2, launched November 3, 1957, was the first animal sent into orbit. This event followed the successful Soviet launch on October 4, 1957, of Sputnik 1, the first man-made satellite ever placed in orbit. Laika was placed in a pressurized compartment within a capsule that weighed 1,103 pounds (500 kilograms). After a few days in orbit, she died, and Sputnik 2 reentered Earth's atmosphere on April 14, 1958. Some sources list the dog as a Russian samoyed laika named "Kudyavka" or "Limonchik."

Goals: • To obtain high resolution images of the surface of Mars
• To characterize the structure and composition of the atmosphere and surface of Mars
• To search for the evidence of life on Mars

Mission: *Voyager 1* and 2
Dates: 1977–
Goals: • Flybys of Jupiter, Saturn, Uranus, and Neptune ("Grand Tour of the Planets") to obtain pictures of the planets, their moons and/or rings
• To obtain information to learn about the atmosphere, magnetic field, and other characteristics of the planets

Mission: Space Transportation System (STS) (space shuttle)
Dates: 1981–2011
Goals: • First reusable spacecraft
• Used to shuttle researchers to and from the International Space Station

Mission: Galileo
Dates: Launched in 1989. Achieved orbit around Jupiter in 1995. Mission ended in 2003
Goals: • To determine the chemical composition of the atmosphere of Jupiter
• To study the atmosphere and surface of four of the moons of Jupiter

Mission: International Space Station
Dates: 1998–
Goals: • First module launched in 1998
• U.S. delivers its final module in 2011
• Scientific laboratory to conduct experiments in space, especially to investigate the effects of weightlessness and microgravity

Mission: Cassini
Dates: Launched in 1997. Reached orbit around Saturn in 2004. Final contact in 2017
Goals: • Flyby of Jupiter enroute to Saturn
• To explore, in depth, the atmosphere, rings, and surface of Saturn and its moon, Titan

Mission: Kepler
Dates: 2009–2018
Goals: • To search for planets outside our solar system (exoplanets)

Mission: Juno
Dates: Launched in 2011. Arrived at Jupiter in 2016. Planned, controlled deorbit in 2021
Goals: • To understand the origin and evolution of Jupiter
• To determine how much water is in Jupiter's atmosphere
• To measure composition, temperature, cloud motions, and other properties of Jupiter's atmosphere
• To map Jupiter's magnetic and gravity fields
• To explore Jupiter's magnetosphere near the planet's poles

Mission: InSight
Dates: Launched in 2018. Anticipated to last one full martian solar orbit until November 2020
Goals: • To explore and study the deep interior (core, mantle, and crust) of Mars to gain an understanding of the formation of rocky planets

What were the first monkeys and chimpanzees in space?

On a Jupiter flight from the United States on December 12, 1958, a squirrel monkey named Old Reliable was sent into space but not into orbit. The monkey drowned during recovery.

On another Jupiter flight, on May 28, 1959, two female monkeys were sent 300 miles (482.7 kilometers) high. Able was a 6-pound (2.7-kilogram) rhesus monkey, and Baker was an 11-ounce (0.3-kilogram) squirrel monkey. Both were recovered alive.

A chimpanzee named Ham was used on a Mercury flight on January 31, 1961. Ham was launched to a height of 157 miles (253 kilometers) into space but did not go into orbit. His capsule reached a maximum speed of 5,857 miles (9,426 kilometers) per hour and landed 422 miles (679 kilometers) downrange in the Atlantic Ocean, where he was recovered unharmed.

On November 29, 1961, the United States placed a chimpanzee named Enos into orbit and recovered him alive after two complete orbits around Earth. Like the Soviets, who usually used dogs, the United States had to obtain information on the effects of space flight on living beings before they could actually launch a human into space.

Who was the first man in space?

Yuri Gagarin (1934–1968), a Soviet cosmonaut, became the first man in space when he made a full orbit of Earth in Vostok 1 on April 12, 1961. Gagarin's flight lasted only 1 hour and 48 minutes, but as the first man in space, he became an international hero.

Who was the first American man in space?

The United States launched the first American into orbit on February 20, 1962. Astronaut John H. Glenn Jr. completed three orbits in Friendship 7 and traveled about 81,000 miles (130,329 kilometers). Prior to this, on May 5, 1961, Alan B. Shepard Jr. became the first American to pilot a spaceflight aboard Freedom 7. This suborbital flight reached an altitude of 116.5 miles (187.45 kilometers).

Who was the first woman in space?

Valentina V. Tereshkova-Nikolaeva (1937–), a Soviet cosmonaut, was the first woman in space. She was aboard the Vostok 6, launched June 16, 1963. She spent three days circling Earth, completing forty-eight orbits. Although she had little cosmonaut training, she was an accomplished parachutist and was especially fit for the rigors of space travel.

Soviet cosmonaut Yuri Gagarin was the first human being to go into space, which he did on the *Vostok 1* in 1961.

Who was the first American woman in space?

The U.S. space program did not put a woman in space until June 18, 1983. Sally

K. Ride (1951–2012) flew aboard the space shuttle Challenger mission STS-7. In 1987, she moved to the administrative side of NASA and was instrumental in issuing the Ride Report, which recommended future missions and direction for NASA. She retired from NASA in August 1987 to become a research fellow at Stanford University after serving on the Rogers Commission that investigated the Challenger disaster in 1986. She was a physics professor at the University of California, San Diego, until 2001, when she founded Sally Ride Science. The company is dedicated to supporting girls' and boys' interests in science, math, and technology by showing that science is fun with a variety of programs.

Who was the first African American in space?

Guion S. Bluford Jr. (1942–) became the first African American to fly in space during the space shuttle Challenger mission STS-8 (August 30–September 5, 1983). Astronaut Bluford, who holds a Ph.D. in aerospace engineering, made a second shuttle flight aboard Challenger mission STS-61-A/Spacelab D1 (October 30–November 6, 1985). The first black man to fly in space was Cuban cosmonaut Arnaldo Tamayo-Mendez (1942–), who was aboard Soyuz 38 and spent eight days aboard the Soviet space station Salyut 6 during September 1980. Dr. Mae C. Jemison (1956–) became the first African American woman in space on September 12, 1992, aboard the space shuttle Endeavour mission Spacelab-J.

What did NASA mean when it said **Voyager 1** and **Voyager 2** would take a "grand tour" of the planets?

Once every 176 years, the giant outer planets—Jupiter, Saturn, Uranus, and Neptune—align themselves in such a pattern that a spacecraft launched from Earth to Jupiter at just the right time might be able to visit the other three planets on the same mission. A technique called "gravity assist" used each planet's gravity as a power boost to point Voyager toward the next planet. The first opportune year for the "grand tour" was 1977. Voyager 1 was launched on September 5, 1977, and Voyager 2 was launched on August 20, 1977. Voyager 1 and Voyager 2 continue to explore space more than forty years after being launched. As of 2018, Voyager 1 was in "interstellar space," and Voyager 2 was in the heliosheath. The heliosheath is the outermost layer of the heliosphere. Both continue to send information back to NASA on a daily basis.

What is the message attached to the Voyager spacecraft?

Voyager 1 and Voyager 2 were unmanned space probes designed to explore the outer planets and then travel out of the solar system. A gold-coated copper phonograph record containing a message to any possible extraterrestrial civilization that they might encounter is attached to each spacecraft. The record contains both video and audio images of Earth and the civilization that sent this message to the stars.

The record begins with 118 pictures. These show Earth's position in the galaxy; a key to the mathematical notation used in other pictures; the Sun; other planets in the solar system; human anatomy and reproduction; various types of terrain (seashore, desert, mountains); examples of vegetation and animal life; people of both sexes and of all ages and ethnic types engaged in a number of activities; structures (from grass huts to the Taj Mahal to the Sydney Opera House) showing diverse architectural styles; and means of transportation, including roads, bridges, cars, planes, and space vehicles.

The pictures are followed by greetings from Jimmy Carter (1924–), who was then president of the United States, and Kurt Waldheim (1918–2007), then secretary general of the United Nations. Brief messages in fifty-four languages, ranging from ancient Sumerian to English, are included, as is a "song" of the humpback whale.

The next section is a series of sounds common to Earth. These include thunder, rain, wind, fire, barking dogs, footsteps, laughter, human speech, the cry of an infant, and the sounds of a human heartbeat and human brainwaves.

The record concludes with approximately 90 minutes of music: "Earth's Greatest Hits." These musical selections were drawn from a broad spectrum of cultures and include such diverse pieces as a Pygmy girl's initiation song; bagpipe music from Azerbaijan; the "Fifth Symphony, First Movement" by Ludwig von Beethoven (1770–1827); and "Johnny B. Goode" by Chuck Berry (1926–2017).

John Glenn (left), the first American in space, is shown here in his Mercury spacesuit; at right is an official portrait of Neil Armstrong, the first man to set foot on the Moon.

It will be tens, or even hundreds of thousands, of years before either Voyager comes close to another star, and perhaps the message will never be heard, but it is a sign of humanity's hope to encounter life elsewhere in the universe.

Who were the first astronauts to walk on the Moon?

Twelve astronauts have walked on the Moon. During the six Apollo missions to the Moon, each Apollo flight had a crew of three. One crew member remained in orbit in the command service module (CSM), while the other two actually landed on the Moon. Apollo 11 landed on the Moon on July 20, 1969. The first man to walk on the Moon was Neil A. Armstrong (1930–2012), followed by Edwin E. (Buzz) Aldrin Jr. (1930–).

What were the first words spoken by an astronaut after touchdown of the lunar module on the Apollo 11 flight and by an astronaut standing on the Moon?

On July 20, 1969, at 4:17:43 P.M. Eastern Daylight Time (20:17:43 Greenwich Mean Time), Neil A. Armstrong and Edwin E. (Buzz) Aldrin Jr. landed the lunar module Eagle on the Moon's Sea of Tranquility, and Armstrong radioed: "Houston, Tranquility Base here. The Eagle has landed." Several hours later, when Armstrong descended the lunar module ladder and made the small jump between the Eagle and the lunar surface, he announced: "That's one small step for man, one giant leap for mankind." The article "a" was missing in the live voice transmission and was later inserted in the record to amend the message to "one small step for a man."

Which manned space flight was the longest?

Dr. Valeri Polyakov (1942–) manned a flight to the space station Mir on January 8, 1994. He returned aboard Soyuz TM-20 on March 22, 1995, making the total time in space equal to 438 days and 18 hours. However, the record for spending the most time in space belongs to Russian cosmonaut Gannady Padalka (1958–). Padalka accumulated 879 cumulative days in space during five separate missions between 1998 and 2015.

What was the mission of the Galileo spacecraft?

Galileo, launched October 18, 1989, required almost six years to reach Jupiter after looping past Venus once and Earth twice. The Galileo spacecraft was designed to make a de-

Which American has spent the most time in space?

Peggy Whitson (1960–) has spent 665 days in space, holding the U.S. record for the most time in space. She attained this goal in September 2017 upon her return from a 288-day mission aboard the International Space Station (ISS). During this mission, she performed four spacewalks, bringing her total number of spacewalks to ten.

tailed study of Jupiter and its rings and moons over a period of years. On December 7, 1995, it released a probe to analyze the different layers of Jupiter's atmosphere. Galileo recorded a multitude of measurements of the planet, its four largest moons, and its mammoth magnetic field. The mission was originally scheduled to continue until the end of 1997, but since it continued to operate successfully, missions exploring Jupiter's moons were added in 1997, 1999, and 2001. Galileo ended on September 21, 2003, when it passed into Jupiter's shadow and disintegrated in the planet's dense atmosphere.

What caused the Challenger accident?

Challenger mission STS-51-L was launched on January 28, 1986, exploding 73 seconds after liftoff. The entire crew of seven was killed, and the space shuttle Challenger was completely destroyed. The consensus of the Rogers Commission (which studied the accident for several months) and participating investigative agencies was that the accident was caused by a failure in the joint between the two lower segments of the right solid rocket motor. The specific failure was the destruction of the seals that are intended to prevent hot gases from leaking through the joint during the propellant burn of the rocket motor. The evidence assembled by the commission indicated that no other element of the space shuttle system contributed to this failure.

Although the commission did not affix blame to any individuals, the public record made clear that the launch should not have been made that day. The weather was unusually cold at Cape Canaveral, and temperatures had dipped below freezing during the night. Test data had suggested that the seals (called O-rings) around the solid rocket booster joints lost much of their effectiveness in the very cold weather.

What was the cause of the Columbia space shuttle disaster?

The Columbia space shuttle was launched on January 16, 2003, on mission STS-107. The mission was a research mission, and the crew had many science experiments ranging from plant growth to a cancer drug study to studying the effects of microgravity on the cardiovascular system. The Columbia space shuttle was lost during its reentry to Earth's atmosphere on February 1, 2003. The investigation of the disaster determined that a piece of foam insulation broke off shortly after liftoff and damaged the thermal protection system on the leading edge of the orbiter's left wing. As the space shuttle descended, superhot gases entered the interior aluminum structure of the orbiter. The internal wing structure was weakened, and eventually, the atmospheric forces tore off the wing. The final communication between Columbia and NASA flight controllers occurred at an altitude of approximately 203,000 feet (61,900 meters) over Texas. Debris from Columbia was later found in Texas, Arkansas, and Louisiana.

How long was the Mir space station in space?

The first component, a 20.4-ton core module, of the Mir space station was launched in February 1986. The core module served as the living and working quarters, including the communications and command center, of the space station. Five more modules were launched and attached to ports on the core module over the next ten years. Mir had self-contained oxygen, power, and water-generation capabilities, allowing cosmonauts and astronauts to spend extended periods of time in space. Scientific investigations aboard Mir included space technology experiments, life science and biological research, astrophysics, and material processing tests. Much of the life science and biological research focused on the effects of microgravity on humans and flora and fauna.

Since some of the astronauts spent many months at a time in Mir, it was possible to study the physiological differences in long and short duration missions. Space technology studies investigated various materials for space use and the effects of the low Earth orbit environment on various materials. The Mir space station was deorbited on March 23, 2001, after more than 86,000 orbits around Earth. The space station broke into several large pieces and thousands of smaller ones over the Pacific Ocean. No injuries occurred.

Which U.S. astronauts spent time aboard Mir?

The collapse of the Soviet Union in 1991 allowed the United States and Russia to enter a new era of space collaboration. NASA made a financial commitment of $400 million to the Russian space program to continue the expeditions to the Mir space station. The space shuttle Atlantis was fitted with a special adaptor to allow it to dock with Mir. Eleven space shuttle flights to Mir occurred between February 1994 and June 1998. In a new spirit of cooperation, Russian cosmonauts began to fly on the shuttle, while American astronauts began to spend time on Mir conducting experiments with their Russian counterparts. The experiences gained by the U.S. astronauts on Mir were beneficial in planning for the ISS. A total of seven U.S. astronauts spent 980 cumulative days aboard Mir.

The Soviets' Mir station is shown at left, and the International Space Station is at right. The ISS is still in service, but the Mir station fell back to Earth in 2001 after fifteen years in orbit.

U.S. Astronauts Who Have Served on Mir

Astronaut	Dates	Total Days
Shannon Lucid (1943–)	March 22, 1996–September 26, 1996	188
C. Michael Foale (1957–)	May 15, 1997–October 5, 1997	145
Andrew Thomas (1951–)	January 22, 1998–June 12, 1998	141
Jerry Linenger (1955–)	January 12, 1997–May 24, 1997	132
John Blaha (1942–)	September 16, 1996–January 22, 1997	128
David Wolf (1956–)	September 25, 1997–January 31, 1998	128
Norman E. Thagard (1943–)	March 14, 1995–July 7, 1995	118

What is the purpose of the ISS?

The ISS orbits Earth at an altitude of about 250 miles (400 kilometers). The ISS travels at 17,500 miles (32,410 kilometers) per hour in an orbit that extends from 52 degrees north latitude to 52 degrees south latitude. Its main purpose is to serve as a laboratory for materials, communications technology, and biological and medical research in an environment that lacks the force of gravity.

When was the first module of the ISS launched?

The ISS is a collaborative partnership program between the space agencies of the United States, Russia, Europe, Japan, and Canada. In addition, 230 astronauts from eighteen countries have visited the ISS to participate in experiments and space exploration. The first module of the ISS was launched by Russia in November 1998. The space shuttle Endeavour brought the first U.S.-built module to the space station in December 1998. A second Russian module arrived in July 2000, followed by the first crew in December 2000. The members of the first crew to live in the space station were Sergei K. Krikalev (1958–) and Yuri P. Gidzenko (1962–) from Russia and William M. (Bill) Shepherd (1949–) from the United States. The space shuttle Discovery delivered the final U.S. module of the ISS in 2011.

How large is the ISS?

The ISS travels at a speed of 5 miles (8 kilometers) per second, orbiting Earth about every 90 minutes. More than twenty different research payloads may be hosted outside the ISS at one time. These include Earth-sensing equipment, materials science payloads, particle physics experiments, and more.

Facts and Figures about the International Space Station

Mass: 925,335 pounds (419,725 kilograms)

Length: 357 feet (109 meters) end to end, or 1 yard (1 meter) shorter than an American football field including the end zones

Solar array wingspan: 240 feet (73 meters) or the same length as the wingspan of an Airbus A380 (the world's largest passenger aircraft)

Pressurized volume: 32,333 cubic feet (916 cubic meters)
Habitable volume: 13,696 cubic feet (388 cubic meters), not including visiting vehicles
Sleeping quarters: 6
Bathrooms: 2
Miles of wire connecting the electrical power system: 8 (13 kilometers)

Astronauts have a gym where they exercise a minimum of 2 hours per day to mitigate the loss of muscle and bone mass in microgravity. The onboard water recovery system decreases the dependence on water delivered by cargo spacecraft by 65 percent from about one gallon per day to about one-third of a gallon per day.

What was the first reusable spacecraft?

The first reusable spacecraft was the space shuttle. Five space shuttles were in the fleet: Columbia, Challenger, Discovery, Atlantis, and Endeavour. The space shuttle missions, beginning with the launch of Columbia in 1982 and ending with the final Atlantis mission in 2011, totaled 135 in number. The purpose of the various space shuttle missions were to transport people to and from space to construct and live in the ISS; to launch, recover, and repair satellites; and to carry out other research.

Did the entire space shuttle travel to and from space?

The space shuttle had three parts: the orbiter, the external tank, and the solid rocket boosters. The orbiter was the only part of the spacecraft that would fly into space. The two solid rocket boosters gave the space shuttle the required lift to leave Earth's gravity. The boosters fell off 2 minutes after launch, landing in the Atlantic Ocean. They were retrieved from the ocean and reused on a subsequent flight. The external tank was a fuel tank. Once the fuel had been used, the external tank would drop off of the orbiter. It would re-enter Earth's atmosphere about 9 minutes after launch and burn up over the Pacific Ocean. The external tank was the only part of the space shuttle that was not reused on subsequent flights.

What was the composition of the tiles on the underside of the space shuttle, and how hot did they get?

The twenty thousand tiles were composed of a low-density, high-purity silica fiber insulator hardened by ceramic bonding. Bonded to a Nomex fiber felt pad, each tile was

Is it possible to see the ISS from Earth?

Yes! As the third brightest object in the sky, it is visible to the naked eye. It looks like a fast-moving plane, only much higher in the sky and traveling much faster. NASA maintains a website (*https://spotthestation.nasa.gov/*) to help spot the ISS in the sky.

directly bonded to the shuttle exterior. The maximum surface temperature can reach up to 922–978 K (649–704°C or 1,200–1,300°F).

What were the liquid fuels used by the space shuttles?

Liquid hydrogen was used as a fuel, with liquid oxygen used to burn it. These two fuels were stored in chambers separately and then mixed to combust the two. Because oxygen must be kept below –297°F (–183°C) to remain a liquid, and hydrogen must be at –423°F (–253°C) to remain a liquid, they were both difficult to handle, but both make useful rocket fuel.

What happened to the space shuttle spacecraft at the completion of the program?

Both Challenger and Columbia were destroyed in tragic accidents. The remaining three space shuttles are on display at museums in the United States. Atlantis is on display at the Kennedy Space Center in Florida. Discovery is on display at the Steven F. Udvar-Hazy Center (a Smithsonian Institution) in Chantilly, Virginia, near Dulles International Airport in Washington, D.C. Endeavour is on display at California Science Center in Los Angeles.

What was the Grand Finale of the Cassini–Huygens mission?

The Cassini–Huygens mission was a joint venture between NASA, the European Space Agency (ESA), and the Italian Space Agency. It consisted of the Cassini orbiter and the Huygens probe. The mission was launched in 1997 and reached Saturn in 2004. The goal of the mission was to study in-depth the planet Saturn, its rings, and its moons. The mission discovered storms in the planet's atmosphere, icy jets from the moon Enceladus, hydrocarbon lakes, and seas that are dominated by liquid ethane and methane on the moon Titan.

The Grand Finale was the final twenty-two orbits of the Cassini–Huygens mission in 2018. During these orbits, the spacecraft traveled in an elliptical path that sent it diving at tens of thousands of miles per hour through the 1,500-mile-wide (2,400-kilometer-wide) space between the rings and the planet. It continued to collect data about the rings until the final orbit. In order to avoid the possibility of Cassini colliding with one of the moons of Saturn, it was decided to send Cassini into Saturn's upper atmosphere, where it incinerated and disintegrated like a meteor.

The space shuttle Discovery is shown here being launched in 1991 from Merritt Island, Florida.

What has the Juno mission revealed about Jupiter?

Juno has revealed many new concepts about Jupiter. It appears that Jupiter is very active and a variable world. Scientists no longer believe it is a well-mixed planet but is churning up material from its depths. Juno's data indicates that storms occur on the surface of Jupiter. As more data is collected, scientists hope to answer more questions about Jupiter and its moons.

What is the goal ofTESS?

TESS, Transiting Exoplanet Survey Satellite, was launched in April 2018 to explore exoplanets and worlds beyond our solar system. TESS will explore and probe to try to find planets or other bodies that can support life. The mission is expected to last two years.

Which planet will the InSight mission investigate?

The InSight (Interior Exploration using Seismic Investigations, Geodesy, and Heat Transport) mission seeks to understand the evolutionary formation of rocky planets, including Earth, by investigating the interior structure and processes of Mars. InSight was launched in May 2018. It will place a geophysical lander on Mars to study the size, thickness, density, and overall structure of the crust, mantle, and core of Mars.

Do plans exist for humans to visit other planets?

NASA worked on a number of initiatives between 2008 and 2016, collectively known as "Journey to Mars." The ultimate goal was to send humans to low-Mars orbit in the early 2030s. A new national space policy in late 2017 directed NASA to return American astronauts to the Moon. Other human exploration journeys have been superseded by the new policy.

Is it possible for interested laypeople to explore the universe?

NASA has an immersive app for PCs, MACs, and mobile devices, NASA's Eyes, that allows anyone to learn about Earth, our solar system, the universe beyond, and the spacecraft exploring space. It is possible to explore planets and exoplanets, tour the solar system, view events on Earth, and more.

Is anyone looking for extraterrestrial life?

A program called SETI (the Search for Extraterrestrial Intelligence) began in 1960, when American astronomer Frank Drake (1930–) spent three months at the National Radio Astronomy Observatory in Green Bank, West Virginia, searching for radio signals coming from the nearby stars Tau Ceti and Epsilon Eridani. Although no signals were detected and scientists interested in SETI have often been ridiculed, support for the idea of seeking out intelligent life in the universe has grown.

Project Sentinel, which used a radio dish at Harvard University's Oak Ridge Observatory in Massachusetts, could monitor 128,000 channels at a time. This project was upgraded in 1985 to META (Megachannel Extraterrestrial Assay) thanks in part to a donation by filmmaker Steven Spielberg (1946–). Project META is capable of receiving 8.4 million channels. NASA began a ten-year search in 1992 using radio telescopes in Arecibo, Puerto Rico, and Barstow, California.

Scientists are searching for radio signals that stand out from the random noises caused by natural objects. Such signals might repeat at regular intervals or contain mathematical sequences. Millions of radio channels and a lot of sky are out there to be examined. As of October 1995, Project BETA (Billion-channel Extraterrestrial Assay) has been scanning a quarter of a billion channels. This new design improves upon Project META three-hundred-fold, making the challenge of scanning millions of radio channels seem less daunting. SETI has since developed other projects, some "piggybacking" on radio telescopes while engaged in regular uses. A program launched in 1999, SETI@home, uses the power of home computers while they are at rest.

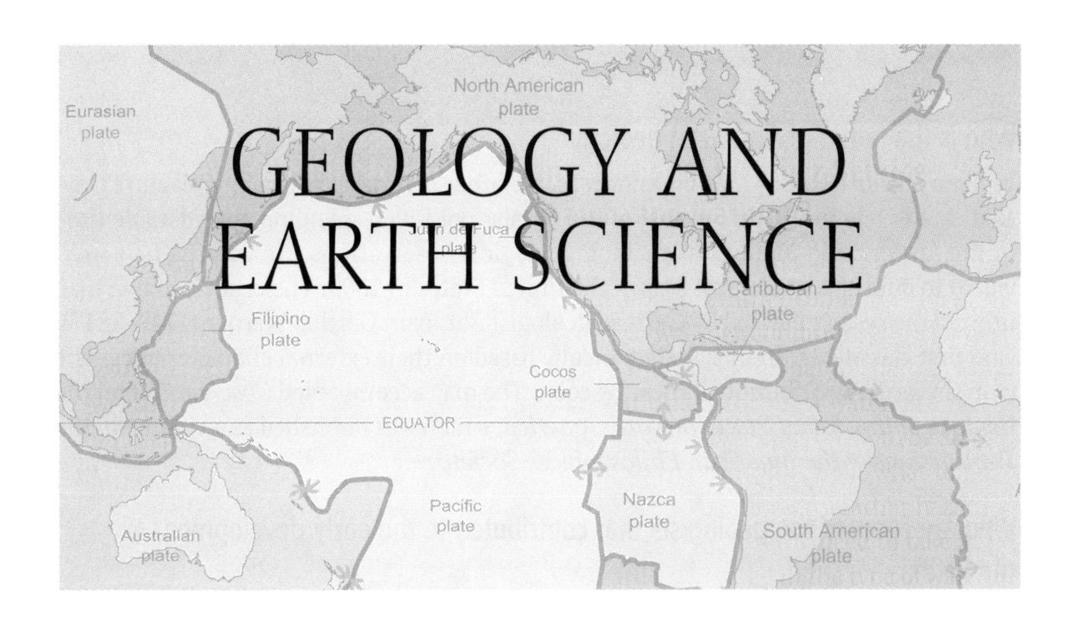

GEOLOGY AND EARTH SCIENCE

INTRODUCTION AND HISTORICAL BACKGROUND

What is geology, and what are some of its subdivisions?

Geology is the study of the composition of Earth and the physical changes in it as well as the materials of which it is made, including the study of rocks, minerals, and sediments. One way to divide the field is physical geology and historical geology. Specialties such as engineering geology, geochemistry, geomorphology, geophysics, glacial geology, hydrology, mineralogy, petroleum geology, volcanology, and related areas are considered subdivisions of physical geology. Paleontology is related to historical geology. Geology is a vast academic discipline that uses a wide variety of tools and techniques, ranging from fieldwork to chemical analysis, to provide new information and insights into this important area of science.

Who is the father of modern geology?

James Hutton (1726–1797), a Scottish farmer and naturalist, is generally recognized as the father of modern geology. He wrote a two-volume treatise in 1795 entitled *Theory of the Earth* in which he proposed that Earth must be much older than the previously accepted age of somewhere around six thousand years. He also proposed that much more time was involved in the erosion of mountains. Furthermore, he proposed that subterranean heat creating metamorphic material was just as important as the process of forming new rocks from sediments laid down at the bottom of the sea, which in turn were raised up to become dry land. Hutton also introduced the idea that current geologic processes and natural laws are sufficient to account for all geologic changes over time. This doctrine was later defined as "uniformitarianism" in 1830 by Charles Lyell (1779–1875).

Who is the father of American geology?

William Maclure (1763–1840) is known as the father of American geology. Maclure traveled extensively in the region east of the Mississippi River, spending considerable time in the Appalachian Mountains making geological observations. He was the first individual to publish a geological map of the United States in 1809. The crudely drawn map utilized the system devised by German geologist Abraham Gottlob Werner (1749–1817), who first classified minerals systematically based on their external characteristics. The map shows the distribution of rocks by color. The map accompanied *Observations on the Geology of the United States of North America*, which was published as a volume in the *Transactions of the American Philosophical Society*.

Who are some of the geologists that contributed to the early development of the field?

Many individuals played a role in developing the field of geology. Some of them and their accomplishments were:

- Nicolas Steno (1638–1686), a Danish geologist, proposed Steno's law, or the principle of superimposition in which in any given rock layer, the bottom rocks are formed first and are the oldest. He made additional contributions to the field of stratigraphy.
- John Playfair (1748–1819), a Scottish geologist, is remembered for his proposal that river valleys were carved by streams as opposed to being formed by cataclysmic upheavals of land with the rivers flowing through at a later time.
- James Hall (1761–1832), also a Scottish geologist, was one of the first individuals to carry out experimental laboratory research in geological investigations.
- Charles Lyell (1779–1875), a British geologist, described geology as the study of Earth's composition, history, structure, and processes in his book *The Principles of Geology*. He was a major advocate for the doctrine of uniformitarianism, which proposed that geological processes that are operating at present on Earth have acted in a similar way and intensity throughout geological time.
- Louis Agassiz (1807–1873), a Swiss geologist and paleontologist, was a major proponent of the concept of Ice Ages, when large masses of ice covered large sections of Earth and the global climate was much colder.
- James Hall (1811–1898), an American geologist, was one of the earliest individuals to propose a theory of mountain building. He was also recognized as an important invertebrate paleontologist.
- James Dwight Dana (1813–1895), an American mineralogist, wrote *Dana's Manual of Mineralogy*, which is still considered one the most important and respected reference works in the field.
- Clarence Edward Dutton (1841–1912), an American geologist, was an early proponent of the theory of isostasy. Isostasy describes how land can rise up, and these iso-

static forces are of major importance in controlling and shaping the topography of Earth's surface.

- Grove Karl Gilbert (1843–1918) was an American geologist who made important contributions in theories of river development and glaciation.
- Thomas Chrowder Chamberlin (1843–1928), an American geologist, also had interests in glacial geology, including the idea of multiple glaciations.
- William Morris Davis (1859–1934) was an American geologist whose primary area of expertise was in the field of geomorphology and is frequently called the "founder of geomorphology." He also had interests in geography and meteorology.

Why was *De re Metallica*, published in 1556, an important book?

This sixteenth-century treatise on mining engineering remained the definitive work for two hundred years. The author, Georg Bauer (1494–1555), collected and mastered the expertise and knowledge of miners about metals and engineering that had not been recorded in any book. The book was published posthumously in 1556 under his Latinized name, Georgius Agricola. Latin scholars encountered great difficulty in translating the work because Agricola had coined many technical terms. The first English translation was published by future U.S. president Herbert Hoover (1874–1964) and his wife Lou Henry Hoover (1874–1944) in 1912. Hoover was a mining engineer before his career in politics. In addition to *De re Metallica,* Agricola produced additional books on geology that were fundamental to the development of the field.

Who was the first woman geologist in the United States who made significant contributions to the field?

Florence Bascom (1862–1945) made many important contributions to geology. She was considered an authority on the rocks of the Piedmont, publishing maps and folios. She also studied and did research on the water resources of the Philadelphia region. Among her many accomplishments:

- First woman to receive a Ph.D. from Johns Hopkins University in 1893 (and the second woman to get a Ph.D. in geology in America [Mary Emilie Holmes was the first in 1888]).
- First woman hired by the U.S. Geological Survey in 1896.
- First woman to present a paper before the Geological Society of Washington in 1901.

American geologist Florence Bascom helped pave the way for women in her field.

- First woman to be elected to the Council of the Geological Society of America (GSA) in 1924.
- First woman officer in the GSA and was elected as its second vice president in 1930.

In the first edition (1906) of *American Men of Science* (title later changed to *American Men and Women of Science*), her entry included four stars, which signified that her colleagues ranked her among the top one hundred leading geologists in America. Bascom also made numerous contributions to teaching and research in geology at Bryn Mawr College, where she founded the Department of Geology in 1901.

What is the mission of the USGS?

The U.S. Geological Survey (USGS) was created by an act of the U.S. Congress in 1879. The main responsibilities of the USGS were to map public lands, examine geological structures, and evaluate mineral resources. Clarence King (1842–1901) was the first director of the USGS from 1879 to 1881. Over the next century, the mission expanded to include the research of groundwater, ecosystems, environmental health, energy, natural hazards, and climate and land use change. The USGS is the sole science agency for the Department of the Interior, and its scientists develop new methods and tools to enable timely, relevant, and useful information about Earth and its processes.

What are the major eras, periods, and epochs in geologic time?

The geologic time scale is divided into geochronological units. Most of the divisions in geologic time are delineated by a major geologic or life form event. Common geologic time divisions are eon, era, period, and epoch. The longest time interval on a geologic time scale is an eon. An era is broken down into periods. Periods may be further broken down into epochs. Modern dating techniques, such as radioactive dating, have given a range of dates as to when the various geologic time periods began.

The Geologic Time Scale
(Based on the Geologic Time Scale of the Geological Society of America

Eon	Era	Period	Epoch	Age
Phanerozoic	Cenozoic	Quaternary	Holocene	10,000 years ago
Phanerozoic	Cenozoic	Quaternary	Pleistocene	2.6 million years ago (mya)–10,000 years ago
Phanerozoic	Cenozoic	Neogene	Pliocene	5.3–2.6 mya
Phanerozoic	Cenozoic	Neogene	Miocene	23–5.3 mya
Phanerozoic	Cenozoic	Paleogene	Oligocene	33.9–23 mya
Phanerozoic	Cenozoic	Paleogene	Eocene	56–33.9 mya
Phanerozoic	Cenozoic	Paleogene	Paleocene	56–66 mya
Phanerozoic	Mesozoic	Cretaceous	Late	66–100 mya
Phanerozoic	Mesozoic	Cretaceous	Early	100–145 mya
Phanerozoic	Mesozoic	Jurassic	Late	145–164 mya

Eon	Era	Period	Epoch	Age
Phanerozoic	Mesozoic	Jurassic	Middle	164–174 mya
Phanerozoic	Mesozoic	Jurassic	Early	174–201 mya
Phanerozoic	Mesozoic	Triassic	Late	201–237 mya
Phanerozoic	Mesozoic	Triassic	Middle	237–247 mya
Phanerozoic	Mesozoic	Triassic	Early	247–252 mya
Phanerozoic	Paleozoic	Permian	Lopingian	252–260 mya
Phanerozoic	Paleozoic	Permian	Guadalupian	260–272 mya
Phanerozoic	Paleozoic	Permian	Cisuralian	272–299 mya
Phanerozoic	Paleozoic	Carboniferous (Pennsylvanian subperiod)	Late	299–307 mya
Phanerozoic	Paleozoic	Carboniferous (Pennsylvanian subperiod)	Middle	307–315 mya
Phanerozoic	Paleozoic	Carboniferous (Pennsylvanian subperiod)	Early	315–323 mya
Phanerozoic	Paleozoic	Carboniferous (Mississippian subperiod)	Late	323–331 mya
Phanerozoic	Paleozoic	Carboniferous (Mississippian subperiod)	Middle	331–347 mya
Phanerozoic	Paleozoic	Carboniferous (Mississippian subperiod)	Early	347–359 mya
Phanerozoic	Paleozoic	Devonian	Late	359–383 mya
Phanerozoic	Paleozoic	Devonian	Middle	383–393 mya
Phanerozoic	Paleozoic	Devonian	Early	393–419 mya
Phanerozoic	Paleozoic	Silurian	Pridoli	419–423 mya
Phanerozoic	Paleozoic	Silurian	Ludlow	423–427 mya
Phanerozoic	Paleozoic	Silurian	Wenlock	427–433 mya
Phanerozoic	Paleozoic	Silurian	Llandovery	433–444 mya
Phanerozoic	Paleozoic	Ordovician	Late	444–458 mya
Phanerozoic	Paleozoic	Ordovician	Middle	458–470 mya
Phanerozoic	Paleozoic	Ordovician	Early	470–485 mya
Phanerozoic	Paleozoic	Cambrian	Furongian	485–497 mya
Phanerozoic	Paleozoic	Cambrian	Epoch 3	497–509 mya
Phanerozoic	Paleozoic	Cambrian	Epoch 2	509–521 mya
Phanerozoic	Paleozoic	Cambrian	Terreneuvian	521–541 mya
Precambrian (Proterozoic)	Neoproterozoic	Ediacaran		541–635 mya
Precambrian (Proterozoic)	Neoproterozoic	Cryogenian		635–850 mya
Precambrian (Proterozoic)	Neoproterozoic	Tonian		850–1000 mya

Eon	Era	Period	Epoch	Age
Precambrian (Proterozoic)	Mesoproterozoic	Stenian		1000–1200 mya
Precambrian (Proterozoic)	Mesoproterozoic	Ectasian		1200–1400 mya
Precambrian (Proterozoic)	Mesoproterozoic	Calymmian		1400–1600 mya
Precambrian (Proterozoic)	Paleoproterozoic	Statherian		1600–1800 mya
Precambrian (Proterozoic)	Paleoproterozoic	Orosirian		1800–2050 mya
Precambrian (Proterozoic)	Paleoproterozoic	Rhyacian		2050–2300 mya
Precambrian (Proterozoic)	Paleoproterozoic	Siderian		2300–2500 mya
Precambrian (Archean)	Neoarchean			2500–2800 mya
Precambrian (Archean)	Mesoarchean			2800–3200 mya
Precambrian (Archean)	Paleoarchean			3200–3600 mya
Precambrian (Archean)	Eoarchean			3600–4000 mya
Precambrian (Hadean)				4000–4500 mya

What are Ice Ages, and when did they occur?

The times when the global climate was colder and large masses of ice covered many continents are referred to as Ice Ages. Ice Ages, or glacial periods, have occurred at irregular intervals for over 2.3 billion years. During an Ice Age, sheets of ice cover large portions of the continents. The exact reasons for the changes in Earth's climate are not known, although some think they are caused by changes in Earth's orbit around the Sun.

The Great Ice Age occurred during the Pleistocene epoch, which began about two million years ago and lasted until ten thousand years ago. At its height, about 27 percent of the world's present land area was covered by ice. In North America, the ice covered Canada and moved southward to New Jersey; in the Midwest, it reached as far south as St. Louis. Small glaciers and ice caps also covered the western mountains. Greenland was covered in ice, as it is today. In Europe, ice moved down from Scandinavia into Germany and Poland; the British Isles and the Alps also had ice caps. Glaciers also covered the northern plains of Russia, the plateaus of Central Asia, Siberia, and the Kamchatka Peninsula.

The glaciers' effect on the United States can still be seen. The drainage of the Ohio River and the position of the Great Lakes were influenced by the glaciers. The rich soil of the Midwest is mostly glacial in origin. Rainfall in areas south of the glaciers formed large lakes in Utah, Nevada, and California. The Great Salt Lake in Utah is a remnant of

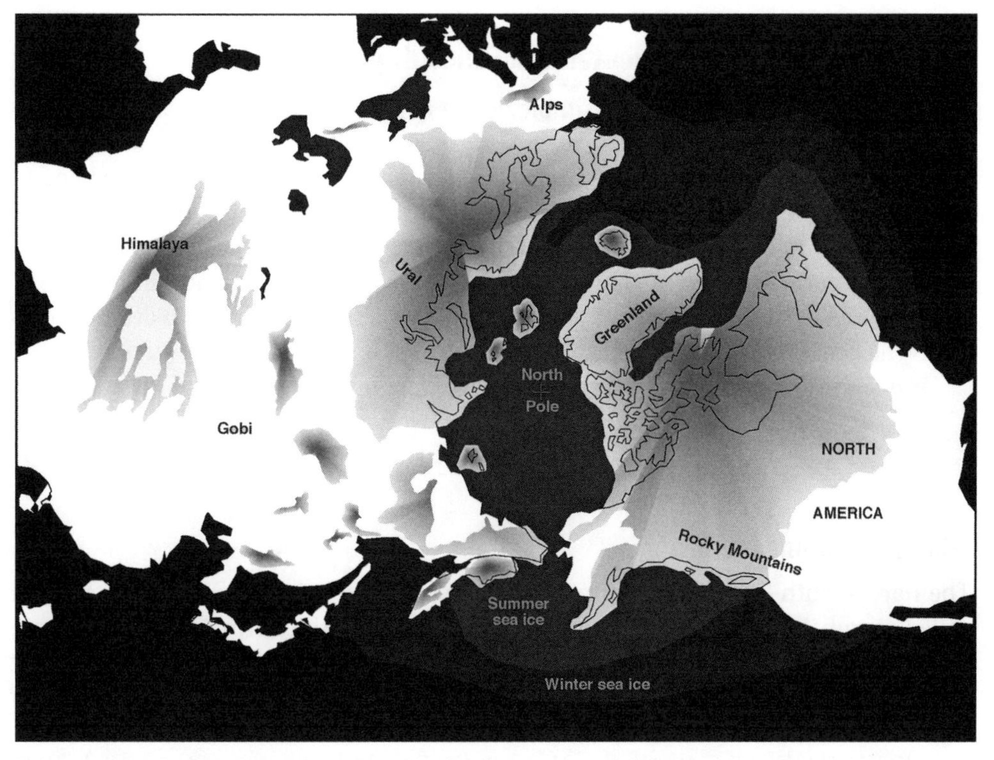

This map of the Northern Hemisphere shows how far glaciation extended during the more recent ice ages.

one of these lakes. The large ice sheets locked up a lot of water; sea level fell about 450 feet (137 meters) below what it is today. As a result, some states, such as Florida, were much larger during the Ice Age.

The glaciers of the last Ice Age retreated about ten thousand years ago. Some believe that the Ice Age is not over yet; the glaciers follow a cycle of advance and retreat many times. Some areas of Earth are still covered by ice, and this may be a time in between glacial advances.

OBSERVATION AND MEASUREMENT

Which direction does a compass point at the North Pole?

At the north magnetic pole, the compass needle would be attracted by the ground and point straight down.

What is magnetic declination?

It is the angle between magnetic north and true north at a given point on Earth's surface. It varies at different points on Earth's surface and at different times of the year.

Who invented the Foucault pendulum?

The Foucault pendulum is a simple device named after French physicist Léon Foucault (1819–1868), also known as Jean Foucault, who conceived an experiment to demonstrate Earth's rotation. The pendulum was introduced in 1851 and consisted of a heavy iron ball suspended from a wire 220 feet (67 meters) long. Sand beneath the pendulum recorded the plane of rotation of the pendulum over time. A Foucault pendulum always swings in the same vertical plane, but on a rotating Earth, this vertical plane slowly changes at a rate and direction dependent on the geographic latitude of the pendulum. For this demonstration and a similar one using a gyroscope, Foucault received the Copley Medal of the Royal Society of London in 1855.

What is the prime meridian?

The north–south lines on a map run from the North Pole to the South Pole and are called meridians. The word "meridian" is derived from the Latin word *meridianus*, meaning "noon." When it is noon at one place on the line, it is noon at any other point on the imaginary line as well. The lines, called longitudes, are used to measure how far east or west a particular place might be from zero degrees of longitude. The prime meridian is the meridian that passes through the Royal Observatory at Greenwich, England. In 1884, it was adopted internationally as zero degrees of longitude. The longitudinal lines are 69 miles (111 kilometers) apart at the equator.

The east–west lines on a map are called parallels and, unlike meridians, are all parallel to each other. They measure latitude, or how far north or south a particular place might be from the equator. One hundred and eighty lines circle Earth, one for each degree of latitude. The degrees of both latitude and longitude are divided into 60 minutes, which are then further divided into 60 seconds each.

When was the first civilian Earth observation satellite launched?

The Earth Resources Technology Satellite (ERTS-1) was launched on July 23, 1972. The satellite was later renamed Landsat 1. Landsat 2, Landsat 3, and Landsat 4 were launched in 1975, 1975, and 1982, respectively. Landsats 1, 2, and 3 orbited Earth at an altitude of 572 miles (920 kilometers), circling Earth every 103 minutes, yielding repeat coverage every eighteen days. Landsat 4, Landsat 5, Landsat 7, and Landsat 8 were launched in 1982, 1984, 1999, and 2013, respectively. These orbited or continue to orbit at an altitude of 438 miles (705 kilometers) above Earth in a 115-mile (185-kilometer) swath above Earth. Each satellite makes a complete orbit every 99 minutes, completes about fourteen full orbits each day, and crosses every point on Earth once every sixteen days. The Landsats use multispectral scanners, which detect visible green and blue wavelengths, and four infrared and near-infrared wavelengths. The results are displayed in "false-color" maps, where the scanner data is represented in shades of easily distinguishable colors—usually, infrared is shown as red, red as green, and green as blue. The differences are especially accurate when multiple wavelengths are compared using multispectral scanners. Even visible light images have proved useful—some of the earliest Landsat images showed that some small Pacific islands were up to 10 miles (16 kilometers) away from their charted positions.

The images provide data that serve as valuable resources for land use/land change research. They are useful in many applications, including forestry, agriculture, geology, regional planning, and education. These scanners can detect differences between soil, rock, water, and vegetation; types of vegetation; states of vegetation (e.g., healthy/unhealthy or underwatered/well-watered); and mineral content. The more than forty years of collected Landsat data have been used for deforestation research, volcanic flow studies, and understanding the effects of natural disasters and man-made disasters, such as tracking oil spills and monitoring mine waste pollution. Since December 2008, the USGS has made all of the Landsat images available without any cost to users worldwide.

What are topographic maps, and why are they important?

A topographic map is a line-and-symbol representation of natural and selected man-made features of a part of Earth's surface plotted to a definite scale. A distinguishing characteristic of a topographic map is the portrayal of the shape and elevation of the terrain by contour lines. They display the three-dimensional ups and downs of the terrain on a two-dimensional surface. Topographic maps usually portray natural features, such as mountains, valleys, plains, lakes, rivers, and vegetation, and man-made features, such as roads, boundaries, transmission towers, and even major buildings. Topographic maps are used by professionals for engineering, environment management, conservation of natural resources, energy exploration, public works design, and general commercial and residential planning. They are also used by individuals for outdoor recreational activities such as camping, hiking, biking, and fishing.

What do the 7.5-minute quadrangle maps represent?

One of the most popular maps of the U.S. Geological Survey (USGS) is the 7.5-minute quadrangle map. Produced at a scale of 1:24,000, each of these maps covers a four-sided area of 7.5 minutes latitude and 7.5 minutes longitude. The United States has been divided into precisely measured quadrangles. Each quadrangle in a state has a unique name. The maps of adjacent quadrangles may be combined to form a large map. More than 54,000 USGS topographic maps cover the forty-eight contiguous states, Alaska, and Hawaii.

What is the Global Positioning System (GPS), and how does it work?

The Global Positioning System (GPS) has three parts: the space part, the user part, and the control part. The space part consists of twenty-four satellites in orbit 11,000 nautical miles (20,300 kilometers) above Earth. The user part consists of a GPS receiver, which may be handheld or mounted in a vehicle. The control part consists of five ground stations worldwide that monitor and maintain the satellites. Using a GPS receiver, an individual can determine his or her location on or above Earth to within about 300 feet (90 meters).

It takes the combined scanning power of twenty-four satellites like this one sending signals to your receiver to make GPS work. Millions of dollars are spent so you don't lose your way driving to visit relatives.

What is Google Earth?

Google Earth is a geobrowser that provides a three-dimensional globe representation of Earth based on satellite imagery. Google Earth was originally developed by Keyhole, Inc., in 2001. It was acquired by Google in 2004. The program maps Earth by superimposing satellite images, aerial photography, and geographic information system (GIS) data onto a 3-D globe, allowing users to see cities and landscapes from various angles. Google Earth is used by individuals and in such fields as real estate, urban planning, defense, and intelligence. Google has extended the focus of the program on exploration and 3-D modeling of cities, street views in cities, oceans, and other water, and 3-D globes of other planets.

PHYSICAL CHARACTERISTICS

What is the mass of Earth?

The mass of Earth is estimated to be 6 sextillion, 588 quintillion short tons (6.6 sextillion short tons), or 5.97×10^{24} kilograms, with Earth's mean density being 5.515 times that of water (the standard). This is calculated from using the parameters of an ellipsoid adopted by the International Astronomical Union in 1964 and recognized by the International Union of Geodesy and Geophysics in 1967.

What are the three layers of Earth?

The planet Earth is made up of three main shells: the very thin, brittle crust, the mantle, and the core. The mantle and core are each divided into two parts: the upper mantle and the lower mantle and the outer core and the inner core. Although the core and mantle are about equal in thickness, the core actually forms only 15 percent of Earth's volume, whereas the mantle occupies 84 percent. The crust makes up the remaining 1 percent.

Which elements are contained in Earth's crust?

The most abundant elements in Earth's crust are listed in the table below. In addition, nickel, copper, lead, zinc, tin, and silver account for less than 0.02 percent, with all other elements comprising 0.48 percent.

Most Common Elements on Earth

Element	Percentage
Oxygen	47.0
Silicon	28.0
Aluminum	8.0
Iron	4.5
Calcium	3.5
Magnesium	2.5
Sodium	2.5
Potassium	2.5
Titanium	0.4
Hydrogen	0.2
Carbon	0.2
Phosphorus	0.1
Sulfur	0.1

What is at the center of Earth?

Geophysicists have held since the 1940s that Earth's interior core is a partly crystallized sphere of iron and nickel that is gradually cooling and expanding. As it cools, this inner core releases energy to an outer core called the fluid core, which is composed of iron, nickel, and lighter elements, including sulfur and oxygen. Another model called the "nuclear earth model" holds that a small core, perhaps 5 miles (8 kilometers) wide, of uranium and plutonium is at the center, surrounded by a nickel–silicon compound. The uranium and plutonium work as a natural nuclear reactor, generating radiating energy in the form of heat, which in turn drives charged particles to create Earth's magnetic field. The traditional model of Earth's core is still dominant; however, scientists have yet to disprove the nuclear earth model.

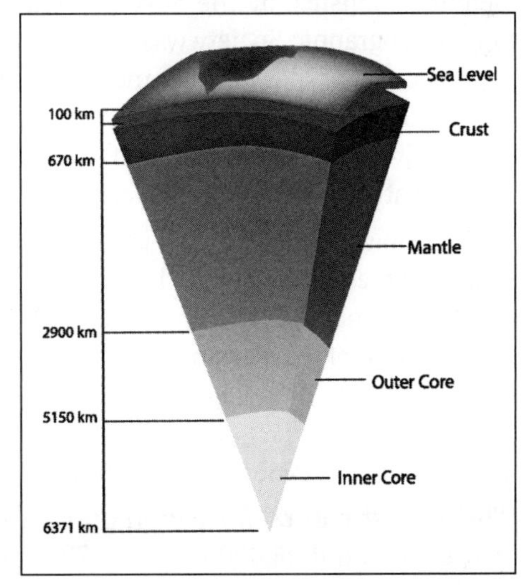

Earth consists of an inner core (sometimes divided into outer and inner cores), a mantle, and the crust.

What are the highest and lowest elevations in the United States?

Denali, previously called Mount McKinley in honor of U.S. president William McKinley (1843–1901), at 20,320 feet (6,194 meters), is the highest point in the United States and North America. Located in central Alaska, it is part of the Alaska Range. Its south peak measures 20,320 feet (6,194 meters) high, and the north peak is 19,470 feet (5,931 meters) high. It boasts one of the world's largest unbroken precipices and is the main scenic attraction at Denali

How does the temperature of Earth change as one goes deeper underground?

Earth's temperature increases with depth. Measurements made in deep mines and drill holes indicate that the rate of temperature increase varies from place to place in the world, ranging from 59 to 167°F (15–75°C) per kilometer in depth. Actual temperature measurements cannot be made beyond the deepest drill holes, which are a little more than 6.2 miles (10 kilometers) deep. Estimates suggest that the temperatures at Earth's center can reach values of 5,000°F (2,760°C) or higher.

National Park. Denali means the "high one" or the "great one" and is a Native American name. Mount Whitney, California, at 14,494 feet (4,421 meters), is the highest point in the continental United States. Death Valley, California, at 282 feet (86 meters) below sea level, is the lowest point in the United States and in the western hemisphere.

What are the highest and lowest points on Earth?

The highest point on land is the top of Mount Everest (in the Himalayas on the Nepal–Tibet border), at 29,035 feet [8,850 meters]) above sea level. This measurement, taken using satellite-based technology, includes the snow and ice layers, which are estimated at between 30 feet (9 meters) to 60 feet (18 meters). The National Geographic Society accepted this height in November 1999. The U.S. National Imagery and Mapping Agency has also accepted 29,035 feet (8,850 meters) as the official height. People are hopeful that ground-penetrating radar will one day be used to determine the snow-pack depth. The height established by the Surveyor General of India in 1954 and accepted by the National Geographic Society was 29,028 feet (8,845 meters), plus or minus 10 feet (3 meters) because of snow. Earlier measurements indicated that the height of Mount Everest was 29,002 feet (8,840 meters). Satellite measurements taken in 1987 indicated that Mount Everest is 29,864 feet (9,102 meters) high, but this measurement was not adopted by the National Geographic Society.

The lowest point on land is the Dead Sea between Israel and Jordan, which is 1,312 feet (399 meters) below sea level. The lowest point on Earth's surface is thought to be in the Mariana Trench in the western Pacific Ocean, extending from southeast of Guam to northwest of the Mariana Islands. It has been measured as 36,198 feet (11,034 meters) below sea level.

What is the deepest natural formation on Earth?

The deepest natural formation on Earth is the famous Mariana Trench, which at its deepest is about 36,000 feet (10.973 meters) or nearly 7 miles (11 kilometers) below sea level in the Pacific Ocean to the southeast of Japan. The Mariana Trench is crescent-shaped, and near the southern tip of the crescent is a small, slot-shaped area

What is the deepest-known place on Earth?

For many years, the deepest-known place on Earth was claimed by the Kola Superdeep Borehole. This well is a man-made well that extends 40,230 feet (12,262 meters) below the surface of Earth. The interesting fact about this well is that it has never been accessed by humans and was completely drilled by machines. The original plan presented in 1970 was to reach 49,000 feet (14,935 meters), but the temperatures recorded that deep were more than initially calculated and the project had to be stopped in 1992 since no drill could work at temperatures above 500°F. This depth is about one-third of the way down to Earth's crust, and it appears impossible to go any further down as of now. In terms of true vertical depth, it is the deepest borehole in the world. It was also the world's longest borehole in terms of measured depth along the well bore until surpassed in 2008 by the 40,318-foot- (12,289-meter-) long Al Shaheen oil well in Qatar and in 2011 by the 40,502-foot- (12,345-meter-) long Sakhalin-I OP-11 well in Russia.

called the Challenger Deep. This area is named after a British Royal Navy ship called the HMS *Challenger*, which measured the original depth by "sounding" techniques. The first attempt at descending into the Challenger Deep was in 1960 by the *Trieste*, which was a special ship called a bathyscaphe invented by Auguste (1884–1962) and Jacques Piccard (1922–2008). The most recent descent occurred in 2012 and was a competition between entrepreneur Richard Branson (1950–) and film director James Cameron (1954–). Cameron won, and his descent took 2 hours and 36 minutes. He spent several hours at the bottom of the trench taking samples and observing various forms of marine life.

LAND

Do the continents move?

In 1912, a German geologist, Alfred Lothar Wegener (1880–1930), theorized that the continents had drifted or floated apart to their present locations and that originally, all the continents had been a single land mass near Antarctica called Pangaea (from the Greek word meaning "all-earth"). Pangaea then broke apart some two hundred million years ago into two major continents called Laurasia and Gondwanaland. These two continents continued drifting and separating until the continents evolved into their present shapes and positions. Wegener's theory was discounted, but it has since been found that the continents do move sideways (not drift) at an estimated 0.75 inches (19 millimeters) annually because of the action of plate tectonics. American geologists William Maurice Ewing (1906–1974) and Harry Hammond Hess (1906–1969) proposed that Earth's crust

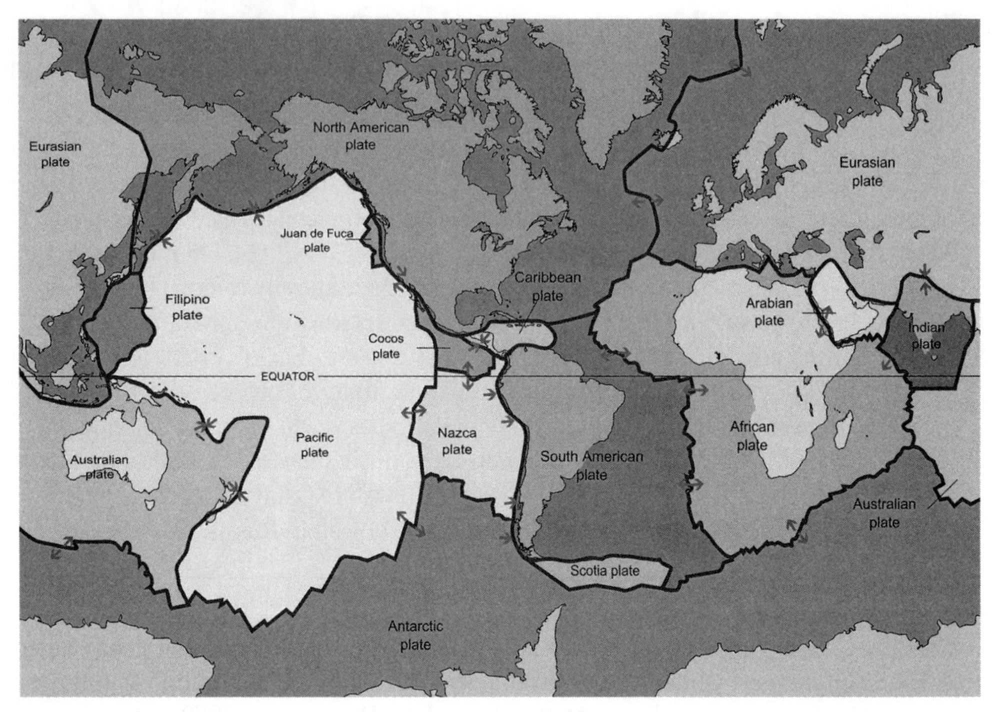

The earth's crust comprises a number of "plates" that move very slowly across the surface of the globe, which explains why islands and continents move over time.

is not a solid mass but is composed of eight major and seven minor plates that can move apart, slide by each other, collide, or override each other. These plates meet in major areas of mountain building, earthquakes, and volcanoes.

How much of Earth's surface is land, and how much is water?

Approximately 30 percent of Earth's surface is land. This is about 57,259,000 square miles (148,300,000 square kilometers). The area of Earth's water surface is approximately 139,692,000 square miles (361,800,000 square kilometers), or 70 percent of the total surface area.

How much of Earth's surface is covered with ice?

About 10 percent of the world's land surface is glaciated, or permanently covered with ice. Approximately 5,712,800 square miles (14,800,000 square kilometers) are covered by ice in the form of ice sheets, ice caps, or glaciers. An ice sheet is a body of ice that blankets an area of land, completely covering its mountains and valleys. Ice sheets have an area of over 19,000 square miles (50,000 square kilometers); ice caps are smaller. Glaciers are larger masses of ice that flow, under the force of gravity, at a rate of 10–1,000 feet (3–300 meters) per year. Glaciers on steep slopes flow faster. The areas of glaciation in some parts of the world are:

Glaciated Land Area

Location	Miles²/Kilometers²
South Polar regions (includes Antarctica)	5,340,000/13,830,000
North Polar regions (includes Greenland)	758,500/1,965,000
Alaska-Canada	22,700/58,800
Asia	14,600/37,800
South America	4,600/11,900
Europe	4,128/10,700
New Zealand	391/1,015
Africa	92/238

How thick is the ice that covers Antarctica?

The average depth of the ice that covers Antarctica is 6,600 feet (2,000 meters) or more than 1 mile thick. In some areas, the ice is as thick as 3 miles (4.8 kilometers). Nearly 90 percent of the world's ice is found in Antarctica.

What causes avalanches?

An avalanche (also called a snowslide) is a rapid flow of great masses of snow down a sloping surface. The snow may be powdery, sliding over compacted, older snow; slabs of snow that roll down the slope; or a combination of ice and snow from the slope, including rocks and other debris. Avalanches are usually triggered in a starting zone from a mechanical failure in the snowpack when the forces on the snow exceed its strength, such as individuals walking, skiing, snowmobiling, and snowboarding on unstable snow. Many avalanches occur naturally without human provocation. Avalanches are not rare or random events, are endemic to any mountain range that accumulates a standing snowpack, and are most common during winter or spring. Colorado, Alaska, and Montana are the states where most avalanche-related deaths occur during the month of February in the United States.

What color is an iceberg, and what fraction of it shows above water?

Most icebergs are blue and white. However, one in a thousand icebergs in Antarctica is emerald green. They are only found in Antarctica because the Northern Hemisphere is not cold enough. These icebergs form when seawater freezes to the bottom of floating ice shelves. The ice looks green from the combination of yellow and blue. Yellow is from the yellowish-brown remains of dead plankton dissolved in seawater and trapped in the ice. Blue is present because although ice reflects virtually all the wavelengths of visible light, it absorbs slightly more red wavelengths than blue. Only one-seventh to one-tenth of an iceberg's mass shows above water.

How much of Earth's surface is permanently frozen?

About one-fifth of Earth's land is permafrost, or ground that is permanently frozen. In the Northern Hemisphere, 24 percent, the equivalent of 9 million square miles (23 million square kilometers), has permafrost underneath it. This classification is based entirely on temperature and disregards the composition of the land. It can include bedrock, sod, ice, sand, gravel, or any other type of material in which the temperature has been below freezing for over two years. Nearly all permafrost is thousands of years old. Permafrost is found in Canada, Russia, northern China, most of Greenland and Alaska, and Antarctica.

Which are the world's largest deserts?

Deserts are distinguished by two general characteristics. First, desert areas receive less than 10 inches of precipitation per year. Second, due to the extreme dryness, little plant or animal life exists in most deserts. Many deserts form a band north and south of the equator at about 20° latitude because moisture-bearing winds do not release their rain over these areas. As the moisture-bearing winds from the higher latitudes approach the equator, their temperatures increase and they rise higher and higher in the atmosphere. When the winds arrive over the equatorial areas and come in contact with the colder parts of Earth's atmosphere, they cool down and release all their water to create the tropical rain forests near the equator. However, other parts of the world, such as Antarctica and the Arctic, are also desert regions.

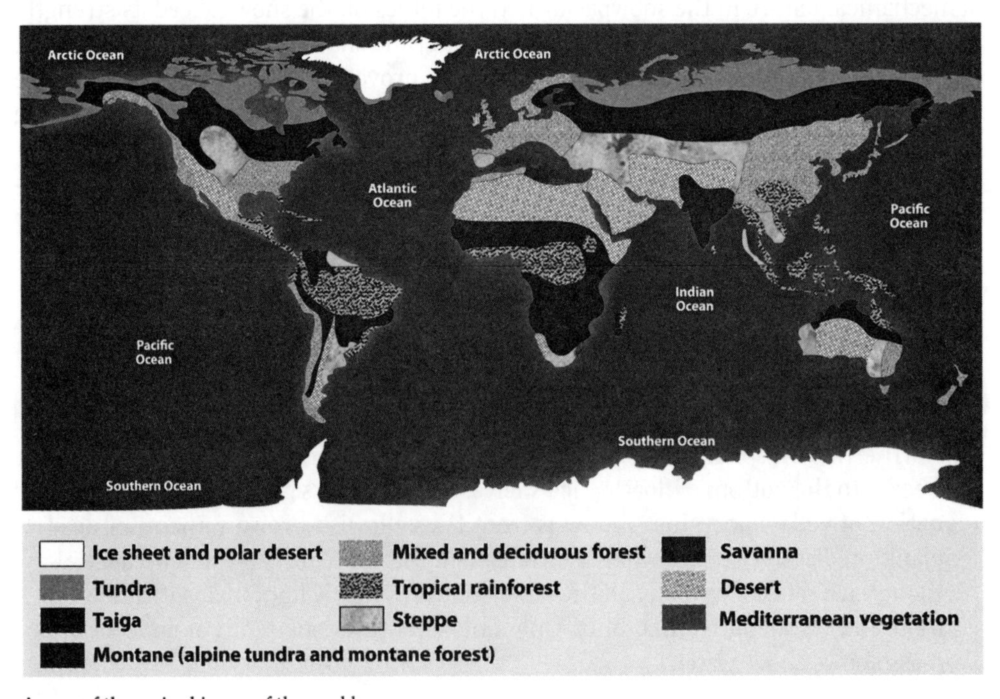

A map of the major biomes of the world.

The Sahara Desert, the world's largest, is three times the size of the Mediterranean Sea. In the United States, the largest desert is the Mojave Desert in southern California, with an area of 15,000 square miles (38,900 square kilometers).

Largest Deserts by Area

Desert	Location	Miles^2/Kilometers^2
Sahara	North Africa	3,500,000/9,065,000
Gobi	Mongolia and China	500,000/1,295,000
Libyan	Libya, SW Egypt, Sudan	425,000/1,100,745
Patagonia	Argentina	300,000/776,996
Kalahari	Southern Africa	275,000/712,247

What are the five longest mountain ranges in the world?

The longest mountain range on Earth is the mid-ocean ridge. It spans 40,389 miles (65,000 kilometers) around the globe, although 90 percent of the ridge is under the ocean. The mid-ocean ridge system of mountains and valleys stretches and crisscrosses around the globe like the seam of a baseball. It is formed by the movement of Earth's tectonic plates.

Five Largest Mountain Ranges

Range	Location (Continent)	Location (Countries)	Length (miles)	Length (kilometers)
Andes	South America	Venezuela, Colombia, Ecuador, Peru, Bolivia, Argentina, Chile	4,500	7,242
Rocky Mountains	North America	United States and Canada	3,750	6,035
Himalayas/ Karakoram/ Hindu Kush	Asia	Bhutan, China, India, Nepal, Pakistan	2,400	3,862
Great Dividing Range	Australia	Australia	2,250	3,621
Trans-Antarctic Mountains	Antarctica	Antarctica	2,200	3,541

How are caves formed?

Water erosion creates most caves found along coastal areas. Waves crashing against the rock over many years wears away part of the rock, forming a cave. Inland caves are also formed by water erosion—in particular, groundwater eroding limestone. As the limestone dissolves, underground passageways and caverns are formed.

What is the difference between spelunking and speleology?

Spelunking, or sport caving, is exploring caves as a hobby or for recreation. Speleology is the scientific study of caves and related phenomena.

What are speleothems?

Speleothems, commonly known as cave formations, are secondary mineral deposits formed in a cave after the cave itself has formed. These secondary mineral deposits are created by the solidification of fluids or from chemical solutions and usually contain calcium carbonate or limestone, but gypsum or silica may also be found. The six most common speleothems are: stalactites, stalagmites, soda straws, cave coral, boxwork, and cave pearls.

How does a stalactite differ from a stalagmite?

A stalactite is a conical or cylindrical calcite formation hanging from a cave roof. It forms from the centuries-long buildup of mineral deposits resulting from the seepage of water from the limestone rock above the cave. This water, containing calcium bicarbonate, evaporates, losing some carbon dioxide, to deposit small quantities of calcium carbonate ($CaCO_3$ carbonate of lime), which eventually forms a stalactite.

A stalagmite is a stone formation that develops upward from the cave floor and resembles an icicle upside down. Formed from water containing calcite that drips from the limestone walls and roof of the cave, it sometimes joins a stalactite to form a column.

Where is the deepest cave in the United States?

Lechuguilla Cave, in Carlsbad Caverns National Park, New Mexico, is the deepest cave in the United States. Its depth is 1,565 feet (477 meters). Unlike most caves in which carbon dioxide mixes with rainwater to produce carbonic acid, Carlsbad Caverns was shaped by sulfuric acid. The sulfuric acid was a result of a reaction between oxygen that was dissolved in groundwater and hydrogen sulfide that emanated from far below the cave's surface.

Where is the longest cave system in the world?

The Mammoth Cave system in Kentucky is the longest cave system in the world. The cave system consists of more than 367 miles (591 kilometers) of subterranean labyrinths. If the second and third longest caves in the world were combined, Mammoth Cave would still be the longest cave in the world.

How long is the Grand Canyon?

The Grand Canyon, cut out by the Colorado River over a period of fifteen million years in the northwest corner of Arizona, is the largest land gorge in the world. It is 4–13

What is a simple way to remember which speleotherm is a stalactite and which is a stalagmite?

A simple way to remember the difference is that the "c" in stalactite stands for "ceiling," while the "g" in stalagmite stands for "ground."

miles (6.4–21 kilometers) wide at its brim, 4,000–5,500 feet (1,219–1,676 meters) deep, and 217 miles (349 kilometers) long, extending from the mouth of the Little Colorado River to Grand Wash Cliffs (and 277 miles, 600 feet [445.88 kilometers] long if Marble Canyon is included).

However, it is not the deepest canyon in the United States; that distinction belongs to Kings Canyon, which runs through the Sierra and Sequoia National Forests near East Fresno, California, with its deepest point being 8,200 feet (2,500 meters). Hell's Canyon of the Snake River between Idaho and Oregon is the deepest U.S. canyon in low-relief territory. Also called the Grand Canyon of the Snake, it plunges 7,900 feet (2,408 meters) down from Devil Mountain to the Snake River.

What is a sinkhole? Which states sustain the most damage from sinkholes?

A sinkhole is an area of ground that has no natural external surface drainage. When it rains, all of the water stays inside the sinkhole and typically drains into the subsurface. Sinkholes can vary from a few feet to hundreds of acres and from less than 1 foot (30 centimeters) to more than 100 feet (30 meters) deep. Sinkholes are common where the rock below the land surface is limestone, carbonate rock, salt beds, or rocks that can be dissolved by groundwater circulating through them. As the rock dissolves, spaces and caverns develop underground. Sinkholes are dramatic because the land usually stays

Mammoth statues have been placed by a tar pit at the La Brea Tar Pits to try to re-create a scene from the Pleistocene Era at this popular Los Angeles tourist attraction.

intact for a while until the underground spaces just get too big. If support for the land above the spaces is lacking, then a sudden collapse of the land surface can occur. The most damage from sinkholes tends to occur in Florida, Texas, Alabama, Missouri, Kentucky, Tennessee, and Pennsylvania.

What are the La Brea Tar Pits?

The tar pits are located in an area of Los Angeles, California, formerly known as Rancho La Brea. Heavy, sticky tar oozed out of the earth there, the scum from great petroleum reservoirs far underground. The pools were cruel traps for uncounted numbers of animals. Today, the tar pits are a part of Hancock Park, where many fossil remains are displayed along with life-sized reconstructions of these prehistoric species.

The tar pits were first recognized as a fossil site in 1875. However, scientists did not systematically excavate the area until 1901. By comparing Rancho La Brea's fossil specimens with their nearest living relatives, paleontologists have a greater understanding of the climate, vegetation, and animal life in the area during the Ice Age. Perhaps the most impressive fossil bones recovered belong to such large, extinct mammals as the imperial mammoth and the saber-toothed cat. Paleontologists have even found the remains of the western horse and the camel, which originated in North America, migrated to other parts of the world, and became extinct in North America at the end of the Ice Age.

ROCKS AND MINERALS

What are the three main groups of rocks?

Rocks can be conveniently placed into one of three groups—igneous, sedimentary, and metamorphic.

- *Igneous rocks*, such as granite, pegmatite, rhyolite, obsidian, gabbro, and basalt, are formed by the solidification of molten magma that emerges through Earth's crust via volcanic activity. The nature and properties of the crystals vary greatly depending in part on the composition of the original magma and partly on the conditions under which the magma solidified. Thousands of different igneous rock types exist. For example, granite is formed by slow cooling of molten material (within Earth). It has large crystals of quartz, feldspar, and mica.
- *Sedimentary rocks*, such as brecchia, sandstone, shale, limestone, chert, and coal, are produced by the accumulation of sediments. These are fine rock particles or fragments, skeletons of microscopic organisms, or minerals leached from rocks that have accumulated from weathering. These sediments are then redeposited underwater and later compressed in layers over time. The most common sedimentary rock is sandstone, which is predominantly quartz crystals.
- *Metamorphic rocks*, such as marble, slate, schist, gneiss, quartzite, and hornsfel, are formed by the alteration of igneous and sedimentary rocks through heat and/or

pressure. One example of these physical and chemical changes is the formation of marble from thermal changes in limestone.

How does a rock differ from a mineral?

Mineralogists use the term "mineral" for a substance that has all four of the following features: it must be found in nature; it must be made up of substances that were never alive (organic); it has the same chemical makeup wherever it is found; and its atoms are arranged in a regular pattern to form solid crystals.

While "rocks" are sometimes described as an aggregate or combination of one or more minerals, geologists extend the definition to include clay, loose sand, and certain limestones.

How and when was coal formed?

Coal is a sedimentary rock. It is formed from the remains of plants that have undergone a series of far-reaching changes, turning into a substance called peat, which subsequently was buried. Through millions of years, Earth's crust buckled and folded, subjecting the peat deposits to very high pressure and changing the deposits into coal. The Carboniferous, or coal-bearing, period occurred about 250 million years ago. Geologists in the United States sometimes divide this period into the Mississippian and the Pennsylvanian periods. Most of the high-grade coal deposits are to be found in the strata of the Pennsylvanian period.

What are the types of coal?

The first stage in the formation of coal converts peat into lignite, a dark-brown type of coal. Lignite is then converted into subbituminous coal as pressure from overlying materials increases. Under still greater pressure, a harder coal called bituminous, or soft, coal is produced. Intense pressure changes bituminous coal into anthracite, the hardest of all coals.

What is mineralogy?

Mineralogy is a branch of geology that studies minerals. It focuses on the identification, distribution, and classification and properties—both physical and chemical—of minerals. Not only does mineralogy provide scientists with information on how Earth was formed, but minerals provide the basic components and building blocks of many of our modern materials. Scientists have identified nearly four thousand different minerals.

What were Abraham Gottlob Werner's contributions to mineralogy?

German mineralogist Abraham Gottlob Werner is considered the father of mineralogy. He wrote the first modern textbook of descriptive mineralogy, *On the External Characters of Fossils*, in 1774. His work devised a method of describing minerals by their external characteristics, including color. Werner worked out an arrangement of colors and color names, illustrated by an actual set of minerals. This simplified approach appealed to a broad audience interested in learning more about geology.

What are the physical traits that are used to identify a mineral?

Minerals are identified by physical traits, including hardness, color, streak, luster, cleavage or fracture, and specific gravity.

Abraham Gottlob Werner proposed the theory of Neptunism, which states that rock formation was the result of the crystallization of minerals in oceans.

- *Hardness* is the ability of a mineral to scratch another mineral.
- *Color* is sometimes a way to identify a mineral, but the color may be the result of impurities.
- *Streak* is related to color but a more reliable test. It is the color of the streak a mineral leaves on an unglazed porcelain plate.
- *Luster* is the appearance when light reflects off the surface of a mineral. Luster can be described in different ways, including metallic luster, vitreous or glassy luster, resinous luster, pearly luster, and greasy luster.

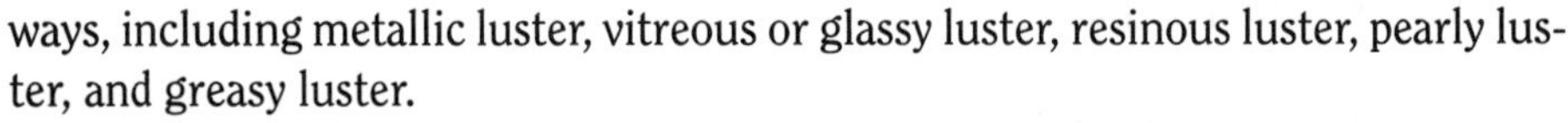

- *Cleavage and fracture* refer to how a mineral breaks when struck with force. It is called cleavage when the minerals break along a smooth plane. Most minerals do not break cleanly but fracture.

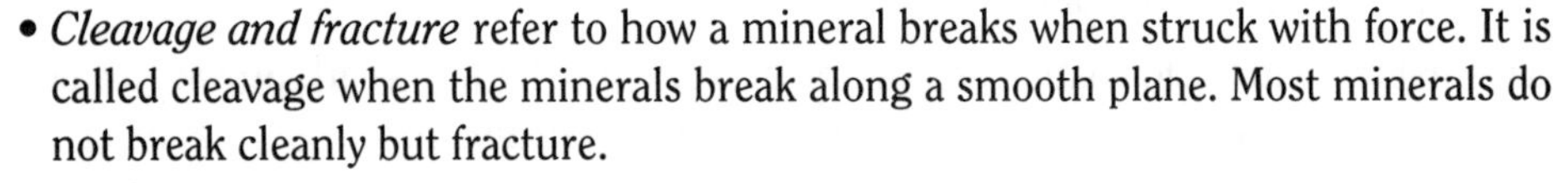

- *Specific gravity* is an indirect measure of density. It is the ratio of the weight of the mineral to an equal volume of water.

What are the most common mineral-forming elements?

Most minerals are compounds of two or more chemical elements. Although more than one hundred chemical elements have been identified, only ten elements account for nearly 99 percent of the weight of Earth's crust. The ten most common mineral-forming compounds are: oxygen, silicon, aluminum, iron, calcium, sodium, potassium, magnesium, hydrogen, and titanium. Since oxygen and silicon make up nearly three-quarters of Earth's crust, silicate minerals—compounds of silicon and oxygen—are the most abundant minerals.

Do any minerals contain just one chemical element?

Yes, minerals that contain just one chemical element are called native elements. Examples of native elements are diamond and graphite, which contain pure carbon. Gold, silver, and copper are also native elements.

What is the softest mineral?

The softest mineral is talc, which is 1 on a hardness scale of 1 to 10. Talc can easily be scratched by a fingernail. Talc is used in many industries, including paper making, plastics, paint and coatings, rubber, food, pharmaceuticals, cosmetics, and ceramics. It is used to coat the insides of inner tubes and rubber gloves during manufacture to prevent the surfaces from sticking. Talc is one of the oldest-known solid lubricants. It is also often used in basketball to keep a player's hands dry, and most tailor's chalk is talc.

When was the Mohs scale introduced?

The Mohs scale is a standard of ten minerals by which the hardness of a mineral is rated. It was introduced in 1812 by German mineralogist Friedrich Mohs (1773–1839). The minerals are arranged from softest to hardest. Harder minerals, with higher numbers, can scratch those with a lower number.

The Mohs Scale

Hardness	Mineral	Comment
1	Talc	Hardness 1–2 can be scratched by a fingernail
2	Gypsum	Hardness 2–3 can be scratched by a copper coin
3	Calcite	Hardness 3–6 can be scratched by a steel pocket knife
4	Fluorite	
5	Apatite	
6	Orthoclase	Hardness 6–7 will not scratch glass
7	Quartz	
8	Topaz	Hardness 8–10 will scratch glass
9	Corundum	
10	Diamond	

What is a strategic mineral and what are some examples?

Strategic minerals are minerals essential to national defense—the supply of which a country uses but cannot produce itself. One-third to half of the eighty minerals used by industry could be classed as strategic minerals. Wealthy countries, such as the United States, stockpile these minerals to avoid any crippling effect on their economy or military strength if political circumstances were to cut off their supplies. As of 2018, thirty-five minerals are considered critical minerals. They are used in many different products and industries, including rechargeable batteries, integrated circuits, steel production and steel products, medicine, and agricultural. Some of the minerals on the list are aluminum, arsenic, cobalt, gallium, lithium, magnesium, platinum group metals, rare earth elements, titanium, and uranium.

What is a geode, and where does it come from?

A geode is a hollow or partially hollow, subspherical body with inward-pointing crystals that grow into an open space. Very simply, a geode is a rock that is rugged and weathered on the outside and colorful and sparkly on the inside. A geode begins when a cavity forms in a rock. These cavities are most common in igneous rock, created by cooling lava or magma. Usually, this happens when a bubble of carbon dioxide and water vapor forms in flows of lava, and as the molten rock cools and the gas dissolves, empty space is left behind. The result of this hollowing is a shell waiting to be filled. Both igneous and sedimentary geodes, which appear to be solid, are very porous, but the openings are microscopic. Eventually, mineral-rich groundwater or rainwater begins to seep through the porous rock and fill the cavity. The water lines a thin crust of minerals inside the cavity. As water flows through the geode, additional mineral layers are deposited in its hollow interior. Over thousands of years, these layers of minerals build crystals that exhibit brilliant colors and eventually fill the cavity. Which minerals end up as crystals varies by location and conditions such as temperature, acidity, and the type of rock the geode is formed from. Quartz crystals are most common in igneous geodes, and silica is more common in geodes that form in sediment. Geodes vary in size from an inch or two in diameter to ones that are many feet long. Many geodes are cut in half, polished, and used as decorations, including bookends.

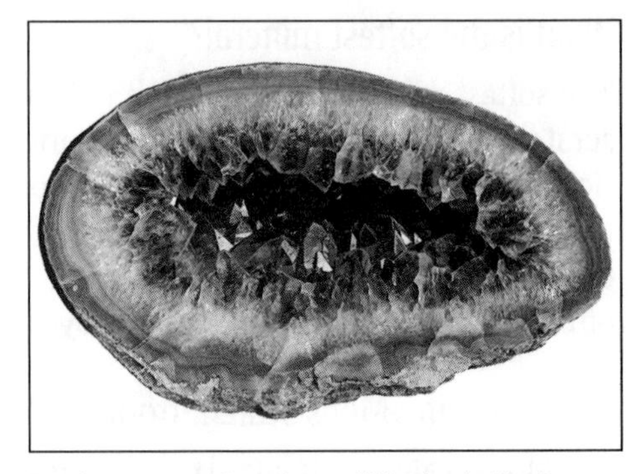

Break open a geode and you will be dazzled by the beautiful crystal formations within it. They form when mineral-rich water seeps into a cavity within a stone.

What is the lightest type of rock?

The lightest type of rock is pumice, which is formed from volcanic lava that cools too quickly to allow dissolved gases to escape, creating a rock filled with holes that is light enough in weight to float on water. Pumice is widely used to make lightweight concrete or insulating cinder blocks. It is also used as an abrasive, especially in polishes, pencil erasers, cosmetic exfoliates, and in the production of stone-washed jeans. Pumice stones are used to dry and remove excess skin, including calluses. Finely ground pumice is added to some toothpastes and hand cleaners, such as Lava soap. In gardening, it is mixed with soil to provide good aeration for plants and is also used as a soilless growing medium in hydroponics.

What rock makes up the White Cliffs of Dover?

The White Cliffs of Dover are part of the British coastline facing the Strait of Dover and France. The cliffs are composed of soft, white chalk, which is a very pure form of limestone and rare geologically. The cliff face, which reaches up to 350 feet (110 meters), owes its striking appearance to its composition of chalk accented by streaks of black flint, which is composed of the remains of sea sponges and planktonic microorganisms. The cliffs stretch for 8 miles (13 kilometers) on both sides of the town of Dover and are England's most spectacular natural feature.

What is the composition of the Rock of Gibraltar?

It is composed of gray limestone, with a dark shale overlay on parts of its western slopes. Located on a peninsula at the southern extremity of Spain, the Rock of Gibraltar is a mountain at the east end of the Strait of Gibraltar, the narrow passage between the Atlantic Ocean and the Mediterranean Sea. "The Rock" is 1,398 feet (425 meters) tall at its highest point.

FOSSILS

What is paleontology?

Paleontology is the study of ancient life primarily by examining and studying fossils, including those of plants and animals—both invertebrates and vertebrates. Fossils are also useful for studying paleoecology, which is the interaction between plants and animals and their environments.

Who is considered the founder of paleontology?

Georges Cuvier (1769–1832), a French naturalist and zoologist, is generally recognized as the founder of paleontology. Paleontology became an established discipline in the eighteenth century as a result of the work of Cuvier in the fields of both comparative anatomy and paleontology comparing living animals with fossils.

French zoologist and naturalist Georges Cuvier, the "founder of paleontology," was instrumental in establishing the field through his contributions to taxonomy, comparative anatomy, and vertebrate fossil studies.

What are fossils, and how are they formed?

Fossils are the remains of animals or plants that were preserved in rock before the beginning of recorded history. It is unusual for complete organisms to be preserved; fossils usually represent the hard parts of animals, such as bones or shells, and leaves, seeds, or woody parts of plants.

Some fossils are simply the bones, teeth, or shells themselves, which can be preserved for a relatively short period of time. Another type of fossil is the imprint of a buried plant or animal that decomposes, leaving a film of carbon that retains the form of the organism.

Some buried material is replaced by silica and other materials that permeate the organism and replace the original material in a process called petrification. Some woods are replaced by agate or opal so completely that even the cellular structure is duplicated. The best examples of this can be found in Petrified Forest National Park in Arizona.

Molds and casts are other very common fossils. A mold is made from an imprint, such as a dinosaur footprint, in soft mud or silt. This impression may harden, then be covered with other materials. The original footprint will have formed a mold, and the sediments filling it will be a cast of the footprint.

What are the basic types of fossils?

Any remains, traces, or imprints of any bacteria, plants, or animals that lived on Earth are fossils. Body fossils and trace fossils are the two principal types of evidence about ancient life. Body fossils are the preserved record of hard (bone, teeth, etc.) or soft body parts. Trace fossils record traces of biological activity, such as footprints, tracks, and burrows. A third type of fossil is geochemical evidence or chemical traces of life, also called chemofossils, which has aided in deciphering the evolution of life before organisms were large enough to leave concrete evidence.

How old are fossils?

The oldest-known fossils are of bacteria that left their impressions approximately 3.5 billion years ago. The oldest animal fossils are of invertebrates that lived approximately seven hundred million years ago. The largest number of fossils come from the Cambrian period of 505–590 million years ago, when living organisms began to develop skeletons and hard parts. Since these parts tended to last longer than ordinary tissue, they were more likely to be preserved in clay and become fossilized.

What areas of the United States are particularly interesting for viewing different types of fossils?

Fossils from the different geologic time periods have been found in many areas of the United States. Some of the oldest fossils from the Precambrian eon (prior to 542 million years ago) may be found in the Grand Canyon. Algae fossils are embedded in the rocks of the Grand Canyon as well as examples of stromatolites, limestone structures formed by cyanobacteria. Glacier National Park in Montana also has examples of fossilized stromatolites in its layers of limestone and dolomite.

Some of the best-known samples of marine fossils from the Paleozoic era (251–542 million years ago) are found in Carlsbad Caverns in New Mexico. These include trilobites, brachiopods, sponges, and bryozoans. Other places to find fossils from the Paleozoic era are the Grand Canyon and the Mojave National Preserve in California. The Guadalupe Mountains National Park in Texas has a collection of fossils from the Permian period

What is amber?

Amber is the fossil resin of trees. The two major deposits of amber are in the Dominican Republic and the Baltics. Amber came from a coniferous tree that is now extinct. Amber is usually yellow or orange in color, semitransparent or opaque with a glossy surface. It is used by both artisans and scientists.

(263 million years ago) from ancient algae to prehistoric gastropods (the ancestors of our snails).

Dinosaur fossils from the Mesozoic era (65.5–251 million years ago) are abundant in the western United States from Montana to New Mexico. Dinosaur National Monument in Utah and Colorado has extensive displays of dinosaur fossils.

Fossils from the most recent geologic time period, the Cenozoic era (65.5 million years ago through today), are found in John Day Fossil Beds in Oregon and the Florissant Fossil Beds in Colorado. Both of these fossil beds have extensive displays of plants and animal fossils.

What were the smallest and largest dinosaurs?

Compsognathus, a carnivore from the late Jurassic period (131 million years ago), was about the size of a chicken and measured, at most, 35 inches (89 centimeters) from the tip of its snout to the tip of its tail. Their average weight was about 6 pounds, 8 ounces (3 kilograms), but they could be as much as 15 pounds (6.8 kilograms).

The largest species for which a whole skeleton is known is Brachiosaurus. A specimen in the Humboldt Museum in Berlin measures 72.75 feet (22.2 meters) long and 46 feet (14 meters) high. It weighed an estimated 34.7 tons (31,480 kilograms). Brachiosaurus was a four-footed, plant-eating dinosaur with a long neck and a long tail and lived between about 121–155 million years ago.

How is petrified wood formed?

Petrified wood is formed when water containing dissolved minerals, such as calcium carbonate ($CaCO_3$) and silicate, infiltrates wood or other structures. The process takes thousands of years. The foreign material either replaces or encloses the organic matter and often retains all of the structural details of the original plant material. Botanists find these types of fossils to be very important since they allow for the study of the internal structure of extinct plants. After a time, wood seems to have turned to stone because the original form and structure are retained. The wood itself does not turn to stone.

How many different kinds of faults have been identified?

Faults are fractures in Earth's crust. Faults are classified as normal, reverse, or strike-slip. Normal faults occur when the end of one plate slides vertically down the end of another. Reverse faults occur when one plate slides vertically up the end of another. A strike-slip fault occurs when two plate ends slide past each other horizontally. In an oblique fault, plate movement occurs vertically and horizontally at the same time.

Where is the San Andreas Fault?

Perhaps the most famous fault in the world, the San Andreas Fault extends from Mexico north through most of California. The San Andreas Fault is not a single fault but rather a system of faults. The northern half of the fault, near San Francisco, has reverse faults and is mostly mountainous. The southern half, near Los Angeles, has mostly normal faults. Land development has made it difficult to see the fault except in a few locations, notably near Lake San Andreas south of San Francisco. The fault was named in honor of Andrew Lawson (1861–1952), a geologist who studied the 1906 San Francisco earthquake.

How does a seismograph work?

A seismograph records earthquake waves. When an earthquake occurs, three types of waves are generated. The first two, the P and S waves, are propagated within Earth, while the third, consisting of Love and Rayleigh waves, is propagated along the planet's surface. The P wave travels about 3.5 miles (5.6 kilometers) per second and is the first wave to reach the surface. The S wave travels at a velocity of a little more than half of the P waves. If the velocities of the different modes of wave propagation are known, the distance between the earthquake and an observation station may be deduced by measuring the time interval between the arrival of the faster and slower waves.

When the ground shakes, the suspended weight of the seismograph, because of its inertia, scarcely moves, but the shaking motion is transmitted to the marker, which leaves a record on the drum.

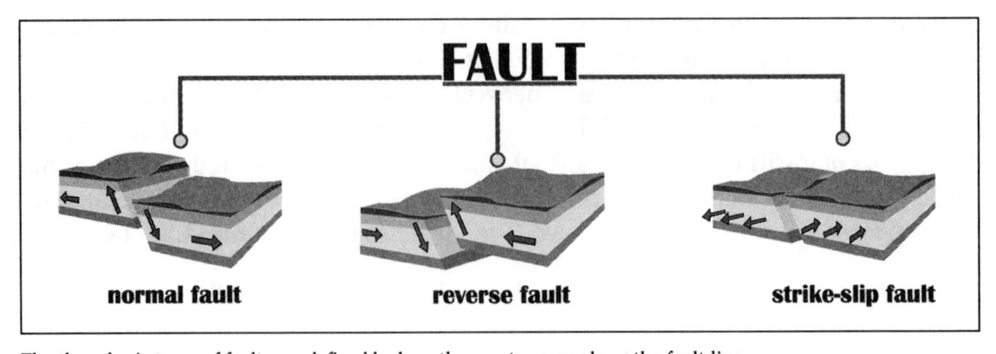

The three basic types of faults are defined by how the crust moves along the fault line.

What is the Richter scale?

The Richter scale measures the magnitude of an earthquake, that is, the size of the ground waves generated at the earthquake's source, on a seismograph. The scale was devised by American geologist Charles W. Richter (1900–1985) in 1935. Every increase of one number means a tenfold increase in magnitude.

Richter Scale

Magnitude	Possible Effects
1	Detectable only by instruments
2	Barely detectable, even near the epicenter
3	Felt indoors
4	Felt by most people; slight damage
5	Felt by all; damage minor to moderate
6	Moderately destructive
7	Major damage
8	Total and major damage

What is the Modified Mercalli Intensity Scale?

The Modified Mercalli Intensity Scale is a means of measuring the intensity of an earthquake. Unlike the Richter scale, which uses mathematical calculations to measure seismic waves, the Modified Mercalli Intensity Scale uses the effects of an earthquake on the people and the structures in a given area to determine its intensity and is primarily used by underwriters and insurance companies. It was invented by Giuseppe Mercalli (1850–1914) in 1902 and modified by Harry Wood (1879–1958) and Frank Neumann (1892–1964) in the 1930s to take into consideration such modern inventions as the automobile and the skyscraper.

The Modified Mercalli Intensity Scale

Scale	Earthquake Effects
I	Only felt by a few under especially favorable circumstances.
II	Felt only by a few sleeping persons, particularly on upper floors of buildings. Some suspended objects may swing.
III	Felt quite noticeably indoors, especially on upper floors of buildings, but may not be recognized as an earthquake. Standing automobiles may rock slightly. Vibration like passing of truck.
IV	During the day, felt indoors by many, outdoors by few. At night, some awakened. Dishes, windows, doors disturbed; walls make creaking sound. Sensation like heavy truck striking building. Standing automobiles rock noticeably.
V	Felt by nearly everyone, many awakened. Some dishes, windows, and so on broken; cracked plaster in a few places; unstable objects overturned. Disturbances of trees, poles, and other tall objects sometimes noticed. Pendulum clocks may stop.

Scale	Earthquake Effects
VI	Felt by all; many frightened and run outdoors. Some heavy furniture moved; a few instances of fallen plaster and damaged chimneys. Damage slight.
VII	Everybody runs outdoors. Damage negligible in buildings of good design and construction; slight to moderate in well-built ordinary structures; considerable in poorly built or badly designed structures; some chimneys broken. Noticed by persons driving cars.
VIII	Damage slight in specially designed structures; considerable in ordinary substantial buildings with partial collapse; great in poorly built structures. Panel walls thrown out of frame structures. Fall of chimneys, factory stacks, columns, monuments, walls. Heavy furniture overturned. Sand and mud ejected in small amounts. Changes in well water. Persons driving cars disturbed.
IX	Damage considerable in specially designed structures; well-designed frame structures thrown out of plumb; great in substantial buildings, with partial collapse. Building shifted off foundation. Ground cracked conspicuously. Underground pipes broken.
X	Some well-built wooden structures destroyed; most masonry and frame structures destroyed with foundations; ground badly cracked. Rails bent. Landslides considerable from riverbanks and steel slopes. Shifted sand and mud. Water splashed, slopped over banks.
XI.	Few, if any, (masonry) structures remain standing. Bridges destroyed. Broad fissures in ground. Underground pipelines completely out of service. Earth slumps and landslips in soft ground. Rails bent greatly.
XII.	Damage total. Waves seen on ground surface. Lines of sight and level distorted. Objects thrown into the air.

Who invented the ancient Chinese earthquake seismometer?

The seismometer, invented by Zhang Heng (78–139 C.E.) around 132 C.E., was a copper-domed urn with dragons' heads circling the outside, each containing a bronze ball. Inside the dome was suspended a pendulum that would swing when the ground shook and knock a ball from the mouth of a dragon into the waiting open mouth of a bronze toad below. The ball made a loud noise and signaled the occurrence of an earthquake. Knowing which ball had been released, one could determine the direction of the earthquake's epicenter (the point on Earth's surface directly above the quake's point of origin).

This clever seismometer from the Eastern Han Dynasty of China could measure the strength and direction of an earthquake.

When did the most severe earthquake in American history occur?

The New Madrid, Missouri, series of earthquakes (a series of quakes starting on December 16, 1811, and lasting until March 1812) is considered to be the most severe earthquake event in U.S. history. It shook more than two-thirds of the United States and was felt in Canada. It changed the level of land by as much as 20 feet (6 meters), altered the course of the Mississippi River, and created new lakes, such as Lake St. Francis west of the Mississippi and Reelfoot Lake in Tennessee. Because the area was so sparsely populated, no known loss of life occurred. Scientists agree that at least three, and possibly five, of the quakes had surface wave magnitudes of 8.0 or greater. The largest was probably a magnitude of 8.8, which is larger than any quake yet experienced in California.

What magnitude was the earthquake that hit San Francisco on April 18, 1906?

The historic 1906 San Francisco earthquake took a mighty toll on the city and surrounding area. Over seven hundred people were killed; the newly constructed $6 million City Hall was ruined; the Sonoma Wine Company collapsed, destroying 15 million gallons (57 million liters) of wine. The quake registered 8.3 on the Richter scale and lasted 75 seconds total. Many poorly constructed buildings built on landfills were flattened, and the quake destroyed almost all of the gas and water mains. Fires broke out shortly after the quake, and when they were finally eliminated, three thousand acres of the city, the equivalent of 520 blocks, were charred. Damage was estimated to be $500 million, and many insurance agencies went bankrupt after paying out the claims.

Another large earthquake hit San Francisco on October 17, 1989. It measured 7.1 on the Richter scale, killed sixty-seven people, and caused billions of dollars worth of damage.

What is a tsunami?

A tsunami is a giant wave set in motion by a large earthquake occurring under or near the ocean that causes the ocean floor to shift vertically. This vertical shift pushes the water ahead of it, starting a tsunami. These are very long waves (100–200 miles [161–322 kilometers]) with high speeds (500 miles per hour [805 kilometers per hour]) that, when approaching shallow water, can grow into a 100-foot- (30.5-meter-) high wave as its wavelength is reduced abruptly. Ocean earthquakes below a magnitude of 6.5 on the Richter scale, and those that shift the sea floor only horizontally, do not produce these destructive waves. The highest recorded tsunami was 1,719 feet (524 meters) high along Lituya Bay, Alaska, on July 9, 1958. Caused by a giant landslip, it moved at 100 miles (161 kilometers) per hour. This wave would have swamped the Petronas Towers in Kuala Lumpur, Malaysia, which are 1,483 feet (452 meters) high.

Where do most tsunamis occur?

Tsunamis occur most frequently in the Pacific Ocean, although they also occur in the Caribbean and Mediterranean seas and the Atlantic and Indian oceans. The Sumatra tsunami, which followed a strong earthquake (9.0 on the Richter scale) in the Indian Ocean on December 26, 2004, is perhaps the most devastating on record. The tsunami had waves up to 100 feet (30 meters) in height. Damage and deaths were recorded in eleven countries, with a total of more than 230,000 people killed. Indonesia (168,000 people killed), Thailand, Malaysia, Bangladesh, India, Sri Lanka, the Maldives, and the eastern coast of Africa in Somalia, Kenya, and Tanzania all recorded damage and deaths from this event.

When was the earliest tsunami warning system tried in the United States?

The first rudimentary system to alert communities of an impending tsunami was attempted in Hawaii in the 1920s. More advanced systems were developed in the wake of the April 1, 1946, tsunami (caused by the 1946 Aleutian Islands earthquake) and May 23, 1960 (caused by the 1960 Valdiva, Chile, earthquake), tsunamis, which caused massive devastation in Hilo, Hawaii.

VOLCANOES

What is the most famous volcano?

The eruption of Mount Vesuvius in Italy in August 79 C.E. is perhaps the most famous historical volcanic eruption. Vesuvius had been dormant for generations. When it erupted, entire cities were destroyed, including Pompeii, Stabiae, and Herculaneum. Pompeii and Stabiae were buried under ashes, while Herculaneum was buried under a mudflow.

What are the different types of volcanoes?

Volcanoes are usually cone-shaped hills or mountains built around a vent connecting to reservoirs of molten rock, or magma, below Earth's surface. At times, the molten rock is forced upward by gas pressure until it breaks through weak spots in Earth's crust. The magma erupts forth, and lava flows or shoots into the air as clouds of lava fragments, ash, and dust. The accumulation of debris from eruptions causes the volcano to grow in size. Volcanoes are categorized into four groups:

- *Cinder cones* are built of lava fragments. They have slopes of 30–40 degrees and seldom exceed 1,640 feet (500 meters) in height. Sunset Crater in Arizona and Paricutin in Mexico are examples of cinder cones.
- *Composite cones* are made of alternating layers of lava and ash. They are characterized by slopes of up to 30 degrees at the summit, tapering off to 5 degrees at the base. Mount Fuji in Japan and Mount St. Helens in Washington are composite cone volcanoes.
- *Shield volcanoes* are built primarily of lava flows. Their slopes are seldom more than 10 degrees at the summit and 2 degrees at the base. The Hawaiian Islands are clusters of shield volcanoes. Mauna Loa is the world's largest active volcano, rising 13,653 feet (4,161 meters) above sea level.

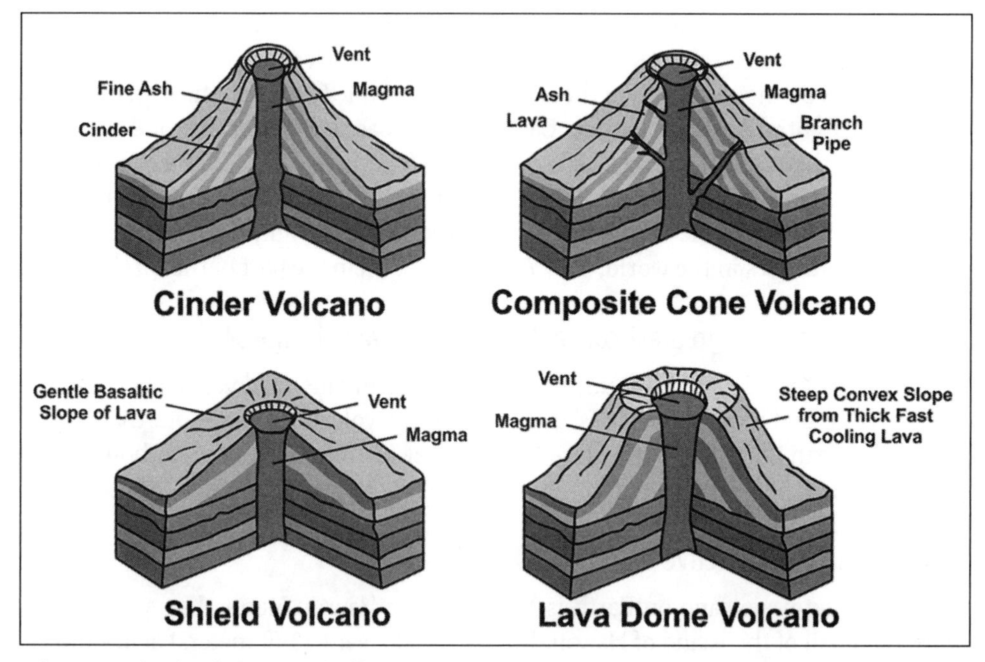

Volcanoes can be classified in several different ways, depending on their structure and how magma is released.

- *Lava domes* are made of viscous, pasty lava squeezed like toothpaste from a tube. Examples of lava domes are Lassen Peak and Mono Dome in California.

Are all craters part of a volcano?

No, not all craters are of volcanic origin. A crater is a nearly circular area of deformed sedimentary rocks with a central, ventlike depression. Some craters are caused by the collapse of the surface when underground salt or limestone dissolves. The withdrawal of groundwater and the melting of glacial ice can also cause the surface to collapse, forming a crater.

How can we determine when ancient volcanic eruptions occurred?

The most common method used to date ancient volcanic eruptions is carbon dating. Carbon dating relies on the rate of radioactive decay of carbon-14. It is used to date eruptions that took place more than two hundred years ago. Charcoal from trees burned during a volcanic eruption is almost pure carbon and is ideal for tracing the tiny amounts of carbon-14.

Where is the Ring of Fire?

The belt of volcanoes bordering the Pacific Ocean is often called the Ring of Fire or the Circle of Fire. Earth's crust is composed of fifteen pieces, called plates, which "float" on the partially molten layer below them. Most volcano and earthquake eruptions and mountain formations occur along the unstable plate boundaries. The Ring of Fire marks the boundary between the plate underlying the Pacific Ocean and the surrounding plates. It runs up the west coast of the Americas from Chile to Alaska (through the Andes Mountains, Central America, Mexico, California, the Cascade Mountains, and the Aleutian Islands), then down the east coast of Asia from Siberia to New Zealand (through the Kamchatka Peninsula, the Kurile Islands, Japan, the Philippines, Sulawesi [formerly Celebes], New Guinea, the Solomon Islands, New Caledonia, and New Zealand). Of the 850 active volcanoes in the world, over 75 percent of them are part of the Circle of Fire.

Which island has the greatest concentration of active volcanoes?

In the early 1990s, 1,133 seamounts (submerged mountains) and volcanic cones were discovered in and around Easter Island. Many of the volcanoes rise more than 1 mile above the ocean floor. Some are close to 7,000 feet (2,134 meters) tall, although their peaks are still 2,500–5,000 feet (760–1,500 meters) below the surface of the sea.

Where is the largest active volcano on Earth?

The largest active volcano on Earth is Mauna Loa in Hawaii. Mauna Loa encompasses more than half of the island of Hawaii. It rises 13,680 feet (4,170 meters) above the Pacific Ocean. In addition, it descends another 16,400 feet (5,000 meters). The sea floor is depressed by Mauna Loa's enormous mass—an additional 26,250 feet (8,000 meters). The combination of height above the ocean, below the ocean, and the depression in the ocean floor yields a total height of 56,000 feet (17,000 meters). Mauna Loa has erupted

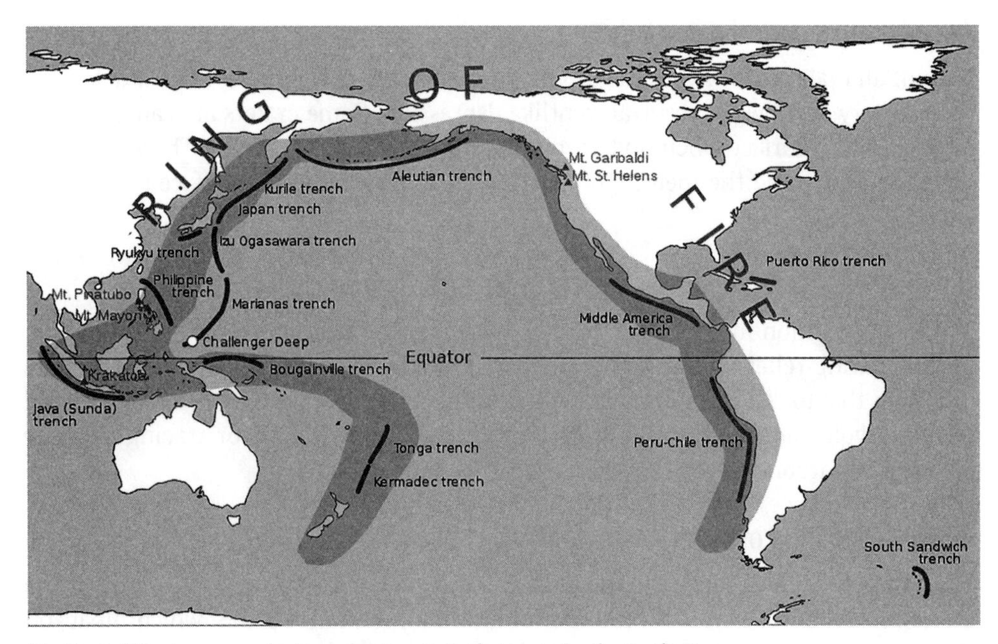

The Ring of Fire is an area of active volcanic activity that encircles the Pacific Ocean.

thirty-three times since 1843, averaging one eruption every five years. The most recent eruption was in 1984. During the 1984 eruption, lava flowed to within 4.5 miles (7.2 kilometers) of the city of Hilo. When Mauna Loa erupts, it produces fast-moving lava, volcanic gas emissions, and potentially destructive earthquakes. It is continuously monitored by the USGS to detect changes that may be indicative of an imminent eruption.

Which volcanoes in the contiguous forty-eight states are considered active and have erupted in the past two hundred years?

Seven major volcanoes in the contiguous forty-eight states are considered active: two in California (Lassen Peak and Mount Shasta), four in Washington (Glacier Peak, Mount Baker, Mount Rainier, and Mount St. Helens), and one in Oregon (Mount Hood).

How long did the 2018 eruption of Kilauea last?

Geologists noticed changes in Kilauea in April 2018. The first cracks and fissures opened in early May 2018. Lava began flowing into the ocean. The eruption continued with more fissures opening, more lava flowing into the ocean, and earthquakes became more frequent, with ash plumes rising many feet into the air. The eruption slowed and was quiet by August 2018. Some facts and statistics about the eruption are:

- More than sixty thousand earthquakes were detected between April 30 and August 4, 2018. The largest volcano had a magnitude of 6.9 on May 4, 2018.
- A total of 13.7 square miles (35.2 square kilometers) of land were inundated by lava.
- 1 billion cubic yards (764,554,858 cubic meters) of lava erupted. This would be enough lava to fill at least 320,000 Olympic-sized swimming pools.
- 875 acres (354 hectares) of new land was created by lava entering the ocean.

When did Mount St. Helens erupt?

Mount St. Helens, located in southwestern Washington State in the Cascade Mountains, erupted on May 18, 1980. Sixty-one people died as a result of the eruption. This was the first known eruption in the forty-eight contiguous United States to claim a human life. Geologists call Mount St. Helens a composite volcano (a steep-sided, often symmetrical cone constructed of alternating layers of lava flows, ash, and other volcanic debris). Composite volcanoes tend to erupt explosively. Mount St. Helens and the other active volcanoes in the Cascade Mountains are a part of the Ring of Fire, the Pacific zone having frequent and destructive volcanic activity. Mount St. Helens erupted again in October 2004. A steam plume billowed 10,000 feet (3,048 meters) into the air. The eruption continued for 2.5 years, building a new lava dome. The new lava dome measured about 125 million cubic yards (95.6 million cubic meters), or a volume equal to nearly two hundred large sports stadiums. The amount of lava that erupted would have been enough to pave seven highway lanes, 3 feet (0.9 meters) thick, from New York City to Portland, Oregon. No loss of life occurred during this eruptive period.

Volcanoes have not only been active in Washington but also in three other U.S. states: California, Alaska, and Hawaii. Lassen Peak is one of several volcanoes in the Cascade Mountains. It last erupted in 1921. Mount Katmai in Alaska had an eruption in 1912, in which the flood of hot ash formed the Valley of Ten Thousand Smokes 15 miles (24 kilometers) away. Hawaii has its famed Mauna Loa, which is the world's largest active volcano, being 60 miles (97 kilometers) in width at its base.

Which volcanoes have been the most destructive?

The five most destructive eruptions from volcanoes since 1700 are as follows:

Volcano	Date of Eruption	Number of Fatalities	Lethal Agent
Mount Tambora, Indonesia	April 5, 1815	92,000	10,000 directly by the volcano; 82,000 from starvation afterward
Krakatoa, Indonesia	August 26, 1883	36,417	90 percent; killed by a tsunami
Mount Pelee, Martinque	August 30, 1902	29,025	Pyroclastic flows
Nevada del Ruiz, Colombia	November 13, 1985	23,000	Mud flow
Unzen, Japan	1792	14,300	70 percent killed by cone collapse; 30 percent by a tsunami

When will Old Faithful erupt again?

We do not know. Future volcanic eruptions could occur within or near Yellowstone National Park for the simple reason that the area has a long volcanic history and because no molten rock, or magma, is beneath the caldera now. USGS (U.S. Geological Survey) scientists monitor Yellowstone for signs of volcanic activity using seismographs to detect earthquakes and GPS to detect ground motion. No evidence exists that a catastrophic eruption at Yellowstone is imminent, and such events are unlikely to occur in the next few centuries. Scientists have also found no indication of an imminent smaller eruption of lava. Yellowstone's two-million-year history of volcanism, the enormous amount of heat that still flows from the ground, the frequent earthquakes, and the repeated uplift and subsidence of the caldera floor also testify to the continuity of magmatic processes beneath Yellowstone and point to the possibility of future volcanism and future earthquake activity.

WATER

What are the major oceans?

Most geographers have recognized four major oceans—Pacific, Atlantic, Indian, and Arctic—for many years. In 2000, the International Hydrographic Organization, an in-

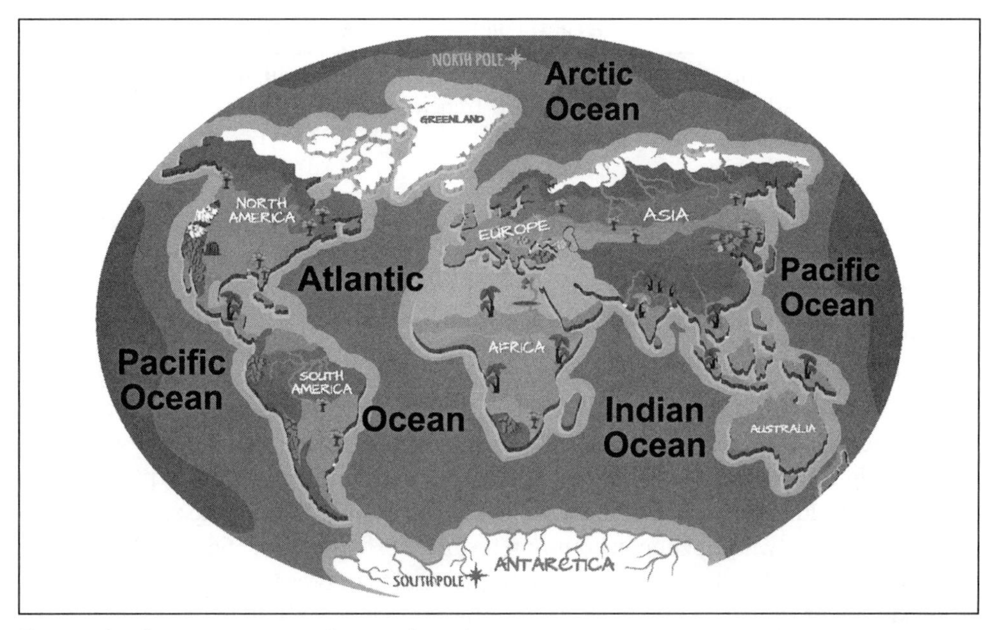

There are four large oceans on our planet, as shown here.

tergovernmental organization dedicated to safety in navigation and support of the marine environment, delimited a fifth ocean called the Southern Ocean, also known as the Antarctic Ocean. The Southern Ocean extends from the coast of Antarctica to 60° south latitude. It encompasses portions of the Atlantic, Indian, and Pacific oceans.

How deep is the ocean?

The average depth of the ocean floor is 13,124 feet (4,000 meters). The average depth of the five major oceans is given below:

Ocean	Average Depth Feet/Meters
Pacific	14,040/4,279
Atlantic	11,810/3,600
Indian	12,800/3,901
Southern	14,450/4,404
Arctic	4,300/1,311

Great variations in depth exist because the ocean floor is often very rugged. The greatest depth variations occur in deep, narrow depressions known as trenches along the margins of the continental plates. The deepest measurements made—36,198 feet (11,034 meters), deeper than the height of the world's tallest mountains—was taken in Mariana Trench east of the Mariana Islands. In January 1960, French oceanographer Jacques Piccard, together with U.S. Navy lieutenant Don Walsh (1931–), took the bathyscaphe *Trieste* to the bottom of the Mariana Trench.

Ocean	Deepest Point	Feet/Meters
Pacific	Mariana Trench	35,840/10,924*
Atlantic	Puerto Rico Trench	28,232/8,605
Indian	Java Trench	23,376/7,125
Arctic	Eurasia Basin	17,881/5,450

*This is the deepest point that the *Trieste* reached, but soundings have occurred as deep as 36,198 feet (11,034 meters).

How far can sunlight penetrate into the ocean?

The ocean is divided into three main zones based on depth and the level of light. The upper level, from 0 to 656 feet (200 meters), is called the epipelagic or, in common terms, the sunlight zone. It is also known as the photic zone since this is where the most sunlight penetrates, allowing for photosynthesis and the growth of plants in the ocean. The middle level is called the mesopelagic zone, or the dysphotic zone. It is often referred to as the twilight zone in common terms. It extends from 656 feet (200 meters) to 3,280 feet (1,000 meters). Some sunlight penetrates to these depths but not enough to sustain plant growth. The bathypelagic zone extends from 3,280 feet (1,000 meters) to the depths of the ocean. It is called the aphotic zone since sunlight cannot penetrate to these depths. In common terms, it is called the midnight zone.

What is the chemical composition of the ocean?

The ocean contains every known naturally occurring element plus various gases, chemical compounds, and minerals. Below is a sampling of the most abundant chemicals.

Common Chemicals in Our Oceans

Chemical	Concentration (parts per million)
Chloride	18,980
Sodium	10,560
Sulfate	2,560
Magnesium	1,272
Calcium	400
Potassium	380
Bicarbonate	142
Bromide	65
Strontium	13
Boron	4.6
Fluoride	1.4

What precious metals are found in the ocean?

Gold and silver are two precious metals that are found in the ocean. One cubic mile of ocean water can contain up to twenty-five pounds of gold and up to forty-five pounds of silver. However, 1 cubic mile of ocean water contains 1,101,117,147,000 gallons (4,168, 181,823,800 liters) of water.

Does ocean water circulate?

Ocean water is in a constant state of movement. Horizontal movements are called currents, while vertical movements are called upwelling and downwelling. Wind, tidal motion, and differences in density due to temperature or salinity are the main causes of ocean circulation. Temperature differentials arise from the equatorial water being warmer than water in the polar regions. In the Northern Hemisphere, the currents circulate in a clockwise direction, while in the Southern Hemisphere, the currents circulate in a counterclockwise direction. In the equatorial regions, the currents move in opposite directions—from left to right in the north and from right to left in the south. Currents moving north and south from equatorial regions carry warm water, while those in polar regions carry cold water.

Major Cold Currents	Major Warm Currents
California	North Atlantic (Gulf Stream)
Humboldt	South Atlantic
Labrador	South Indian
Canary	South Pacific
Benguela	North Pacific
Falkland	Monsoon
West Australian	Okhotsk

What causes waves in the ocean?

The most common cause of surface waves is air movement (the wind). Waves within the ocean can be caused by tides, interactions among waves, submarine earthquakes or volcanic activity, and atmospheric disturbances. Wave size depends on wind speed, wind duration, and the distance of water over which the wind blows. The longer the distance the wind travels over water, or the harder it blows, the higher the waves. As the wind blows over water, it tries to drag the surface of the water with it. The surface cannot move as fast as air, so it rises. When it rises, gravity pulls the water back, carrying the falling water's momentum below the surface. Water pressure from below pushes this swell back up again. The tug-of-war between gravity and water pressure constitutes wave motion. Capillary waves are caused by breezes of less than two knots. At thirteen knots, the waves grow taller and faster than they grow longer, and their steepness causes them to break, forming whitecaps. For a whitecap to form, the wave height must be one-seventh the distance between wave crests.

Why is the ocean salty?

Oceans cover about 70 percent of Earth's surface, and about 97 percent of all water on and in Earth is saline. Basically, all of the salt in the oceans comes from rocks on land.

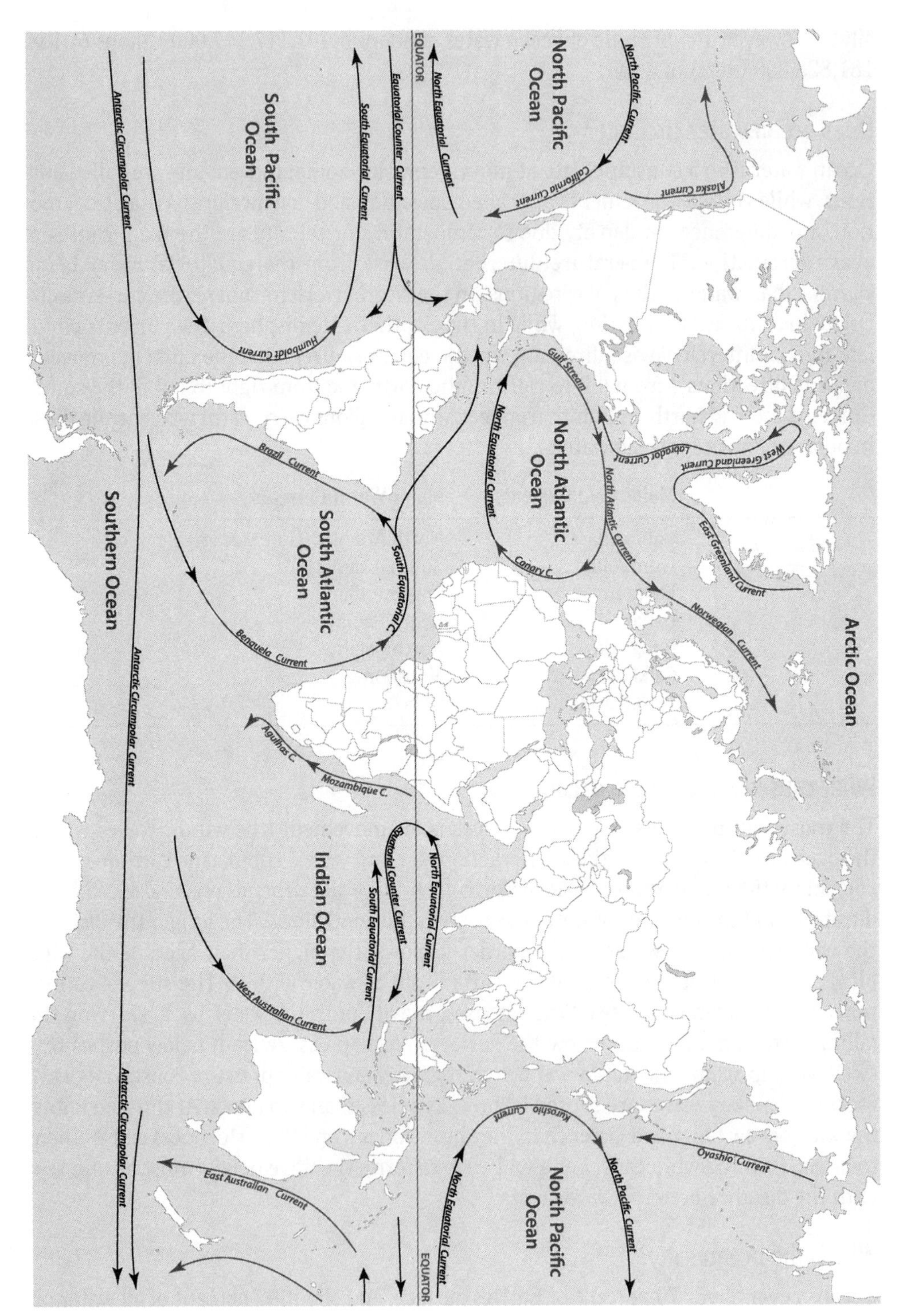

 The waters in our oceans circulate through the actions of large warm and cold currents.

Why is the sea blue?

The colors of the sea have no single cause. What is seen depends in part on when and from where the sea is observed. Eminent authority can be found to support almost any explanation. Some explanations include absorption and scattering of light by pure water; suspended matter in seawater; the atmosphere; and color and brightness variations of the sky. For example, one theory is that when sunlight hits seawater, part of the white light, composed of different wavelengths of various colors, is absorbed, and some of the wavelengths are scattered after colliding with the water molecules. In clear water, red and infrared light are greatly absorbed but blue is least absorbed so that the blue wavelengths are reflected out of the water. The blue effect requires a minimum depth of 10 feet (3 meters) of water.

The rain that falls on the land contains some dissolved carbon dioxide from the surrounding air. This causes the rainwater to be slightly acidic due to carbonic acid. The rain physically erodes the rock, and the acids chemically break down the rocks and carry salts and minerals along in a dissolved state as ions. The ions in the runoff are carried to the streams and rivers and then to the ocean. Many of the dissolved ions are used by organisms in the ocean and are removed from the water. Others are not used up and are left for long periods of time where their concentrations increase over time. The two ions that are present in seawater are chloride and sodium. These two make up over 90 percent of all dissolved ions in seawater.

How salty is seawater?

The concentration of salt in seawater (salinity) is about thirty-five parts per thousand. In other words, about 3.5 percent of the weight of seawater comes from the dissolved salts. In 1 cubic mile of seawater, the weight of the salt, as sodium chloride, would be about 120 million tons. According to one estimate from the National Oceanographic and Atmospheric Administration (NOAA), if the salt in the ocean could be removed and spread evenly over Earth's land surface, it would form a layer more than 500 feet (166 meters) thick, about the height of a forty-story office building.

Is the Dead Sea really dead?

Since the Dead Sea, on the boundary between Israel and Jordan, is the lowest body of water on Earth's surface, any water that flows into it has no outflow. It is called "dead" because its extreme salinity makes any animal or vegetable life impossible except bacteria. Fish introduced into the sea by the Jordan River or by smaller streams die instantly. The only plant life consists primarily of halophytes (plants that grow in salty or alkaline soil). The concentration of salt increases toward the bottom of the lake. The water also has such a high density that bathers float on the surface easily.

What are tides, and where are the world's highest tides?

Tides are the periodic rise and fall of the ocean surface and the alternate submersion and exposure of the intertidal zone along coasts. Tides are caused by the combined effects of the gravitational forces exerted by the Moon and Sun and the rotation of Earth. The Bay of Fundy (New Brunswick, Canada) has the world's highest tides. They average about 45 feet (14 meters) high in the northern part of the bay, far surpassing the world average of 2.5 feet (0.8 meters). The shape and size of the bay, as well as the increasing up-bay shallowness, cause these very high tides.

What are riptides, and why are they so dangerous?

At points along a coast where waves are high, a substantial buildup of water is created near the shore. This mass of water moves along the shore until it reaches an area of lower waves. At this point, it may burst through the low waves and move out from shore as a strong surface current moving at an abnormally rapid speed known as a rip current or riptide. Swimmers who become exhausted in a rip current may drown unless they swim parallel to the shore.

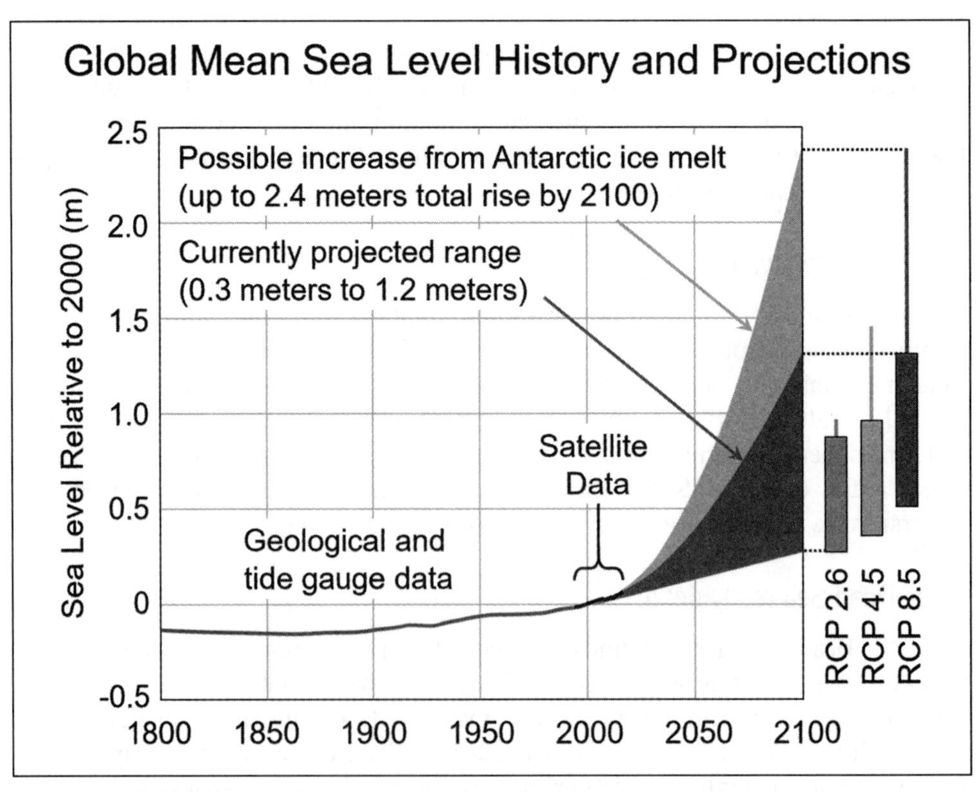

This 2017 graph published by the U.S. Global Change Research Program indicates that our oceans could rise by as much as 2.4 meters (7.87 feet) by 2100. Note: RCP stands for the Representative Concentration Pathway—the trajectory of the concentration of greenhouse gases.

How much would sea level change if glaciers melted?

The USGS predicts that if the glaciers in Antarctica melted, sea level would rise approximately 240 feet (73 meters). Furthermore, the USGS predicts that if all of the glacier ice on Earth (Antarctica, Greenland, Alaska, and the glaciers in the temperate areas) were to melt, sea level would rise approximately 265 feet (80 meters), causing flooding to every coastal city on the planet.

What is the bearing capacity of ice on a lake?

The following chart indicates the maximum safe load. It applies only to clear lake ice that has not been heavily traveled. For early winter slush ice, ice thickness should be doubled for safety.

Ice Thickness (inches/centimeters)	Examples	Maximum Safe Load (tons/kilograms)
2/5	One person on foot	
3/7.6	Group in single file	
7.5/19	Car or snowmobile	2/907.2
8/20.3	Light truck	2.5/1,361
10/25.4	Medium truck	3.5/1,814.4
12/30.5	Heavy truck	9/7,257.6
15/38		10/9,072
20/50.8		25/22,680

Where are the five largest lakes in the world located?

Largest Lakes in the World

Lake (Location)	Area (miles^2/km^2)	Length (miles/kilometers)	Depth (feet/meters)
Caspian Sea (Eurasia)	143,244/370,922	760/1,225	3,363/1,025
Superior (North America)	31,700/82,103	350/560	1,330/406
Victoria (Africa)	26,828/69,464	250/360	27/85
Huron (North America)	23,000/59,570	206/330	750/229
Michigan (North America)	22,300/57,757	307/494	923/281

Which of the U.S. Great Lakes is the largest?

Lake Superior is the largest of the Great Lakes. The North American Great Lakes form a single watershed with one common outlet to the sea—the St. Lawrence Seaway. The total volume of all five basins is 6,000 trillion gallons (22.7 trillion liters), equivalent to about 20 percent of the world's freshwater. Only Lake Michigan lies wholly within the United States's borders; the others share their boundaries with Canada. Some believe that Lake Huron and Lake Michigan are two lobes of one lake since they are the same elevation and are connected by the 120-foot- (36.5-meter-) deep Strait of Mackinac, which is 3.6–5 miles (6–8 kilometers) wide. Gauge records indicate that they both have similar water level regimes and mean long-term behavior so that hydrologically, they act as one lake. Historically, they were considered two by the explorers who named them, but this is considered a misnomer by some.

Lake	Surface Area (miles^2/km^2)	Length (miles/kilometers)	Maximum Depth (feet/meters)	Volume of Water (cubic miles/ cubic kilometers
Superior	31,700/82,103	350/560	1,330/406	2,935/12,234
Michigan	22,300/57,757	307/494	923/281	1,180/4,918
Huron	23,000/59,570	206/330	750/229	850/3,543
Erie	9,910/25,667	241/388	210/64	116/484
Ontario	7,340/19,011	193/311	802/244	393/1,638

What is an aquifer?

An aquifer is basically a reservoir for groundwater. Porous, water-bearing layers of sand, gravel, and rock exist below Earth's surface. Aquifers are always underlain by relatively impermeable layers of rock or clay that keep water from seeping out at the bottom. Aquifers can yield an economically significant amount of water. The High Plains Aquifer, perhaps the largest aquifer in the world, underlies 174,000 square miles (450,000 square kilometers) of land stretching across eight states (Wyoming, South Dakota, Colorado, Nebraska, Kansas, Oklahoma, New Mexico, and Texas). Billions of gallons of water are withdrawn from it each day, mostly for agriculture. Even an aquifer of this size, however, can lose more water through pumping than it gains through replenishment from rain. In 2007, the High Plains Aquifer stored about 2.9 billion acre-feet of water—the equivalent of 2.9 billion acres of water a foot deep. This is a lot of water, but it is some 270 million acre-feet less than the aquifer contained in 1950.

Which are the longest rivers in the world?

The two longest rivers in the world are the Nile in Africa and the Amazon in South America. However, which is the longest is a matter of some debate. Measuring the length of a river is complicated since it is often difficult to pinpoint the exact beginning (source) and end (mouth) of a river. Many rivers meander or are dammed, or the mouth grows

Have the Great Lakes ever been completely frozen?

The greatest maximum ice coverage on the Great Lakes was 94.7 percent, which occurred in 1979. Both Lake Superior and Lake Erie have completely frozen. Lake Superior had 100 percent ice coverage in 1996. Lake Erie had 100 percent ice coverage in 1978, 1979, and 1996.

or recedes. A further complication is that the small tributaries may be dry at certain seasons of the year. Official estimates indicate that the length of the Amazon is between 3,903 miles (6,259 kilometers) and 4,225 miles (6,800 kilometers). Official estimates indicate that the length of the Nile is between 3,347 miles (5,499 kilometers) and 4,180 miles (6,690 kilometers). According to the USGS, the longest river in the United States is the Missouri River at 2,540 miles (4,088 kilometers).

What is an estuary?

Estuaries are places where freshwater streams and rivers flow into the sea. Known as brackish water, the salinity of such areas is less than that of the open ocean but greater than that of a typical river. Organisms living in or near estuaries have special adaptations. Estuaries are rich sources of invertebrates, such as clams, shrimps, and crabs, as well as fish, such as striped bass, mullet, and menhaden.

Some estuaries are entirely freshwater, such as streams and rivers that flow in the Great Lakes. In many cases, the streams and rivers flowing into the lake have different physical and chemical properties than the lake water.

How much salt is in brackish water?

Brackish water has a saline (salt) content between that of freshwater and seawater. It is neither fresh nor salty but somewhere in between. Brackish waters are usually regarded as those containing 0.5–30 parts per thousand salt.

What is a watershed?

A watershed is an area of land that contains a common set of streams and rivers that all drain into a single, larger body of water such as a larger river, a lake, or an ocean. For example, the Mississippi River watershed is an enormous watershed in which all the tributaries to the Mississippi that collect rainwater eventually drain into the Mississippi, which eventually drains into the Gulf of Mexico.

What and where is the Continental Divide of the Americas?

The Continental Divide, also known as the Great Divide, is a continuous ridge of peaks in the Rocky Mountains that marks the watershed separating easterly flowing waters from westerly flowing waters in North America. To the east of the Continental Divide, water drains into Hudson Bay or the Mississippi River before reaching the Atlantic Ocean. To the west, water generally flows through the Columbia River or the Colorado River on its way to the Pacific Ocean.

What is the world's highest waterfall?

Angel Falls, named after the explorer and bush pilot Jimmie Angel (1899–1956), on the Carrao River tributary in Venezuela is the highest waterfall in the world. It has a total height of 3,212 feet (979 meters), with its longest unbroken drop being 2,648 feet (807 meters).

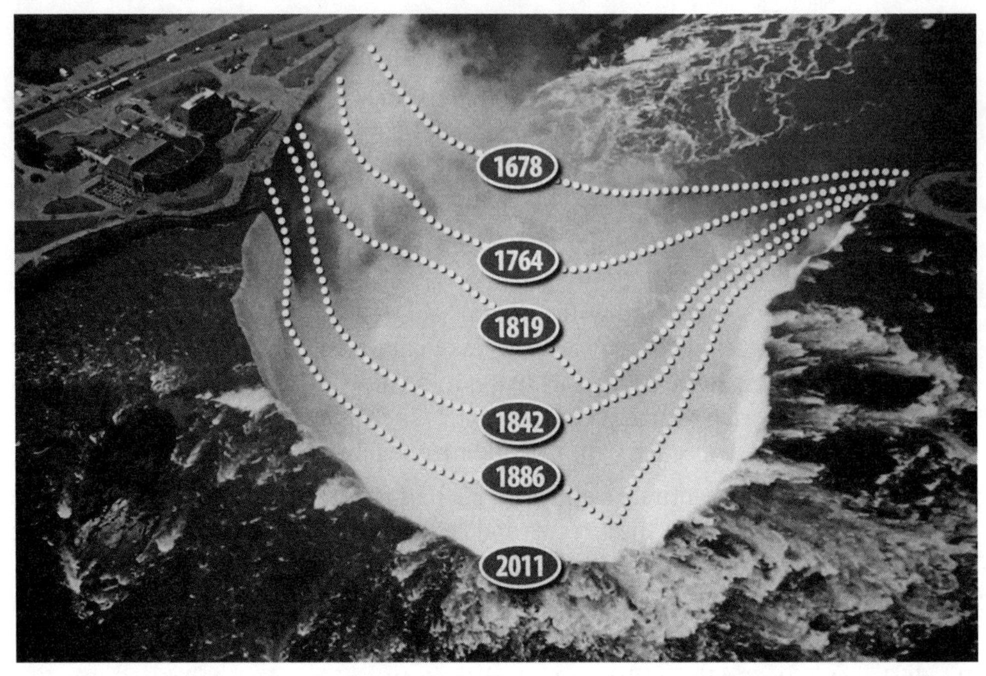

The extent of erosion of Horseshoe Falls since the late seventeenth century is illustrated here. Eventually, the falls will disappear entirely, but not for thousands of years.

It is difficult to determine the height of a waterfall because many are composed of several sections rather than one straight drop. The highest waterfall in the United States is Yosemite Falls on a tributary of the Merced River in Yosemite National Park, California, with a total drop of 2,425 feet (739 meters). Yosemite Falls has three sections: Upper Yosemite Falls is 1,430 feet (435 meters), Cascade Falls (middle portion) is 675 feet (205 meters), and Lower Yosemite Falls is 320 feet (97 meters).

When will Niagara Falls disappear?

The water dropping over Niagara Falls digs great plunge pools at the base, undermining the shale cliff and causing the hard limestone cap to cave in. Niagara has eaten itself 7 miles (11 kilometers) upstream since it formed ten thousand years ago. At this rate, it will disappear into Lake Erie in 22,800 years. The Niagara River connects Lake Erie with Lake Ontario and marks the United States–Canada boundary (New York–Ontario).

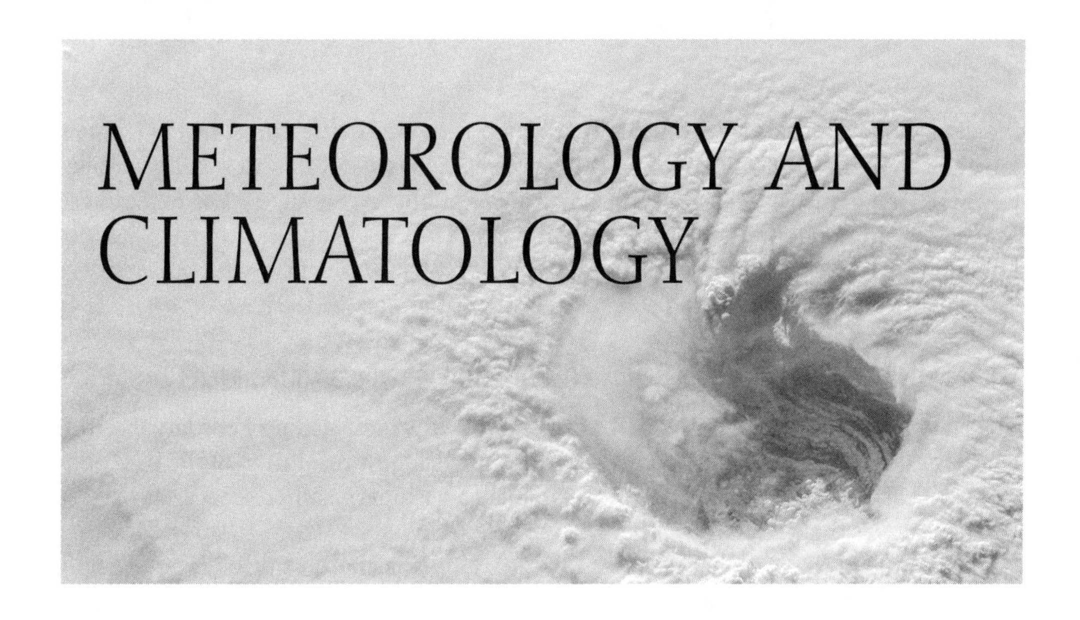

METEOROLOGY AND CLIMATOLOGY

INTRODUCTION AND HISTORICAL BACKGROUND

What is the difference between weather and climate?

Weather is the current condition of the atmosphere. The scientific term for the study of weather and how changes in the weather may be forecasted is meteorology. Climate is the long-term average weather for a particular place or region. Often, a thirty-year average, although also calculated for ten-year averages, climate describes the average weather pattern of a locale. Climatic elements include precipitation, temperature, humidity, sunshine, wind velocity, and other measures of weather such as fog, frost, and other storms. Climatology is the study of the world's climates and how they are changing over time.

What is the purpose of the National Oceanic and Atmospheric Administration (NOAA)?

NOAA, an agency within the U.S. Department of Commerce, is responsible for monitoring conditions on land and in the seas that have an effect on our weather, climate, and environment. NOAA is heavily involved in atmospheric research and weather forecasting, but the agency is also interested in fostering the economic and environmental health of the country, as well as the safety of its citizens, through scientific management of oceanic, coastal, and mainland resources.

When did modern weather forecasting begin?

On May 14, 1692, a weekly newspaper, *A Collection for the Improvement of Husbandry and Trade*, gave a seven-day table with pressure and wind readings for the comparable dates of the previous year. Readers were expected to make up their own forecasts from the data. Other journals soon followed with their own weather features. In 1771, a new journal called the *Monthly Weather Paper* was completely devoted to weather prediction. The first daily newspaper weather report was published on August 31, 1848, in the *Daily News* in London. The first daily weather forecast was published in the *Times* of London in 1860. The first broadcast of weather forecasts was done by the University of Wisconsin's station 9XM in Madison, Wisconsin, on January 3, 1921.

What is the National Weather Service (NWS), and when was it founded?

The National Weather Service is part of the NOAA. It was started on February 9, 1870, when President Ulysses S. Grant (1822–1885) signed a joint resolution of the U.S. Congress authorizing the secretary of war to establish a national weather service. Its first official name was the National Weather Bureau. Cleveland Abbe (1838–1916) was the first chief scientist of the National Weather Bureau. Abbe was chosen since he was the only person in the country with any practical experience with weather forecasting based on scientific principles.

The National Weather Bureau was renamed the U.S. Weather Bureau in 1891 and became the National Weather Service in 1967. Its mission is to provide weather, hydrologic, and climate forecasts and warnings for the citizens of the United States, its territories, adjacent waters, and ocean areas in order to protect life and property and for the enhancement of the national economy. The NWS has forecasting centers in 122 locations around the country, including U.S. territories Guam, American Samoa, and Puerto Rico.

Do specific criteria exist to prompt the National Weather Service to issue an advisory, watch, or warning?

Certain advisories, watches, or warnings are issued when specific conditions are met, while others will vary by location. Examples of when some common National Weather Service advisories, watches, and warnings are issued are:

- Hurricane watch: Issued when a hurricane poses a possible threat to specific coastal areas, generally within 36 hours.
- Hurricane warning: Issued when sustained winds of 74 miles (119 kilometers) per hour or higher, associated with a hurricane, are expected in a specified coastal area in 24 hours or less. A hurricane warning may remain in effect even when winds are below hurricane force when dangerously high water and/or exceptionally high waves continue to persist.
- Severe thunderstorm warning: Issued when either a severe thunderstorm is indicated by radar or a spotter reports a thunderstorm producing hail 0.75 inches (2 centimeters) in diameter and/or winds equal to or exceeding 58 miles (93 kilometers) per hour. Lightning frequency is not a criterion for issuing a severe thunderstorm warning. It is advisable to seek safe shelter immediately when a severe thunderstorm warning is issued.

What is the difference between a National Weather Service statement, advisory, watch, and warning?

The National Weather Service will issue a *statement* as a "first alert" of the possibility of severe weather. An *advisory* is issued when weather conditions are not life threatening, but individuals need to be alert to weather conditions. A weather *watch* is issued when conditions are more favorable than usual for dangerous weather conditions, e.g., tornadoes and violent thunderstorms. A watch is a recommendation for planning, preparation, and increased awareness (i.e., to be alert for changing weather, listen for further information, and think about what to do if the danger materializes). A *warning* is issued when a particular weather hazard is either imminent or has been reported. A warning indicates the need to take action to protect life and property. The type of hazard is reflected in the type of warning (e.g., tornado warning, blizzard warning).

- Tornado watch: Issued when conditions are favorable for the development of tornadoes in the area. They normally are issued well in advance of the actual occurrence of severe weather for a duration of 4–8 hours.
- Tornado warning: Issued when a tornado is indicated by radar or sighted by spotters. People in the affected area should seek safe shelter immediately. They are usually issued for around 30 minutes.
- Wind advisory: Issued when sustained winds occur of 25–39 miles (40–63 kilometers) per hour and/or gusts up to 57 miles (92 kilometers) per hour.
- Winter weather advisory: Issued when a low-pressure system produces a combination of winter weather (snow, freezing rain, sleet) that presents a hazard but does not meet warning criteria.
- Winter storm watch: Issued when the potential exists for heavy snow or significant ice accumulations, usually at least 24–36 hours in advance. The criteria for this watch can vary from place to place.
- Winter storm warning: Issued when a winter storm is producing or is forecast to produce heavy snow or significant ice accumulations. The criteria for this warning can vary from place to place.

What was Aristotle's contribution to the study of weather?

Greek philosopher Aristotle released his book *Meteorologica* around 340 B.C.E., which was the first comprehensive text written on the subject. It was this work that gave us the term "meteorology." In Aristotle's time, the word "meteor" referred not just to extraterrestrial rocks entering the atmosphere but to anything in the sky, including clouds, rain, and snow. Many of the theories expressed in Aristotle's work are based on mythology and other misplaced notions of what causes weather. He believed that hurricanes resulted from a "moral conflict" between "evil" and "good" winds.

What contribution did Theophrastus make to the study of weather?

Aristotle's student Theophrastus of Eresus (c. 372–287 B.C.E.) continued his mentor's study of weather with his *On Weather Signs*, a book that became the last word on weather. It was consulted through the twelfth century as a predictor of weather, including when rain, wind, and storms were coming. The work of Theophrastus was a mix of well-reasoned scientific observation and superstition.

What was Benjamin Franklin's contribution to the study of meteorology?

Benjamin Franklin, who is said to have discovered electricity by flying a kite in a storm, made the important discovery that low-pressure systems caused the atmosphere to circulate in a rotating pattern. He made this discovery in 1743 after unsuccessfully attempting to view an eclipse on October 21. A storm was occurring in Philadelphia at the time, but he later learned that the skies were clear in Boston that day. He then found out that the storm that had been in Philadelphia traveled to Boston the next day. From this information, he surmised that the storm was traveling in a clockwise manner from southwest to northeast. Franklin thus concluded that the low-pressure system was causing the storm to move in this manner.

Which of America's founding fathers were interested in meteorology?

Among his many other interests ranging from agriculture to architecture, law, and politics, Thomas Jefferson (1743–1826) was also fascinated by the weather. Jefferson was offended by French naturalist Georges Louis Leclerc de Buffon's (1707–1788) assertion that Americans were negatively impacted by their climate, making them somehow inferior to Europeans. To prove him wrong, Jefferson and his friend and fellow Founding Father James Madison (1751–1836) decided to study the weather in earnest. Jefferson made daily observations from his Virginia home at Monticello from 1772 to 1778, and Madison followed his lead from 1784 to 1802. While it might seem painfully obvious today, it was Madison who broke with English logic that said that temperature read-

An engraving by artist Alfred Jones showing Benjamin Franklin flying a kite appeared on the U.S. $10 bill from the 1860s to the 1890s.

ings should be done indoors; he took the unheard-of step of placing his thermometer outside. Today, universities are using Madison's measurements of temperature and precipitation for comparative studies on climate change.

Who was named the first official meteorologist of the United States?

James P. Espy (1785–1860) was most noted as the author of *The Philosophy of Storms* (1841). A year after this book's publication, the U.S. Congress named him the federal government's meteorologist. He is credited with giving the first accurate description of how thermodynamics plays a role in cloud formation, also explaining the dynamics of low-pressure systems.

Polymath Sir Francis Galton is credited with developing the first weather map back in 1862.

Who prepared the first weather map?

The first weather map was prepared by Sir Francis Galton (1822–1911) in 1862. His initial map was prepared by requesting detailed information about the weather in the month of December 1861 from weather stations throughout the British Isles. He plotted the data he collected, drawing the world's first weather map.

When was the first weather satellite launched?

The first weather satellite, the Television and Infrared Observation Satellite (TIROS I), was launched by NASA on April 1, 1960. Although the images were not of the same resolution as we have now, they were able to reveal the organization and structure of clouds and storms. One of its accomplishments was to see a previously undetected tropical storm near Australia. The information was conveyed to the people so they could prepare for the approaching storm. It operated for seventy-seven days until mid-June 1960, when an electrical fire caused it to cease operating.

OBSERVATION, MEASUREMENT, AND PREDICTION

How is rainfall measured?

Agencies like the National Weather Service use very accurate devices that measure rainfall to the nearest one-hundredth of an inch (.25 millimeters). The devices, known as rain gauges or tipping-bucket gauges, collect rainwater at a point unaffected by local buildings or trees that may interfere with the rain.

An anemometer is a device for measuring wind speeds.

How is snowfall measured?

Snowfall is measured in a very practical and low-tech way: with a ruler. To get a good average indication of snowfall in a selected area, the National Weather Service takes measurements from several locations, instead of just one, and then averages them out. In places where heavy amounts of snow often occurs, tall poles are erected that can measure snow when it accumulates up to several feet (meters) deep.

What is an anemometer?

An anemometer is a device, invented by Robert Hooke, that measures wind speed. The most commonly used anemometer uses three or four small cups that spin around a central pole. As the cups spin, a switch within the central pole detects each time a magnet in a cup swings by. This generates an electronic pulse that has been calibrated to calculate wind speed, which is then transmitted to a weather station.

How is wind direction measured?

A wind vane is the most common instrument used to discover wind direction. Wind vanes look like windmills mounted on a pole that allows them to rotate toward the direction of the incoming wind. Older wind vanes frequently had a rooster or other farm animal mounted on top. Newer methods of measuring wind directions may use Doppler sodar (sound radar) or lidar (light radar).

What is the Beaufort scale?

The Beaufort scale was devised in 1805 by British admiral Sir Francis Beaufort (1774–1857) to help mariners in handling ships. It uses a series of numbers from 0 to 17 to indicate wind speeds and applies to both land and sea.

Beaufort Number	Name	Wind Speed (mph/kph)
0	Calm	< 1/< 1.5
1	Light air	1–3/1.5–4.8
2	Light breeze	4–7/6.4–11.3
3	Gentle breeze	8–12/12.9–19.3

Beaufort Number	Name	Wind Speed (mph/kph)
4	Moderate breeze	13–18/21–29
5	Fresh breeze	19–24/30.6–38.6
6	Strong breeze	25–31/40.2–50
7	Moderate gale	32–38/51.5–61.1
8	Fresh gale	39–46/62.8–74
9	Strong gale	47–54/75.6–86.9
10	Whole gale	55–63/88.5–101.4
11	Storm	64–73/103–117.5
12–17	Hurricane	74+/119.1+

Who invented the hygrometer?

A hygrometer is an instrument that measures humidity or the amount of water vapor in the air. Leonardo da Vinci (1452–1519) is often credited with constructing the first hygrometer. Hygrometers come in two types: 1) dry and wet bulb psychrometers and 2) mechanical hygrometers. The first type uses dry and wet bulb thermometers to compare temperature changes resulting from humidity; the second type uses either organic material (blonde human hairs), which expand or contract based on humidity levels, or semiconductors made of lithium chloride or other substances, whose degrees of electrical resistance change according to humidity. Francesco Folli (1624–1685) improved upon da Vinci's hygrometer in 1664. Swiss physicist and geologist Horace Bénédict de Saussure (1740–1799) designed the first mechanical hygrometers.

What is a barometer, and who invented it?

A barometer is a device that measures air pressure. A standard barometer consists of a glass tube filled with mercury (a liquid metal) that is inserted into a reservoir, which also contains mercury. When the surrounding air pressure exerts more weight on the reservoir than the mercury in the tube does, the mercury level rises, and vice versa. Evangelista Torricelli (1608–1647) invented the barometer in 1644.

What is the Fujita and Pearson Tornado Scale?

The Fujita and Pearson Tornado Scale was developed in 1971 by University of Chicago professor T. Theodore Fujita (1920–1998) and Allen Pearson (1925–2016), who was then the director of the National Severe Storms Forecast Center. It ranked tornadoes by their

What are thermographs and hygrothermographs?

A thermograph is a device that records changes in temperature over time by drawing a line over a rotating chart. A hygrothermograph does the same thing, except it also indicates levels of humidity.

wind speed, path, length, and width. Tornadoes are not assessed based on actual wind speed and damage but, rather, the scale determines wind speed based on damage. Sometimes known simply as the Fujita Scale, the rankings ranged from F0 (very weak) to F6 (inconceivable).

- F0—Light damage: damage to trees, billboards, and chimneys.
- F1—Moderate damage: mobile homes pushed off their foundations and cars pushed off roads.
- F2—Considerable damage: roofs torn off, mobile homes demolished, and large trees uprooted.
- F3—Severe damage: even well-constructed homes torn apart, trees uprooted, and cars lifted off the ground.
- F4—Devastating damage: houses leveled, cars thrown, and objects become flying missiles.
- F5—Incredible damage: structures lifted off foundations and carried away; cars become missiles. Less than 2 percent of tornadoes are in this category.
- F6—An F6 tornado has never been recorded, but we surmise that the damage would be devastating.

Fujita and Pearson Tornado Scale

Scale	Speed (mph/kph)	Path Length (miles/km)	Path Width
0	< 72/< 116	< 1.0/< 1.6	< 17 yards/< 15.5 meters
1	73–112/117–180	1.0–3.1/1.6–4.99	18–55 yards/16.5–50.3 meters
2	113–157/182–253	3.2–9.9/5.15–15.93	56–175 yards/51.2–160 meters
3	158–206/254–332	10.0–31.0/16–49.89	176–556 yards/160.9–508.4 meters
4	207–260/333–418	32.0–99.0/51.5–159.33	0.34–0.9 miles/0.55–1.45 km
5	261–318/420–512	100–315/160–506.94	1.0–3.1 miles/1.6–4.99 km
6	319–380/513–612	316–999/508.55–1,607.73	3.2–9.9 miles/5.15–15.9 km

How does the Enhanced Fujita Scale differ from the original Fujita Scale?

The National Weather Service adapted the Enhanced Fujita (EF) Scale on February 1, 2007, to rate tornadoes. The enhanced scale has six categories, EF0 to EF5, representing increasing levels of damage. It was revised to better estimate wind speeds by considering different types of construction and low-populated areas with few structures. The enhanced scale offers more detailed descriptions of potential damages by using twenty-eight damage indicators based on different building structures and vegetation.

Enhanced Fujita Scale

Scale	Wind Speed (mph/kph)	Damages
EF0	65–85/105–137	Tree branches break off, trees with shallow roots fall over; house siding and gutters damaged; some roof shingles peel off or other minor roof damage.

Scale	Wind Speed (mph/kph)	Damages
EF1	86–110/137–177	Mobile homes overturned; doors, windows, and glass broken; severe damage to roofs.
EF2	111–135/178–217	Large tree trunks split and big trees fall over; mobile homes destroyed and homes on foundations are shifted; cars lifted off the ground; roofs torn off; some lighter objects thrown at the speed of missiles.
EF3	136–165/218–265	Trees broken and debarked; mobile homes completely destroyed, houses on foundations lose stories, and buildings with weaker foundations are lifted and blown distances; commercial buildings such as shopping malls are severely damaged; heavy cars are thrown and trains are tipped over.
EF4	166–200/266–322	Frame houses leveled; cars thrown long distances; larger objects become dangerous projectiles.
EF5	>200/>322	Homes are completely destroyed and even steel-reinforced buildings are severely damaged; objects the size of cars are thrown distances of 300 feet (90 meters) or more. Total devastation.

What is Doppler radar?

Doppler radar measures frequency differences between signals bouncing off objects moving away from or toward it. By measuring the difference between the transmitted and received frequencies, Doppler radar calculates the speed of the air in which the rain, snow, ice crystals, and even insects are moving. It can then be used to predict speed and direction of wind and amount of precipitation associated with a storm. The National Weather Service has installed a series of NEXRAD (Next Generation Radar) Doppler radar systems throughout the country. They are especially helpful in measuring the speed of tornadoes and other violent thunderstorms.

How accurate is *The Old Farmer's Almanac* in predicting weather?

The Farmer's Almanac, published in Lewiston, Maine, and *The Old Farmer's Almanac*, published in Dublin, New Hampshire, both

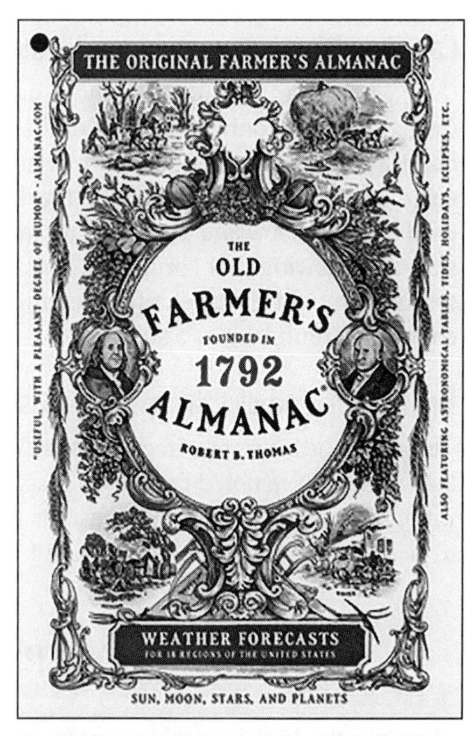

The Old Farmer's Almanac is a compilation of tips and forecasts, recipes, folklore, sports, and astronomy. Each annual also includes general predictions about the year's upcoming weather.

How can the temperature be determined from the frequency of cricket chirps?

Listen for the chirping of either katydids or crickets, then count the number of chirps you hear in 1 minute. For the following equations, "C" equals the number of chirps per minute:

- For katydids, Fahrenheit temperature = 60 + (C – 19)/3
- For crickets, Fahrenheit temperature = 50 + (C – 50)/4

make predictions about the weather for the coming year when they are published. Each book claims to have secret formulas to predict the weather. *The Old Farmer's Almanac* claims it has an accuracy rate of 80 percent. Prior to modern forecasting techniques using radar, satellites, and computer simulation, many people relied on the almanacs for long-range weather forecasts. However, many of the predictions in the almanacs are very general in nature; for example, a five-day period in November may be predicted to be sunny/cool without a range of expected temperatures or whether it will be sunny each of the five days.

Can groundhogs accurately predict the weather?

Over a sixty-year period, groundhogs have accurately predicted the weather (i.e., when spring will start) only 28 percent of the time on Groundhog Day, February 2. Groundhog Day was first celebrated in Germany, where farmers would watch for a badger to emerge from winter hibernation. If the day was sunny, the sleepy badger would be frightened by his shadow and duck back for another six weeks' nap; if it was cloudy, he would stay out, knowing that spring had arrived. German farmers who immigrated to Pennsylvania brought the celebration to America. Finding no badgers in Pennsylvania, they chose the groundhog as a substitute.

Do trees that predict the weather exist?

Observing the leaves of a tree may be an old-fashioned method of predicting the weather, but farmers have noted that when maple leaves curl and turn bottom up in a blowing wind, rain is sure to follow. Woodsmen claim they can tell how rough a winter is going to be by the density of lichens on a nut tree.

TEMPERATURE

What are the highest and lowest recorded temperatures on Earth?

For ninety years, the highest temperature in the world was an observation of 136°F (58°C) at Al Aziziyah (el-Aziia), Libya, on September 13, 1922. However, the World

Is a halo around the Sun or Moon a sign of rain or snow approaching?

The presence of a ring around the Sun or, more commonly, the Moon in the night sky indicates very high ice crystals composing cirrostratus clouds. The brighter the ring, the greater the odds of precipitation and the sooner it may be expected. Rain or snow will not always fall, but two times out of three, precipitation will start to fall within 12–18 hours. These cirroform clouds are a forerunner of an approaching warm front and an associated low-pressure system.

Meteorological Organization spent two years (2010–2012) reviewing the data and rejected this claim of the highest recorded temperature. Some of the factors leading to this decision included problematic instrumentation, an inexperienced observer, and an observation site not representative of the desert surroundings. The official highest temperature recorded in the world was 134°F (56.7°C) in Death Valley, California, on July 10, 1913.

The lowest temperature was –128.6°F (–89.6°C) at Vostok Station in Antarctica on July 21, 1983. The record cold temperature for an inhabited area was –90.4°F (–68°C) in Oymyakon, Siberia (population 4,000) on February 6, 1933. This temperature tied with the readings at Verkhoyansk, Siberia, on January 3, 1885, and February 5 and 7, 1892. The lowest temperature reading in the United States was –79.8°F (–62.1°C) on January 23, 1971, in Prospect Creek, Alaska; for the contiguous forty-eight states, the coldest temperature was –69.7°F (–56.5°C) in Rogers Pass, Montana, on January 20, 1954.

What is the hottest place in the United States?

The most inhospitably hot and dry place in America is Death Valley in eastern California. Also the lowest place at 280 feet below sea level, Death Valley regularly experiences temperatures in the 120s Fahrenheit (high 40s to low 50s Celsius) and higher, and it sees temperatures above 100°F (37.8°C) between 140 and 160 days every year, even in the shade. The record high here has been tallied at 134°F (56.7°C). Annual rainfall in Death Valley is less than two inches (50 millimeters).

What is a heat wave?

A heat wave is a period of two days in a row when apparent temperatures on the National Weather Service heat index exceed 105–110°F (40–43°C). Heat waves can be extremely dangerous. According to the National Weather Service, 175–200 Americans die from heat in a normal summer. Between 1936 and 1975, as many as fifteen thousand Americans died from problems related to heat. In metropolitan areas, large concentrations of buildings, parking lots, and roads create an "urban heat island" in cities with higher temperatures than the surrounding open, rural areas.

What is the heat index?

The index is a measure of what hot weather feels like to the average person for various temperatures and relative humidities. Heat exhaustion and sunstroke are inclined to happen when the heat index reaches 105°F (40°C). The chart below provides the heat index for some temperatures and relative humidities.

How Relative Humidity Affects Temperature

Humidity Temp (°F/°C)	0% Feels Like (°F/°C)	10%	20%	30%	40%	50%	60%	70%	80%	90%	100%
70/21	67/19	67/20	68/20	68/20	69/20	69/21	70/21	70/21	70/21	71/21	71/22
75/24	72/22	73/23	73/23	74/23	74/23	75/24	75/24	75/24	76/24	76/25	77/25
80/27	78/25	78/26	79/26	79/26	80/27	81/27	82/28	83/28	84/29	86/30	89/32
85/29	80/27	81/27	82/28	83/28	84/29	86/30	89/32	93/34	97/36	102/39	108/42
90/32	84/29	85/30	86/30	88/31	91/33	95/35	100/38	106/41	113/45	122/50	132/55
95/35	87/31	89/32	91/33	94/35	99/37	105/41	113/45	123/50	134/57		
100/38	91/33	94/35	97/36	102/39	109/43	118/48	129/54				
105/41	95/35	99/37	104/40	112/44	121/50	134/57					
110/43	99/37	104/40	112/44	122/50	136/58						
115/46	103/39	110/43	121/49	134/57							
120/49	105/41	116/46	130/54								

Is the world actually getting warmer?

Two sustained periods of warming have occurred in history, one beginning around 1910 and ending around 1945 and the most recent beginning in about 1976. The yearly global temperature has increased at an average rate of 0.13°F (0.07°C) per decade since 1880. Since 1980, the average rate of increase has been twice as great. The following chart indicates the top ten warmest years on record since 1880 according to NOAA:

Warmest Years on Record

Year	°F/°C above Twentieth-Century Average
2016	1.69/0.94
2015	1.62/0.90
2017	1.51/0.84
2018	1.42/0.79
2014	1.33/0.74
2010	1.26/0.70
2013	1.21/0.67
2005	1.19/0.66
2009	1.15/0.64
1998	1.13/0.63
2012	1.12/0.62

According to NASA, 2018 was the fourth warmest year on record (1.42 °F [0.79 °C] above the twentieth-century average). Different methodologies yield differences in re-

Why was 1816 known as the year without a summer?

The eruption of Mount Tambora, a volcano in Indonesia, in 1815 threw billions of cubic yards of dust over 15 miles (24 kilometers) into the atmosphere. Because the dust penetrated the stratosphere, wind currents spread it throughout the world. As a consequence of this volcanic activity, in 1816, normal weather patterns were greatly altered. Some parts of Europe and the British Isles experienced average temperatures 2.9–5.8°F (1.6–3.2°C) below normal. In New England, heavy snow fell between June 6 and June 11, and frost occurred in every month of 1816. Crop failures occcured in Western Europe and Canada as well as in New England. By 1817, the excess dust had settled, and the climate returned to more normal conditions.

sults between NOAA and NASA. They both agree that the five warmest years on record have all occurred since 2010.

Which place has the maximum amount of sunshine in the United States?

Yuma, Arizona, has an annual average of 90 percent of sunny days, or over 4,000 sunny hours per year. St. Petersburg, Florida, had 768 consecutive sunny days from February 9, 1967, to March 17, 1969. On the other extreme, the South Pole has no sunshine for 182 days annually, and the North Pole has none for 176 days per year.

Does the North Pole have the same average temperature as the South Pole?

One might think that the northern and southern extremes of the planet would have comparable annual temperatures. In actuality, the South Pole is much colder. This is because giant glaciers have built up on the continent of Antarctica, where the South Pole is located. The inland ice sheets of Antarctica are 8,036 feet (2,450 meters) thick on average, and this thick layer of ice keeps the South Pole about 50°F (23.6°C) colder than the North Pole.

A 2013 photograph of the Mount Tambora caldera proves the volcano is still active. Its major eruption in 1815 threw so much dust into the atmosphere that it cooled the entire planet.

How warm can it get in Antarctica?

Antarctica is not always an icebox perpetually below freezing. During the summer months, it can often get into the 40s and 50s Fahrenheit (4–14°C). The warmest temperature on record thus far on the continent is 59°F (15°C) at Vanda Station on January 5, 1974.

What is the lowest temperature possible?

According to physics, the lowest temperature possible is zero degrees Kelvin (–459.67°F or –273.15°C), which is also known as absolute zero. This is the temperature at which molecular activity stops and it is not possible for any more heat energy to be lost. Such an extreme does not exist naturally on Earth, except in some university laboratories. The lowest that air temperatures can go on the planet is about –130°F (–90°C). An unverified reading of –132°F (–91°C) was reported in 1997 at Vostok Station in Antarctica.

What is wind chill, and who coined the term?

The term "wind chill" was coined by Paul A. Siple (1908–1969) to describe variations in human comfort caused by wind at cold temperatures. He specifically defined wind chill as the measure of the quantity of heat that the atmosphere is capable of absorbing within 1 hour from an exposed surface 1 meter square. Siple was the youngest member of Rear Admiral Richard Byrd's (1888–1957) Antarctica expedition from 1928 to 1930. He first used the term "wind chill" in his 1939 dissertation "Adaptions of the Explorer to the Climate of Antarctica" at Clark University in Worcester, Massachusetts. Siple continued to conduct experiments that involved the relationship between temperature and wind speed and the freezing of water and related areas.

THE ATMOSPHERE

What is the composition of Earth's atmosphere?

Earth's atmosphere, apart from water vapor and pollutants, is composed of 78 percent nitrogen, 21 percent oxygen, less than 1 percent each of argon and carbon dioxide, and traces of hydrogen, neon, helium, krypton, xenon, methane, and ozone. Earth's original atmosphere was probably composed of ammonia and methane; twenty million years ago, the air started to contain a broader variety of elements. The atmosphere weighs approximately five million billion tons. It exerts an average of 14.7 pounds per square inch (PSI) of pressure on the surface of the planet.

What are the layers of Earth's atmosphere?

The atmosphere, the "skin" of gas that surrounds Earth, consists of six layers that are differentiated by temperature:

- The *troposphere* is the lowest level; it averages about 7 miles (11 kilometers) in thickness, varying from 5 miles (8 kilometers) at the poles to 10 miles (16 kilometers) at the equator. Most clouds and weather form in this layer. Temperature decreases with altitude in the troposphere.

- The *stratosphere* ranges between 7 and 30 miles (11–48 kilometers) above Earth's surface. The ozone layer, important because it absorbs most of the Sun's harmful ultraviolet radiation, is located in this band. Temperatures rise slightly with altitude to a maximum of about 32°F (0°C).
- The *mesosphere* (above the stratosphere) extends from 30 to 55 miles (48–85 kilometers) above Earth. Temperatures decrease with altitude to –130°F (–90°C).
- The *thermosphere* (also known as the heterosphere) is between 55 and 435 miles (85–700 kilometers) above Earth. Temperatures in this layer range up to 2,696°F (1,475°C).

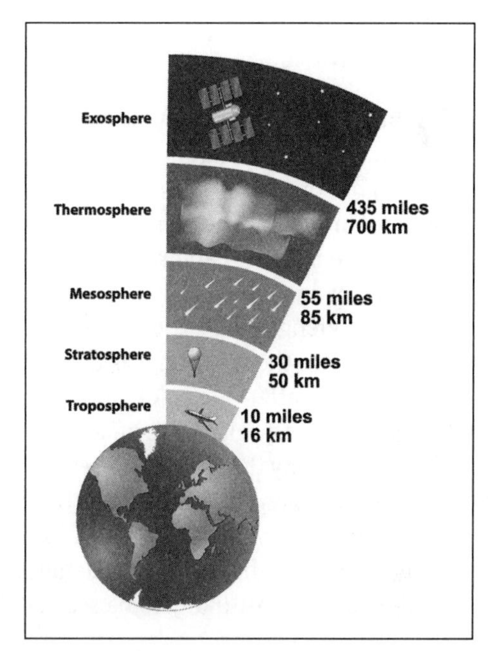

Meteorologists distinguish between five layers of Earth's atmosphere.

- The *exosphere*, beyond the thermosphere, applies to anything over 435 miles (700 kilometers) above Earth. In this layer, temperature no longer has any meaning.
- The *ionosphere* is a region of the atmosphere that overlaps the others, reaching from 30 to 250 miles (48–402 kilometers). In this region, the air becomes ionized (electrified) from the Sun's ultraviolet rays. This area affects the transmission and reflection of radio waves. It is divided into three regions: the D region (at 35–55 miles [56–88 kilometers]), the E region (Heaviside–Kennelly layer, 55–95 miles [88–153 kilometers]), and the F region (Appleton–Barnett layer, 95–250 miles [153–402 kilometers]).

What is the ozone layer, and who discovered it?

The ozone layer is part of the stratosphere, a layer of Earth's atmosphere that lies about 10–30 miles (16–48 kilometers) above the surface of Earth. Ozone (O_3) is like regular gaseous oxygen (O_2) with an extra oxygen atom attached to it. It is created when short-wavelength ultraviolet radiation interacts with O_2 molecules. The energy from the radiation breaks the molecules apart, which then recombine into ozone. The ozone layer is important because it protects life on Earth from harmful ultraviolet radiation. While it does not absorb *all* of this radiation (otherwise, it would be impossible for you to get a tan!), it prevents about 80 percent of it from reaching life on Earth. Too much ultraviolet radiation can cause skin cancers. In 1913, French physicists Henri Buisson (1873–1944) and Charles Fabry (1867–1945) theorized that an ozone layer existed in the upper

Why is the sky blue?

The sunlight interacting with Earth's atmosphere makes the sky blue. In outer space, the astronauts see blackness because outer space has no atmosphere. Sunlight consists of light waves of varying wavelengths, each of which is seen as a different color. The minute particles of matter and molecules of air in the atmosphere intercept and scatter the white light of the Sun. A larger portion of the blue color in white light is scattered, more so than any other color because the blue wavelengths are the shortest. When the size of atmospheric particles are smaller than the wavelengths of the colors, selective scattering occurs—the particles only scatter one color, and the atmosphere will appear to be that color. Blue wavelengths especially are affected, bouncing off the air particles to become visible. This is why the Sun looks yellow (yellow equals white minus blue). At sunset, the sky changes color because as the Sun drops to the horizon, sunlight has more atmosphere to pass through and loses more of its blue wavelengths. The orange and red, having the longer wavelengths and making up more of the sunlight at this distance, are most likely to be scattered by the air particles.

atmosphere. It was confirmed in a series of measurements of ultraviolet radiation levels that were recorded by Walter Noel Hartley (1847–1913) and Marie Alfred Cornu (1841–1902) from 1879 to 1881.

Is ozone beneficial or harmful to life on Earth?

Ozone, a form of oxygen with three atoms instead of the normal two, is highly toxic; less than one part per million of this blue-tinged gas is poisonous to humans. In Earth's upper atmosphere (stratosphere), it is a major factor in making life on Earth possible. About 90 percent of the planet's ozone is in the ozone layer. The ozone belt shields and filters Earth from excessive ultraviolet (UV) radiation generated by the Sun. Scientists predict that a diminished or depleted ozone layer could lead to increased health problems for humans, such as skin cancer, cataracts, and weakened immune systems. Increased UV can also lead to reduced crop yields and disruption of aquatic ecosystems, including the marine food chain. While beneficial in the stratosphere, near ground level, it is a pollutant that helps form photochemical smog and acid rain.

What is the ozone hole?

The ozone hole is not technically a "hole" where no ozone is present but is actually a region with extremely low amounts of ozone in the stratosphere over the Antarctic. It occurs annually at the beginning of Southern Hemisphere spring (August to October). Satellite instruments provide daily images of ozone over the Antarctic region. From the historical record, the total ozone values of less than 220 Dobson units were not observed prior to 1979. Readings of less than 220 Dobson units are due to catalyzed ozone loss from chlorine, bromine, and related compounds.

Why does a hole exist in the ozone layer?

The ozone layer is not evenly distributed around Earth. It is thicker around the equator and nearby latitudes and thinner as one progresses north and south. At the poles, the atmosphere is thinner, including the ozone layer. In addition, because ozone is dependent upon the interaction of sunlight and oxygen, naturally less ozone is at the poles. Since 1975, scientists believe that more than 33 percent of the ozone layer has disappeared. The reduction in ozone at any given time during the year also has a seasonal factor. Scientists also know that chlorofluorocarbons (CFCs), which are used for air conditioning, aerosol sprays, and halon in fire extinguishers, along with methane (CH_4), affect the ozone layer.

How large is the Antarctic ozone hole?

The term "hole" is widely used in popular media when reporting on ozone. However, the concept is more correctly described as a low concentration of ozone that occurs from August through October (springtime in the Southern Hemisphere). It was not observed until 1979. The first scientific article on ozone depletion in the Antarctic was published in *Nature* in 1985. The largest single-day ozone hole recorded by satellite was 11.5 million square miles (29.9 million square kilometers) on September 9, 2000. The largest ozone hole ever observed, 11.4 million square miles (29.6 million square kilometers), occurred on September 24, 2006. The single-day maximum ozone hole area for 2018 was

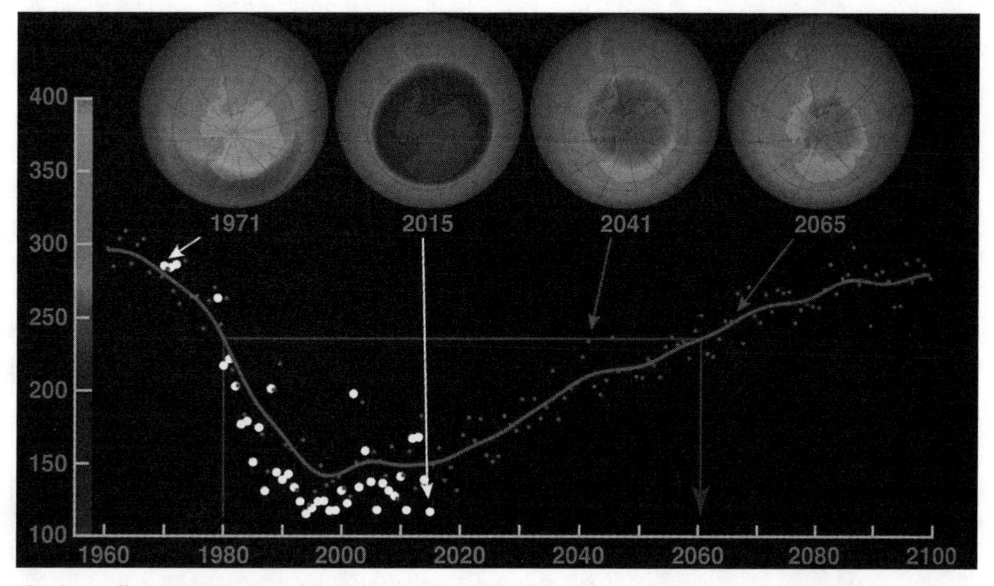

Thanks to efforts to reduce harmful chemicals from being released into the atmosphere, the hole in the protective ozone layer has started to shrink and should recover completely in the next fifty to seventy years.

9.6 million square miles (24.8 million square kilometers). The 2018 ozone hole was slightly larger than the average for 2013–2017 but not as large as it would have been if chlorine levels hadn't decreased thanks to the Montréal Protocol.

How was the ozone hole discovered?

The famous meteorologist Gordon Miller Bourne Dobson (1889–1976) was the first to make accurate measurements of the ozone layer in the 1920s, but it was not until 1979 that the depletion of ozone was observed at the South Pole by the Nimbus 7 satellite. Today, a network of Dobson spectrophotometers has been set up around the world to monitor the changes in the ozone.

What is the jet stream, and when was it discovered?

The jet stream was discovered by World War II bomber pilots flying over Japan and the Mediterranean Sea. Their planes were capable of cruising at over 30,000 feet (9,144 meters), and the pilots discovered the effects of the jet stream on their aircraft. The jet stream is a flat and narrow band of air that moves more rapidly than the surrounding air. It is located high in the atmosphere and affects the movements of storms and air masses closer to the ground. The currents of air flow from west to east and are usually a few miles deep, up to 100 miles (160 kilometers) wide, and well over 1,000 miles (1,600 kilometers) in length. The air current flows at over 57.5 miles (92 kilometers) per hour, sometimes moving as fast as 230 miles (386 kilometers) per hour.

One polar jet stream is located in each hemisphere. Each moves between 30 and 70° latitude, occurs at altitudes of 25,000–35,000 feet (7,620–10,668 meters), and achieve maximum speeds of over 230 miles (368 kilometers) per hour. The subtropical jet streams (again, one per hemisphere) wander between 20 and 50° latitude. They are found at altitudes of 30,000–45,000 feet (9,144–13,715 meters) and have speeds of over 345 miles (552 kilometers) per hour.

What is a front?

A front is any boundary existing between differing masses of air—that is, masses that have different overall temperatures and humidity. Fronts are categorized into four types:

What is the Coriolis effect?

The nineteenth-century French engineer Gaspard Gustav de Coriolis discovered that the rotation of Earth deflects streams of air. Because Earth spins to the east, all moving objects in the Northern Hemisphere tend to turn somewhat to the right of a straight path, while those in the Southern Hemisphere turn slightly left. The Coriolis effect explains the lack of northerly and southerly winds in the tropics and polar regions; the northeast and southeast trade winds and the polar easterlies all owe their westward deflection to the Coriolis effect.

warm, cold, stationary, and occluded. A warm front is a front that is advancing over a colder mass of air, while a cold front is just the opposite. Stationary fronts are boundaries between warm and cold air that are in relative equilibrium but can still move back and forth by as much as several hundred miles. An occluded front occurs whenever a cold front, instead of merely overtaking warm air, actually separates and breaks apart warm masses of air.

What is barometric pressure, and what does it mean?

Barometric, or atmospheric, pressure is the force exerted on a surface by the weight of the air above that surface as measured by an instrument called a barometer. Pressure is greater at lower levels because the air's molecules are squeezed under the weight of the air above, so while the average air pressure at sea level is 14.7 pounds per square inch (1,013.53 hecto Pascals), at 1,000 feet (304 meters) above sea level, the pressure drops to 14.1 pounds per square inch (972.1 hecto Pascals), and at 18,000 feet (5,486 meters), the pressure is 7.3 pounds (503.32 hecto Pascals), about half of the figure at sea level. Changes in air pressure bring weather changes. High-pressure areas bring clear skies and fair weather; low-pressure areas bring wet or stormy weather. Areas of very low pressure have serious storms, such as hurricanes.

Where is the polar vortex usually located?

The polar vortex is an area of low pressure that is a mass of very cold, swirling air. It is usually found over the polar regions—both the south and the north. If the winds of the arctic polar vortex weaken, the shape of the polar vortex may become distorted, sending blasts of very cold air to more southern regions of the United States.

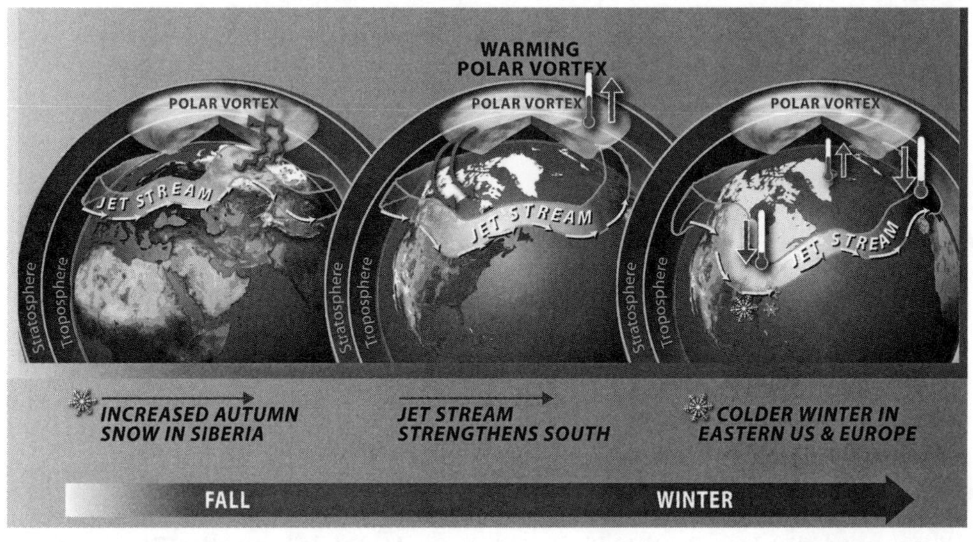

Global warming has increased the frequency of polar vortexes during the winter months, a phenomenon that has hit North America particularly hard with sub-freezing temperatures and heavy snowfalls.

Where do haboobs occur?

Haboobs, derived from the Arabic word *habb*, meaning "to blow," are violent dust storms with strong winds of sand and dust. They are most common in the Sahara region of Africa and the deserts of southwestern United States, Australia, and Asia.

What is wind shear?

Wind shear refers to rapid changes in wind speed and/or direction over short distances and is usually associated with thunderstorms. It is especially dangerous to aircraft.

What is a microburst?

Microbursts are downbursts of air with a diameter of 2.5 miles (4 kilometers) or less. Often associated with thunderstorms, they can generate winds of hurricane forces that change direction abruptly.

What conditions result in microclimates?

Microclimates are small-scale regions where the average weather conditions are measurably different from the larger, surrounding region. Differences in temperature, precipitation, wind, or cloud cover can produce microclimates. Frequent causes of microclimates are differences in elevation, mountains that alter wind patterns, shorelines, and man-made structures, such as buildings, that can alter the wind patterns. Microclimates can be found anywhere from a protected corner in the backyard to larger areas, such as several miles from the shore near an inland lake.

CLOUDS

What is a cloud?

Clouds are collections of trillions of rain droplets suspended in the atmosphere. These droplets consist of water or ice crystals that condense around a nucleus—usually a speck of dust or other small particle, such as pollen, volcanic ash, mineral flakes, or other organic or inorganic material. To form, clouds require air that has cooled to the point where the air is saturated with moisture or ice crystals; that is, when the humidity reaches 100 percent. Each cloud droplet is only a few micrometers in diameter, and it takes a million such droplets to form a single raindrop.

Who was the first person to classify clouds?

French naturalist Jean-Baptiste Pierre Antoine de Monet de Lamarck (1744–1829) was the first individual to propose a system for classifying clouds in 1802. However, his work did not receive wide recognition. A year later, amateur British meteorologist Luke Howard (1772–1864) developed a cloud-classification system that has been generally accepted and is still in use today. Howard's system distinguished clouds according to their general appearance ("heap clouds" versus "layer clouds") and their height above ground. Latin names and prefixes are used to describe these characteristics. The shape names are *cirrus* (curly or fibrous), *stratus* (layered), and *cumulus* (lumpy or piled). The prefixes denoting height are *cirro* (high clouds with bases above 20,000 feet [6,000 meters]) and *alto* (midlevel clouds from 6,000 to 20,000 feet [1,800–6,000 meters]). Low clouds do not have a prefix. Nimbo and nimbus are also added as a name or prefix to indicate that the cloud produces precipitation.

What are the four major cloud groups and their types?

1. High Clouds—composed almost entirely of ice crystals. The bases of these clouds start at 20,000 feet (6,096 meters) and reach 40,000 feet (12,192 meters).
   - *Cirrus* (from the Latin "lock of hair")—thin, featherlike crystal clouds in patches or narrow bands. The large ice crystals that often trail downward in well-defined wisps are called "mares' tails."
   - *Cirrostratus*—a thin, white cloud layer that resembles a veil or sheet. This layer can be striated or fibrous. Because of the ice content, these clouds are associated with the halos that surround the Sun or Moon.
   - *Cirrocumulus*—thin clouds that appear as small, white flakes or cotton patches and may contain supercooled water.

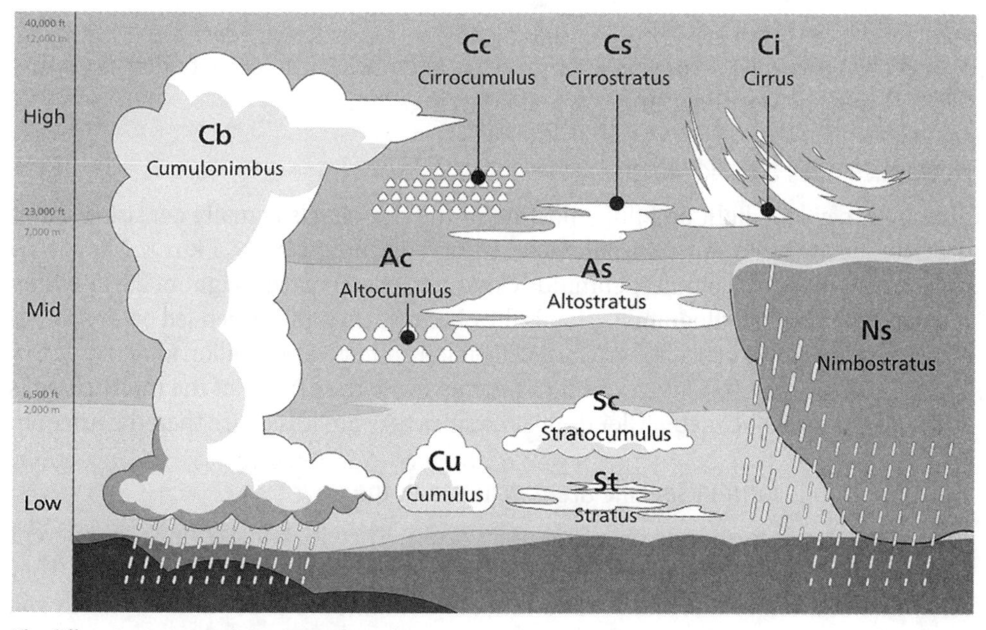

The different categories of clouds are shown here.

2. Middle Clouds—composed primarily of water. The height of the cloud bases range from 6,500 to 20,000 feet (2,000–6,096 meters).
   - *Altostratus*—appears as a bluish or grayish veil or layer of clouds that can gradually merge into altocumulus clouds. The Sun may be dimly visible through it, but flat, thick sheets of this cloud type can obscure the Sun.
   - *Altocumulus*—a white or gray layer or patches of solid clouds with rounded shapes.
3. Low Clouds—composed almost entirely of water that may at times be supercooled; at subfreezing temperatures, snow and ice crystals may be present as well. The bases of these clouds start near Earth's surface and climb to 6,500 feet (2,000 meters) in the middle latitudes.
   - *Stratus*—gray, uniform, sheetlike clouds with a relatively low base, or they can be patchy, shapeless, low, gray clouds. Thin enough for the Sun to shine through, these clouds bring drizzle and snow.
   - *Stratocumulus*—globular, rounded masses that form at the top of the layer.
   - *Nimbostratus*—seen as a gray or dark, relatively shapeless, massive cloud layer containing rain, snow, and ice pellets.
4. Clouds with Vertical Development—contain supercooled water above the freezing level and grow to great heights. The cloud bases range from 1,600 to 20,000 feet (488–6,096 meters).
   - *Cumulus*—detached, fair-weather clouds with relatively flat bases and dome-shaped tops. These usually do not have extensive vertical development and do not produce precipitation.
   - *Cumulonimbus*—unstable, large, vertical clouds with dense, boiling tops that bring showers, hail, thunder, and lightning.

How much does a cloud weigh?

Although they look light and fluffy floating in the sky, clouds actually consist of water droplets and are heavy. A 1 cubic kilometer (1 km^3, meaning 1 km × 1 km × 1 km or 0.6 miles × 0.6 miles × 0.6 miles) cumulus cloud has an estimated weight of 2.211 billion pounds (1.003 billion kilograms). Clouds float because they are suspended on dry air. An equal volume (1 km^3) of dry air weighs 2.220 billion pounds (1.007 billion kilograms). Dry air has a density of 1.007 kilograms/cubic meter, while the density of the moist cloud is 1.003 kilograms/cubic meter. Clouds float because they are less dense than the drier air.

Do some clouds form in specific areas or times of the day?

Lenticular clouds form only over mountain peaks. They look like a stack of different layers of cloud matter. Noctilucent clouds are the highest in Earth's atmosphere. They form only between sunset and sunrise between the latitudes of 50° and 70° north and south of the equator.

What is cloud seeding?

Cloud seeding is an attempt to modify the weather to produce precipitation. A cloud contains water droplets and ice crystals. Precipitation occurs when the ice crystals grow large enough to fall as rain, snow, hail, or other precipitation. Unless the conditions are right for ice crystals to grow, precipitation does not occur. Cloud seeding is an attempt to convert the supercooled droplets of liquid water in a cloud to ice crystals. Dry ice (solid carbon dioxide) and silver iodide are the substances most often used to seed clouds to transform the water droplets to ice crystals. Seeding is most efficient when the seeding agent is dropped from an airplane onto the top of a cloud.

How much of Earth is usually covered by clouds?

At any given time, about one-half of Earth is covered by clouds.

What are the cloudiest cities in the United States?

One recent study identifies the five cloudiest cities in the United States as Seattle, Washington; Portland, Oregon; Buffalo, New York; Pittsburgh, Pennsylvania; and Cleveland, Ohio. In these cities, over half the days of a year, clouds cover more than three-quarters of the sky.

Cloudiest U.S. Cities

City	Days of Heavy Clouds	Percentage of Cloudy Days
Seattle, WA	226	62
Portland, OR	222	61
Buffalo, NY	208	57
Pittsburgh, PA	203	56
Cleveland, OH	202	55

PRECIPITATION

What is the shape of a raindrop?

Although a raindrop has been illustrated as being pear shaped or tear shaped, high-speed photographs reveal that a large raindrop has a spherical shape with a hole not quite through it (giving it a doughnutlike appearance). Water surface tension pulls the drop into this shape. As a drop larger than 0.08 inches (2 millimeters) in diameter falls, it will become distorted. Air pressure flattens its bottom, and its sides bulge. If it becomes larger than 0.25 inches (6.4 millimeters) across, it will keep spreading crosswise as it falls and will bulge more at its sides, while at the same time, its middle will thin into a bowtie shape. Eventually in its path downward, it will divide into two smaller, spherical drops.

How fast does rain fall?

The speed of rainfall varies with drop size and wind speed. A typical raindrop in still air falls about 600 feet (182 meters) per minute or about 7 miles (11 kilometers) per hour. Large raindrops can reach speeds of 16–20 miles (26–32 kilometers) per hour. Very small raindrops may fall slowly, at no more than 1 mile (1.6 kilometers) per hour.

Where does it rain the most in the United States?

Mount Wai'ale'ale, on the island of Kauai in Hawaii, receives an average of 460 inches (1,168 centimeters) of rain a year. That amount is over 38 feet (about 12 meters) of rain per year.

Where is the driest place on Earth?

Probably the driest place on the planet is the Atacama Desert, which is located in Chile near the Pacific Coast. The average annual rainfall there, specifically in the town of Arica, is about 0.02 inches (0.05 centimeters). Meteorologists believe that this never-ending drought is the result of the Humboldt Current, which blocks rain from reaching the Atacama Desert. Some parts of the Atacama Desert have not seen a drop of rain in centuries.

The Atacama Desert in Chile is the driest place on Earth. Annual rainfall is about 0.02 inches (0.05 centimeters).

What is the dew point?

The dew point is the temperature at which air is full of moisture and cannot store any more. When the relative humidity is 100 percent, the dew point is either the same as or lower than the air temperature. If a fine film of air contacts a surface and is chilled to below the dew point, then actual dew is formed. This is why dew often forms at night or early morning: as the temperature of the air falls, the amount of water vapor the air can hold also decreases. Excess water vapor then condenses as very small drops on whatever it touches. Fog and clouds develop when sizable volumes of air are cooled to temperatures below the dew point.

How is hail formed?

Hail is precipitation consisting of balls of ice. Hailstones usually are made of concentric, or onionlike, layers of ice alternating with partially melted and refrozen snow, structured around a tiny, central core. It is formed in cumulonimbus or thunderclouds when freezing water and ice cling to small particles in the air, such as dust. The winds in the clouds blow the particles through zones of different temperatures, causing them to accumulate additional layers of ice and melting snow and to increase in size.

How large can hailstones become?

The average hailstone is about 0.25 inches (0.64 centimeters) in diameter. However, hailstones weighing up to 7.5 pounds (3.4 kilograms) are reported to have fallen in Hyderabad, India, in 1939, although scientists think these huge hailstones may be several stones that partly melted and stuck together. On April 14, 1986, hailstones weighing 2.5 pounds (1 kilogram) were reported to have fallen in the Gopalgang district of Bangladesh.

The largest hailstone ever recorded in the United States fell in Coffeyville, Kansas, on September 3, 1970. It measured 5.57 inches (14.15 centimeters) in diameter and weighed 1.67 pounds (0.75 kilogram).

Hail Size Estimates

Description	Size (inches/centimeters)
Pea Size	0.25/.064
Penny/Dime	0.75/1.91
Nickel	0.88/2.24
Quarter	1.00/2.54
Half Dollar or Susan B. Anthony Dollar	1.25/3.18
Ping Pong Ball	1.50/3.81
Golf Ball	1.75/4.45
Hen Egg	2.00/5.08
Tennis Ball	2.50/6.35
Baseball	2.75/6.99
Grapefruit	4.00/10.16

When does frost form?

A frost is a crystalline deposit of small, thin ice crystals formed on objects that are at freezing or below freezing temperatures. This phenomenon occurs when atmospheric water vapor condenses directly into ice without first becoming a liquid; this process is called sublimation. Usually frost appears on clear, calm nights, especially during early autumn, when the air above Earth is quite moist.

What is the difference between freezing rain, sleet, and hail?

Freezing rain is rain that falls as a liquid but turns to ice on contact with a freezing object to form a smooth ice coating called glaze. Usually, freezing rain only lasts a short time because it either turns to rain or to snow. Sleet is frozen or partially frozen rain in the form of ice pellets. Sleet forms when rain falls from a warm layer of air, passes through a freezing air layer near Earth's surface, and forms hard, clear, tiny ice pellets that can hit the ground so fast that they bounce off with a sharp click. Hail is a larger form of sleet.

How does snow form?

Snow is not frozen rain. Snow forms by sublimation of water vapor—the turning of water vapor directly into ice without going through the liquid stage. High above the ground, chilled water vapor turns to ice when its temperature reaches the dew point. The result of this sublimation is a crystal of ice, usually hexagonal. Snow begins in the form of these tiny, hexagonal ice crystals in the high clouds; the young crystals are the seeds from which snowflakes will grow. As water vapor is pumped up into the air by updrafts, more water is deposited on the ice crystals, causing them to grow. Soon, some of the larger crystals fall to the ground as snowflakes.

How much water is contained in snow?

Snow can range from light and fluffy to heavy, dense, and slushy. On average, every 10 inches of snow is equivalent to one inch of water. Heavy, wet snow has a high water con-

Are any two snowflakes identical?

Some snowflakes may have strikingly similar shapes, but these twins are probably not molecularly identical. In 1986, cloud physicist Nancy Knight believed she found a uniquely cloned pair of crystals on an oil-coated slide that had been hanging from an airplane. This pair may have been the result of breaking off from a star crystal or were attached side by side, thereby experiencing the same weather conditions simultaneously. Unfortunately the smaller aspects of each of the snow crystals could not be studied because the photograph was unable to capture possible molecular differences. So, even if the human eye may see twin flakes, on a minuscule level, these flakes are different.

tent; 4–5 inches (10–12 centimeters) may contain 1 inch (2.5 centimeters) of water. A dry, powdery snow might require 15 inches (38 centimeters) of snow to equal 1 inch (2.5 centimeters) of water.

What is the record for the greatest snowfall in the United States?

The record for the most snow in a single storm is 189 inches (480 centimeters) at Mount Shasta Ski Bowl in California from February 13 to 19, 1959. For the most snow in a 24-hour day, the record goes to Silver Lake in Colorado on April 14 to 15, 1921, with 76 inches (193 centimeters) of snow. The year record goes to Paradise, an area of Mount Rainier in Washington, with 1,224.5 inches (3,110 centimeters) from February 19, 1971, to February 18, 1972. The highest average annual snowfall is 241 inches (612 centimeters) in Blue Canyon, California. In March 1911, Tamarack, California, had the deepest snow accumulation—over 37.5 feet (11.4 meters).

Is it ever too cold to snow?

No matter how cold the air gets, it still contains some moisture, and this can fall out of the air in the form of very small snow crystals. Very cold air is associated with no snow because these invasions of air from northerly latitudes are associated with clearing conditions behind cold fronts. Heavy snowfalls are associated with relatively mild air in advance of a warm front. The fact that snow piles up, year after year, in Arctic regions illustrates that it is never too cold to snow.

What is a "white-out"?

An official definition for "white-out" does not exist. It is a colloquial term that can describe any condition during snowfall that severely restricts visibility. That may mean a blizzard, snow squall, and the like If you get some sunlight in the mix, that makes the situation even worse—it's like driving in fog with your headlights on high beam. The light gets backscattered right into your eyes, and you can't see.

ATMOSPHERIC PHENOMENA

What is lightning?

Lightning is an electrical discharge occurring in the atmosphere accompanied by a vivid flash of light. During a thunderstorm, a positive charge builds in the upper part of a cloud, and a negative charge builds in the lower part of the cloud. The difference between the positive and negative charges increases, generating an electrical field, until the electrical charge jumps from one area to another. Lightning may travel from cloud to ground, cloud to air, cloud to cloud, or stay within a cloud. Nearly 90 percent of all lightning strikes never reach the ground; they merely jump from cloud to cloud or cloud layer to cloud layer. The main types of lightning are:

- Streak lightning: A single or multiple zigzagging line from cloud to ground.
- Forked lightning: Lightning that forms two branches simultaneously.
- Sheet lightning: A shapeless flash covering a broad area.
- Ribbon lightning: Streak lightning blown sideways by the wind to make it appear like parallel successive strokes.
- Bead or chain lightning: A stroke interrupted or broken into evenly spaced segments or beads.
- Ball lightning: A rare form of lightning in which a persistent and moving, luminous, white or colored sphere is seen. It can last from a few seconds to several minutes, and it travels at about a walking pace. Spheres have been reported to vanish harmlessly or to pass into or out of rooms—leaving, in some cases, signs of their passage, such as a hole in a window pane. Sphere dimensions vary but are most commonly from 4 to 8 inches (10–20 centimeters) in diameter.
- Heat lightning: Lightning seen along the horizon during hot weather and believed to be a reflection of lightning occurring beyond the horizon.

When a volcano erupts it can produce lightning as a result of charged particles forming from the ejected materials and from charges building up as the ash cloud moves through the atmosphere. The results are quite spectacular!

How hot is lightning?

The temperature of the air around a bolt of lightning is about 54,000°F (30,000°C), which is six times hotter than the surface of the Sun, yet many times, people survive being struck by a bolt of lightning. American park ranger Roy Sullivan (1912–1983), for example, was hit by lightning seven times between 1942 and 1977. In cloud-to-ground lightning, its energy seeks the shortest route to Earth, which could be through a person's shoulder, down the side of the body, or through the leg, down to the ground. As long as the lightning does not pass across the heart or spinal column, the victim usually does not die.

How many volts are in lightning?

A stroke of lightning discharges anywhere from 10 to 100 million volts of electricity. An average lightning stroke has 30,000 amperes.

What is the color of lightning?

The atmospheric conditions determine the color of lightning. Blue lightning within a cloud indicates the presence of hail. Red lightning within a cloud indicates the presence of rain. Yellow or orange lightning indicates a large concentration of dust in the air. White lightning is a sign of low humidity in the air.

How long is a lightning stroke?

The visible length of the streak of lightning depends on the terrain and can vary greatly. In mountainous areas where clouds are low, the flash can be as short as 300 yards (273 meters), whereas in flat terrain, where clouds are high, the bolt can measure as long as 4 miles (6.5 kilometers). The usual length is about 1 mile (1.6 kilometers), but streaks of lightning up to 20 miles (32 kilometers) have been recorded. The stroke channel is very narrow—perhaps as little as half an inch (1.27 centimeters). It is surrounded by a "corona envelope" or a glowing discharge that can be as wide as 10 to 20 feet (3–6 meters) in diameter. The speed of lightning can vary from 100 to 1,000 miles (161–1,610 kilometers) per second for the downward leader track; the return stroke is 87,000 miles (140,070 kilometers) per second (almost half the speed of light).

How many times does lightning strike Earth each year?

About twenty million bolts of lightning are generated in the atmosphere every year, and during any particular second, about 100–125 lightning strokes are occurring on our planet. Thunderstorms are very common in Earth's atmosphere, with about 1,500–2,000 such storms being active at any given time.

Does lightning ever strike twice in the same place?

Lightning can and often does strike twice in the same place. Since lightning bolts head for the highest and most conductive point, that point often receives multiple strikes of lightning in the course of a storm. In fact, tall buildings, such as the Empire State Building in New York, can be struck several times during the same storm. During one storm, lightning struck the Empire State Building twelve times. Designed as a lightning rod for the surrounding area, the Empire State Building is struck by lightning about one hundred times per year during multiple storms.

Why are lightning rods important?

The lightning rod was invented by Benjamin Franklin in around 1750 following his experiment with the kite and the key. In 1752, Franklin tied a metal key to the end of a kite string and flew it during a thunderstorm. Franklin suspected that lightning was a nat-

How is the distance from a lightning flash calculated?

Count the number of seconds between seeing a flash of lightning and hearing the sound of the thunder. Divide the number by five to determine the number of miles away that the lightning flashed.

ural form of electricity. He knew that if lightning was electricity, it would be attracted to the metal key. When sparks jumped from the metal key, he understood that an electrical current had traveled from the electrified air above down the kite string to the key. This experiment confirmed that lightning is an electrical phenomenon. A lightning rod is often placed on the top of buildings to attract lightning bolts. They are designed to provide a safe path to ground the electricity so that it does not damage the building. In recent years, lightning rods have become even more important because the metal pipes that used to be installed for indoor plumbing and could serve as lightning rods are being replaced by nonconductive PVC pipes.

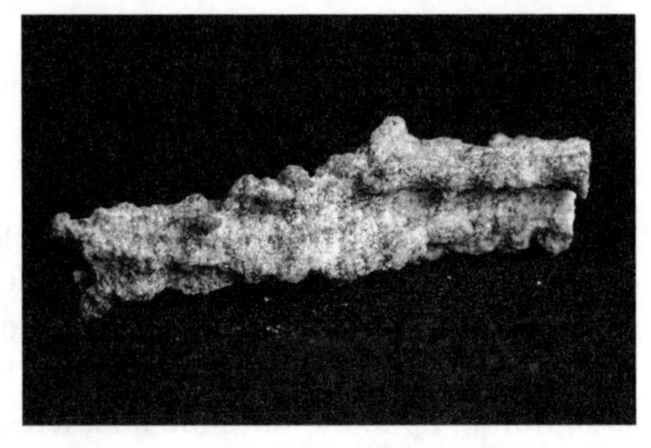

Fulgurites form when lightning strikes silica-rich land. They can often be found in sandy areas such as beaches and deserts, but also in other sandy soils that lack clay.

What are fulgurites?

Fulgurites (from the Latin word *fulgur*, meaning "lightning") are petrified lightning, created when lightning strikes an area of dry sand. The intense heat of the lightning melts the sand surrounding the stroke into a rough, glassy tube, forming a fused record of its path. These tubes may be 0.5–2 inches (1.5–5 centimeters) in diameter and up to 10 feet (3 meters) in length. They are extremely brittle and break easily. The inside walls of the tube are glassy and lustrous, while the outside is rough, with sand particles adhering to it. Fulgurites are usually tan or black in color, but translucent white ones also have been found.

How does lightning strike people?

The five ways that lightning may strike people are direct strike, side flash, ground current, conduction, and streamers. As the term indicates, in a direct strike, the lightning strikes the person directly. The heat of a direct strike may cause burns on the skin, and the current moves through the body through either the cardiovascular system or the nervous system. In a side flash (also called a side splash) strike, the lightning strikes a taller object near the victim, and a portion of the current jumps from the taller object to the victim. Side flashes often occur when a person has taken shelter near a tree during a storm. Ground current is when lightning strikes an object, for example, a tree, and the energy from the strike travels outward from the strike in and along the ground surface. Ground current can travel along garage floors with conductive materials. Ground current is responsible for the most lightning-related injuries and fatalities. In a conduction strike, lightning travels along wires and other metal surfaces, including plumbing. Streamers are the least common type of lightning to cause injuries. Streamers develop as the downward-moving lightning strike leader approaches the ground. Typically, only one of the streamers makes contact with the leader as it approaches the ground and provides the path for the bright return stroke; however, when the main channel discharges, so do all the other streamers in the area. If a person is in one of the streamers, they may be injured.

How many fatalities occur each year in the United States due to lightning?

During the past thirty years, from 1978 to 2017, an average of forty-four people were killed each year by lightning. The ten-year average from 2008 to 2017 decreased to only twenty-seven people per year. An estimated 252 people were injured by lightning strikes in the United States each year during the period 2008–2017.

What causes thunder?

Thunder is created when lightning rapidly heats a section of air. As the air expands, it compresses, releasing the energy as a sound wave that we hear as a loud boom or clap of thunder.

What is a thunderstorm, and how many occur in a year?

Thunderstorms are localized atmospheric phenomena that produce heavy rain, thunder, lightning, and sometimes hail. They are formed in cumulonimbus clouds (big and bulbous) that rise many miles into the sky. Most of the southeastern United States has over forty days of thunderstorm activity each year, and about one hundred thousand thunderstorms occur across the country annually.

How loud is thunder?

A clap of thunder can be as loud as 120 decibels, which is comparable to the noise at a rock concert, a chainsaw, or a pneumatic drill.

What is Saint Elmo's fire?

Saint Elmo's fire has been described as a corona from electric discharge produced on high-grounded metal objects, chimney tops, and ship masts. Since it often occurs during thun-

How far away can thunder be heard?

Thunder is the crash and rumble associated with lightning. It is caused by the explosive expansion and contraction of air heated by the stroke of lightning. This results in sound waves that can be heard easily 6 to 7 miles (9.7–11.3 kilometers) away. Occasionally, such rumbles can be heard as far away as 20 miles (32.2 kilometers). The sound of great claps of thunder is produced when intense heat and the ionizing effect of repeated lightning occur in a previously heated air path. This creates a shock wave that moves at the speed of sound.

derstorms, the electrical source may be lightning. Another description refers to this phenomenon as weak static electricity formed when an electrified cloud touches a high exposed point. Molecules of gas in the air around this point become ionized and glow. The name originated with sailors who were among the first to witness the display of spearlike or tufted flames on the tops of their ships' masts. Saint Elmo (d. 303) (which is a corruption of Saint Ermo) is the patron saint of sailors, so they named the fire after him.

What is a rainbow, and how do they occur?

Rainbows are colorful bands of light that are formed when water particles in the air reflect sunlight. As sunlight enters the drops and droplets, the different wavelengths of colors that compose sunlight are refracted at different wavelengths to produce a spectrum of color. To see a rainbow, you must be standing with the Sun behind you and the raindrops in front of you. The Sun needs to be less than 42° above the horizon to obtain the correct angle so that the light waves are properly reflected. The light is refracted as it enters a raindrop, reflects off the inside of the back of the raindrop, and is refracted again as it leaves.

What is the order of colors in a rainbow?

Red, orange, yellow, green, blue, indigo, and violet are the colors of the rainbow, but these are not necessarily the sequence of colors that an observer might see. Rainbows are formed when raindrops reflect sunlight. As sunlight enters the drops, the different wavelengths of the colors that compose sunlight are refracted at different lengths to produce a spectrum of color. Each observer sees a different set of raindrops at a slightly different angle. Drops at different angles from the observer send different wavelengths (i.e., different color) to the observer's eyes. Since the color sequence of the rainbow is the result of refraction, the color order depends on how the viewer sees this refraction from the viewer's angle of perception.

Is it possible for a monochromatic rainbow to occur?

Rare reports of all-red or all-white rainbows have indeed occurred.

When does the green flash phenomenon occur?

On rare occasions, the Sun may look bright green for a moment as the last tip of the Sun is setting. This green flash occurs because the red rays of light are hidden below the horizon and the blue rays are scattered in the atmosphere. The green rays are seldom seen because of dust and pollution in the lower atmosphere. It may best be seen when the air is cloudless and when a distant, well-defined horizon exists, as on an ocean.

How often does an aurora appear?

Because it depends on solar winds (electrical particles generated by the Sun) and sunspot activity, the frequency of an aurora cannot be determined. Auroras usually appear two days after a solar flare (a violent eruption of particles on the Sun's surface) and reach

Iceland is an ideal place to see the Aurora Borealis as is shown here at the scenic Kirkjufell mountain.

their peak two years into the eleven-year sunspot cycle. The auroras, occurring in the polar regions, are broad displays of usually colored light at night. The northern polar aurora is called Aurora Borealis or Northern Lights, and the southern polar aurora is called the Aurora Australis.

STORMY WEATHER

Is it true that the United States has the worst weather of any nation?

Yes, the United States has the worst weather of any country. Subjected to hurricanes, flooding, drought, heat and cold waves, blizzards, and the worst tornado activity on the planet, the United States experiences more weather disasters than any other nation.

What is the Storm of the Century?

One candidate for the title "Storm of the Century" was a storm that arrived in 1993. It was a blizzard that struck the American East Coast, killing 318 people, including forty-eight at sea. Fifty percent of the American population was affected in some way by the storm. The storm reached from Maine to Florida, where half a foot of snow fell in the Florida Panhandle, and even Daytona Beach saw freezing temperatures. Winds near Key West raged at up to 109 miles (175 kilometers) per hour. Meanwhile, Mount LeConte in Tennessee received 56 inches (142 centimeters) of snow. The 1993 storm ranged far be-

TYPICAL JANUARY-MARCH WEATHER ANOMALIES AND ATMOSPHERIC CIRCULATION DURING MODERATE TO STRONG EL NIÑO & LA NIÑA

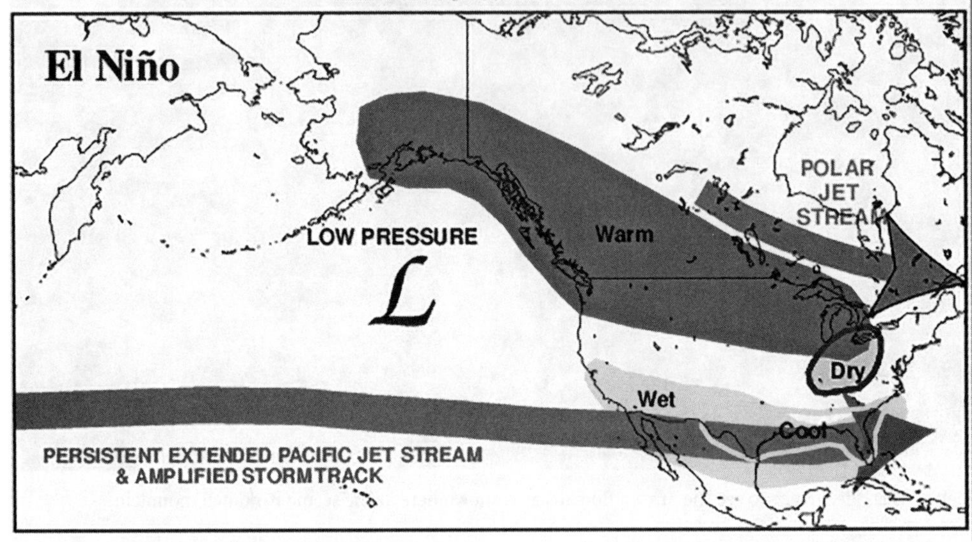

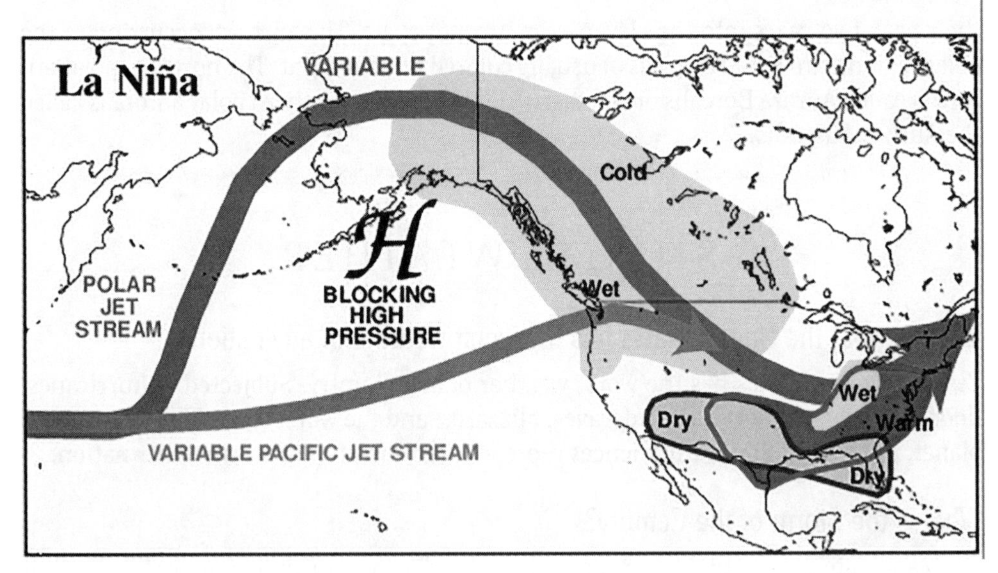

Pressure systems and jet stream patterns move in a certain way every few years to dramatically affect weather in what are called La Niña and El Niño patterns.

yond U.S. borders, extending north to Canada and south to Central America. At its peak, it reached the strength of a category 3 hurricane, and by the time it was over, it had dumped 44 million acre-feet (about 14.3 trillion gallons or 54.3 trillion liters) of water onto the ground. Add to this several killer tornadoes, and perhaps, the 1993 storm deserves the twentieth century's title of "Storm of the Century."

What is El Niño?

El Niño is the unusual warming of the surface waters of large parts of the tropical Pacific Ocean. Occurring around Christmastime, it is named after the Christ child. El Niño occurs erratically every three to seven years. It brings heavy rains and flooding to Peru, Ecuador, and southern California and a milder winter with less snow to the northeastern United States. Studies reveal that El Niño is not an isolated occurrence but is instead part of a pattern of change in the global circulation of the oceans and atmosphere. The 1982–1983 El Niño was one of the most severe climate events of the twentieth century in both its geographical extent as well as in its degree of warming (14°F or 8°C).

What is La Niña?

La Niña is the opposite of El Niño. It refers to a period of cold surface temperatures of the tropical Pacific Ocean. La Niña winters are especially harsh, with heavy snowfall in the northeastern United States and rain in the Pacific Northwest.

What was the greatest disaster in U.S. history?

The greatest natural disaster in the United States occurred when a hurricane struck Galveston, Texas, on September 8, 1900, and killed between eight and twelve thousand people.

What is a blizzard?

According to the U.S. National Weather Service, a winter storm is considered a blizzard when wind speeds reach 35 miles (56 kilometers) per hour and poor visibility of less than 0.25 miles (400 meters) occurs. Snow does not need to be falling at the time, but blowing and drifting should occur, with drifts exceeding 10 inches (25 centimeters) deep.

What is a Nor'Easter?

A Nor'Easter is a storm along the eastern coast of North America that affects the region, with northeasterly winds having speeds up to 75 miles (121 kilometers) per hour or more. Such storms evolve when low-pressure systems accumulate humid air from the Atlantic Ocean or the Gulf of Mexico and combine it with cold, dry air coming down from Canada in conjunction with a strong jet stream. The system rotates counterclockwise, bringing strong rainstorms in the South and, in the winter, snow to the Northeast. Nor'Easters have wreaked havoc on the United States several times. A February 1969 Nor'Easter dumped 70 inches (178 centimeters) of snow on Rumford, Maine, and 164 inches (416.5 centimeters) on Pinkham Notch, New Hampshire.

What is a hurricane, and what is the origin of the term?

A hurricane is a tropical storm that forms in the Atlantic Basin and has winds of 74 miles (112 kilometers) per hour or more. Hurricanes typically occur in the North At-

What is a Siberian Express?

This term describes storms that are severely cold and cyclonic; they descend from northern Canada and Alaska to other parts of the United States.

lantic Ocean and Caribbean Sea during the months of July, August, and September, when warm surface ocean temperatures exceed 80°F (26.5°C), providing energy that feeds into the storm. Seawater evaporates into the air, creating clouds, while the Coriolis effect causes the clouds to rotate. For a hurricane to develop, the wind speeds in the upper and lower elevations of the storm must be similar. If a significant difference exists in these speeds, the resulting wind shear will cause the hurricane to become unstable, with clouds and winds opposing each other rather than working together in a gigantic swirl that increases in speed. The term "hurricane" is derived from Hurican, the Carib god of evil, which was derived from the Mayan god Hurakan. Hurakan was one of the Mayan creator gods who blew his breath across the chaotic water and brought forth dry land.

How are hurricanes classified?

The Saffir/Simpson Hurricane Damage-Potential Scale assigns numbers 1 through 5 to measure the disaster potential of a hurricane's winds and its accompanying storm surge. The purpose of the scale, developed in 1971 by Herbert Saffir (1917–2007) and Robert Simpson (1912–2014), is to help disaster agencies gauge the potential significance of these storms in terms of assistance.

Saffir/Simpson Hurricane Scale Ranges

Scale Number (Category)	Barometric Pressure (in inches)	Winds (miles per hour)	Surge (in feet)	Damage
1	> 28.94	74–95	4–5	Minimal
2	28.50–28.91	96–110	6–8	Moderate
3	27.91–28.47	111–130	9–12	Extensive
4	27.17–27.88	131–155	13–18	Extreme
5	< 27.17	> 155	> 18	Catastrophic

Damage categories:

- *Minimal*—No real damage to building structures. Some tree, shrubbery, and mobile home damage. Coastal road flooding and minor pier damage.
- *Moderate*—Some roof, window, and door damage. Considerable damage to vegetation, mobile homes, and piers. Coastal and low-lying escape routes flood 2–4 hours before center of storm arrives. Small craft can break moorings in unprotected areas.

- *Extensive*—Some structural damage to small or residential buildings. Mobile homes destroyed. Flooding near coast destroys structures and floods of homes 5 feet (1.5 meters) above sea level as far inland as 6 miles (9.5 kilometers).
- *Extreme*—Extensive roof, window, and door damage. Major damage to lower floors of structures near the shore and some roof failure on small residences. Complete beach erosion. Flooding of terrain 10 feet (3 meters) above sea level as far as 6 miles (9.5 kilometers) inland, requiring massive residential evacuation.

The eye of Hurricane Florence stands out sharply in this photo taken by an astronaut on the International Space Station in 2018.

- *Catastrophic*—Complete roof failure to many buildings; some complete building failure, with small utility buildings blown away. Major damage to lower floors of all structures 19 feet (5.75 meters) above sea level located within 500 yards (547 meters) of the shoreline. Massive evacuation of residential areas on low ground 5–10 miles (8–16 kilometers) from shoreline may be required.

What is the eye of a hurricane?

The eye of a hurricane is a region of relative calm in the middle of the swirling storm. Hurricane eyes range in size from 4–40 miles (7–74 kilometers) in diameter. The eye can be so free of clouds that sunshine can be seen. The more intense the hurricane, the smaller the eye tends to be. The eye is surrounded by the "eye wall," an apt phrase because it is literally a circular wall that can reach 7 miles (11.3 kilometers) into the sky. Once past the eye wall, the hurricane resumes, with winds blowing as fast as 150 miles (278 kilometers) per hour or more.

How do hurricanes get their names?

Since 1950, hurricane names have been officially selected from library sources and are decided on during the international meetings of the World Meteorological Organization (WMO). The names are chosen to reflect the cultures and languages found in the Atlantic, Caribbean, and Hawaiian regions. When a tropical storm with rotary action and wind speeds above 39 miles (63 kilometers) per hour develops, the National Hurricane Center near Miami, Florida, selects a name from one of the six listings for Region 4 (Atlantic and Caribbean area). Letters Q, U, X, Y, and Z are not included because of the scarcity of names beginning with those letters. The next six years' names for the Atlantic hurricane season are:

Hurricane Names 2020 to 2025

2020	2021	2022	2023	2024	2025
Arthur	Ana	Alex	Arlene	Alberto	Andrea
Bertha	Bill	Bonnie	Bret	Beryl	Barry
Cristobal	Claudette	Colin	Cindy	Chris	Chantal
Dolly	Danny	Danielle	Don	Debby	Dorian
Edouard	Elsa	Earl	Emily	Ernesto	Erin
Fay	Fred	Fiona	Franklin	Florence	Fernand
Gonzalo	Grace	Gaston	Gert	Gordon	Gabrielle
Hanna	Henri	Hermine	Harold	Helene	Humberto
Isaias	Ida	Ian	Idalia	Isaac	Imelda
Josephine	Julian	Julia	Jose	Joyce	Jerry
Kyle	Kate	Karl	Katia	Kirk	Karen
Laura	Larry	Lisa	Lee	Leslie	Lorenzo
Marco	Mindy	Martin	Margot	Michael	Melissa
Nana	Nicholas	Nicole	Nigel	Nadine	Nestor
Omar	Odette	Owen	Ophelia	Oscar	Olga
Paulette	Peter	Paula	Philippe	Patty	Pablo
Rene	Rose	Richard	Rina	Rafael	Rebekah
Sally	Sam	Shary	Sean	Sara	Sebastien
Teddy	Teresa	Tobias	Tammy	Tony	Tanya
Vicky	Victor	Virginie	Vince	Valerie	Van
Wilfred	Wanda	Walter	Whitney	William	Wendy

During which hurricane season were the most names retired?

Once a hurricane has done a great deal of damage and caused loss of life, its name is retired from the six-year list cycle. Five hurricane names—Dennis, Katrina, Rita, Stan, and Wilma—were retired in the 2005 hurricane season. This was the greatest number of names retired in one season since the practice of retiring names of very destructive or deadly storms was started in 1954. Thirty-seven names have been retired since 2000.

Year	Hurricane Names Retired
2000	Keith
2001	Allison
	Iris
	Michelle
2002	Isidore
	Lili
2003	Fabian
	Isabel
	Juan
2004	Charley
	Frances

Year	Hurricane Names Retired
	Ivan
	Jeanne
2005	Dennis
	Katrina
	Rita
	Stan
	Wilma
2006	No names retired
2007	Dean
	Felix
	Noel
2008	Gustav
	Ike
	Paloma
2009	No names retired
2010	Igor
	Tomas
2011	Irene
2012	Sandy
2013	Ingrid
2014	No names retired
2015	Erika
	Joaquin
2016	Matthew
	Otto
2017	Harvey
	Irma
	Maria
	Nate

Which hurricane in the United States was the most destructive?

The economic loss caused by Hurricane Katrina is estimated at $100–$150 billion, making it the costliest natural disaster in the United States. The death toll for Katrina was less than two thousand, which is significant although nowhere near the death toll of at least eight thousand during the Galveston, Texas, hurricane in 1900. Katrina made landfall in Plaquemines Parish, Louisiana, as a category 3 hurricane with winds of 125 miles (201 kilometers) per hour on August 29, 2005. The coastal areas of Louisiana (including New Orleans), Mississippi, and Alabama suffered extensive damage.

How does a cyclone differ from a hurricane or a tornado?

All three wind phenomena are rotating winds that spiral in toward a low-pressure center as well as upward. Their differences lie in their size, wind velocity, rate of travel, and duration. Generally, the faster the winds spin, the shorter (in time) and smaller (in size) the event becomes.

Damage to property in New Orleans and surrounding areas was well over $100 billion from Hurricane Katrina, and thousands lost their homes, cars, and possessions.

A cyclone has rotating winds from 10 to 60 miles (16–97 kilometers) per hour, can be up to 1,000 miles (1,600 kilometers) in diameter, travels about 25 miles (40 kilometers) per hour, and lasts from one to several weeks.

A hurricane (or typhoon, as it is called in the Pacific Ocean area) has winds that vary from 75 to 200 miles (120–320 kilometers) per hour, moves between 10 to 20 miles (16–32 kilometers) per hour, can have a diameter up to 600 miles (960 kilometers), and can exist from several days to more than a week.

A tornado can reach a rotating speed of 300 miles (400 kilometers) per hour, travels between 25 to 40 miles (40–64 kilometers) per hour, and generally lasts only minutes, although some have lasted for five to six hours. Its diameter can range from 300 yards (274 meters) to 1 mile (1.6 kilometers), and its average path length is 5 miles (8 kilometers), with a maximum of 300 miles (483 kilometers).

Typhoons, hurricanes, and cyclones tend to breed in low-altitude belts over the oceans, generally around 5° to 15° latitude north or south. A tornado generally forms several thousand feet above Earth's surface, usually during warm, humid weather; many times, it is in conjunction with a thunderstorm. Although tornadoes can occur in many places, they mostly appear on the continental plains of North America (i.e., from the Plains states eastward to western New York and the southeastern Atlantic Coast states).

What is storm surge?

Storm surge is an abnormal rise in seawater level caused solely by a storm. It is caused primarily by the strong winds in a hurricane or tropical storm. Some of the factors that contribute to the amount of storm surge produced by a given storm are the storm intensity and size, the forward speed of the storm, the angle of approach to the coast, the shape of the coastline, and the width and slope of the ocean bottom. Storm surge tends to be greatest in larger storms with stronger winds. A storm that moves onshore perpendicular to the coast is more likely to produce higher storm surge than a storm that moves onshore parallel to the coast. On an open coast, a faster storm will produce higher storm surge, but areas of bays and sounds have a higher storm surge in slower moving storms. Finally, higher storm surge occurs with wide, gently sloping continental shelves found in the areas along the Gulf Coast. The area of the east coast of Florida has narrow, steeply sloping continental shelves and much lower storm surge.

How long do most tornadoes last?

Most tornadoes have short life spans, lasting less than 10 minutes, though they may be as short as only several seconds. Some may last an hour or longer. Tornadoes during the early to mid-1900s were often reported to be longer lived, though many climatologists believe these may have been a series of tornadoes instead of one single event. An average tornado will travel 5 miles (8 kilometers) during its life span. The tornado of March 18, 1925, traveled 219 miles (352 kilometers) through Missouri, Illinois, and Indiana at an average speed of 60–73 miles (97–117 kilometers) per hour.

When do most tornadoes happen?

Tornadoes can happen at any time of the year and at any time during the day or night. Tornadoes are most likely to happen in the United States from March through August. A general northward shift in the "tornado season" occurs from late winter through midsummer. Regionally, tornado activity increases in the central Gulf Coast as early as February, then shifts to the southeastern Atlantic Coast states in March and April before moving to the southern Plains in May and early June. The greatest amount of tornado activity is in June and July in the northern Plains and upper Midwest. Statistically, most tornadoes strike between 4:00 P.M. and 9:00 P.M.

The United States is prone to tornado strikes because of its geography and weather patterns. There are typically more than a thousand tornado reports in America annually.

How many tornadoes occur in the United States each year?

According to NOAA, the 1991–2010 annual average of tornadoes for the United States is 1,253. Compiling an actual average is difficult since reporting methods have changed over the last several decades, so the officially recorded tornado climatologies are believed to be incomplete. Some tornadoes, especially ones that cause little or no damage in remote areas, may not be reported. The three-year average for 2015–2017 is 1,143 tornadoes per year. The largest single outbreak of tornadoes occurred on April 3 and 4, 1974; 148 tornadoes were recorded in this period in the Plains and Midwestern states, which was called the "Super Outbreak." Six of these tornadoes had winds greater than 260 miles (420 kilometers) per hour, and some of them were the strongest ever recorded.

How many fatalities occur each year in the United States due to tornadoes?

During the past thirty years, from 1988 to 2017, an average of sixty-nine people were killed each year by tornadoes. The ten-year average, from 2008 to 2017, for deaths due to tornadoes was 101 people per year. These averages include the 2011 tornado season, which was one of the most active and most destructive since modern record keeping began in 1950. The tornado that hit Joplin, Missouri, on May 22, 2011, is ranked as the seventh most deadly tornado event in the United States. A total of 158 fatalities occurred during the Joplin tornado.

Have wind speeds during tornadoes been accurately measured?

No, tornado wind speeds have been scientifically estimated using Doppler radar and video observations, but no attempts to physically measure wind speeds using an anemometer have been successful. Many severe tornadoes will destroy an anemometer before it records the wind speed during a tornado. Furthermore, they may occur in many random locations without equipment in place to measure wind speed.

What is the biggest-known tornado?

The tornado on May 22, 2004, in Hallam, Nebraska, holds the record for the peak width of any tornado at nearly 2.5 miles (4 kilometers) across.

Do tornadoes in the Northern and Southern hemispheres rotate in the same direction?

In general, tornadoes in the Northern Hemisphere rotate counterclockwise (cyclonically), while those in the Southern Hemisphere rotate clockwise (anticyclonically). Oc-

Who are storm chasers?

Storm chasers are scientists and amateur storm enthusiasts who track and intercept severe thunderstorms and tornadoes. Two reasons for storm chasing are 1) to gather data to use in researching severe storms and 2) to provide a visual observation of severe storms indicated on remote radar stations. In addition, television personnel will chase storms to produce a dramatic storm video. Storm chasing can be an extremely dangerous activity in which strong winds, heavy rain, hail, and lightning threaten one's safety. Most individuals who chase storms are trained in the behavior of severe storms.

casionally, anticyclonic tornadoes have been observed in the Northern Hemisphere. Typically, anticyclonic tornadoes in the Northern Hemisphere are weaker twisters associated with weak storm cells or sometimes appearing as waterspouts. In 1998, a tornado spinning anticyclonically was observed near Sunnyvale, California. Rarer, but still possible, is a supercell that generates both cyclonic and anticyclonic tornadoes.

What causes a flood?

Flooding results when more water enters an environment than can be easily absorbed into the soil or drained away in rivers and streams. Flooding is usually caused by intense rainfalls that dump many inches of water onto an area over a short period of time, or they can also be caused by ocean swells and storm surges initiated by hurricanes and tropical storms. Tsunamis also cause flooding. The 2004 tsunamis that resulted in an undersea earthquake in the Indian Ocean killed about 238,000 people in eleven surrounding countries. Most of these people died from the initial landfall of the waves and resulting floods. In addition, floods can also be caused when a dam or levee breaks.

BIOLOGY

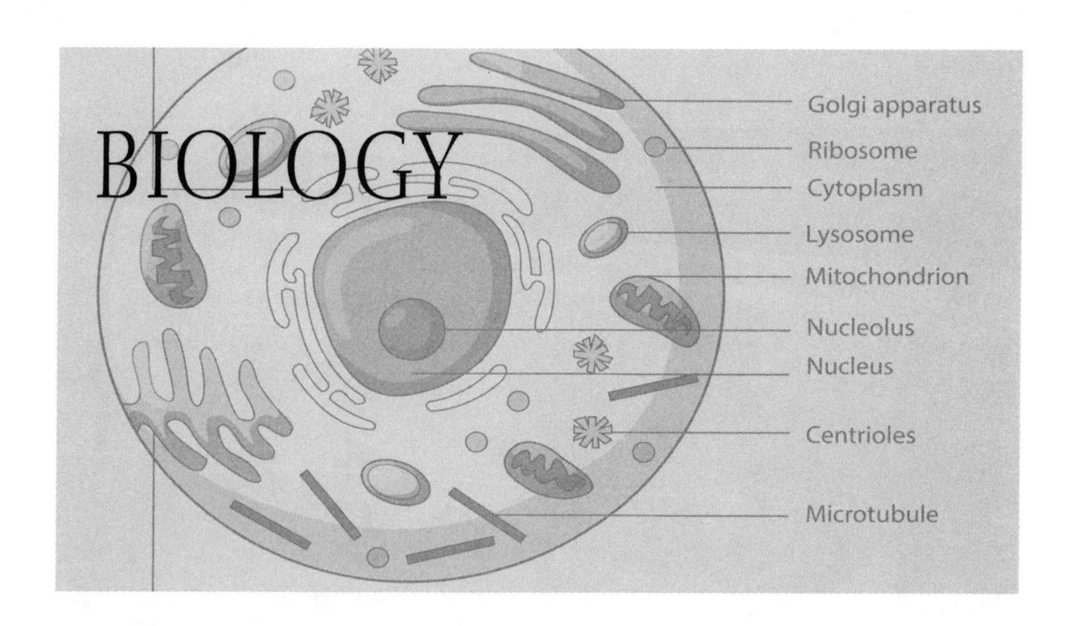

INTRODUCTION AND HISTORICAL BACKGROUND

What is biology, and who coined the term?

Biology is the scientific study of life. It is mainly concerned with the study of living systems—from animal to plant and everything in between—and includes the study of various organisms' cells, metabolism, reproduction, growth, activity of systems, and response to the stimuli in their environment. French biologist Jean-Baptiste Pierre Antoine de Monet de Lamarck is credited with coining the term "biology," which comes from the Greek words *bios*, meaning "life," and *logy*, meaning "study of," in 1802.

How many species of biological organisms have been discovered?

It is estimated that 1.3 million named species exist in the world, approximately half of which are insects. The largest group of insects is the beetle (the order Coleoptera), with approximately three hundred thousand species. In contrast, only 4,500 species of mammals exist. Biologists believe that many more species have not yet been discovered. Most biologists estimate that eight to twelve million species of organisms in the world exist, many of which have not yet been discovered. Some biologists estimate the number of species to be far greater, as high as a trillion. The vast majority of these species (about 78 percent) are bacteria.

Who is considered the "father of microbiology"?

Antonie van Leeuwenhoek (1632–1723), a Dutch fabric merchant and civil servant, discovered bacteria and other microorganisms in 1674 when he looked at a drop of pond water through a glass lens. Early, single-lens instruments produced magnifications of

Robert Koch (left) and Louis Pasteur are considered to be the scientists who established the field of bacteriology in the nineteenth century.

fifty to three hundred times that of real size (approximately one-third of the magnification produced by modern light microscopes). Primitive microscopes provided a perspective into the previously unknown world of small organisms, which van Leeuwenhoek called "animalcules" in a letter he wrote to the Royal Society of London. Because of these early investigations, van Leeuwenhoek is considered to be the "father of microbiology."

What period of time has come to be known as the golden age of microbiology?

The era known as the golden age of microbiology began in 1857 with the work of Louis Pasteur and Robert Koch and lasted about sixty years. During this period of time, many important scientific discoveries were made. Joseph Lister's (1827–1912) practice of treating surgical wounds with a phenol solution led to the advent of aseptic surgery. The advancements Paul Ehrlich (1854–1915) made to the theory of immunity synthesized the "magic bullet," an arsenic compound that proved effective in treating syphillis in humans.

In 1884, Elie Metchnikoff (1845–1916), an associate of Pasteur, published a report on phagocytosis. The report explained the defensive process in which the body's white blood cells engulf and destroy microorganisms. In 1897, Masaki Ogata reported that rat fleas transmitted bubonic plague, ending the centuries-old mystery of how plague was transmitted. The following year, Kiyoshi Shiga (1871–1957) isolated the bacterium responsible for bacterial dysentery. This organism was eventually named *Shigella dysenteriae*.

Who were the founders of modern bacteriology?

German bacteriologist Robert Koch (1843–1910) and French chemist Louis Pasteur (1822–1895) are considered the founders of bacteriology. In 1864, Pasteur devised a way to slowly heat foods and beverages to a temperature that was high enough to kill most of the microorganisms that would cause spoilage and disease but would not spoil or curdle the food. This process is called pasteurization.

By demonstrating that tuberculosis was an infectious disease caused by a specific species of *Bacillus*, Koch, in 1882, set the groundwork for public-health measures that would go on to significantly reduce the occurrences of many diseases. His laboratory procedures, methodologies for isolating microorganisms, and four postulates for determining agents of disease gave investigators valuable insights into the control of bacterial infections.

What are some well-known diseases and infectious agents, and who discovered them?

During the sixty years between 1857 and 1917, numerous laboratory investigations occurred dealing with microorganisms, including their isolation and identification. The following chart identifies a number of diseases, their infectious agents, the scientists who discovered them, and the year in which they were discovered.

Diseases and Their Infectious Agents

Disease	Infectious Agent	Discoverer	Year Discovered
Anthrax	*Bacillus anthracis*	Robert Koch (1843–1910)	1876
Gonorrhea	*Neisseria gonorrhoeae*	Albert L. S. Neisser (1855–1916)	1879
Malaria	*Plasmodium malariae*	Charles-Louis Alphonse Laveran (1845–1922)	1880
Wound infections	*Staphylococcus aureus*	Sir Alexander Ogston (1844–1929)	1881
Erysipelas	*Streptococcus pyogenes*	Friedrich Fehleisen (1854–1924)	1882
Tuberculosis	*Mycobacterium tuberculosis*	Robert Koch (1843–1910)	1882
Cholera	*Vibrio cholerae*	Robert Koch (1843–1910)	1883
Diphtheria	*Carynebacterium diphtheriae*	Edwin Klebs (1834–1913) and Friedrich Loeffler (1852–1915)	1883–1884
Typhoid fever	*Salmonella typhi*	Karl Eberth (1877–1952) and Georg Gaffky (1850–1918)	1884
Bladder infections	*Escherichia coli*	Theodor Escherich (1857–1911)	1885
Salmonellosis	*Salmonella enteritidis*	August Gärtner (1848–1934)	1888

Disease	Infectious Agent	Discoverer	Year Discovered
Tetanus	*Clostridium tetani*	Shibasaburo Kitasato (1853–1931)	1889
Gas gangrene	*Clostridium perfringens*	William Henry Welch (1850–1934) and George Henry Falkiner Nuttall (1862–1937)	1892
Plague	*Yersinia pestis*	Alexandre Yersin (1863–1943) and Shibasaburo Kitasato (1853–1931)	1894
Botulism	*Clostridium botulinum*	Émile Van Ermengem (1851–1932)	1897
Shigellosis	*Shigella dysenteriae*	Kiyoshi Shiga (1871–1957)	1898
Syphilis	*Treponema pallidum*	Fritz R. Schaudinn (1871–1906) and P. Erich Hoffmann (1868–1959)	1905
Whooping cough	*Bordetella pertussis*	Jules Bordet (1870–1961) and Octave Gengou (1875–1957)	1906

Who discovered that penicillin was effective against bacterial infections?

Scottish physician Alexander Fleming (1881–1955) decided to pursue a career in scientific research rather than practice medicine. He was the first individual to discover penicillin's use as an antibacterial agent. In 1928, Fleming was researching staphylococci at St. Mary's Hospital in London. As part of his investigation, he had spread staphylococci on several Petri dishes before going on vacation. When he returned, he noticed a green-yellow mold contaminating one of his Petri dishes and that the staphylococci had failed to grow near the mold. He identified the mold as *Penicillium notatum*. Further investigations showed that staphylococci and other microorganisms are killed by *Penicillium notatum*. However, it was not until 1940 that penicillin was first isolated and produced in concentrated form by Howard Florey (1898–1968) and Ernst Chain (1906–1979). In 1945, Fleming, Florey, and Chain shared the Nobel Prize in Physiology or Medicine for their work on penicillin.

What was the first virus to be isolated in a laboratory?

In 1935, Wendell Stanley (1904–1971) of the Rockefeller Institute (known today as Rockefeller University) prepared an extract of the tobacco mosaic virus and purified it. The purified virus precipitated in the form of crystals. During this investigation, Stanley was able to demonstrate that viruses can be regarded as chemical matter rather than as living organisms. The purified crystals retained the ability to infect healthy tobacco plants, thus characterizing them as viruses, not merely chemical compounds derived from a virus. Subsequent studies showed that the tobacco mosaic virus consisted of a protein and a nucleic acid. Further studies showed that this virus consisted of RNA (ribonucleic acid) surrounded by a protein coat. Stanley was awarded the Nobel Prize in Chemistry in 1946 for his discovery.

CLASSIFICATION

What is taxonomy?

Taxonomy is a general term for the practice and science of classification. In a wider, more general sense, it may refer to a classification of things or concepts as well as to the principles underlying such a classification. In biology, taxonomy is the science of defining and naming groups of organisms on the basis of shared characteristics. Organisms are grouped together into taxa (singular: taxon), and these groups are given a taxonomic rank. Groups of a given rank can be aggregated to form a supergroup of higher rank, thus creating a taxonomic hierarchy. The principal ranks in modern classification are domain, kingdom, phylum (division is sometimes used in botany in place of phylum), class, order, family, genus, and species.

Who devised one of the earliest systems for animal and plant classification?

Swedish botanist Carolus Linnaeus is regarded as the father of taxonomy, as he developed a system known as Linnaean taxonomy for categorization of organisms and binomial nomenclature for naming organisms. His classification system divided organisms into two major groups: plants and animals. He categorized the organisms by perceived physical differences and similarities. Every plant and animal was given two scientific names (binomial method) in Latin, one for the species and the other for the group or genus within the species. This system of nomenclature (naming) continues to be used today.

What is systematics?

Systematics is the area of biology devoted to the classification of organisms. While the Linnaean system of classification was based on physical traits, systematics now includes the similarities of DNA, RNA, and proteins across species as criteria for classification.

Swedish botanist and zoologist Carolus Linnaeus is best remembered for formalizing the system of binomial nomenclature for naming organisms.

What is the difference between a domain and a kingdom?

A domain is a taxonomic category above the kingdom level. The three domains are Bacteria, Archaea, and Eukarya, which are the major categories of life. Essentially, domains are superkingdoms. A kingdom is a taxonomic group that contains one or more phyla. The four traditional kingdoms of Eukarya are Protista, Fungi, Plantae, and Animalia.

Who first proposed establishing a third kingdom in addition to plants and animals?

German zoologist Ernst Haeckel (1834–1919) proposed establishing a third kingdom—Protista—for simple organisms that did not appear to fit into either the plant or animal kingdom in 1866. The term "protist" is derived from the Greek word *protistos*, meaning "the very first."

What classification systems were proposed in the twentieth century?

In 1969, American plant ecologist Robert Harding Whitaker (1920–1980) proposed a system of classification based on five different kingdoms. The groups suggested by Whitaker were the bacteria group Prokaryote (originally called Monera), Protista, Fungi (for multicellular forms of nonphotosynthetic heterotrophs and single-celled yeasts), Plantae, and Animalia. This classification system is still widely accepted.

A six-kingdom system of classification, as proposed in 1977 by American microbiologist and biophysicist Carl Woese (1928–2012), included Archaebacteria and Eubacteria (both for bacteria), Protista, Fungi, Plantae, and Animalia. In 1981, Woese further proposed a classification system based on three domains (a level of classification higher than a kingdom): Bacteria, Archaea, and Eukarya. The domain Eukarya is further subdivided into four kingdoms: Protista, Fungi, Plantae, and Animalia. The system involving three domains is the most current classification scheme that is generally accepted by biologists.

What is cladistics?

Cladistics is one of the newest approaches to classification. It is often defined as a set of concepts and methods for determining cladograms, which portray branching patterns of evolution. Overall, it is a method of classification of organisms (e.g., a group of Gram-negative bacteria) according to the proportion of measurable characteristics that they have in common. Thus, the organisms are grouped together based on whether or not they have one or more shared characteristics that are unique and that come from the group's last common ancestor; in addition, the characteristics cannot be present in more distant ancestors.

CELLS

What is a cell?

A cell is the basic structural and functional unit of all forms of life. It is a membrane-bound unit that contains hereditary material (DNA) and cytoplasm. The study of microscopic cell structure is cytology and was established as a branch of biology in 1892 by the specialized investigations of Oscar Hertwig (1849–1922).

What is the cell theory?

The cell theory is the concept that all living things are made up of essential units called "cells." Diverse forms of life exist as single-celled organisms. More complex organisms, including plants and animals, are multicellular—cooperatives of many kinds of specialized cells that could not survive for long on their own. All cells come from preexisting cells and are related by division to earlier cells that have been modified in various ways during the long evolutionary history of life on Earth. Everything an organism does occurs fundamentally at the cellular level.

What is the origin of the term "cell"?

The term "cell" was first used by Robert Hooke, a British scientist who described cells he observed in a slice of cork in 1665. Using a microscope that magnified thirty times, Hooke identified little chambers or compartments in the cork that he called *cellulae*, a Latin word meaning "little rooms," because they reminded him of the cells inhabited by monks. It is from this word that we got the modern term "cell." He calculated that one square inch of cork would contain 1,259,712,000 of these tiny chambers or cells!

Which scientists made important discoveries associated with the cell?

Once Robert Hooke observed a cell, other scientists began to research and study cells.

- Henri Dutrochet (1776–1847) proposed in 1824 that animals and plants had similar cell structures.
- Robert Brown (1773–1858) discovered the cell nucleus in 1831. Matthias Schleiden (1804–1881) named the nucleolus (the structure within the nucleus now known to be involved in the production of ribosomes) around that time.
- Matthias Schleiden and Theodor Schwann (1810–1882) described a general cell theory in 1839, the former stating that cells were the basic unit of plants and Schwann extending the idea to animals.
- Robert Remak (1815–1865) was the first to describe cell division in 1855.
- Heinrich Wilhelm von Waldeyer-Hartz (1836–1921) observed and named chromosomes in the nucleus of a cell in 1888.
- Walther Flemming (1843–1905) was the first individual to follow chromosomes through the entire process of cell division in 1879 and published his work in 1882.

What is the difference between prokaryotic and eukaryotic cells?

French biologist Edouard Chatton (1883–1947) first proposed terms *procariotique* and *eucariotique* (French for prokaryotic and eukaryotic, respectively) in 1937. Prokaryotic, meaning "before nucleus," was used to describe bacteria, while eukaryotic, meaning "true nucleus," was used to describe all other cells. Today, the terms are more well defined: eukaryotic cells are much more complex than prokaryotic cells, having compartmentalized interiors and membrane-contained organelles, which are small structures within cells that perform specific functions within the cytoplasm. The major feature of a eukaryotic cell is its membrane-bound nucleus, which is the part of the cell that contains genetic information. Prokaryotic cells do not have a nuclear membrane. The cells also differ in size. Eukaryotic cells are generally much larger and more complex than prokaryotic cells. In fact, most eukaryotic cells are one hundred to one thousand times the volume of typical prokaryotic cells. The following chart lists the general differences between these two types of cells.

Characteristic	Prokaryotic Cells	Eukaryotic Cells
Organisms	Eubacteria and archaebacteria	Protista, fungi, plants, animals
Cell size	Usually 1–10 μ across	Usually 10–100 μ across
Membrane-bound organelles	No	Yes
Ribosomes	Yes	Yes
Mode of cell division	Cell fission	Mitosis and meiosis
DNA location	Nucleoid	Nucleus
Membranes	Some	Many
Cytoskeleton	No	Yes
Metabolism	Anaerobic or aerobic	Aerobic

What are organelles?

Organelles—frequently called "little organs"—are found in all eukaryotic cells; they are specialized, membrane-bound, cellular structures that perform a specific function. Eukaryotic cells contain several kinds of organelles, including the nucleus, mitochondria, chloroplasts, endoplasmic reticulum, and Golgi apparatus.

What is the cytoplasm?

All of the organelles of a cell are suspended in a watery medium called the cytoplasm, which is also the location of many of the cell's chemical reactions. The cytoplasm includes everything within the plasma or cell membrane except the nucleus. Everything within the nuclear membrane is called the nucleoplasm.

What are the largest and smallest organelles in a cell?

The largest organelle in a cell is the nucleus. The next largest organelle would be the chloroplast, which is substantially bigger than a mitochondrion. The smallest organelle in a cell is the ribosome.

What cell structures are unique to plant cells, and which ones are unique to animal cells?

The chloroplast, central vacuole, tonoplast, cell wall, and plasmodesmata are only found to occur in plant cells. Lysosomes and centrioles are found only in animal cells.

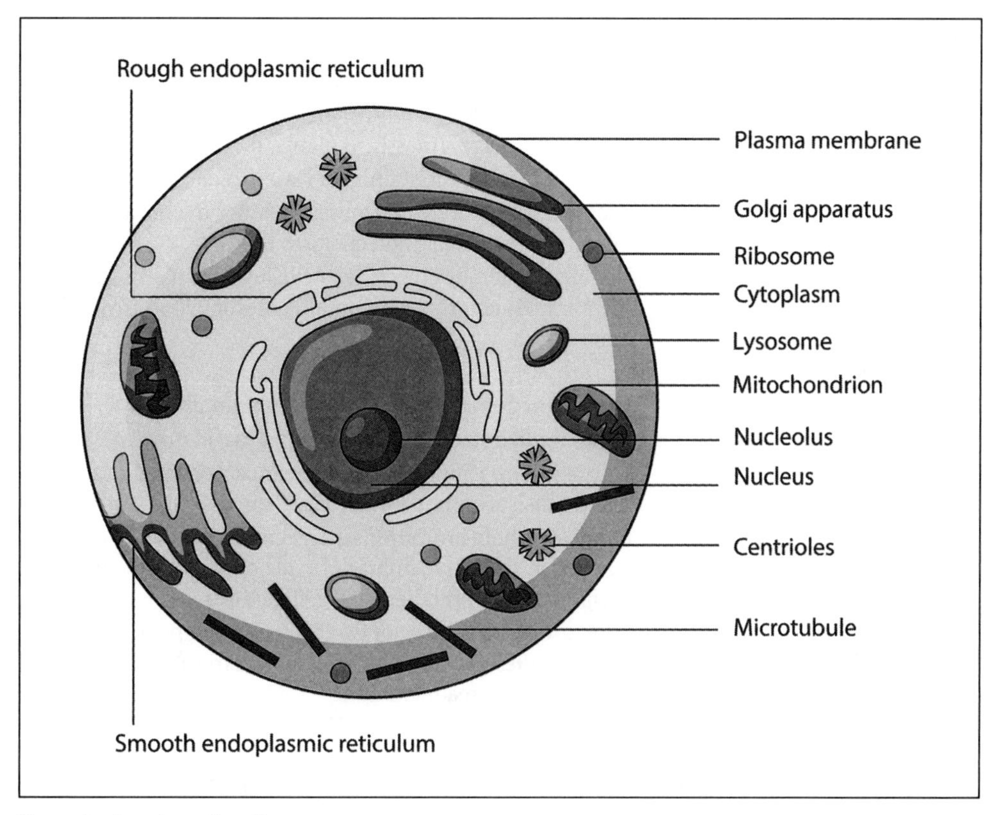

The parts of a eukaryotic cell.

What are the major components of the eukaryotic cell?

The major components of the eukaryotic cell are the cell nucleus, cytoplasmic organelles, and the cytoskeleton.

Structure	Description
Cell nucleus	Contains the cell's genetic information in the form of chromosomes
Nucleus	Large structure surrounded by double membrane
Nucleolus	Special body within nucleus; consists of RNA and protein
Chromosomes	Composed of a complex of DNA and protein known as chromatin; resemble rodlike structures after cell division
Cytoplasmic organelles	
Plasma membrane	Membrane boundary of living cell
Endoplasmic reticulum (ER)	Network of internal membranes extending through cytoplasm
Smooth endoplasmic reticulum	Lacks ribosomes on the outer surface
Rough endoplasmic reticulum	Ribosomes stud outer surface
Ribosomes	Granules composed of RNA and protein; some are attached to ER and some are free in cytoplasm

Structure	Description
Golgi complex	Stacks of flattened membrane sacs
Lysosomes	Membranous sacs (in animals)
Vacuoles	Membranous sacs (mostly in plants, fungi, and algae)
Microbodies (e.g., peroxisomes)	Membranous sacs containing a variety of enzymes
Mitochondria	Sacs consisting of two membranes; inner membrane is folded to form cristae and encloses matrix
Plastids (e.g., chloroplasts)	Double membrane structure enclosing internal thylakoid membranes; chloroplasts contain chlorophyll in thylakoid membranes
Cytoskeleton	
Microtubules	Hollow tubes made of subunits of tubulin protein
Microfilaments	Solid, rodlike structures consisting of actin protein
Centrioles	Pair of hollow cylinders located near center of cell; each centriole consists of nine microtubule triplets (9×3 structure)
Cilia	Relatively short projections extending from surface of cell; covered by plasma membrane; made of two central and nine peripheral microtubules ($9 + 2$ structure)
Flagella	Long projections made of two central and nine peripheral microtubules ($9 + 2$ structure); extend from surface of cell; covered by plasma membrane

How do the cells of bacteria, plants, and animals compare to each other?

	Bacterium	Plant (Eukaryote)	Animal (Eukaryote)
Cell wall	Present (protein polysaccharide)	Present (cellulose)	Absent
Plasma membrane	Present	Present	Present
Flagella and cilia	May be present	Absent except in sperm of a few species	Frequently present
Endoplasmic reticulum	Absent	Usually present	Usually present
Ribosome	Present	Present	Present
Microtubule	Absent	Present	Present
Centriole	Absent	Absent	Present
Golgi apparatus	Absent	Present	Present
Cytoskeleton	Absent	Present	Present
Nucleus	Absent	Present	Present
Mitochondrion	Absent	Present	Present
Chloroplast	Absent	Present	Absent
Nucleolus	Absent	Present	Present
Chromosome	A single circle of naked DNA	Multiple; DNA-protein complex	Multiple; DNA-protein complex

	Bacterium	Plant (Eukaryote)	Animal (Eukaryote)
Microbody	Absent	Present	Present
Lysosome	Absent	Absent	Present
Vacuole	Absent	Usually a large single vacuole	Absent

What is the major function of the nucleus?

The nucleus is the information center for the cell and the storehouse of the genetic information (DNA) that directs all of the activities of a living eukaryotic cell. It is usually the largest organelle in a eukaryotic cell and contains the chromosomes.

What are the main functions of the plasma membrane?

The main purpose of the plasma membrane is to provide a barrier that keeps cellular components inside the cell while simultaneously keeping unwanted substances from entering the cell. The membrane allows essential nutrients to be transported into the cell and aids in the removal of waste products from the cell. The specific functions of a membrane depend on the kinds of phospholipids and proteins present in the plasma membrane.

What are stem cells, and why is stem cell research important?

Stem cells are undifferentiated cells—meaning they do not have a specific function—that are capable, under certain conditions, of producing cells that can become a specific type of tissue. In adult humans, stem cells are found in bone marrow and other tissues, such as fat. However, most stem cell research is focused on stem cells present in fetal tissue, which, in a laboratory setting, can divide indefinitely and be stimulated into becoming a variety of different cell types. The potential benefits of stem cells have made the investigation of these cells an exciting research field. In fact, stem cells have been used to repair heart tissue or to renew injured structures, like the spinal cord. They could also be used as cell models for drug testing, thereby increasing the speed at which drugs are approved or finding new cures.

What is the cell cycle, and how is it controlled?

The life cycle of a single eukaryotic cell is known as the cell cycle. The cycle has two major phases: interphase and mitosis. When a cell is not dividing, it is in interphase; for

How thick is the plasma membrane?

The plasma membrane is only about 8 nanometers thick. It would take over eight thousand plasma membranes to equal the thickness of an average piece of paper.

example, a mature neuron conducting an impulse in the brain is in interphase. Interphase is broken down into the G_1 and G_2 phases—periods of growth during which a cell increases in size, complexity, and protein content. The G_1 phase prepares the cell for DNA synthesis (known as the S phase); G_2 prepares the cell for both mitosis and the synthesis of proteins. Although many cells eventually divide, it is an interesting observation that some cells remain in interphase almost indefinitely.

What is mitosis, and what are the stages of this process?

Mitosis involves the replication of DNA and its separation into two new daughter cells, which are genetically identical to the parent cell. While only four phases of mitosis are often listed, the entire process is actually comprised of six phases:

- Interphase: Involves extensive preparation for the division process.
- Prophase: The condensation of chromosomes; the nuclear membrane disappears; formation of the spindle apparatus; chromosomes attach to spindle fibers.
- Metaphase: Chromosomes, attached by spindle fibers, align along the midline of a cell.
- Anaphase: The entromere splits and chromatids move apart.

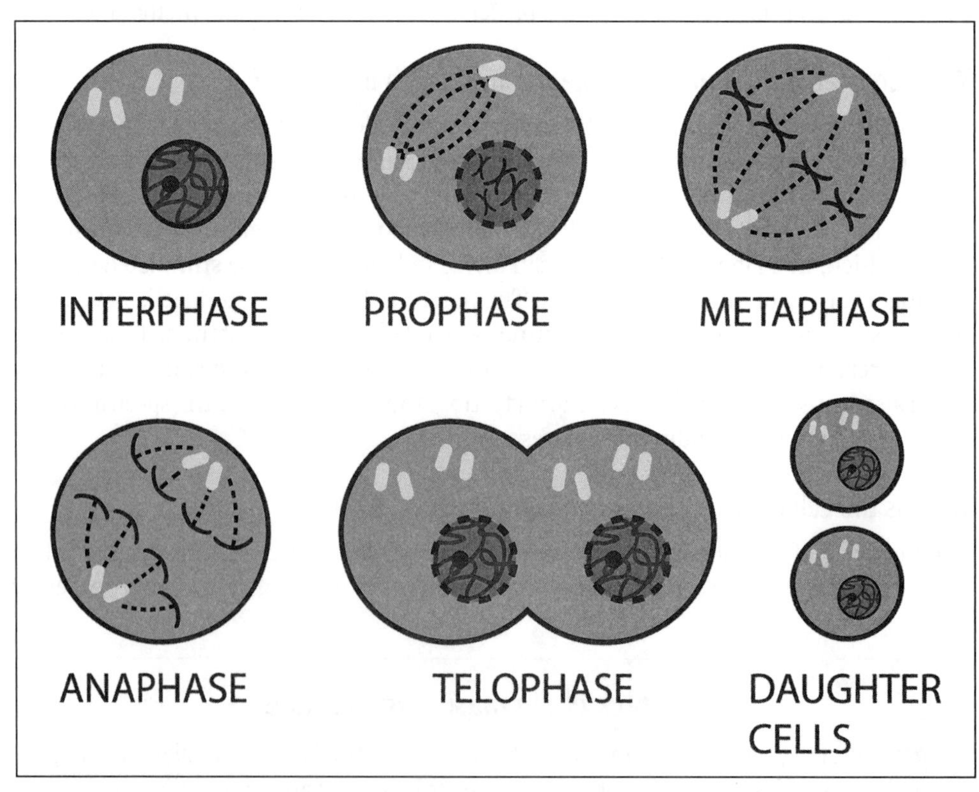

Cell division involves several stages to allow for the successful replication of DNA into the two resulting daughter cells.

Can artificial cells be made?

Research in progress at the National Aeronautical and Space Administration (NASA) is focused on cells as a means to deliver medicine in outer space; these cells are able to withstand dehydration and thus can be safely stored for long periods. Artificial cells are made of a polymer that acts like a cell membrane, but the polymer is stronger and more manageable than real membranes. These polymers are called polymersomes and can be made to cross-link with other polymers. Researchers feel that many different kinds of molecules can be encapsulated within these polymersomes and then delivered to specific target organs. An example would be an artificial blood cell that not only delivers oxygen but also medication.

- Telophase: The nuclear membrane reforms around newly divided chromosomes.
- Cytokinesis: The division of cytoplasm, cell membranes, and organelles occur. In plants, a new cell wall forms.

Do all cells undergo cell division?

No. Some cells, such as human brain cells, exist for a lifetime in a nondividing state.

Do all cells divide at the same rate?

No. All cells do not divide at the same rate. Cells that require frequent replenishing, such as skin or intestinal cells, may only take approximately 12 hours to complete a cell cycle. Other cells, such as liver cells, remain in a resting state (interphase) for up to a year before undergoing division.

How do cells communicate with each other?

Cells communicate with each other via small, signaling molecules that are produced by specific cells and received by target cells. This communication system operates on both a local and long-distance level. The signaling molecules can be proteins, fatty acid derivatives, or gases. Nitric oxide is an example of a gas that is part of a locally based signaling system and is able to signal for a human being's blood pressure to be lowered. Hormones are long-distance signaling molecules that must be transported via the circulatory system from their production site to their target cells. Plant cells, because of their rigid cell walls, have cytoplasmic bridges called plasmodesmata that allow cell-to-cell communication. Animals use gap junctions to transfer material between adjacent cells.

What is cell cloning, and how is it used in scientific research?

Cell cloning is the process by which an exact copy is made of a cell. This cellular process is known as mitosis and is required for the growth and repair of multicellular organisms. Different types of cells in the body differ in their ability to perform mitosis. Some cells, like skin cells, produce clones quite often. Others, like those of the nervous system, will not reproduce after they have reached maturity and have differentiated. The scientific purpose of cloning is to produce many copies of certain types of cells that can then be used for a variety of purposes, like basic research or the growth of replacement organs.

Can cells ever change functions?

The more specialized function that a cell performs, the less likely it is for the cell to change jobs within an organism. However, some cells have unspecialized functions and are able to adapt to the changing needs of the body. In mammals, a good example of cells with adapting functions would be bone marrow cells, which are responsible for producing different types of cells in the blood. Bone marrow cells produce red blood cells and five types of white blood cells. Slime molds of the kingdom Protista have cells that are capable of drastically changing their function. The cellular adaptations that can occur in slime molds allow them to change from single-celled amoebas to multicellular, reproductive spore producers.

What is cell culture?

Cell culture is the cultivation of cells (outside the organism) from a multicellular organism. This technique is very important to biotechnology processes because most research programs depend on the ability to grow cells outside the parent organism. Cells grown in culture usually require very special conditions (e.g., specific pH, temperature,

What are the oldest living cultured human cells?

The oldest living cultured human cells are the HeLa cell line. They were the first line of human cells to survive in a test tube (called *in vitro*) and have been a standard for investigating and understanding many human biological processes. All HeLa cells are derived from Henrietta Lacks (1920–1951), a thirty-one-year-old woman from Baltimore, Maryland, who died of cervical cancer. This cell culture led to advancements in many areas of biology and medicine, including cancer and HIV/AIDS and was also useful in the development of a polio vaccine in the 1950s. In addition, using this cell line, scientists discovered that 80–90 percent of cervical cancers (carcinomas) contain human papilloma virus DNA. In 2013, scientists did the first detailed sequence of the genome of a HeLa line. Along with discovering many abnormalities in the structure and number of chromosomes, they found other unexpected results. For example, they discovered that regions of the chromosomes in each cell were arranged in the wrong order and had either extra or too few copies of the genes. This was a sign of a newly discovered phenomenon (or concept) called chromosome shattering—one that is associated with about 2–3 percent of all cancers.

nutrients, and growth factors). Cells can be grown in a variety of containers ranging from a simple Petri dish to large-scale cultures in roller bottles, which are bottles that are rolled gently to keep a culture medium flowing over the cells.

Why do cells die?

Cells die for a variety of reasons, many of which are not deliberate. For example, cells can starve to death, asphyxiate, or die from trauma. Cells that sustain some sort of damage, such as DNA alteration or viral infection, frequently undergo programmed cell death. This process eliminates cells with a potentially lethal mutation or limits the spread of the virus. Programmed cell death can also be a normal part of embryonic development. Frogs undergo cell death that results in the elimination of tissues, allowing a tadpole to morph into an adult frog.

What is programmed cell death?

Apoptosis, or programmed cell death, is a process by which cells deliberately destroy themselves. The process follows a natural sequence of events controlled by the genes within a nucleus. First, the chromosomal DNA breaks into fragments, followed by the breakdown of the nucleus. The cell then shrinks and breaks up into pieces (called vesicles) that are absorbed (phagocytosis) by macrophages and neighboring cells. While programmed cell death may initially seem counterproductive, it plays an important part in maintaining the life and health of all living organisms. During human embryonic development, apoptosis removes the unnecessary webbing between the fingers and toes of the fetus, and it is also vital to the development and organization of both the immune and nervous systems.

VIRUSES

What is a virus?

A virus is an infectious, protein-coated fragment of DNA or RNA. Viruses replicate by invading host cells and taking over the cell's "machinery" for DNA replication. Viral particles can then break out of the cells, spreading disease.

Are viruses living organisms?

All living things have six common characteristics: 1) they adapt to their environment; 2) they have a cell makeup; 3) they have metabolic processes that help them obtain and use energy; 4) they move in response to their environment; 5) they grow and develop; and 6) they reproduce. Thus, viruses are not living organisms, as they cannot grow or reproduce (replicate) on their own. They need a host cell in which to grow and reproduce. In other words, they are inert outside their living host cell. British biologist Sir Peter Medwar (1915–1987), the 1960 Nobel Prize recipient in Physiology or Medicine, described viruses as "a piece of bad news wrapped in a protein." He was referring to the fact that viruses cause influenza, smallpox, hepatitis, yellow fever, polio, rabies, AIDS, and many other diseases.

What is the structure of a virus?

Viruses consist of strands of the genetic material (nucleic acid) either as DNA or RNA, but not both, surrounded by a protein coat called a capsid. The capsid protects the genome and is often subdivided into individual protein particles called capsomeres. The capsomeres are the features that create the shape of the virus.

What are the three main shapes of viruses?

A virus usually has one of three main shapes: 1) helical, resembling a wound spring, such as those of the tobacco mosaic virus; 2) icosahedral, a multifaceted virus, such as the herpes simplex; and 3) complex, which, as the name implies, can be a combination of shapes, such as the T-4 bacteriophage.

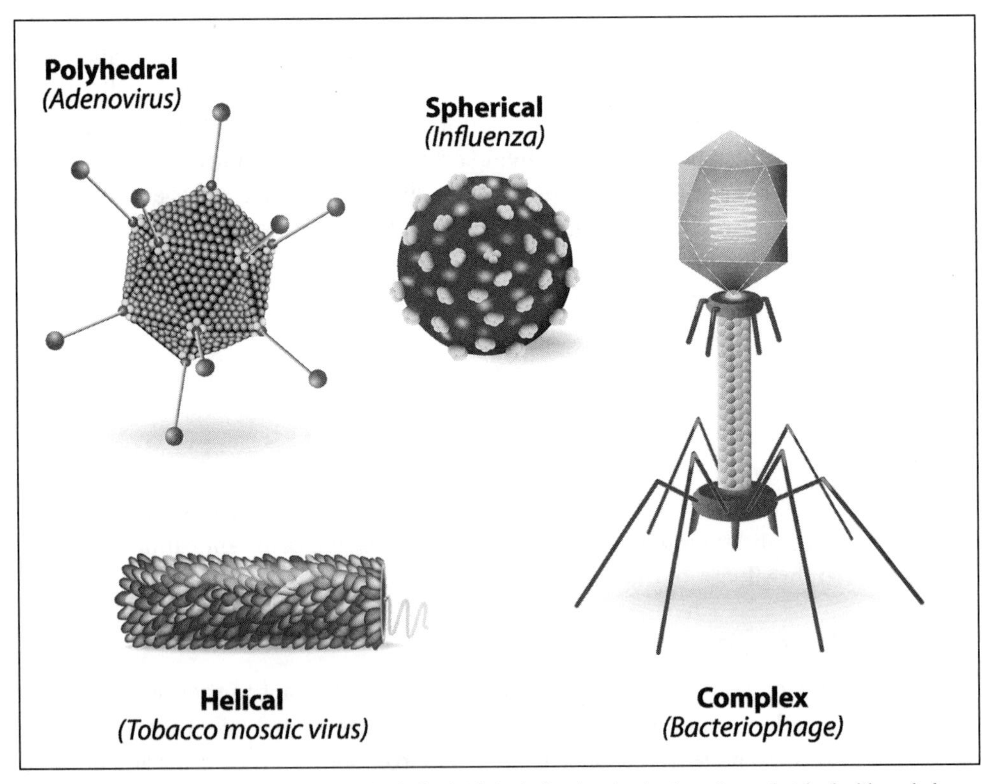

Viruses come in three basic forms: 1) polyhedral, which includes the classic virus shape that looks like a dodecahedron (top, left), a geometric shape with 12 sides such as the adenovirus, or spherical, such as in the case of the influenza virus; 2) helical, resembling a wound spring, such as those of the tobacco mosaic virus; and 3) complex, which, as the name implies, can be a combination of shapes, such as the T-4 bacteriophage, which consists of a geometric head and a tail.

How did viruses originate?

The most widely accepted hypothesis is that viruses are bits of nucleic acid that "escaped" from cells. According to this view, some viruses trace their origin to animal cells, some to plant cells, and others to bacterial cells. The variety of origins may explain why viruses are species-specific, that is, why some viruses only infect species that they are closely related to or the organisms from which they originated. This hypothesis is supported by the genetic similarity between a virus and its host cell.

Where are viruses found?

Viruses lie dormant in any environment (land, soil, or air) and on any material. They infect every type of cell—plant, animal, bacterial, and fungal.

What is the average size of a virus?

The smallest viruses are about 17 nanometers in diameter, and the largest viruses are up to 1,000 nanometers (1 micrometer) in length. By comparison, the bacterium *Escherichia coli* is 2,000 nanometers in length, a cell nucleus is 2,800 nanometers in diameter, and an average eukaryotic cell is 10,000 nanometers in length.

What is the difference between a virus and a retrovirus?

The difference between a virus and a retrovirus is a function of how each replicates its genetic material. A virus has a single strand of genetic material—either DNA or RNA. A retrovirus consists of a single strand of RNA. Once a retrovirus enters a cell, it collects nucleotides and assembles itself as a double strand of DNA that splices itself into the host's genetic material. Retroviruses were first identified by David Baltimore (1938–) and Howard Temin (1934–1994). They were awarded the Nobel Prize in Physiology or Medicine in 1975 for their discovery.

What was the first retrovirus to be discovered?

Dr. Robert Gallo (1937–) discovered the first retrovirus, the human T-cell lymphotropic virus (HTLV), in 1979. The second human retrovirus to be discovered was human immunodeficiency virus, HIV.

What is a bacteriophage?

A bacteriophage, also called a phage, is a virus that infects bacteria. The term "bacteriophage" means "bacteria eater" (from the Greek word *phagein*, meaning "to devour"). Phages consist of a long nucleic acid molecule (usually DNA) coiled within a polyhedral-shaped protein head. Many phages have a tail attached to the head. Fibers extending from the tail may be used to attach the virus to the bacterium.

When were bacteriophages first discovered?

Bacteriophages were discovered in the early 1900s by Frederick W. Twort (1877–1950) and Félix d'Hérelle (1873–1949). In 1915, Twort reported observing a filterable agent that destroyed bacteria growing on solid media, and d'Hérelle independently confirmed the discovery in 1917. It was actually d'Herelle who named the agent "bacteriophage." However, very few scientists accepted the findings on the growth and infectious nature of bacteriophages. It was not until 1934 that Martin Schlesinger (d. 1936) further characterized bacteriophages, establishing their unique place in the microbial world.

When did scientists first observe a virus infecting a bacterial cell?

It may seem like something out of a science fiction movie, but in 2013, scientists observed, for the first time, the detailed changes in a virus's structure—called T-7—as it infected a bacterium, *Escherichia coli*. The researchers observed that while the virus searched for prey, it extended one or two of six ultrathin fibers that are normally folded at the base of the virus's "head." When the prey is spotted, the T-7 virus extends the feeler-fibers, essentially walking around the cell until it finds the perfect spot to infect the prey's cell. Like a minitransformer, the T-7 virus changes its structure, injecting some of its proteins through a protein path in the bacterium's cell membrane—thus allowing the virus to easily send its genetic material into the prey. Once the transfer of the viral DNA is in the prey's cell, the pathway in the membrane seals up—and the infection is complete. Although this experiment only included the T-7 virus and *Escherichia coli*, scientists believe this may be typical of how many viruses attack other organisms.

What is the difference between a virus and a viroid?

Viroids are small fragments of nucleic acid (RNA) without a protein coat. They are usually associated with plant diseases and are several thousand times smaller than a virus.

What is a prion?

Prions are abnormal forms of natural proteins. Stanley Prusiner (1942–) first used the term "prion" in place of the expression "proteinaceous infectious particle" when describing an infectious agent. Scientists still have not discovered exactly how prions work. Current research shows that a prion is composed of about 250 amino acids, but no nucleic acid component has been found. Prions appear to accumulate in lysosomes. In the brain, it is possible that the filled lysosomes burst and damage cells. As diseased cells die, the prions contained in the cells are released and are able to attack other cells—thus, like viruses, prions are considered infectious agents. It is thought that prions are responsible for the group of brain diseases known as transmissible spongiform encephalopathies (TSEs)—in-

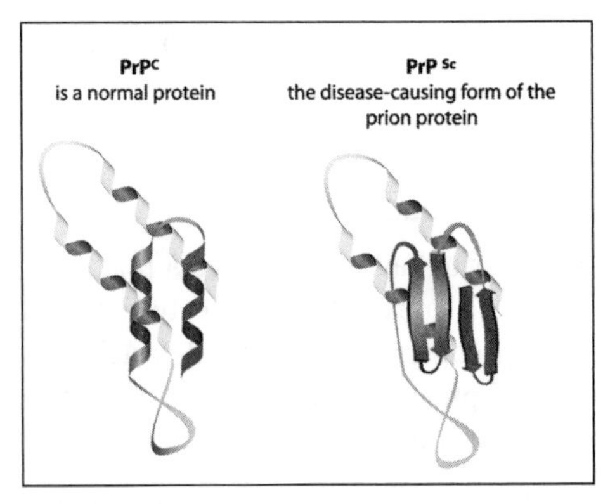

One peculiar property of prions is "misfolded" proteins. Scientists do not understand how prions work, but they seem to be proteins that have become abnormal for some reason.

cluding bovine spongiform encephalopathy (mad cow disease) when it occurs in cattle and Creutzfeldt-Jakob disease when it occurs in humans.

What is the difference between a plasmid and a prion?

A plasmid is a small, circular, self-replicating DNA molecule separate from the bacterial chromosome. Plasmids do not normally exist outside the cell and are generally beneficial to the bacterial cell. Plasmids are often used to pick up foreign DNA for use in genetic engineering. A prion is an infectious form of a protein or malformed protein that may increase in number by converting related proteins to more prions.

Why is influenza considered to be one of the most lethal viruses?

The influenza virus is probably the most lethal virus in human history. In fact, the symptoms of human influenza were described by Greek physician Hippocrates (c. 460–c. 377 B.C.E.) about 2,400 years ago. Since that time, many outbreaks have occurred. It is thought that Christopher Columbus (1451–1506) brought the virus to the Antilles in 1493, causing almost the entire population to die. Influenza was also responsible for fifty million deaths during a worldwide outbreak in 1918; furthermore, the 1957 Asian flu and the 1968 Hong Kong flu killed millions of people. According to statistics, seasonal outbreaks around the world often average between 250,000 and 500,000 deaths from the virus.

BACTERIA

What are the main components of a bacterial cell?

The major components of a bacterial cell are the plasma membrane, cell wall, and a nuclear region containing a single, circular DNA molecule. Plasmids—small, circular pieces of DNA that exist independently of the bacterial chromosome—are also present in a bacterial cell. In addition, some bacteria may have flagella, which aid in movement; pili or fimbriae, which are short, hairlike appendages that help bacteria adhere to vari-

Why is it difficult to treat viral infections with medications?

Antibiotics are not effective against viral infections because viruses lack the structures (for example, a cell wall in bacteria) with which antibiotics interfere and are effective in fighting off a bacterial infection. It is difficult to treat viral infections with medications without affecting the host cell, as viruses are dependent on the host cell's machinery during replication. Eventually, it is the body's immune system that fights off viral infections.

ous surfaces, including the cells that they infect; or a capsule of slime around the cell wall that protects it from other microorganisms.

How many genes are in a typical bacterial cell?

The bacterium *Escherichia coli* has about five thousand genes.

Do bacteria all have the same shape?

No, not all bacteria have the same shape. The spherical bacteria, called cocci (such as *Staphylococcus aureus*), occur singularly in some species and as groups in other species. Spherical bacteria have the ability to stick together and form a pair (diplococci) or stick together and form long chains called streptococci. In irregularly shaped clumps or clusters of bacteria, they are called staphylococci. Rod-shaped bacteria are called bacilli and can occur as single rods (*Bacillus subtilis*) or as long chains of rods (*Salmonella typhi*). Helical-shaped bacteria include spirilla (*Spirillum volutans*) and corkscrew-shaped spirochetes (*Borrelia garinii*).

Where are bacteria found?

Bacteria inhabit every place on Earth—including places where no other organism can survive. Bacteria have been detected as high as 20 miles (32 kilometers) above Earth and 7 miles (11 kilometers) deep in the waters of the Pacific Ocean. They are found in extreme environments, such as the Arctic tundra, boiling hot springs, and our bodies. Heat-tolerant bacteria have been found at a gold mine in South Africa at a level of 2.17 miles (3.5 kilometers) below Earth's surface, where the temperature in the mine was 149°F (65°C).

What is the Gram stain, and why is it important?

One of the most valuable tools for identifying specific bacteria is the Gram stain, developed in 1884 by Hans Christian Gram (1853–1938). It distinguishes between two different kinds of bacterial cell walls. Bacteria are stained with a violet dye and iodine, rinsed in alcohol, and then stained again with a red dye. The structure of the cell wall determines the staining response. Gram-positive bacteria have cell walls with a large amount of peptidoglycan that traps the violet stain. Gram-negative bacteria have less peptidoglycan and the violet dye is easily rinsed from Gram-negative bacteria, but the cells retain the red dye. Gram-negative bacteria are commonly more resistant than Gram-positive species to antibiotics because the outer membrane impedes entry of the drugs. Gram stain results can be combined with other information on a bacterium's cellular structure and biochemical characteristics to allow scientists to identify an unknown type of bacteria.

How do bacteria reproduce?

Bacteria reproduce asexually by binary fission—a process in which one cell divides into two similar cells. First the circular, bacterial DNA replicates, then a transverse wall is formed by an ingrowth of both plasma membrane and the cell wall.

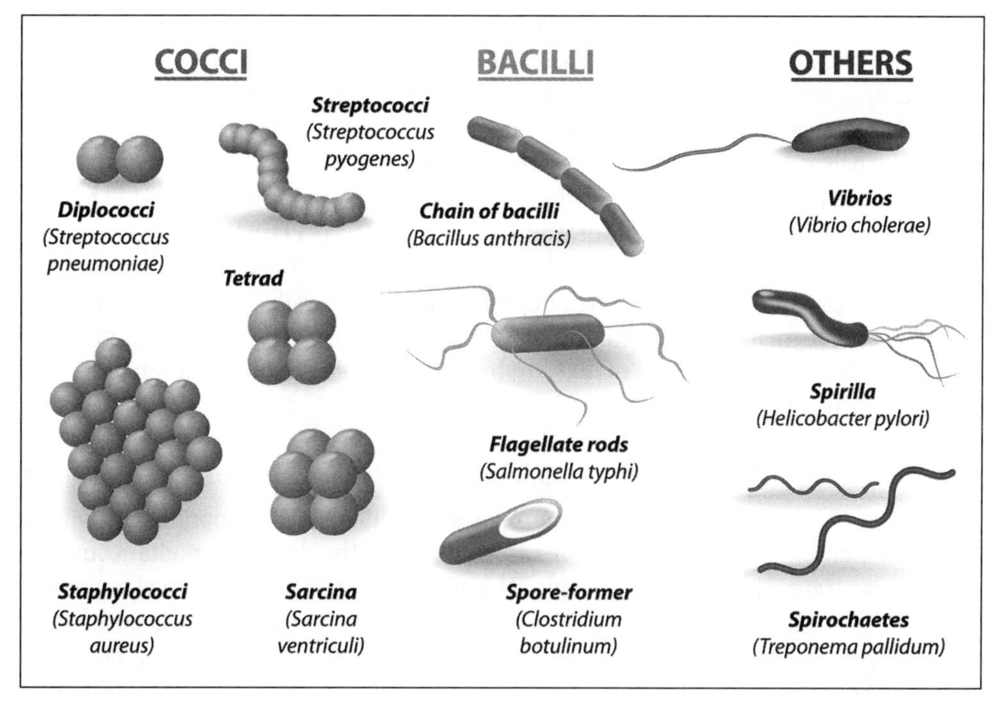

The main bacteria groups are cocci (spherical), bacilli (rod), spirilla (spiral), vibrios (comma), and spirochaetes (corkscrew).

How quickly do bacteria reproduce?

Bacteria can reproduce very rapidly in favorable environments in both laboratory cultures and natural habitats. The time required for a bacterial population to double is called generation time. For example, under optimal conditions, *Escherichia coli* can divide every 17 minutes. A laboratory culture started with a single cell can produce a colony of 10^7 (10,000,000) to 10^8 (100,000,000) bacteria in about 12 hours.

Generation Time for Selected Bacteria

Bacterium	Generation Time (in minutes)
Escherichia coli	17
Salmonella typhimurium	24
Staphylococcus aureus	32
Clostridium botulinum	35
Streptococcus lactis	48
Lactobacillus acidophilus	66

How are bacteria cultured?

Bacteria are usually cultured in Petri dishes that contain a culture medium, usually nutrient agar. Petri dishes were developed in 1887 by Julius Richard Petri (1852–1921), a member of Robert Koch's laboratory. The top of the dish is larger than the bottom so that when the dish is closed, a strong seal is created, preventing contamination of the culture. Agar was developed as a culture media for bacteria by Koch. He was interested in the isolation of bacteria in pure culture. Because isolation was difficult in liquid media, he began to study ways in which bacteria could be grown on solid media. After sterile, boiled potatoes proved unsatisfactory, a better alternative was suggested by Fannie E. Hesse (1850–1934), the wife of Walther Hesse (1846–1911), who was one of Koch's assistants. She suggested that agar, which she had used to thicken sauces, jams, and jellies, be used to solidify liquid nutrient broth. Agar is generally inexpensive and, once jelled, does not melt until reaching a temperature of 212°F (100°C). If 1–2 grams of agar are added to 100 milliliters of nutrient broth, it produces a solid medium that is not degraded by most bacteria.

How did the discovery of bacteria impact the theory of spontaneous generation?

The theory of spontaneous generation proposes that life can arise spontaneously from nonliving matter. One of the first scientists to challenge the theory of spontaneous generation was Italian physician Francesco Redi (1626–1698). Redi performed an experiment to show that meat placed in covered containers (either glass-covered or gauze-covered) remained free of maggots, while meat left in an uncovered container eventually became infested with maggots from flies laying their eggs on the meat. After the discovery of microorganisms by Antonie van Leeuwenhoek, the controversy surrounding spontaneous generation was renewed, as it had been assumed that food became spoiled by organisms arising spontaneously within food. In 1776, Lazzaro Spallanzani (1729–1799) showed that no growth occurred in flasks that were boiled after sealing. The controversy over the theory of spontaneous generation was finally solved in 1861 by Louis Pasteur. He showed that the microorganisms found in spoiled food were similar to those found in the air. He concluded that the microorganisms that caused food to spoil were from the air and did not spontaneously arise.

What are Koch's postulates?

Robert Koch was the first to identify that various microorganisms are the cause of disease. His four basic criteria of bacteriology, known as Koch's postulates, are still considered fundamental principles of bacteriology. The characteristics are as follows: 1) the organism must be found in tissues of animals that have been infected with the disease rather than in disease-free animals; 2) the organism must be isolated from the diseased animal and grown in a pure culture or *in vitro*; 3) the cultured organism must be able to be transferred to a healthy animal, which will show signs of the disease after having been exposed to the organism; and 4) the organism must be able to be isolated from the infected animal. Koch was awarded the Nobel Prize in Physiology or Medicine in 1905 for his research on tuberculosis.

How do bacteria compare to viruses?

Bacteria vs. Viruses

Characteristic	Bacteria	Viruses
Able to pass through bacteriological filters	No	Yes
Contain a plasma membrane	Yes	No
Contain ribosomes	Yes	No
Possess genetic material	Yes	Yes
Require a living host to multiply	No	Yes
Sensitive to antibiotics	Yes	No
Sensitive to interferon	No	Yes

How has the classification of bacteria evolved?

Early systems for classifying bacteria were based on structural and morphological characteristics displayed by the organisms—for example, bacterial shape, size, and the presence or absence of various elements within the bacterial cell. They were later classified by their biochemical and physiological traits, such as the best temperatures and pH ranges for growth, respiration, fermentation, and the types of carbohydrates used as an energy source. Other classifications were based on stains (using a dye to see bacteria under the microscope), such as the Gram stain. With newer technology, genetic and molecular characteristics are now used to understand the evolution of bacteria and their relationship to other organisms.

What groups are identified in the domain Bacteria?

Biologists recognize at least a dozen different groups of bacteria.

Major Group	Gram Reaction	Characteristics	Examples
Actinomycetes	Positive	Produce spores and antibiotics; live in soil environment	*Streptomyces*
Chemoautotroph	Negative	Live in soil environment; important in the nitrogen cycle	*Nitrosomonas*
Cyanobacteria	Negative	Contain chlorophyll and are capable of photosynthesis; live in aquatic environment	*Anabaena*
Enterobacteriaceae	Negative	Live in intestinal and respiratory tracts; ability to decompose materials; do not form spores; pathogenic	*Escherichia, Salmonella, Vibrio*

Major Group	Gram Reaction	Characteristics	Examples
Gram-positive cocci	Positive	Live in soil environment; inhabit the skin and mucous membranes of animals; pathogenic to humans	*Streptococcus, Staphylococcus*
Gram-positive rod	Positive	Live in soil environment, or animal intestinal tracts; anaerobic; disease-causing	*Clostridia, Bacillus*
Lactic acid bacteria	Positive	Important in food production, especially dairy products; pathogenic to animals	*Lactobacillus, Listeria*
Myxobacteria	Negative	Move by secreting slime and gliding; ability to decompose materials	*Chondromyces*
Pseudomonas	Negative	Aerobic rods and cocci; live in soil environment	*Pseudomonas*
Rickettsia and Chlamydia	Negative	Very small, intracellular parasites; pathogenic to humans	*Rickettsia, Chlamydia*
Spirochete	Negative	Spiral-shaped; live in aquatic environment	*Treponema, Borrelia*

What are Archaebacteria?

Archaebacteria (domain Archaea) are primitive bacteria that often live in extreme environments. This domain includes the following:

- Thermophiles ("heat lovers") live in very hot environments, including the hot sulfur springs of Yellowstone National Park, which reach temperatures ranging from 140 to 176°F (60–80°C).
- Halophiles ("salt lovers") live in locations with high concentrations of salinity, such as the Great Salt Lake in Utah, which have salinity levels that range from 15 to 20 percent. Seawater normally has a level of salinity of 3 percent.

Why do some bacteria cause disease and others do not?

All strains of bacteria possess genetic differences. These differences are not sufficient for them to be considered as separate species, but each strain is distinctive. For example, many different strains of *Escherichia coli* (frequently referred to as *E. coli*) exist. Some, such as *Escherichia coli* 0157:H7, cause serious diseases, while others live in the intestine and can be considered beneficial because they aid digestion. In fact, although billions of bacteria exist in the world, less than 1 percent of all bacteria cause disease.

- Methanogens, which obtain their energy by using hydrogen gas (H_2) to reduce carbon dioxide (CO_2) to methane gas (CH_4).

What is the most abundant group of organisms?

The Eubacteria are the most abundant group of organisms on Earth. More living species of Eubacteria inhabit the human mouth than the total number of mammals living on Earth.

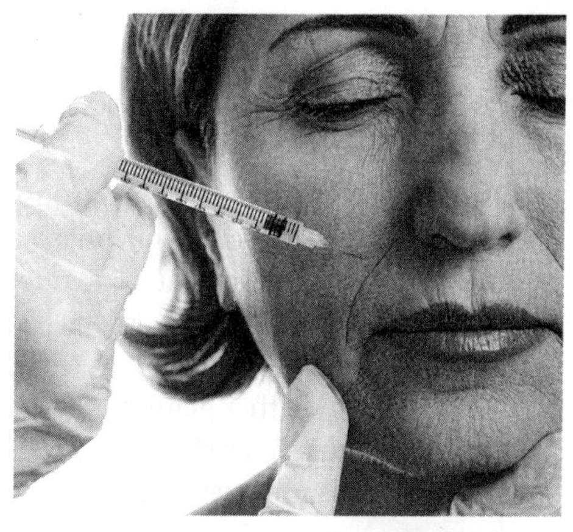

Those magical Botox injections dermatologists give people to reduce wrinkles contain a form of botulinum toxin. The toxin blocks nerve signals, causing muscles to relax and ease wrinkling for about three to six months.

How dangerous is *Clostridium botulinum*?

The bacterium *Clostridium botulinum* is very dangerous. It can grow in food products and produce a toxin called botulinum, the most toxic substance known. Microbiologists estimate that one gram of this toxin can kill fourteen million adults! This bacterium can withstand boiling water (212°F [100°C]) but is killed in 5 minutes in 248°F (120°C). This tolerance makes *Clostridium botulinum* a serious concern for people who can vegetables at home. If home canning is not done properly, this bacterium will grow in the anaerobic conditions of the sealed container and create extremely poisonous food. The endospores of *Clostridium botulinum* can germinate in poorly prepared canned goods, so individuals should never eat food from a can that appears swollen, as it is a sign that the can has become filled with gas released during germination. Consuming food from a can containing endospores that have undergone germination can lead to nerve paralysis, severe vomiting, and even death. Two major antiserums have been developed for botulism, but their effectiveness depends on how much is ingested and how long the toxin has been in the body.

What is Botox?

Botox, the trade name for botulinum toxin type A, is a protein produced by the bacterium *Clostridium botulinum*—yes, the same acute toxin that causes botulism. In this case, the botulinum toxin is purified, sterilized, and then converted to a form that can be injected and used in a medical environment.

PROTISTS

What are protists?

Protists are an extremely diverse group of predominantly microscopic organisms. All protists are eukaryotic, and while many are unicellular and have only one nucleus, they may be multicellular, have multiple nuclei, or form colonies of identical, unspecialized cells. Although most are microscopic, some are much larger, reaching lengths of over 1 inch (just over 2 millimeters). In early, traditional classifications, they were listed on the basis of being neither plant nor animal nor fungus. Further evidence suggested that protists exhibit characteristics of the plant, animal, and fungi kingdoms—but their classification is still highly debated.

What are the major groups of organisms in the kingdom Protista?

Few taxonomists agree on how to classify the protists, but they may be conveniently divided into seven general groups that share certain characteristics of locomotion, nutrition, and reproduction. Some of the general groupings of protists are listed below:

- Sarcodinas: Amoebas and related organisms that have no permanent locomotive structure
- Algae: Single-celled and multicellular organisms that are photosynthetic
- Diatoms: Photosynthetic organisms with hard shells formed of silica
- Flagellates: Organisms that propel themselves through water with flagella
- Sporozoans: Nonmotile parasites that spread by forming spores
- Ciliates: Organisms that have many short, hairlike structures on their cell surface associated with locomotion
- Molds: Heterotrophs with restricted mobility that have cell walls composed of carbohydrates

How do protists move?

Protists move in several different ways. Because they are exclusively found in aquatic or very moist environments, they have certain appendages to remain motile. Two of the most common ways are with cilia and flagella, both used by cells to move through watery environments. Cilia move back and forth, while flagella undulate in a whiplike motion, moving in the same direction as the cell's axis. Other protists use pseudopods (or "false feet") to move and are large, lobe-shaped extensions of the organism. However, not all protists move. Some are sessile—attached using certain structures (usually stalks) that adhere to a substrate. Other protists are both sessile and mobile; for example, many of the brown algae have free-floating sperm, whereas mature algae are attached to rock or other substrate and are not mobile.

How do protists reproduce?

Protists reproduce either asexually—by buddings, binary fission, or mitosis—or sexually, depending on their environment, life cycle, and type. During binary fission, the organism's DNA replicates and the cells divide; during budding, the organism produces a smaller bud of itself that will grow into an individual protist identical to the original. Some protists use a variety of reproductive methods. For example, paramecia reproduce using binary fission, but after so many hundreds of times essentially splitting apart, they use sexual reproduction to exchange genetic material. Scientists do not know what triggers this urge to sexually reproduce, but it may have to do with surrounding environmental stresses.

Which protists are indicators of polluted water?

Of the many protists that are found in aquatic environments, the euglenoids, typically among the largest algal cells, appear to be the best bioindicators. Euglenoids are unicellular flagellates and live in aquatic habitats than can vary in pH and light. Some euglenoids are capable of photosynthesis, while others do not carry on photosynthesis and are predatory. In general, though, because their population thrives under high nutrient levels—and human pollution often carries the decomposing organics needed by euglenoids—they are useful bioindicators of polluted water.

What is a slime mold?

A slime mold is an organism that uses spores to reproduce and resembles a gelatinous slime. They were once thought to be fungi but are now considered protists. They are divided into two groups: 1) Myxogastria or true slime molds and 2) Dictyosteliomycota or cellular slime molds. One cellular slime mold, *Dictyostelium discoideum*, has been studied as a model for the developmental biology of complex organisms. Under the best conditions, this organism lives as individual, amoeboid cells. When food is scarce, the cells stream together into a moving mass resembling a slug that can change into a stalk with a spore-bearing body at its top. This structure then releases spores that can grow into a new amoeboid cell. The development from identical, free-living cells to a multicellular organism simulates many of the properties of more complex and complicated organisms.

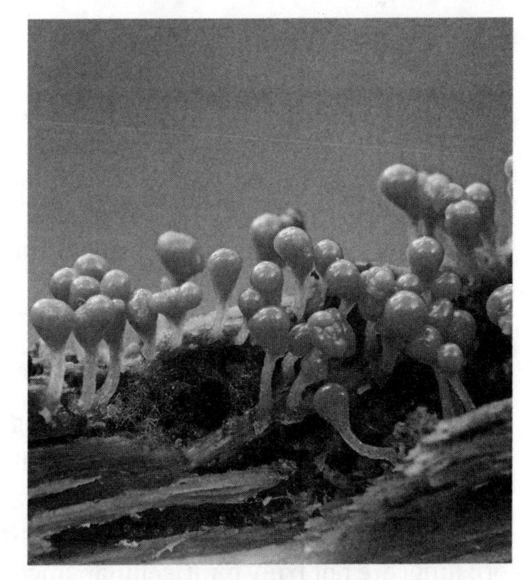

Some slime molds are less slimy looking than others, such as this *Trichia decipiens*, which is native to Asia.

How did the protist *Phytophthora infestans* influence Irish history?

Phytophthora infestans, one of the potato's most lethal pathogens, caused the late blight of potato disease. This pathogen was responsible for the Irish potato famine of 1845–1849. *Phytophthora infestans* causes the leaves and stem of the potato plant to decay, eventually causing the tuber to stop growing. In addition, the tubers are attacked by the pathogen and rot. It has been estimated that during the potato famine, 1.5 million Irish people emigrated from their country and moved to various parts of the world, but most immigrated to the United States. An estimated four hundred thousand people perished during the famine due to malnutrition.

What are diatoms?

Diatoms are microscopic algae of the phylum Chrysophyta in the kingdom Protista. Almost all diatoms are single-celled algae and dwell in both fresh- and saltwater. They are abundant in the cold waters of the northern Pacific and Antarctic oceans. Diatoms are yellow or brown in color and are an important food source for marine plankton and many small animals. Diatoms have hard cell walls. These "shells" are made from silica that has been extracted from the water. When diatoms die, the glassy shells, called frustules, sink to the bottom of the sea and harden into rock called diatomite. Diatoms are familiar to gardeners since diatomaceous earth is often used to control garden insect pests.

FUNGI

What is the scientific study of fungi called?

The scientific study of fungi (plural for fungus) is called mycology, from the Greek word *mycote*, meaning "fungus."

What characteristics do all fungi share?

In the earliest classification systems, fungi were classified as plants. The first classification system to recognize fungi as a separate kingdom was proposed in 1784. Researchers identified four characteristics shared by all fungi: fungi lack chlorophyll; the cell walls of fungi contain the carbohydrate chitin (the same tough material a crab shell is made of); fungi are not truly multicellular since the cytoplasm of one fungal cell mingles with the cytoplasm of adjacent cells; and fungi are heterotrophic eukaryotes (unable to produce their own food from inorganic matter), while plants are autotrophic eukaryotes.

What organisms are included in the kingdom Fungi?

Members of the kingdom Fungi range from single-celled yeasts to *Armillaria ostoyea*, a species that covers 2,220 acres (890 hectares)! Also included are mushrooms that are commonly consumed, the black mold that forms on stale bread, the mildew that grows on damp shower curtains, rusts, smuts, puffballs, toadstools, shelf fungi, and the death cap mushroom, *Amanita phalloides*. Of the bewildering variety of organisms that live on the planet Earth, perhaps the most unusual and peculiarly different one from human beings are fungi. Fungi are able to rot timber, attack living plants, spoil food, and afflict humans with athlete's foot and even worse maladies. Fungi also decompose dead organisms, fallen leaves, and other organic materials. In addition, they produce antibiotics and other drugs, make bread rise, and ferment beer and wine.

Where are fungi found?

Fungi grow best in dark, moist habitats, but they can be found wherever organic material is available. Moisture is necessary for their growth, and they can obtain water from the atmosphere as well as from the medium upon which they live. When the environment becomes very dry, fungi survive by going into a resting stage or by producing spores that are resistant to drying. The optimum pH for most species is 5.6, but some fungi can tolerate and grow in pH ranging from 2 to 9. Certain fungi can grow in concentrated salt solutions or sugar solutions, such as jelly or jam, that prevent bacterial growth. Fungi also thrive over a wide temperature range. Even refrigerated food may be susceptible to fungal invasion.

How old are fungi?

Although not as old as bacteria (fossil evidence suggests that bacteria may be at least 3.5 billion years old), fungi have been on Earth for hundreds of millions of years. The earliest fungi fossil evidence is from the Ordovician period, about 460–455 million years ago. Some scientists suggest that fungi may have played an important part in the colonization of vascular plants on land about 425 million years ago. Even when dinosaurs

What famous children's author studied and drew illustrations of fungi?

Beatrix Potter (1866–1943), perhaps best known for having written *The Tale of Peter Rabbit* in 1902, began drawing and painting fungi in 1888. She eventually completed a collection of almost three hundred detailed watercolors, which are now in the Armitt Library in Ambleside, England. In 1897, she prepared a scientific paper on the germination of *Aaricineae* spores for a meeting of the Linnean Society of London. Although her findings were originally rejected, experts now consider her ideas correct.

were on the planet some sixty-five million years ago, fungi were there also with some ninety-four-million-year-old fossils displaying mushroom-forming fungi similar to those that exist today.

How many species of fungi have been identified?

Scientists have identified between seventy and eighty thousand species of fungi. Approximately two thousand new species are discovered and identified each year. Some mycologists estimate that about 1.5 million species of fungi exist worldwide, placing them second only to insects in the number of different species.

Since fungi lack chlorophyll, which is necessary to produce their own food, how do they obtain food?

Fungi are saprophytes that absorb nutrients from wastes and dead organisms. Instead of taking food inside its body and then digesting it as an animal would, a fungus digests food outside its body by secreting strong, hydrolytic enzymes onto the food. In this way, complex organic compounds are broken down into simpler compounds that the fungus can absorb through the cell wall and cell membrane.

How many kinds of mushrooms are edible, and what are some of the more common ones?

Approximately two hundred varieties of mushrooms are edible, and about seventy species are poisonous. Some edible mushrooms are cultivated commercially. More than 844 million pounds (382,832 metric tons) of commercially grown mushrooms are produced in the United States each year. The following are some of the most popular edible mushrooms:

Edible Mushrooms

Common Name	Scientific Name
Black truffle	*Tuber melanosporum*
Chanterelle	*Cantharellus cibarius*
King bolete	*Boletus edulis*
Morels	*Morchella esculenta* (and others)
Oyster mushroom	*Pleurotus ostreatus*
Porcino	*Boletus edulis*
Shiitake	*Lentinula edodes*
Straw mushroom	*Volvariella volvacea*
Trompette des morts	*Craterellus corrucopioides*

What is unusual about *Amanita* mushrooms?

Some of the most poisonous mushrooms belong to the genus *Amanita*. Toxic species of this genus have been called such names as "death angel" (*Amanita phalloides*) and "destroying angel" (*Amanita virosa*). Ingestion of a single cap can kill a healthy, adult

Fungi come in a wide variety of forms, including this collection of edible mushrooms. Mmmmm!

human! Even ingesting a tiny bit of the amatoxin—the toxin present in species of this genus—may result in liver ailments that will last the rest of a person's life.

What antidote is available for mushroom poisoning?

No effective antidote for human poisoning by mushrooms has been discovered. The toxins produced by mushrooms accumulate in the liver and lead to irreversible liver damage. Unfortunately, poisoning may not become apparent for several hours after ingesting a toxic mushroom. When the symptoms do present, they often resemble typical food poisoning. Liver failure becomes apparent three to six days after ingesting the poisonous mushroom. Oftentimes, a liver transplant may be the only possible treatment.

How were fungi involved in World War I?

During World War I, the Germans needed glycerol to make nitroglycerin, which is used in the production of explosives, such as dynamite. Before the war, the Germans had imported their glycerol, but this impact was prevented by the British naval blockade during the war. German scientist Carl Neuberg (1877–1956) knew that trace levels of glycerol are produced when *Saccharomyces cerevisiae* is used during the alcoholic fermentation of sugar. He sought and developed a modified fermentation process in which the yeast would produce significant quantities of glycerol and less ethanol. The production of glycerol was improved by adding 3.5 percent sodium sulfite at pH 7 to the fermentation process, which blocked one chemical reaction in the metabolic pathway.

Neuberg's procedure was implemented with the conversion of German beer breweries to glycerol plants. The plants produced one thousand tons of glycerol per month. After the war ended, the production of glycerol was not in demand, so it was suspended.

Why is *Neurospora* an important fungus?

The pink bread molds of the genus *Neurospora* have long served as powerful laboratory models used to study genetics, biochemistry, and molecular biology. Scientists first demonstrated the concept that one gene produces a corresponding protein by studying *Neurospora*. Its ease of growth and the extensive genetic information available for this organism makes it a convenient model for the study of many processes found in higher plants and animals. Among the fungi, it is second only to yeast as a basic laboratory model organism.

What fungus plays an important role in human organ transplantation?

The fungus *Tolypocladium inflatum* is the source of cyclosporine, a medication that suppresses the immune reactions that cause organ transplant rejections. Cyclosporine does not cause the undesirable side effects that other immune-suppressing medications do. This remarkable drug became available in 1979, making it possible to resume organ transplants, which had essentially been abandoned. As a result of cyclosporine, successful organ transplants are commonplace today.

Why are fungi important in recycling?

Fungi are the ultimate recyclers on Earth. As the primary decomposers in the biosphere, they break down organic matter, including dead plants and other vegetation. As

Which fungus may have played a role in the Salem witch trials?

The Salem, Massachusetts, witch hunts of 1692 may have initially been caused by an infestation of a microbiological poison. The fungus *Claviceps purpurea*, commonly known as rye smut, produces the poison ergot. When ingested, this poison produces symptoms similar to the ones observed in the girls accused of being witches in Salem in 1692. Historians and biologists have reviewed environmental conditions in New England from 1690 to 1692 and have found that conditions were perfect for an occurrence of rye smut overgrowth. The weather conditions during those years were particularly wet and cool. Rye grass had replaced wheat as the principal grain because wheat had become seriously infected with wheat rust during long periods of cold and damp weather. The symptoms of ergot poisoning include convulsions, pinching or biting sensations, and stomach ailments as well as temporary blindness, deafness, and muteness.

fungi actively decompose materials, carbon, nitrogen, and the mineral components present in the organic compounds are released—all of which can also be recycled, with carbon dioxide released into the atmosphere and minerals returned to the soil. It is estimated that, on average, the top 8 inches (20 centimeters) of fertile soil contain nearly 11,023 pounds (5 metric tons) of fungi and bacteria per 2.47 acres (1 hectare). Without fungi acting as decomposers, dead organic matter would overpower the world, and life on Earth would eventually become impossible.

American chestnut trees almost became extinct because of the fungus *Cryphonectria parasitica*, which caused a blight that killed almost all of them.

How many species of fungi are plant pathogens?

More than eight thousand species of fungi cause diseases in plants. In fact, most diseases found in both cultivated and wild plants are caused by fungi. Some pathogenic fungi grow and multiply in their host plants; others grow and multiply on dead organic matter and host plants. Fungi that are pathogenic to plants can occur below the soil surface, at the soil surface, and throughout the body of a plant. Fungi are responsible for leaf spots, blights, rusts, smuts, mildews, cankers, scabs, fruit rots, galls, wilts, and tree diebacks and declines as well as root, stem, and seed rots.

Which native U.S. tree nearly became extinct due to a fungus?

The American chestnut (*Castena dentata*) was devastated by the chestnut blight. The American chestnut once made up almost half the population of hardwood forests in central and southern Pennsylvania, New Jersey, and southern New England. In its entire range, the species dominated deciduous forests, making up almost one-quarter of the trees. By the mid-1900s, the fungus *Cryphonectria parasitica* caused a disease commonly known as chestnut blight and destroyed nearly every mature specimen of the American chestnut tree in the United States. The blight most often attacks older trees, infecting a tree's layers of living bark and the adjacent layers of wood; this often creates cankers that cause the bark on the trunk to crack. The fungus continues to kill the cells that carry the food made in the leaves to other parts of a tree, and nutrients are not able to reach various parts of the tree. Although the fungus leaves the roots unaffected, allowing a tree to send up new sprouts and saplings, within a number of years, the "older" bark and wood of those trees can also become infected. The American Chestnut Foundation is dedicated to restoring the American chestnut to its native range through genetic studies and improvements in plant-breeding techniques.

Which other tree species has been affected by a fungus?

The fungus *Ophiostoma ulmi*, which causes Dutch elm disease, has caused widespread disease to elm trees. The fungus spores (carried by bark beetles) enter the cells in the outermost wood of trees. As the fungus grows, the tree's cells become plugged, and water and nutrients are not able to move from the roots to the top of a tree—eventually causing the tree to die. Dutch elm disease is believed to have originated in the Himalayas and traveled to Europe from the Dutch East Indies in the late 1800s. It emerged in Holland just after World War I and was first identified in 1930 in Cincinnati, Ohio—thought to have been carried on elm logs imported from Europe. By 1940, the disease had spread to nine states; by 1950, it was found in seventeen states and had spread into southern Canada. Today, it is found wherever elm trees grow throughout North America.

What cheeses are associated with fungi?

The unique flavor of cheeses such as Roquefort, Camembert, and Brie is produced by the action of members of the genus *Penicillium*. Roquefort is often referred to as "the king of cheeses"; it is one of the oldest and best known in the world. This "blue cheese" has been enjoyed since Roman times and was a favorite of Charlemagne (742–814), king of the Franks and emperor of the Holy Roman Empire. Roquefort is made from sheep's milk that has been exposed to the mold *Penicillium roqueforti* and aged for three months or more in the limestone caverns of Mount Combalou, near the village of Roquefort in southwestern France. This is the only place true Roquefort can be aged. It has a creamy, rich texture and is pungent, piquant, and salty. It has a creamy, white interior with blue veins; the cheese is held together with a snowy-white rind. True Roquefort is authenticated by the presence of a red sheep on the emblem of the cheese's wrapper.

Penicillium camemberti give Camembert and Brie cheeses their special qualities. Napoleon Bonaparte (1769–1821) is said to have christened Camembert cheese with its name; supposedly the name comes from the Norman village where a farmer's wife first served it to Napoleon. This cheese is formed of cow's milk cheese and has a white, downy rind and a smooth, creamy interior. When perfectly ripe and served at room temperature, the cheese should ooze thickly. Although Brie is made in many places, Brie from the region of the same name east of Paris is considered one of the world's finest cheeses by connoisseurs. Similar to Camembert, it has a white, surface-ripened rind and a smooth, buttery interior.

What is the role of yeast in beer production?

Beer is made by fermenting water, malt, sugar, hops, yeast (species *Saccaromyces spp.*), salt, and citric acid. Each ingredient has a specific role in the brewing of beer. Malt is produced from a grain—usually barley—that has sprouted, been dried in a kiln, and ground into a powder. Malt gives beer its characteristic body and flavor. Hops is made from the fruit that grows on the herb *Humulus lupulus* (a member of the mulberry family). The fruit is picked when ripe and is then dried; the ingredient gives beer a slightly bitter flavor. Yeast is used for the fermentation process.

Making beer is a complex process. One method begins by mixing and mashing malted barley with a cooked cereal grain, such as corn. This mixture, called "wort," is filtered before hops is added to it. The wort is then heated until it is completely soluble. The hops is removed, and after the mixture is cooled, yeast is added. The beer ferments for eight to eleven days at temperatures that range between 50° and 70°F (10°–21°C). The beer is then stored and kept at a state that is close to freezing. During the next few months, the liquid takes on its final character before carbon dioxide is added for effervescence. The beer is then refrigerated, filtered, and pasteurized in preparation for bottling or canning.

What are truffles?

Truffles, a delight of gourmets, are arguably the most prized edible fungi. Found mainly in Western Europe, they grow near the roots of trees (particularly oak but also chestnut, hazel, and beech) in open woodlands. Unlike typical mushrooms, truffles develop 3–12 inches (7.6–30.5 centimeters) underground, making them difficult to find. Truffle hunters use specially trained dogs and pigs to find the flavorful morsels. Both animals have a keen sense of smell and are attracted by the strong, nutlike aroma of truffles. In fact, trained pigs are able to pick up the scent of a truffle from 20 feet (6.1 meters) away. After catching a whiff of a truffle's scent, the animals rush to the origin of the aroma and quickly root out the precious prize. Once the truffle is found, the truffle hunter (referred to in French as *trufficulteur*) carefully scrapes back the earth to reveal the fungus. Truffles should not be touched by human skin, as doing so can cause the fungus to rot.

What do truffles look like?

A truffle has a rather unappealing appearance—round and irregularly shaped with a thick, rough, wrinkled skin that varies in color from almost black to off-white. The fruiting bodies present on truffles are fragrant, fleshy structures that usually grow to about the size of a golf ball; they range from white to gray or brown to nearly black in color. Nearly seventy varieties of truffles are known, but the most desirable is the black truffle—known as the black diamond—that grows in France's Périgord and Quercy regions as well as Italy's Um-

Yeast is what makes bread, beer, and wine possible in our cuisine.

bria region. The flesh of the black diamond appears to be black, but it is actually dark brown and contains white striations. The flesh has an aroma that is extremely pungent. The next most popular is the white truffle (actually off-white or beige) of Italy's Piedmont region. Both the aroma and flavor of this truffle are earthy and garlicky. Fresh truffles are available from late fall to midwinter and can be stored in the refrigerator for up to three days. Dark truffles are generally used to flavor foods such as omelets, polentas, risottos, and sauces. White truffles are usually served raw; they are often grated over foods such as pasta or dishes containing cheese, as their flavors are complementary. They are also added at the last minute to cooked dishes.

Why is the yeast ***Saccharomyces cerevisiae*** important in genetic research?

Biologists have studied *Saccharomyces cerevisiae*, a yeast used by bakers and brewers, for many decades because it offers valuable clues into the understanding of how more advanced organisms work. For example, yeast and humans share a number of similarities in their genetic composition. The DNA present in certain regions of yeast contain stretches of DNA subunits that are nearly identical to those in human DNA. These similarities indicate that humans and yeast both have similar genes that play a critical role in cell function. In 1996, a consortium of scientists from the United States, Canada, Europe, and Japan completed the genome of *Saccharomyces cerevisiae*—the first eukaryotic organism to be completely sequenced. They found that the genome is composed of about 12,156,677 base pairs and 6,275 genes, of which 5,800 are believed to be true functional genes that are organized on sixteen chromosomes. With their rapid generation time of 100 minutes, yeasts continue to be the organism of choice to provide new and significant insights into the functioning of eukaryotic cells and systems.

What are lichens?

Lichens are organisms that grow on rocks, tree branches, and bare ground. They are composed of two different entities living together in a symbiotic relationship: 1) a population of either algal or cyanobacterial cells that are single or filamentous and 2) fungi. Lichens do not have roots, stems, flowers, or leaves. The fungal component of the lichen is called the mycobiont (from the Greek words *mykes*, meaning "fungus," and *bios*, meaning "life"), and the photosynthetic component is called the photobiont (from the Greek words *photo*, meaning "light," and *bios*, meaning "life"). The scientific name given to the lichen is the name of the fungus and is most often an ascomycete. As the fungus has no chlorophyll, it cannot manufacture its own food but can absorb food from algae. Lichens and algae enjoy a symbiotic relationship. Lichens can often be found growing around and on top of algae, providing the algae protection from the Sun, thus decreasing the loss of moisture. Fungi and algae were the first organisms recognized as having a symbiotic relationship. A unique feature of this relationship is that it is so perfectly developed and balanced that the two organisms behave as a single organism.

What is the relationship between lichens and pollution?

Lichens are extremely sensitive to pollutants in the atmosphere and can be used as bioindicators of air quality. They absorb minerals from the air, from rainwater, and directly from their substrate. Lichen growth has been used as an indicator of air pollution, especially sulfur dioxide. Pollutants are absorbed by lichens, causing the destruction of their chlorophyll, which leads to a decrease in the occurrence of photosynthesis and changes in membrane permeability. Lichens are generally absent in and around cities, even though suitable substrates exist; the reason for this is the polluted exhaust from automobiles and industrial activity. They are beginning to disappear from national parks and other relatively remote areas that are becoming increasingly contaminated by industrial pollution. The return of lichens to an area frequently indicates a reduction in air pollution.

Lichens are also used to assess radioactive pollution levels in the vicinity of uranium mines, environments where nuclear-powered satellites have crashed, former nuclear bomb-testing sites, and power stations that have incurred accidents. Following the Chernobyl, Ukraine, nuclear power station disaster in 1986, arctic lichens as far away as Lapland were tested and showed levels of radioactive dust that were as much as 165 times higher than had been previously recorded.

What is the relationship between lichens and pollution?

Lichens are extremely sensitive to pollutants in the atmosphere and can be used as bioindicators of air quality. They absorb [illegible] from the air, from rainwater, and [illegible] [illegible] lichen growth has been used as an indicator of air pollution. [illegible] pollutants [illegible] by lichens, causing the destruction of [illegible] which leads to a decrease [illegible] photosynthesis and [illegible] around cities, [illegible] exhaust from automobiles and industrial [illegible] [illegible] national parks [illegible]

[illegible]

GENETICS

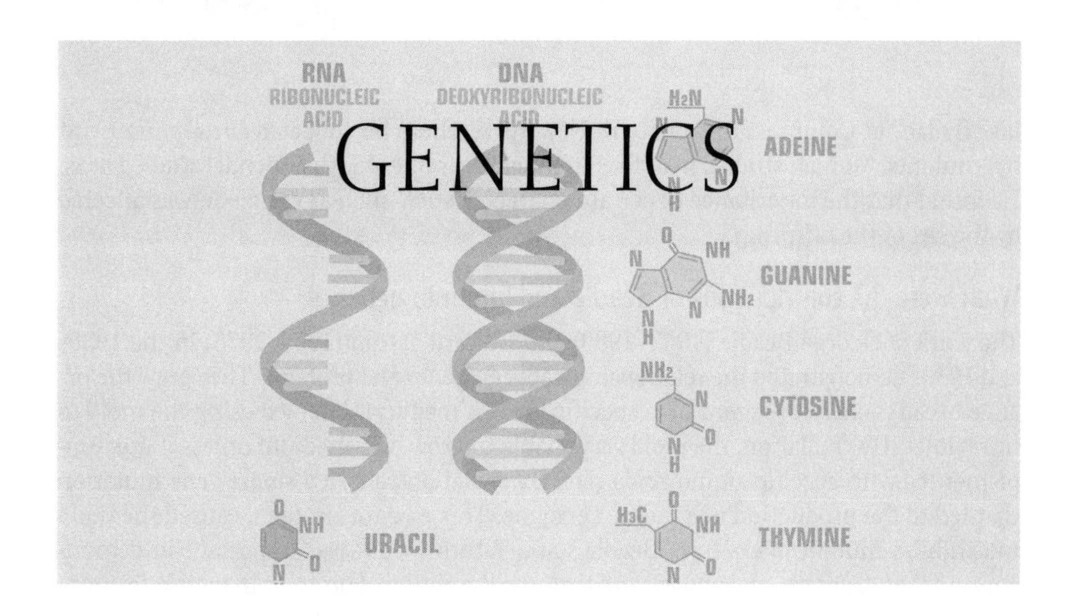

INTRODUCTION AND HISTORICAL BACKGROUND

Who was Mendel?

Gregor Mendel (1822–1884), an Austrian monk and biologist, is considered the founder of genetics. Mendel was the first to demonstrate the transmission of distinct physical characteristics from generation to generation through his work with the garden pea, *Pisum sativuum*. He studied peas of distinct and recognizable plant varieties, including the characteristics of height, flower color, pea color, pea shape, pod color, and the position of flowers on the stem. Mendel's theory was called particulate inheritance, in which certain inherited characteristics were carried by what he called elements, which eventually became the now well-known name "genes."

What is meant by the modern era of genetics?

Mendel's work was really not appreciated until advances in cytology enabled scientists to better study cells. In 1900, Hugo deVries (1848–1935) of Holland, Carl Correns (1864–1933) of Germany, and Erich von Tschermak (1871–1962) of Austria examined Mendel's original 1866 paper and repeated the experiments. In the following years, chromosomes were discovered as discrete structures within the nucleus of a cell. In 1917, Thomas Hunt Morgan (1866–1945), a fruit fly geneticist at Columbia University, extended Mendel's findings to the structure and function of chromosomes. This and subsequent findings in the 1950s were the beginning of the modern era of genetics.

What was Thomas Hunt Morgan's contribution to genetics?

Thomas Hunt Morgan won the Nobel Prize in Physiology or Medicine in 1933 for his discoveries of the role played by chromosomes in heredity. He is perhaps most noted for his "fly lab" at Columbia University in New York, where he collected *Drosophila* (fruit fly) mutants. Morgan studied fruit flies in much the same way that Mendel studied peas. He found that the inheritance of certain characteristics, such as eye color, was affected by the sex of the offspring.

What were the contributions of Beadle and Tatum to genetics?

The work of George Beadle (1903–1989) and Edward Tatum (1909–1975) in the 1930s and 1940s demonstrated the relationship between genes and proteins. They grew the orange bread mold *Neurospora* on a specific growth medium. After exposing the mold to ultraviolet (UV) radiation, the mold was unable to grow on a medium unless it was supplemented with specific amino acids. The UV radiation caused a single gene mutation that led to the production of a mutant enzyme. This enzyme, in turn, caused the mold to exhibit a mutant phenotype. Beadle's and Tatum's work was important in demonstrating that genes control phenotypes through the action of proteins in metabolic pathways. In other words, one gene affects one enzyme. Since Beadle and Tatum were awarded their Nobel Prize in 1958, scientists have discovered that the relationship between genes and their proteins is much more complex.

How was DNA shown to be the genetic material for all cellular organisms?

The proof that the material basis for a gene is DNA came from the work of Oswald T. Avery (1877–1955), Colin M. MacLeod (1909–1972), and Maclyn McCarty (1911–2005) in a paper published in 1944. This group of scientists followed the work of Frederick Griffith (1879–1941) in order to discover what causes nonlethal bacteria to transform to a lethal strain. Using specific enzymes, all parts of the *S* (lethal) bacteria were degraded, including the sugarlike coat, the proteins, and the RNA. The degradation of these substances by enzymes did not affect the transformation process. Finally, when the lethal bacteria were exposed to DNase, an enzyme that destroys DNA, all transformation activity ceased. The transforming factor was DNA.

What other scientists provided additional evidence that DNA was the genetic material?

In 1952, Alfred D. Hershey (1908–1997) and Martha C. Chase (also known as Martha Epstein; 1927–2003) further carried out investigations to prove that DNA rather than protein was the carrier of genetic material. Bacteriophages (viruses that infect and replicate in bacteria) were 1) grown with radioactive sulfur, which is incorporated into the protein coat of the phage or 2) grown with radioactive phosphorus, which is incorporated into the DNA of the phage. Hershey and Chase demonstrated that phosphorus (contained in DNA), not sulfur (contained in protein), was incorporated into the genetic material of bacteriophages that had multiplied within the bacterial cells.

What were some early ideas on evolution?

While some Greek philosophers had theories about the gradual evolution of life, Plato and Aristotle were not among them. In the 1700s, "natural theology" (the explanation of life as the manifestation of the creator's plan) held sway in Europe. This idea was the motive force behind the work of Carolus Linnaeus, who was the first to classify all known living things by kingdom. Also popular prior to the work of Charles Darwin (1809–1882) were the theories of "special creation" (creationism), "blending inheritance" (that offspring were always the mixture of the traits of their two parents), and "acquired characteristics."

Although the ideas of evolution and natural selection were not original to Charles Darwin, his *On the Origin of Species* was key to shifting the discussion toward a scientific explanation of how species arise.

What were Charles Darwin's contributions to evolutionary theory?

Charles Darwin first proposed a theory of evolution based on natural selection in his treatise *On the Origin of Species*. The publication of *On the Origin of Species* ushered in a new era in our thinking about the nature of man. The intellectual revolution it caused and the impact it had on man's concept of himself and the world were greater than those caused by the works of Isaac Newton and other individuals. The effect was immediate—the first edition sold out on the day of publication (November 24, 1859). *Origin* has been referred to as "the book that shook the world." Every modern discussion of man's future, the population explosion, the struggle for existence, the purpose of man and the universe, and man's place in nature rests on Darwin.

The work was a product of his analyses and interpretations of his findings from his voyages on the HMS *Beagle*. In Darwin's day, the prevailing explanation for organic diversity was the story of creation in the book of Genesis in the Bible. *Origin* was the first publication to present scientifically sound, well-organized evidence for the theory of evolution. Darwin's theory was based on natural selection in which the best, or fittest, individuals survive more often than those who are less fit. If a difference exists in the genetic endowment among these individuals that correlates with fitness, the species will change over time and will eventually resemble more closely (as a group) the fittest individuals. It is a two-step process: the first consists of the production of variation and the second of the sorting of this variability by natural selection, in which the favorable variations tend to be preserved.

DNA AND RNA

What are nucleic acids?

Deoxyribonucleic acid (DNA) and ribonucleic acid (RNA) are nucleic acids. They are molecules comprised of monomers (structural units of a polymer), known as nucleotides. These molecules may be relatively small (as in the case of certain kinds of RNA) or quite large (a single DNA strand may have millions of monomer units). Individual nucleotides and their derivatives are important in living organisms. For example, ATP, the molecule that transfers energy to cells, is built from a nucleotide as are a number of other molecules crucial to metabolism.

What is DNA, and what term was originally used for DNA?

DNA (deoxyribonucleic acid) is the genetic material for all cellular organisms. The elucidation of the structure of DNA is considered the most important molecular discovery of the twentieth century. DNA was originally called nuclein because it was first isolated from the nuclei of cells in 1869. In the 1860s, Johann Friedrich Miescher (1844–1895), a Swiss biochemist working in Germany at the University of Tubingen laboratory of Felix Hoppe-Seyler (1825–1895), was given the task of researching the composition of white blood cells. He found a good source of white blood cells from the used bandages that he obtained from a nearby hospital. He washed off the pus and isolated a new molecule from the nuclei of the white blood cells. He called the isolated substance nuclein, which we now call DNA.

Where is DNA found in a cell?

In addition to the nuclear DNA of eukaryotic cells, mitochondria (an organelle found in both plant and animal cells) and chloroplasts (found in plant and algal cells) both contain DNA. Mitochondrial DNA contains genes essential to cellular metabolism, while chloroplast DNA contains genetic information essential to photosynthesis.

How much DNA is in a typical human cell?

If the DNA in a single human cell were stretched out and laid end to end, it would measure approximately 6.5 feet (2 meters) long. The average human body contains 10–20 billion miles (16–32 billion kilometers) of DNA distributed among trillions of cells. If the total DNA in all the cells from one human were unraveled, it would stretch to the Sun and back more than five hundred times.

What are the component molecules of DNA?

The full name of DNA is deoxyribonucleic acid, with the "nucleic" part coming from the location of DNA in the nuclei of eukaryotic cells. DNA is actually a polymer (long strand) of nucleotides. A nucleotide has three component parts: a phosphate group, a five-carbon sugar (deoxyribose), and a nitrogen base. If you visualize DNA as a ladder, the sides of the ladder are made of the phosphate and deoxyribose molecules, and the rungs are made of two different nitrogen bases. The nitrogen bases are the crucial part of the molecule with regard to genes. Specific sequences of nitrogen bases make up a gene.

What are the nitrogenous bases of DNA?

The nitrogenous bases have a ring of nitrogen and carbon atoms with various functional groups attached. Nitrogenous bases come in two types. They differ in their structure: thymine and cytosine (pyrimidines) have a single-ring structure, while adenine and guanine (purines) have double-ring structures. When James Watson (1928–) and Francis Crick (1916–2004) were imagining how the bases would join together, they knew that the pairing had to be such that the molecule always had a uniform diameter. It therefore became apparent that a double-ring base must always be paired with a single-ring base on the opposite strand.

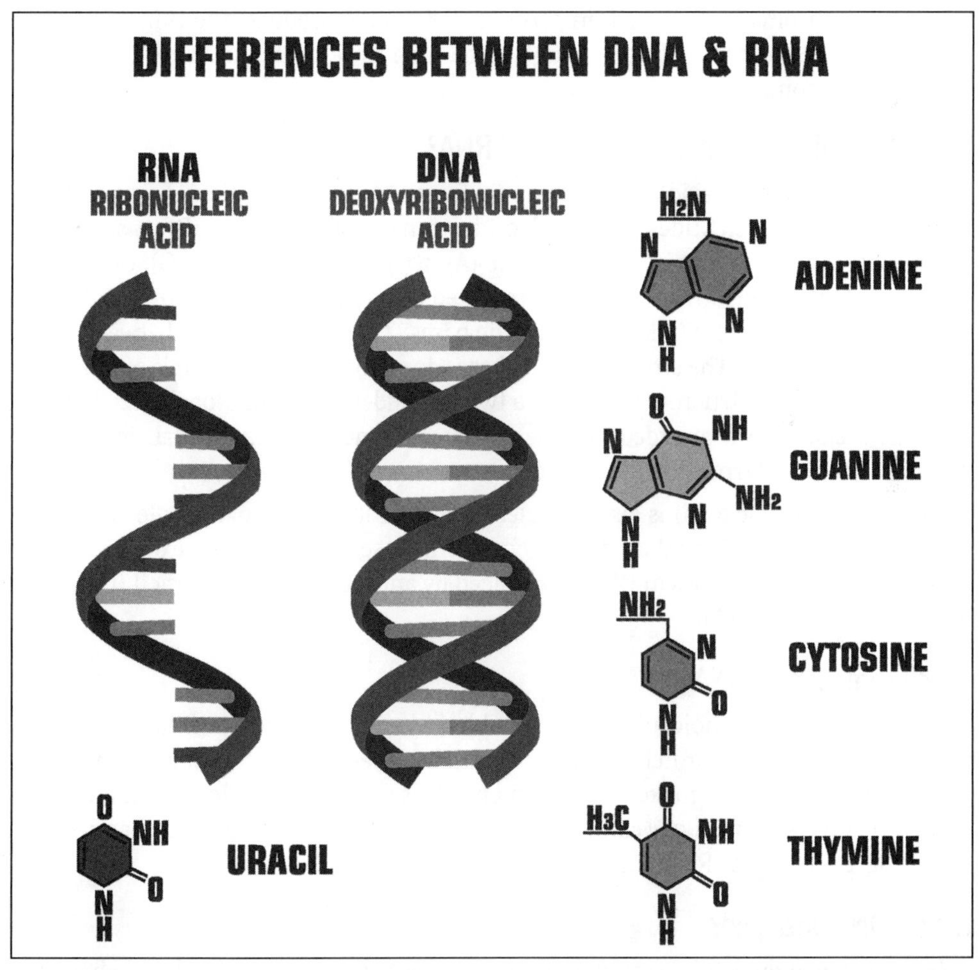

The basic difference between DNA and RNA is that DNA has a double helix structure and RNA is just one helix. Also, the thymine in DNA is replaced by uracil in RNA.

How is a DNA molecule held together?

Although DNA is held together by several different kinds of chemical interactions, it is still a rather fragile molecule. The nitrogen bases that constitute the "rungs" of the ladder are held together by hydrogen bonds. The "sides" of the ladder (the phosphate and deoxyribose molecules) are held together by a type of covalent bond called a phosphodiester bond. Because part of the DNA molecule is polar (the outside of the ladder) and the rungs (nitrogen bases) are nonpolar, other interactions—called hydrostatic interactions—occur between the hydrogen and oxygen atoms of DNA and water. The internal part of the DNA tends to repel water, while the external sugar-phosphate molecules tend to attract water. This creates a kind of molecular pressure that glues the helix together.

How is DNA organized in the nucleus?

Within the nucleus, DNA is organized with proteins, called histones, into a fibrous material called chromatin. As a cell prepares to divide or reproduce, the thin chromatin fibers condense, becoming thick enough to be seen as separate structures, which are called chromosomes.

What is the difference between DNA and RNA?

DNA (deoxyribonucleic acid) is a nucleic acid formed from a repetition of simple building blocks called nucleotides. The nucleotides consist of phosphate (PO_4), sugar (deoxyribose), and a base that is either adenine (A), thymine (T), guanine (G), or cytosine (C). In a DNA molecule, this basic unit is repeated in a double helix structure made from two chains of nucleotides linked between the bases. The links are either between A and T or between G and C. The structure of the bases does not allow other kinds of links. The famous double helix structure resembles a twisted ladder. The 1962 Nobel Prize in Physiology or Medicine was awarded to James Watson, Francis Crick, and Maurice Wilkins (1916–2004) for determining the molecular structure of DNA.

RNA (ribonucleic acid) is also a nucleic acid, but it consists of a single chain, and the sugar is ribose rather than deoxyribose. The bases are the same except that the thymine (T), which appears in DNA, is replaced by another base called uracil (U), which links only to adenine (A).

Which came first—DNA or RNA?

The first information molecule had to be able to reproduce itself and carry out tasks similar to those done by proteins. However, proteins, even though bigger and more complicated than DNA, can't make copies of themselves without the help of DNA and RNA. Therefore, RNA was the likely candidate for the first information molecule because scientists have found that RNA, unlike DNA, can replicate and then self-edit.

How is DNA unzipped?

DNA is unzipped during its replication process; the two strands of the double helix are separated, and a new complementary DNA strand is synthesized from the parent strands.

Also, during DNA transcription, one DNA strand, known as the template strand, is transcribed (copied) into a mRNA strand. In order for the two strands of DNA to separate, the hydrogen bonds between the nitrogen bases must be broken. DNA helicase (an enzyme) breaks the bonds. However, the enzyme does not actually unwind the DNA; special proteins first separate the DNA strands at a specific site on the chromosome. These are called initiator proteins.

What is needed for DNA replication?

DNA replication is a complex process requiring more than a dozen enzymes, nucleotides, and energy. Eukaryotic cells have multiple sites called origins of replication; at these sites, enzymes unwind the helix by breaking the double bonds between the nitrogen bases. Once the molecule is opened, separate strands are kept from rejoining by DNA-stabilizing proteins. DNA polymerase molecules read the sequences in the strands being copied and catalyze the addition of complementary bases to form new strands.

Is DNA always double-stranded?

No. DNA is not always double-stranded. Certain viruses have only a single strand of DNA. Furthermore, at temperatures greater than 176°F (80°C), eukaryotic DNA will become single-stranded. Single-stranded eukaryotic DNA does not always have the characteristic secondary helical structure and can form a hairpin, stem, or cross.

How does DNA correct its own errors?

Spontaneous damage to DNA occurs at a rate of one event per billion nucleotide pairs per minute. Assuming this rate in a human cell, DNA is damaged every 24 hours in ten thousand different sites in the body. DNA has a number of quality-control mechanisms. DNA polymerase (the enzyme that catalyzes DNA replication) has a proofreading function that immediately corrects 99 percent of these errors during replication. Those errors that pass through are corrected by a mismatch repair system. When a mismatch base (such as A–G) is detected, the incorrect strand is cut, and the mismatch is removed. The gap is then filled in with the correct base, and the DNA is resealed. However, some errors can be permanent.

How fast is DNA copied in prokaryotes and eukaryotes?

In prokaryotes, about one thousand nucleotides can be copied per second, so all of the 4.7Mb of *Escherichia coli* can be copied in about 40 minutes. Since the eukaryotic genome is immense compared to the prokaryotic genome, one might think that the eukaryotic DNA replication would take a very long time. However, actual measurements show that the chromosomes in eukaryotes have multiple replication sites per chromosome. Eukaryotic cells can replicate about five hundred to five thousand bases per minute; the actual time to copy the entire genome would depend on the size of their genome.

Is DNA always copied exactly?

Considering how many cells are in the human body and how often it occurs, DNA replication is fairly accurate. Spontaneous damage to DNA is low, occurring at the rate of one to one hundred mutations per ten billion cells in bacteria. The rate for eukaryotic genes is higher, about one to ten mutations per million gametes. The rate of mutation can vary according to different genes in different organisms.

What is polymerase chain reaction?

Polymerase chain reaction, or PCR, is a laboratory technique that amplifies or copies any piece of DNA very quickly without using cells. The DNA is incubated in a test tube with a special kind of DNA polymerase, a supply of nucleotides, and short pieces of synthetic, single-stranded DNA that serve as primers for DNA synthesis. With automation, PCR can make billions of copies of a particular segment of DNA in a few hours. Each cycle of the PCR procedure takes only about 5 minutes. At the end of the cycle, the DNA segment—even one with hundreds of base pairs—has been doubled. A PCR machine repeats the cycle over and over. PCR is much faster than the days it takes to clone a piece of DNA by making a recombinant plasmid and letting it replicate within bacteria.

PCR was developed by the biochemist Kary Mullis (1944–) in 1983 at Cetus Corporation, a California biotechnology firm. In 1993, Mullis, along with Michael Smith (1932–2000), won the Nobel Prize in Chemistry for the development of PCR.

What is a mutation?

A mutation is an alteration in the DNA sequence of a gene. Mutations are a source of variation to a population, but they can have detrimental effects in that they may cause diseases and disorders. One example of a disease caused by a mutation is sickle cell disease, in which a change occurs in the amino acid sequence (valine is substituted for glutamic acid) of two of the four polypeptide chains that make up the oxygen-carrying protein known as hemoglobin. However, not all mutations are bad. Although many individuals may use the term "mutant" in a negative or disparaging manner, mutations are important to a population's gene pool because of the variation that is contributed. Variations cannot occur without mutations.

When was RNA discovered?

By the 1940s, it was known that another kind of nucleic acid existed other than DNA, this one called RNA. Phoebus Levene (1869–1940), a Russian-born chemist, further refined the work of Albrecht Kossel (1853–1927). Kossel was awarded a Nobel Prize in 1910 for determining the composition of nuclein. At the time of Kossel's work, it was not clear that DNA and RNA were different substances. In 1909, Levene isolated the carbohydrate portion of nucleic acid from yeast and identified it as the pentose sugar ribose. In 1929, he

Two giants in early research into DNA and RNA were Albrecht Kossel (left) and Phoebus Levene. Kossel won a Nobel Prize for determining the composition of nucleic acids, and Levene figured out that DNA was composed of adenine, thymine, guanine, cytosine, deoxyribose, and a phosphate group.

succeeded in identifying the carbohydrate portion of the nucleic acid isolated from the thymus of an animal. It was also a pentose sugar, but it differed from ribose in that it lacked one oxygen atom. Levene called the new substance deoxyribose. These studies defined the chemical differences between DNA and RNA by their sugar molecules.

What are the three primary types of RNA?

The three primary types of RNA are 1) messenger RNA or mRNA, a single strand copied from a DNA strand that carries the genetic code from the DNA to the site of protein synthesis on the ribosomes; 2) transfer RNA or tRNA, the translation molecule that recognizes the nucleic acid message and converts it into polypeptide language; and 3) ribosomal RNA or rRNA, the most abundant type of RNA, which participates with mRNA in protein synthesis on the ribosome. Each tRNA molecule carries a specific anticodon, picks up a specific amino acid, and conveys the amino acid to the appropriate codon on mRNA in the production of the polypeptide.

What is a codon?

A codon is the three-unit sequence (AUG, ACG, etc.) of mRNA nucleotides that codes for a specific amino acid. Only twenty amino acids but sixty-four ($4 \times 4 \times 4$) possible codon sequences exist. Each codon codes for only one amino acid, but each amino acid may have more than one matching codon.

What is the genetic code?

The genetic code is a chart showing the relationship between each of the possible mRNA codons and their associated amino acids. The codons are grouped according to the amino acids for which they code. For example, two of the mRNA codons (UGU and UGC) can be associated with the production of the amino acid cysteine. Stop and start codons also exist. It is interesting to note that the genetic code contains only one start codon—AUG (and also codes for the amino acid methionine)—but three stop codons—UAA, UGA, and UAG.

What is transcription and translation?

In a cell, inherited information flows from DNA to RNA to protein. The two main stages of information flow are transcription and translation. In transcription, a gene (DNA) provides the instructions for synthesizing messenger RNA (mRNA) molecules. In translation, the information encoded in mRNA determines the order of amino acids that are joined to form a polypeptide on the ribosome.

GENES AND CHROMOSOMES

What is a gene, and what are its components?

The term "gene" describes a sequence of DNA that is a discrete unit of hereditary information. In addition to this information, each gene also contains a promoter region, which indicates where the coding information actually begins, and a terminator, which delineates the end of the gene.

What is the average size of a gene?

The average size of a vertebrate gene is about thirty thousand base pairs. Bacteria, because their sequences contain only coding material, have smaller genes of about one thousand base pairs each. Human genes are in the twenty to fifty thousand base pairs range, although sizes greater than one hundred thousand have been suggested as well.

How can the entire genome fit into the nucleus of a cell?

The average nucleus has a diameter of 5-6 μm (micrometers), and eukaryotic DNA has a length of 1-2 μm. In order to fit DNA into a nucleus, DNA and proteins are tightly packed to form threads called chromatin. These threads are so thick that they actually become visible with a light microscope.

How are genes controlled?

Genes are controlled by regulatory mechanisms that vary by whether the organism is a prokaryote or a eukaryote. Bacteria (prokaryote) genes can be regulated by DNA binding proteins that influence the rate of transcription or by global regulatory mechanisms that result in an organism's response to specific environmental stimuli, such as heat shock. This is particularly important in bacteria.

Gene control in eukaryotes depends on a complex set of regulatory elements that turn genes off and on at specific times. Among these regulatory elements are DNA binding proteins as well as proteins that, in turn, control the activity of the DNA binding proteins.

What is a jumping gene, and who discovered it?

A jumping gene is a gene that can move from one location to another on a chromosome or can even "jump" from one chromosome to another. Another name for a jumping gene is a transposon. Barbara McClintock (1902–1992), who worked on the cytogenetics of maize during the 1950s at Cold Spring Harbor Laboratory in New York, discovered that certain mutable genes were transferred from cell to cell during development of the corn kernel. McClintock made this inference based on observations of changing patterns of coloration in maize kernels over many generations of controlled crosses. She was awarded the Nobel Prize in Physiology or Medicine in 1983 for her work.

What is the difference between a gene and a chromosome?

The human genome contains twenty-three distinct, physically separate units called chromosomes. Arranged linearly along the chromosomes are tens of thousands of genes. The term "gene" refers to a particular part of a DNA molecule defined by a specific sequence of nucleotides. It is the specific sequence of the nitrogen bases that encodes a gene. The human genome contains about three billion base pairs, and the length of genes varies widely.

When were chromosomes first observed?

Chromosomes were observed as early as 1872, when Edmund Russow (1841–1897) described seeing items that resembled small rods during cell division; he named the rods *Stäbchen*. Edouard van Beneden (1846–1910) used the term *bâtonnet* in 1875 to describe nuclear duplication. The following year, 1876, Edouard Balbiani (1825–1899) de-

What is the p53 gene?

Discovered in 1979, p53—sometimes referred to as "The Guardian Angel of the Genome"—is a gene that, when a cell's DNA is damaged, acts as an "emergency brake" to halt the resulting cycle of cell division that can lead to tumor growth and cancer. It also acts as an executioner, programming damaged cells to self-destruct before their altered DNA can be replicated. However, when it mutates, p53 can lose its suppressive powers or have the devastating effect of actually promoting abnormal cell growth. Indeed, p53 is the most commonly mutated gene found in human tumors. Scientists have discovered a compound that could restore function to a mutant p53. Such a discovery may lead to the development of anticancer drugs targeting the mutant p53 gene.

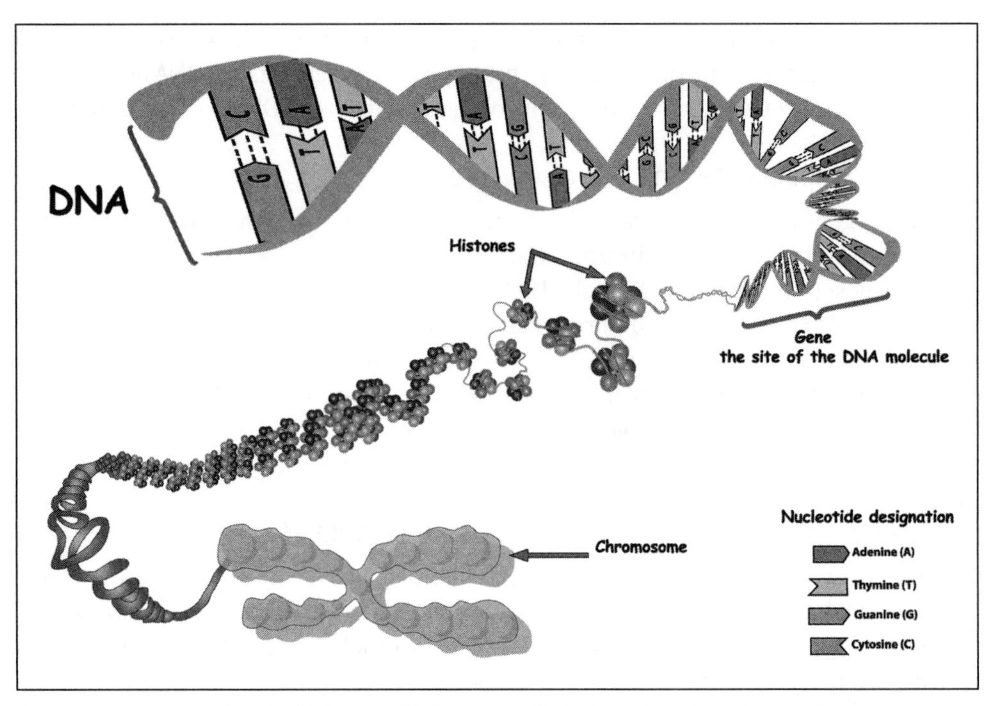

Chromosomes are long strands of DNA assembled upon a scaffold of proteins called histones. Chromosomes contain numerous, individual genes.

scribed that at the time of cell division, the nucleus dissolved into a collection of *bâtonnet étroits* ("narrow little rods"). Walther Flemming discovered that the chromosomal "threads," or *Fäden,* split longitudinally during mitosis.

How are chromosomes assembled?

Chromosomes are assembled on a scaffold of proteins (histones) that allow DNA to be tightly packed. Five major types of histones exist, and all have a positive charge. The positive charges of the histones attract the negative charges on the phosphates of DNA, thus holding the DNA in contact with the histones. These thicker strands of DNA and proteins are called chromatin. Chromatin is then packed to form the familiar structure of a chromosome. During mitosis, chromosomes acquire characteristic shapes that allow them to be counted and identified.

How many chromosomes are in a human body cell?

A human being normally has forty-six chromosomes (twenty-three pairs) in all but the sex cells. Half of each pair is inherited from the mother's egg and the other from the father's sperm. When the sperm and egg unite in fertilization, they create a single cell, or zygote, with forty-six chromosomes. When cell division occurs, the forty-six chromosomes are duplicated; this process is repeated billions of times over, with each of the cells containing the identical set of chromosomes. Only the gametes, or sex cells, are different. In their cell division, the members of each pair of chromosomes are separated and distributed to different cells. Each gamete has only twenty-three chromosomes.

What organisms have the most and fewest number of chromosomes?

Ophioglossum reticulatum, a species of fern also known as Adders-tongue, has the largest number of chromosomes with more than 1,260 (630 pairs). To date, the organism with the fewest number of chromosomes is the male Australian ant, *Myrmecia pilosula*, with one chromosome per cell (male ants are generally haploid—that is, they have half the number of normal chromosomes, while the female ant has two chromosomes per cell). Most bacteria also have a single chromosome that consists of a circular DNA molecule and associated proteins. Some of the more common animals and plants have different numbers of chromosomes. Dogs (*Canis lupus familiaris*) have seventy-eight; cats (*Felis catus*) have thirty-eight; mice (*Mus musculus*) have forty; fruit flies (*Drosophila melanogaster*) have eight; mosquitoes (*Aedes aegypti*) have six; potatoes (*Solanum tuberosum*) have forty-eight; and yeast (*Saccharomyces cerevisiae*) has thirty-two. Chromosome number is not necessarily correlated with the complexity of the organism.

What are sex chromosomes, and when is the sex of an organism determined?

Sex (X and Y) chromosomes are found in mammals. In humans, the number of chromosomes is forty-six (twenty-three pairs); of these, twenty-two pairs are autosomes, and the remaining pair of chromosomes is known as sex chromosomes, or the X and Y chromosomes. Each human cell normally contains two of these chromosomes: the XX lead-

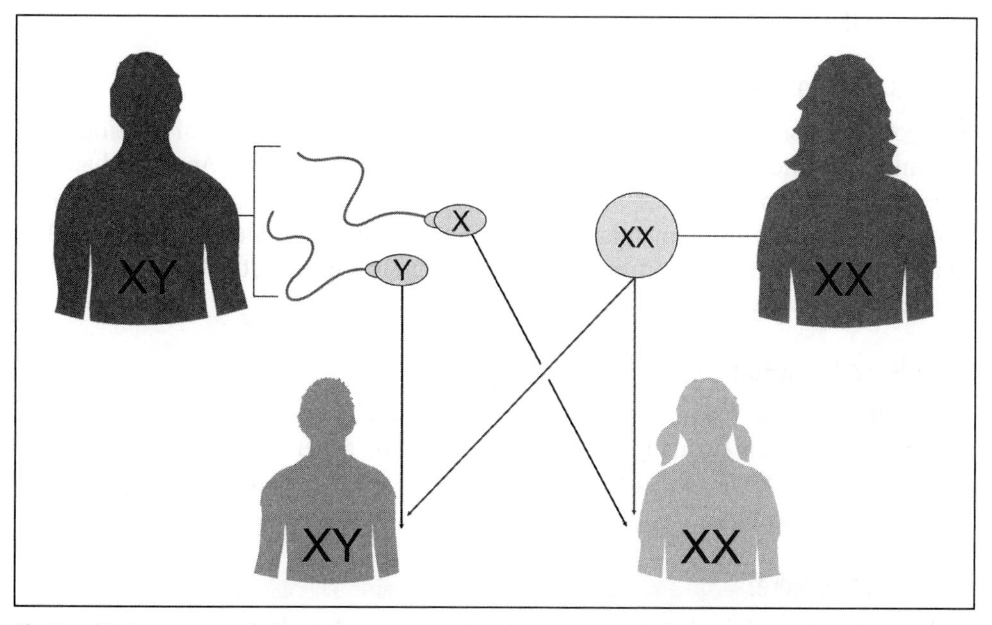

The X and Y chromosomes determine sex in mammals. A female has two X chromosomes, and a male has one X and one Y. The sex of a fertilized egg is set by whether the sperm from the male carries an X or Y.

ing to the female, and the XY leading to the male. The Y chromosome, found only in males, has the sex-determining gene, SRY. The SRY determines whether or not an individual will develop testes and produce testosterone that is responsible for male characteristics. No correlate gene on the X chromosome for the formation of ovaries is known. Therefore, it is the absence of the Y chromosome that determines a female. The sex of a new organism is determined at the instant the egg is fertilized by the sperm.

How can chromosomes become damaged?

Chromosomes can become damaged physically (in their appearance) and molecularly (in the specific DNA sequences they contain). Chromosomes can break randomly or due to exposure to ionizing radiation, which acts like a miniature cannonball, blasting the strands of DNA. Other factors, such as physical trauma or interactions with chemicals, can also cause breaks. When a chromosome breaks, the broken part may then reattach to another chromosome, a process called translocation. Sometimes, the broken chromosomes may attach to each other, forming a ring. Any type of physical or chemical change is known as a mutation.

Can people have missing or extra chromosomes?

Yes, people can live with this chromosomal abnormality, depending on which chromosomes are copied or missing. For example, Down syndrome (an extra copy of chromosome 21) and Turner syndrome (a female with only one X chromosome) are common chromosome abnormalities. Furthermore, almost 1 percent of all conceptions are triploid, which means that three copies of each chromosome are present, but over 99 percent of these die before birth.

What is a karyotype?

A karyotype is a snapshot of the genome. It can be used to detect extra or missing chromosomes, chromosomal rearrangements, or chromosomal breaks. Any cell that contains a nucleus can be used to make a karyotype. However, white blood cells seem to work best for preparing human karyotypes. After the cells are cultured, they are killed by using a drug that halts mitosis. The chromosomes are then stained, observed, and studied, and a size-order chart is produced.

What is a chromosome map?

A chromosome map lists the sequence of genes found on a particular chromosome. Chromosome maps are usually determined by breeding experiments in which the ratio of the offspring with certain combinations of traits indicates how far apart those traits are on the chromosomes.

What is a telomere?

At the end of eukaryotic chromosomes lies a unique structure known as a telomere. Experiments have shown that without telomeres, the chromosome structure could be

What was the first artificial chromosome created?

The first artificial or synthetic chromosome was reported in *Nature* in 1983. These earliest molecules were composed of 55,000 message-carrying units of deoxyribonucleic acid in a yeast. Naturally occurring yeast chromosomes contain from 150,000 to one million DNA message units. These first synthetic chromosomes were able to perform most, but not all, of the basic function of a chromosome in yeast. They were assembled from deoxyribonucleic acid derived from diverse sources and attached to a basic "backbone" of DNA. In 2014, the first complete and functional artificial chromosome was built by attaching synthetic strands of DNA together in a sequence based on the known genome of brewer's yeast. Scientists involved with the Synthetic Yeast Genome Project (Sc2.0) hope to build the first fully synthetic eukaryotic organism by 2020. They will swap all sixteen yeast chromosomes for engineered ones. As of 2018, they had constructed six of the sixteen chromosomes of the yeast's genome.

damaged, i.e., the DNA of the chromosome tends to stick to other pieces of DNA, and enzymes (deoxyribonucleases) are more likely to degrade or digest the ends of the chromosomes. Human telomeres have specific repetitive DNA sequences that may be repeated from 250 to 1,500 times.

What is the relationship between telomeres and cancer?

Increased telomerase activity can increase longevity of cells, but it is also implicated in cancer formation. Almost 90 percent of cancer cells have been found to have enhanced telomerase activity, and a cancer cell usually divides about eighty times before a tumor mass becomes large enough to be detected. In contrast, normal human cells usually divide thirty to fifty times before telomeres become too short and doubling stops.

BIOTECHNOLOGY AND GENETIC ENGINEERING

What is biotechnology?

Biotechnology is the use of a living organism to produce a specific product. It includes any technology associated with the manipulation of living systems for industrial purposes. In its broadest sense, biotechnology includes the fields of chemical, pharmaceutical, and environmental technology as well as engineering and agriculture.

What were some of the earliest uses of biotechnology?

The earliest examples of biotechnology involved using yeast and bacteria to make bread, vinegar, and alcoholic beverages, such as beer and wine. Archeological evidence indicates that beer was used as rations for workers in ancient Mesopotamia in 3100–3000 B.C.E.

How can biotechnology be used to manufacture vaccines?

Vaccine development is risky using conventional methods because vaccines must be manufactured inside living organisms, and the diseases themselves are extremely dangerous and infectious. Using genetic engineering, specific pathogen proteins that trigger antibody production are isolated and inserted into a bacterial or fungal vector. The vector organisms are then cultured to produce large quantities of the protein.

What is genetic engineering?

Genetic engineering, also popularly known as molecular cloning or gene cloning, is the artificial recombination of nucleic acid molecules in a test tube; their insertion into a virus, bacterial plasmid, or other vector system; and the subsequent incorporation of the molecules into a host organism, where they are able to propagate. The construction of such molecules has also been termed gene manipulation because it usually involves the production of novel genetic combinations by biochemical means. Genetic engineering techniques include cell fusion and the use of recombinant DNA or gene splicing.

In cell fusion, the tough, outer membranes of sperm and egg cells are stripped off by enzymes, then the fragile cells are mixed and combined with the aid of chemicals or viruses. The result may be the creation of a new life form from two species.

Recombinant DNA techniques, also called gene splicing, transfer a specific DNA sequence from one organism to another through the use of small, circular pieces of DNA and enzymes. The recombinant DNA process begins with the isolation and fragmentation of suitable DNA strands. After these fragments are combined with vectors, they are carried into bacterial cells, where the DNA fragments are "spliced" onto plasmid DNA that has been opened up. These hybrid plasmids are then mixed with host cells to form transformed cells. Since only some of the transformed cells will exhibit the desired characteristic or gene activity, the transformed cells are separated and grown individually in cultures. This methodology has been successful in producing large quantities of hormones (such as insulin) for the biotechnology industry. It is more difficult to transform animal and plant cells, yet the technique exists to make plants resistant to diseases and to make animals grow larger.

What was the first commercial use of genetic engineering?

Recombinant DNA technology was first used commercially to produce human insulin from bacteria. In 1982, genetically engineered insulin was approved for use by diabetics. Insulin is normally produced by the pancreas, and the pancreas of slaughtered animals, such as swine or sheep, was used as a source of insulin. To provide a reliable source of human insulin, researchers isolated the human insulin gene, made multiple copies of the gene, and inserted it into bacteria. When the bacteria were cultured, the gene was multiplied many times during each cell division. Insulin was then isolated from the bacterial culture.

Corn is the most genetically modified crop in the United States, with nearly 90 percent of all corn grown in America being GMO. Other heavily modified crops include soy, cottonseed, sugar beets, alfalfa, canola, and papaya.

What are some fields that use genetic engineering techniques?

- *Agriculture:* Crops having larger yields, disease- and drought resistancy; bacterial sprays to prevent crop damage from freezing temperatures; and livestock improvement through changes in animal traits.
- *Industry:* Use of bacteria to convert old newspaper and wood chips into sugar; oil- and toxin-absorbing bacteria for oil spill or toxic waste cleanups; and yeasts to accelerate wine fermentation.
- *Medicine:* Alteration of human genes to eliminate disease (experimental stage); faster and more economical production of vital human substances to alleviate deficiency and disease symptoms (but not to cure them) such as insulin, interferon (cancer therapy), vitamins, human growth hormone ADA, antibodies, vaccines, and antibiotics.
- *Research:* Modification of gene structure in medical research, especially cancer research.
- *Food processing:* Rennin (enzyme) in cheese production.

What are some examples of genetic engineering in plants?

Genetically engineered plants include transgenic crop plants that are resistant to herbicides used in weed control. These transgenic crops carry genes for resistance to herbicides such that all plants in a field are killed with the exception of the modified plant. Transgenic soybeans, corn, cotton, canola, papaya, rice, and tomatoes are used by many farmers in the United States. Plants resistant to predatory insects have also been genetically engineered.

What is bioinformatics?

Bioinformatics is the field of science in which biology, computer science, and information technology merge into a single discipline. The ultimate goal of the field is to enable the discovery of new biological insights as well as to create a global perspective from which unifying principles in biology can be discerned. Bioinformatics includes three important subdisciplines: 1) the development of new algorithms and statistics with which to assess relationships among members of the large data sets; 2) the analysis and interpretation of various types of data, including nucleotide and amino acid sequences, protein domains, and protein structures; and 3) the development and implementation of tools that enable efficient access and management of different types of information.

What is bioprospecting?

Bioprospecting involves the search for possible new plant or microbial strains, particularly from the world's largest rain forests and coral reefs. These organisms are then used to develop new phytopharmaceuticals. Some controversy is going on as to who owns the resources of these countries: the countries in which the resources reside or the company that turns them into valuable products.

What is biopharming?

Biopharming is a relatively new technology in which transgenic plants are used to "grow" pharmaceuticals. In this technique, scientists bioengineer medically important proteins into a corn plant, thus spurring the corn plant to produce large amounts of a particular medicinal protein. This is less expensive than using a microbial fermentation chamber.

What is a biopesticide?

A biopesticide is a chemical derived from an organism that interferes with the metabolism of another species. An example is the BT toxin *Bacillus thuringiensis*, which interferes with the absorption of food in insects but does not harm mammals.

What is bioterrorism?

Bioterrorism is the use of biological substances or toxins with the goal of causing harm to humans. Biotechnology can be used to manufacture biological weapons, such as large amounts of anthrax spores. However, biotechnology can also be used positively to identify bioweapons. A new faster method of PCR, called continuous flow PCR, uses a biochip and requires only nanoliter amounts of DNA to detect a bioweapon.

What is bioethics?

Bioethics is the study of the ethical aspects of biotechnology. Specific concerns include 1) access to personal genetic information; 2) privacy and confidentiality of genetic information; 3) the effects of personal genetic information on the individual and how society perceives that individual; 4) use of genetic information in reproductive issues; 5) philosophical implications regarding how genes can affect behavior and enhancement of genes; and 6) commercialization issues concerning patents on genes and gene products.

What was the Human Genome Project?

The Human Genome Project (HGP) began in 1990 as a thirteen-year effort and was completed in 2003. The goals of the HGP were:

- Identify all of the approximately thirty to forty thousand genes in human DNA
- Determine the sequence of the three billion chemical pairs of human DNA
- Store this information in public databases
- Improve tools for data analysis
- Transfer related technologies to the private sector
- Address the ethical, legal, and social issues that may stem from the project

At London's Wellcome Collection one can find this bookshelf, which is filled with the printout of the full human genome.

What is proteomics?

Proteomics is the study of proteins encoded by a genome. This field extends the Human Genome Project and is a far more complex study than finding where genes are located on chromosomes. Proteins are dynamic molecules that can change according to the needs of a cell, and complete understanding of cell metabolism requires that scientists understand all of the proteins involved as well as their genes.

How much information does the human genome contain?

The amount of genetic information in the human genome is equivalent to the amount of information in two hundred telephone books of one thousand pages each! If the twenty-three chromosomes in one human cell were removed, unwound, and placed end to end, the DNA would stretch more than 5 feet (1.5 meters) and be fifty-trillionths of an inch wide.

When was the first genome sequenced?

The first eukaryotic organism to be completely sequenced was *Saccharomyces cerevisiae*, a yeast, in 1996. An international consortium of scientists from the United States, Canada, Europe, and Japan found that the genome is composed of about 12,156,677 base pairs and 76,275 genes organized on sixteen chromosomes. Only about 5,800 are believed to be true functional genes.

When did scientists sequence the wheat genome?

The first fully annotated sequence of the wheat genome was published in 2018. The wheat genome is five times longer than the human genome and forty times longer than the rice genome. It is also very complex, consisting of three distinct subgenomes and containing many repetitive sequences. More than two hundred scientists from twenty countries collaborated for more than twelve years to determine the coordinates for all of the 108,000 genes on the chromosomes for the wheat plant. Not only have the scientists located the exact position of the wheat genes, they also know their genomic sequence, which contains all the information about how genes are regulated in time and space. Scientists hope that plant breeders will be able to use information about the wheat genome to improve the nutritional value and disease resistance of wheat. They may be able to breed plants with bigger grains or more grains per stalk, provide more protein, or be more resistant to heat and/or drought. Since wheat is the most widely grown crop in the world, hope exists that this new knowledge will reduce the risk of food scarcity.

Which individuals were awarded the Nobel Prize for their work on sequencing of DNA?

Frederick Sanger and Walter Gilbert (1932–) shared the Nobel Prize in Chemistry in 1980 for contributions in determining the nitrogen base sequences of DNA. This method was later referred to as the Sanger sequencing method for reading DNA. Sanger won the Nobel Prize in Chemistry in 1958 for his work on the structure of proteins (especially insulin), making him one of only four individuals to receive two Nobel prizes.

What are the types of genetic testing?

More than one thousand genetic tests are available and in use to identify changes in chromosomes, genes, or proteins. The three methods used for genetic testing are:

1. Molecular genetic tests, which study single genes or short lengths of DNA to identify variations or mutations that lead to a genetic disorder.
2. Chromosomal genetic tests, which analyze whole chromosomes or long lengths of DNA to see whether large genetic changes exist, such as an extra copy of a chromosome, that cause a genetic condition.
3. Biochemical genetic tests, which study the amount or activity level of proteins. Abnormalities in either can indicate changes to the DNA that result in a genetic disorder.

Who was the first individual to find the gene for breast cancer?

Mary-Claire King (1946–) determined that in 5–10 percent of women with breast cancer, the cancer is the result of a mutation of a gene on chromosome 17, the BRCA1 (Breast Cancer 1). She announced her findings at the American Society of Human Genetics annual meeting in 1990. The BRCA1 gene is a tumor suppressor gene and is also linked to ovarian cancer. A research team led by Mark Skolnick (1946–) succeeded in pinpointing the exact location on chromosome 17 in September 1994.

What is gene therapy?

Gene therapy involves replacement of an "abnormal" disease-causing gene with a "normal" gene. The normal gene is delivered to target cells using a vector, which is usually a virus that has been genetically engineered to carry human DNA. The virus genome is altered to remove disease-causing genes and insert therapeutic genes. Target cells are infected with the virus. The virus then integrates its genetically altered material containing the human gene into the target cell, where it should produce a functional protein product.

Which is the first gene therapy approved by the Food and Drug Administration in the United States?

In 2018, the Food and Drug Administration (FDA) approved a "living drug" called tisagenlecleucel (pronounced tis-agen-LEK-loo-sell) to treat acute lymphoblastic leukemia (ALL) in children and young adults who stopped responding to chemotherapy. Tisagenlecleucel is a type of chimeric antigen receptor T cell (CAR-T) therapy. First, an inactivated form of HIV is packed with snippets of custom-designed DNA. Second, T cells are harvested from the patient's blood and infected with the virus. The virus rewrites the genetic code to destroy cancer cells. Once the infected T cells have multiplied, they are infused back into the patient, where they attack the cancer cells. While risks are associated with CAR-T therapy, when successful, patients are spared months and years of the side effects associated with chemotherapy and can eliminate the need for a bone marrow transplant.

What were some of the early techniques for gene editing?

Genome, or simply gene, editing techniques and methods were first developed toward the end of the twentieth century. The earliest method used to edit genes was homologous recombination. Homologous recombination was developed in the late 1970s. It involves the exchange (or recombination) of genetic information between two similar (homologous) strands of DNA. Scientists generate and isolate DNA fragments with similarities to the genome to be edited. The isolated fragments are injected into individual cells or taken up by cells using special chemicals. Once inside a cell, the DNA fragments are able to recombine with the cell's DNA to replace the targeted portion of the genome. While potentially successful, the homologous recombination is inefficient and prone to error. Some estimate that the technique has as low as a one-in-a-million probability of successful editing.

During the 1990s, researchers began to use zinc-finger nucleases (ZFNs) for genome editing. The structures of ZFNs are engineered from naturally occurring proteins. These proteins are engineered to bind to specific DNA sequences in the genome. Once bound to the targeted DNA sequence, the ZFNs cut the genome at the specified locations, allowing scientists to either delete the target DNA sequence or replace it with a new DNA sequence via homologous recombination. The rate of success for genome editing with

ZFNs is about 10 percent. The technique is difficult and time-consuming, and a new ZFN must be engineered for each new target DNA sequence.

In 2009, a new class of proteins called Transcription Activator-Like Effector Nucleases (TALENs) became available for genome editing. TALENs are similar to ZFNs in terms of how efficiently they can create edits to the genome. TALENs are easier to engineer than ZFNs.

What is CRISPR technology?

Clustered Regularly Interspaced Short Palindromic Repeats (CRISPR) is a new technique for genome editing. Researchers create a short, synthetic RNA template that matches a target DNA sequence in the genome. The guide RNA directs Cas9 protein to the targeted DNA sequence. Cas9 cuts the genome at the location of the targeted DNA sequence. It can either make deletions in the genome and/or be engineered to insert new DNA sequences. Some scientists estimate that CRISPR is six times more efficient than ZFNs or TALENs in creating targeted mutations to the genome. Using CRISPR technology, genomic projects may be completed much more quickly at a much lower cost than using other technologies.

Who developed CRISPR?

CRISPR technology was developed over a period of twenty years in a dozen cities around the world by teams of researchers. Francisco Mojico (1963–) is credited with being the first researcher to recognize the palindromic, repeated sequence of thirty bases separated by spacers of roughly thirty-six bases while studying an archaeal microbe. He published his first paper describing the palindromic repeats in 1993. Twenty years later, in 2013, Feng Zhang (1982–) published a paper reporting mammalian genome editing using CRISPR–Cas9.

How does the CRISPR–Cas9 system work in nature?

CRISPR is a natural product. The CRISPR system has two components—the clustered regularly interspaced short palindromic repeats array and the Cas-9 (CRISPR-associated) proteins. Working together as a system, they help bacteria defend against attacking viruses, known as bacteriophages. The bacteria capture fragments of DNA from invading viruses and use them to create DNA segments known as CRISPR arrays. These arrays allow the bacteria to "remember" the viruses. If the virus attacks again, the bacteria produce RNA segments from the CRISPR arrays to target the viruses' DNA. The Cas-9 proteins destroy the matching viral DNA by cutting it, thus disabling the virus.

What are some of the anticipated uses of CRISPR?

Great hope exists that CRISPR technology will be used to develop treatments for many diseases, including cancer, HIV, blindness, chronic pain muscular dystrophy, Lyme disease, malaria, and Huntington's disease. Researchers are studying the potential of editing out disease-causing mutations in human embryos. Not only does hope exist to treat serious diseases in humans, but researchers are investigating CRISPR techniques to reduce disease in crops by removing the genes that make certain crops vulnerable to disease. An experiment in 2018 repaired a mutation by removing a disease-causing gene from a human embryo. Researchers hope to start clinical trials once the method is proven safe.

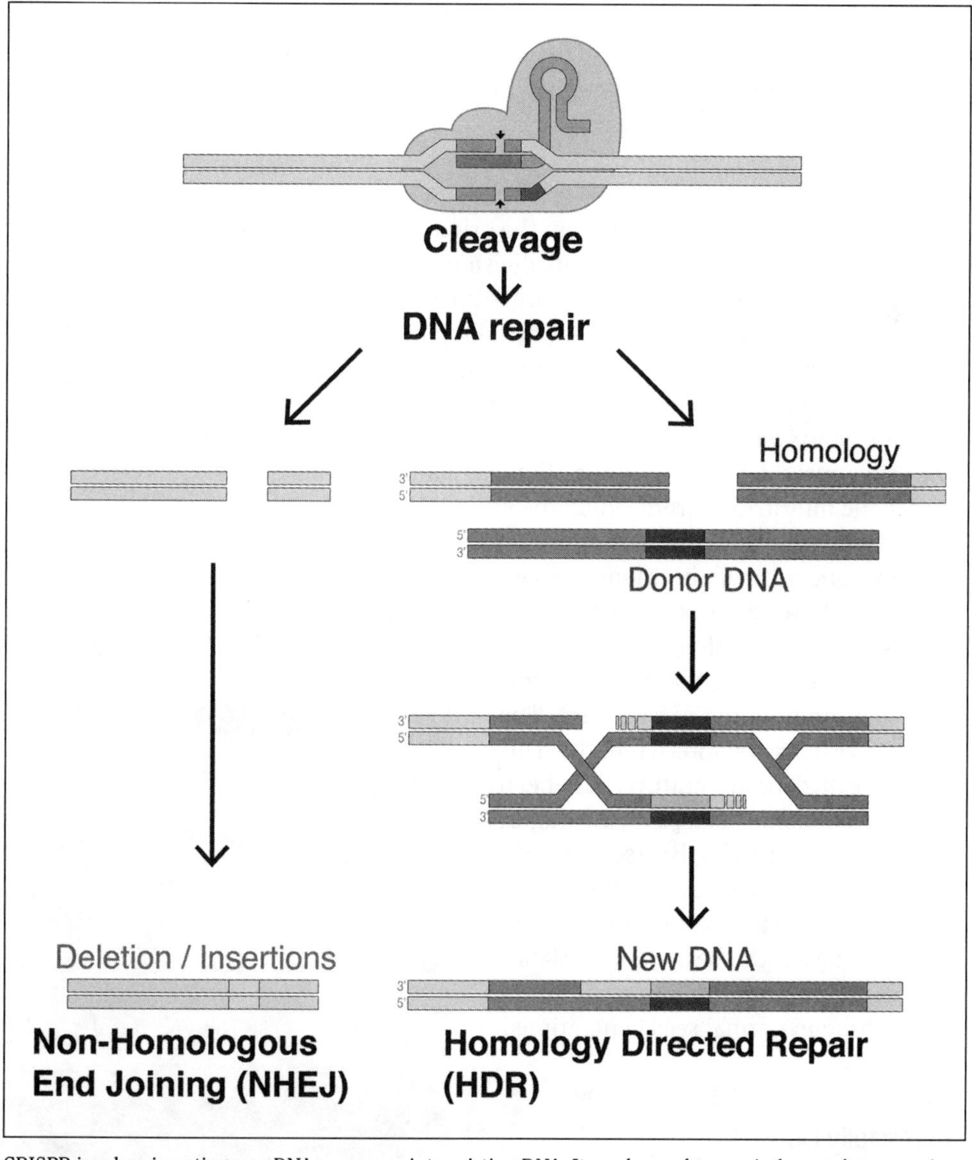

CRISPR involves inserting new DNA sequences into existing DNA. It can be used to repair damaged genes or insert genes that are normally there but are missing. The technology could prove useful for treating a variety of genetic diseases.

What is the biological basis for DNA fingerprinting?

British geneticist Alec Jeffreys (1950–) formulated the method of DNA fingerprinting, also known as DNA typing or DNA profiling, based on the fact that unique genetic differences exist between individuals. Most DNA sequences are identical, but out of one hundred base pairs (of DNA), two people will generally differ by one base pair. Since human DNA contains three billion base pairs, one individual's DNA will differ from another's by three million base pairs. To examine an individual's DNA fingerprint, the DNA sample is cut with a restriction endonuclease, and the fragments are separated by gel electrophoresis. The fragments are then transferred to a nylon membrane, where they are incubated in a solution containing a radioactive DNA probe that is complementary to specific polymorphic sequences.

What is cloning?

A clone is a group of cells derived from the original cell by fission (one cell dividing into two cells) or by mitosis (cell nucleus division, with each chromosome splitting into two). Cloning perpetuates an existing organism's genetic makeup. Gardeners have been making clones of plants for centuries by taking cuttings of plants to make genetically identical copies. For plants that refuse to grow from cuttings or for the animal world, modern scientific techniques have greatly extended the range of cloning. The technique for plants starts with taking a cutting of a plant that best satisfies the criteria for reproductive success, beauty, or some other standard. Since all of the plant's cells contain the genetic information from which the entire plant can be reconstructed, the cutting can be taken from any part of the plant. Placed in a culture medium having nutritious chemicals and a growth hormone, the cells in the cutting divide, doubling in size every six weeks until the mass of cells produces small, white, globular points called embryoids. These embryoids develop roots, or shoots, and begin to look like tiny plants. Transplanted into compost, these plants grow into exact copies of the parent plant. The whole process takes eighteen months. This process, called tissue culture, has been used to make clones of oil palm, asparagus, pineapples, strawberries, Brussels sprouts, cauliflower, bananas, carnations, ferns, and others. Besides making highly productive copies of the best plant available, this method controls viral diseases that are passed through normal seed generations.

Dolly the sheep was the first successfully cloned mammal.

What was the first mammal to be successfully cloned?

The first mammal cloned from adult cells was Dolly, a ewe born in July 1996. Dolly was born in a research facility in Scotland. Ian Wilmut (1944–) led the team of biologists who removed a nucleus from a mammary cell of an adult ewe and transplanted it into an enucleated egg extracted from a second ewe. Electrical pulses were administered to fuse the nucleus with its new host. When the egg began to divide and develop into an embryo, it was transplanted into a surrogate mother ewe. Dolly was the genetic twin of the ewe that donated the mammary cell nucleus. On April 13, 1998, Dolly gave birth to Bonnie—the product of a normal mating between Dolly and a Welsh mountain ram. This event demonstrates that Dolly was a healthy, fertile sheep, able to produce healthy offspring. Dolly died on February 14, 2003.

GENETICS AND EVOLUTION

What is meant by Mendelian inheritance?

Mendelian inheritance refers to genetic traits carried through heredity; the process was studied and described by Austrian monk Gregor Mendel. Mendel was the first to deduce correctly the basic principles of heredity. Mendelian traits are also called single gene or monogenic traits because they are controlled by the action of a single gene or gene pair. More than 4,300 human disorders are known or suspected to be inherited as Mendelian traits, encompassing autosomal dominant (e.g., neurofibromatosis), autosomal recessive (e.g., cystic fibrosis), and sex-linked dominant and recessive conditions (e.g., color blindness and hemophilia).

Overall, incidence of Mendelian disorders in the human population is about 1 percent. Many nonanomalous characteristics that make up human variation are also inherited in Mendelian fashion.

Gregor Mendel was an Augustinian friar living in what is now the Czech Republic city of Brno. His experiments with breeding peas led to the founding principles of genetics.

Why was Mendel successful in his genetics work while others were not?

Using a simple organism like the garden pea (*Pisum sativuum*), Mendel was able to control pollination among his experimental plants and, most importantly, he used true breeding plants with easily observable characteristics (e.g., flower color, height). He kept meticulous records and discovered consistent ratios involving thousands of plant-breeding experiments over eleven years. Interestingly, Mendel, while a student at the University of Vienna, studied physics and math under Christian Doppler, who later discovered the Doppler effect. Mendel's interest in statistics helped in determining the ratios and laws for which he is famous.

How does phenotype relate to genotype?

The phenotype is the physical manifestation of the genotype and the polypeptides it codes for. Consider the rose; the color and shape of the petals (the phenotype) are the result of chemical reactions within the cells of each petal. Polypeptides synthesized from the directions encoded within the cell's genes (the genotype) are part of those reactions. Different versions of a gene will produce different polypeptides, which in turn will cause different molecular interactions and ultimately a different phenotype.

What is the relationship between probability and genetics?

Probability is a branch of mathematics used to predict the likelihood of an event occurring. It is also an important tool for understanding inheritance patterns of specific traits. Two rules are applied to genetic inheritance: 1) The rule of multiplication is used when determining the probability of any two events happening simultaneously. For example, if the probability of having dimples is 1/4, and the probability of having a male child is 1/2, then the probability of having a boy with dimples is $1/4 \times 1/2$, or 1/8. 2) The rule of addition is used when determining the outcome of an event that can occur in two or more ways. For example, if we were to consider the probability of having a male child or of having a child with dimples, then the probability would be $1/2 + 1/4$, or 3/4. In other words, chances are 3 out of 4 that a given birth will produce a male child or a child with dimples.

Why are some species more commonly used for genetic studies than others?

Species with a relatively small genome, with a short generation time from seed to seed, that are adaptable to living in captivity are appealing as experimental organisms. Even though many of these species bear little physical resemblance to humans, they do share part of our genome and therefore can answer some of the questions we have about genetic inheritance and gene expression.

Species Commonly Used in Genetics Research

Species	Kingdom	Genome Size (in million base pairs)
Arabidopsis thaliana (plant)	Plantae	120
Neurospora (orange bread mold)	Fungi	40
Escherichia coli (bacteria)	Monera	4.64
Drosophila melanogaster (fruit fly)	Animalia	170
Caenorhabditis elegans (roundworm)	Animalia	97

How has genetics been linked to the Salem witch trials held in 1692 in Salem, Massachusetts?

It is believed that some of the early English colonists who settled in New England may have had Huntington's disease. Huntington's disease is an autosomal dominant disor-

What is a pedigree?

A pedigree is a genetic history of a family, which shows the inheritance of traits through several generations. Information that can be obtained from a pedigree includes birth order, sex of children, twins, marriages, deaths, stillbirths, and pattern of inheritance of specific genetic traits.

der characterized by late-onset symptoms (age forty to fifty) such as mild behavioral and neurological changes; as the disease progresses, psychiatric problems develop that frequently lead to insanity. Early descriptions of the odd behavior included names such as "that disorder" and "Saint Vitus's dance" to describe involuntary muscle jerks and twitches. Many of the witches who were on trial for possession by spirits may have had Huntington's disease, which causes uncontrollable movements and odd behavior.

What is eugenics?

Sir Francis Galton, a cousin of Charles Darwin, founded eugenics. After reading Darwin's work on natural selection, Galton thought that the human species could be improved by artificial selection—the selective breeding for desirable traits. This method is often used in domesticated animals. In Galton's plan, those with desirable traits would be encouraged to have large families, while those with undesirable traits would be kept from breeding. However, Galton's theory overlooked two important points: the importance of environmental factors on phenotypic expression and the difficulty of removing recessive traits from the gene pool. Recessive alleles can be passed from one individual to the next as part of a heterozygous genotype, thereby escaping detection for generations. Galton's work was enthusiastically adopted in both the United States and Europe. In the United States between 1900 and 1930, eugenics gave rise to changes in federal immigration laws and the passage of state laws requiring the sterilization of "genetic defectives" and certain types of criminals. In Europe, eugenics became a cornerstone of the Nazi movement.

What is evolution?

Although it was originally defined in the nineteenth century as "descent with modification," evolution is currently described as the change in frequency of genetic traits (also known as the allelic frequency) within populations over time.

How and when did different forms of life evolve on Earth?

We do not really know how life evolved on Earth. One of the reasons is the minute size (single cells) of the first organisms, which makes it difficult to detect them in ancient rocks. In addition, most of the oldest rocks have been exposed to the heat and pressure of geologic activity over time, making detection impossible by erasing all traces of that life. The following is only one interpretation of how early life on Earth developed (all years are approximations):

- 3.6 billion years ago, simple cells (prokaryotes) evolved.
- 3.4 billion years ago, stromatolites began the process of photosynthesis.
- 2 billion years ago, complex cells (eukaryotes) developed.
- 1 billion years ago, multicellular life began.
- 600 million years ago, simple animals evolved in the oceans.
- 570 million years ago, arthropods (ancestors of insects, arachnids, and crustaceans) began to become more widespread.
- 550 million years ago, complex animals began to evolve.

What is Lamarckian evolution?

French biologist Jean-Baptiste Pierre Antoine de Monet de Lamarck is credited as the first person to propose a theory that attempts to explain how and why evolutionary change occurs in living organisms. The mechanism Lamarck proposed is known as "the inheritance of acquired characteristics," meaning that what individuals experience during their lifetime will be passed along to their offspring as genetic traits. This is sometimes referred to as the theory of "use and disuse." A classic example of this would be the giraffe's neck. Lamarckian evolution would predict that as giraffes stretch their necks to reach higher branches on trees, their necks grow longer. As a result, this increase in neck length will be transmitted to egg and sperm such that the offspring of giraffes whose necks have grown will also have long necks. While Lamarck's idea was analytically based on available data (giraffes have long necks and give birth to offspring with long necks as well), he did not know that, in general, environmental factors do not change genetic sequences in such a direct fashion.

It was French biologist Jean-Baptiste Lamarck who first developed the theory of evolution, not Charles Darwin.

What were the *Beagle* voyages?

The HMS *Beagle* was a naval survey ship that left England in December 1831 to chart the coastal waters of Patagonia, Peru, and Chile. On a voyage that would last five years, Darwin's job as unpaid companion to the captain onboard the *Beagle* allowed him to satisfy his interests in natural history. On its way to Asia, the ship spent time in the Galapagos Islands off the coast of Ecuador; Darwin's observations there caused him to generate his theory of natural selection.

What is the significance of Darwin's finches?

In his studies on the Galapagos Islands, Charles Darwin observed patterns in animals and plants that suggested to him that species changed over time to produce new species. Darwin collected several species of finches. The species were all similar, but each had developed beaks and bills specialized to catch food in a different way. Some species had heavy bills for cracking open tough seeds. Others had slender bills for catching insects. One species used twigs to probe for insects in tree cavities. All the species resembled one species of South American finch. In fact, all the plants and animals of the Galapagos Islands were similar to those of the nearby (600 miles [1,000 kilometers] away) coast of South America. Darwin felt that the simplest explanation for this similarity was that a few species of plants and animals from South America must have migrated to the Galapagos Islands. These few plants and animals then changed during the years they lived in their new home, giving rise to many new species. Evolutionary theory proposes that species change over time in response to environmental challenges.

How did geology influence Darwin?

While traveling aboard the HMS *Beagle,* Charles Darwin read the *Principles of Geology* by Charles Lyell. Catastrophism was the popular theory of the time about the forces driving geological change. Lyell's theory suggested that geologic change was not solely the result of random catastrophes. Rather, he proposed that geologic formations were most often the result of everyday occurrences like storms, waves, volcanic eruptions, and earthquakes that could be observed within an individual lifetime. This idea, that the same geologic processes at work today were also present during our evolutionary past, is known as uniformitarianism.. This conclusion also led Lyell and, before him, James Hutton to suggest that Earth must be much older than the previously accepted age of six thousand years because these uniform processes would have required many millions of years to generate the structures he observed. Reading Lyell's work gave Darwin a new perspective as he traveled through South America and sought a mechanism by which he could explain his thoughts on evolution.

Who was Alfred Russel Wallace?

Alfred Russel Wallace (1823–1913) was a naturalist whose work was presented with Charles Darwin's at the Linnaean Society of London in 1858. After extensive travels in the Amazon basin, Wallace independently came to the same conclusions as Darwin on the significance of natural selection in driving the diversification of species. Wallace also worked as a natural history specimen collector in Indonesia. Wallace, like Darwin, also read the work of Thomas Malthus (1766–1834). During an attack of malaria in Indonesia, Wallace made the connection between the Malthusian concept of the struggle for existence and a mechanism for change within populations. From this, Wallace wrote the essay that was eventually presented with Darwin's work in 1858.

Who was "Darwin's bulldog"?

Thomas Huxley (1825–1895) was a staunch supporter of Darwin's work; in fact, Huxley wrote a favorable review of Darwin's *On the Origin of Species* that appeared soon after its publication. When the firestorm of controversy began after the appearance of Darwin's book, Huxley was ready and able to defend Darwin, whose chronic public reticence about his theories was at that time exacerbated by illness. Huxley's defense of Darwin was so vigorous during a debate with Bishop Samuel Wilberforce (1805–1873) at the British Association for the Advancement of Science in 1860 that he earned the title "Darwin's bulldog."

Naturalist, biologist, anthropologist, and explorer Alfred Russel Wallace deserves equal credit with Darwin for simultaneously theorizing that natural selection drives the mechanism of evolution.

Who coined the phrase "survival of the fittest"?

Although frequently associated with Darwinism, this phrase was coined by Herbert Spencer (1820–1903), a British sociologist. It is the process by which organisms that are less well adapted to their environment tend to perish and better adapted organisms tend to survive.

What is the Darwin–Wallace theory?

The Darwin–Wallace theory can be summarized as the following: Species as a whole demonstrate descent with modification from common ancestors, and natural selection is the sum of the environmental forces that drive those modifications. The modifications or adaptations make the individuals in the population better suited to survival in their environment, more "fit," as it were.

The four postulates presented by Darwin in *On the Origin of Species* are as follows: 1) individuals within species are variable; 2) some of these variations are passed on to offspring; 3) in every generation, more offspring are produced than can survive; and 4) the survival and reproduction of individuals are not random—the individuals who survive and go on to reproduce the most are those with the most favorable variation. They are naturally selected. It follows logically from these that the characteristics of the population will change with each subsequent generation until the population becomes distinctly different from the original; this process is known as evolution.

Why is evolution a theory?

A scientific theory is an explanation of observed phenomena that is supported by the available scientific data. The term "theory" is used as an indication that the explanation will be modified as new data becomes available. For example, the Darwin–Wallace theory was proposed prior to the discovery of the molecular nature of genetics but has since been expanded to encompass that information as well.

What is gradualism?

The Darwin–Wallace theory of evolution is based on gradualism—the idea that speciation occurs by the gradual accumulation of new traits. This would allow one species to gradually evolve into a different-looking one over many, many generations, which is the scale of evolutionary time.

What is punctuated equilibrium?

Punctuated equilibrium is a model of macroevolution first detailed in 1972 by Niles Eldredge (1942–) and Stephen Jay Gould (1941–2002). It can be considered either a rival or supplementary model to the more gradual-moving model of evolution posited by neo-Darwinism. The punctuated equilibrium model essentially asserts that most of geological history shows periods of little evolutionary change, followed by short (geologically speaking, a few million years) periods of rapid evolutionary change. Gould and Eldredge's work has been buttressed by the discovery of the Hox genes that control embryonic development. Hox genes are found in all vertebrates and many other species as well; they control the placement of body parts in the developing embryo. Relatively minor mutations in these gene sequences could result in major body changes for species in a short period of time, thereby giving rise to new forms of organisms and therefore new species.

Which scientific disciplines provide evidence for evolution?

Although information from any area of natural science is relevant to the study of evolution, several in particular directly support the work of Darwin and Wallace. Paleobiology,

What is social Darwinism?

Social Darwinism is a social movement—one of a number of perversions of the Darwin–Wallace theory. These movements attempt to use evolutionary mechanisms as excuses for social change. Followers of social Darwinism believe that the "survival of the fittest" applies to socioeconomic environments as well as evolutionary ones. By this reasoning, the weak and the poor are "unfit" and should be allowed to die without societal intervention. This idea has nothing to do with Charles Darwin and Alfred Russel Wallace but was promoted by Herbert Spencer and is related to the works of Thomas Malthus, whose work did indeed inspire Darwin. Although social Darwinism has faded as a movement, it did help to spur the eugenics movement of Nazi Germany as well as a number of laws and policies in the United States in the twentieth century.

geology, and organic chemistry provide insight on how living organisms have evolved. Ecology, genetics, and molecular biology also demonstrate how living species are currently changing in response to their environments and therefore undergoing evolution.

What is the value of fossils to the study of evolution?

Fossils are the preserved remains of once living organisms. The value of fossils comes not only from the information they give us about the structures of those animals. The placement of common fossils in the geologic layers also gives researchers a method for dating other, lesser known samples.

What is the Oparin–Haldane hypothesis?

In the 1920s, while working independently, Alexandr Oparin (1894–1980) and John Haldane (1892–1964) both proposed scenarios for the "prebiotic" conditions on Earth (the conditions that would have allowed organic life to evolve). Although they differed on details, both models described an early Earth with an atmosphere containing ammonia and water vapor. Both also surmised that the assemblage of organic molecules began in the atmosphere and then moved into the seas. The steps of the Oparin–Haldane hypothesis are described below.

1. Organic molecules, including amino acids and nucleotides, are synthesized abiotically (without living cells).
2. Organic building blocks in the prebiotic soup are assembled into polymers of proteins and nucleic acids.
3. Biological polymers are assembled into a self-replicating organism that feeds on the existing organic molecules.

Who verified the Oparin–Haldane hypothesis?

In 1953 (the year that Watson and Crick published their famous paper on DNA structure), Stanley Miller (1930–2007), a graduate student in the lab of Harold Urey, built an apparatus that mimicked what was then thought to be the atmosphere of early Earth, a reducing atmosphere containing methane, ammonia, and hydrogen. In the closed chamber, Miller boiled water, then exposed it to electric shocks and cooled it. After the apparatus ran for a period of days, Miller tested the water and found several amino acids that are the building blocks of proteins. Eventually,

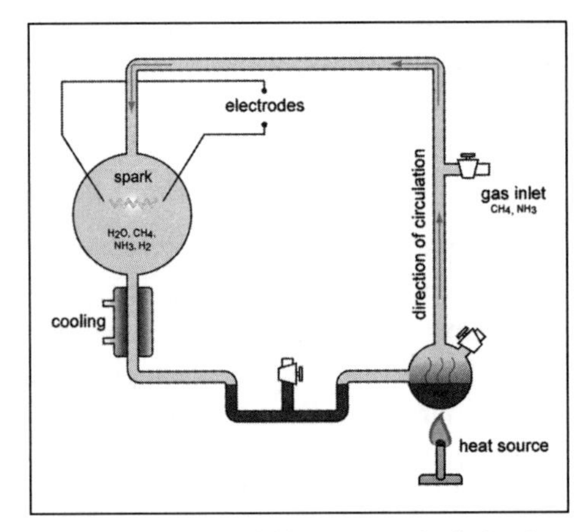

Stanley Miller and Harold Urey created a device that simulated conditions on Earth before life emerged. They showed that water, ammonia, hydrogen, and methane, mixed with a bit of heat and electricity (from lightning), could react to form the basic building blocks of life.

scientists replicating the Miller–Urey experiment were able to generate other types of amino acids as well as nucleotides (basic units of DNA) and sugars.

How did humans evolve?

Evolution of the *Homo* lineage of modern humans (*Homo sapiens*) has been proposed to originate from a hunter of nearly 5 feet (1.5 meters) tall, *Homo habilis*, who is widely presumed to have evolved from an australopithecine ancestor. Near the beginning of the Pleistocene epoch (two million years ago), *Homo habilis* is thought to have transformed into *Homo erectus* (Java Man), who used fire and possessed culture. Middle Pleistocene populations of *Homo erectus* are said to show steady evolution toward the anatomy of *Homo sapiens* (Neanderthals, Cro-Magnons, and modern humans) 120,000–40,000 years ago. Premodern *Homo sapiens* built huts and made clothing.

What was the Scopes (monkey) trial?

John T. Scopes (1900–1970), a high school biology teacher, was brought to trial by the State of Tennessee in 1925 for teaching the theory of evolution. He challenged a recent law passed by the Tennessee legislature that made it unlawful to teach in any public school any theory that denied the divine creation of man. He was convicted and sentenced, but the decision was reversed later and the law repealed in 1967.

In the early twenty-first century, pressure against school boards still affects the teaching of evolution. Recent drives by anti-evolutionists either have tried to ban the teaching of evolution or have demanded "equal time" for "special creation" as described in the biblical book of Genesis. This has raised many questions about the separation of church and state, the teaching of controversial subjects in public schools, and the ability of scientists to communicate with the public. The gradual improvement of the fossil record, the result of comparative anatomy, and many other developments in biological science have contributed toward making evolutionary thinking more palatable.

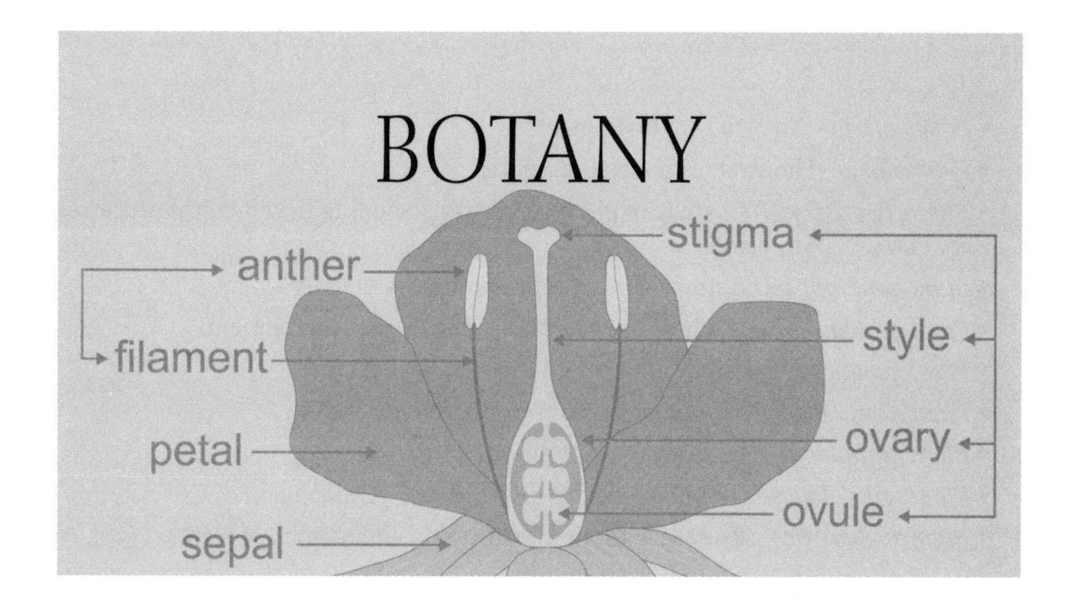

INTRODUCTION AND HISTORICAL BACKGROUND

What are the general characteristics of a plant?

A plant is a multicellular, eukaryotic organism with cellulose-rich cell walls and chloroplasts that has starch as the primary carbohydrate food reserve. Plants are primarily terrestrial, autotrophic (capable of making their own food) organisms. Most plants contain chlorophylls *a* and *b* and xanthophylls (yellow pigments) as well as carotenes (orange pigments).

Why are plants important?

Plants are essential for creating and sustaining life on Earth—both human and animal. Plants provide food and shelter for all life. They provide clothing and medicines for human life. They also play a vital role in removing carbon dioxide and producing the oxygen we breathe through photosynthesis. They help with the water cycle and reduce soil erosion.

What are the major subdivisions of botany?

The major subdivisions of botany are:

- *Agronomy*—The application of plant science to crop production.
- *Bryology*—The study of mosses and liverworts.
- *Economic botany*—The study of the utilization of plants by humans.
- *Ethnobotany*—The study of the use of plants by indigenous peoples.
- *Forestry*—The study of forest management and the utilization of forest products.
- *Horticulture*—The study of ornamental plants, vegetables, and fruit trees.

- *Paleobotany*—The study of fossil plants.
- *Palynology*—The study of pollen and spores.
- *Phytochemistry*—The study of plant chemistry, including the chemical processes that take place in plants.
- *Plant anatomy*—The study of plant cells and tissues.
- *Plant ecology*—The study of the role plants play in the environment.
- *Plant genetics*—The study of genetic inheritance in plants.
- *Plant morphology*—The study of plant forms and life cycles.
- *Plant pathology*—The study of plant diseases.
- *Plant physiology*—The study of plant function and development.
- *Plant systematics*—The study of the classification and naming of plants.

Who is known as the founder of botany?

The ancient Greek Theophrastus is known as the father of botany. His two works on botany, *On the History of Plants* and *On the Causes of Plants*, were so comprehensive that 1,800 years passed before any new significant botanical information was discovered. He integrated the practice of agriculture into botany and established theories regarding plant growth and the analysis of plant structure. He related plants to their natural environment and identified, classified, and described 550 different plants.

What contributions did John and William Bartram make to botany?

John Bartram (1699–1777) was the first American-born botanist. He and his son, William Bartram (1739–1823), traveled throughout the American colonies observing the flora and fauna. Although John Bartram never published his observations, he was considered the authority on American plants. In 1791, his son William published his notes on American plants and animals as *Bartram's Travels*.

What is *Gray's Manual*?

Gray's Manual of Botany, first published in 1848 by Asa Gray (1810–1888) under the title *Manual of the Botany of Northern United States*, was one of the first guides to the identification of plants of eastern North America. The publication contained keys and thorough descriptions of plants. The eighth, and centennial, edition was largely rewritten and expanded by Merritt Lyndon Fernald (1873–1950) and published in 1950. This edition was corrected and updated by R. C. Rollins and reprinted in 1987 by Dioscorides Press.

What was the significance of *De Materia Medica*?

De Materia Medica ("About medicinal materials") was written by Greek physician Dioscorides (c. 40–90 C.E.) in the first century C.E. The manuscript included the names and uses of the six hundred plants that, at the time, were known to have medicinal properties. The purpose of the publication was to improve medical service in the Roman Empire. In addition to its medical use, it became the book most often used for plant classification in the Western world for nearly 1,500 years. During the fifteenth and sixteenth centuries, European botanists and physicians used *De Materia Medica* to formulate their "herbals"—illustrated books on the presumed medicinal uses of plants.

American botanist and horticulturalist Luther Burbank is credited with developing plant breeding into a modern science.

Who developed plant breeding into a modern science?

Luther Burbank (1849–1926) developed plant breeding as a modern science. His breeding techniques included crosses of plant strains native to North America and foreign strains. He obtained seedlings that were then grafted onto fully developed plants for an appraisal of hybrid characteristics. His keen sense of observation allowed him to recognize desirable characteristics, enabling him to select only varieties that would be useful. One of his earliest hybridization successes was the Burbank potato, from which more than eight hundred new strains and varieties of plants—including 113 varieties of plums and prunes—were developed. More than twenty of these plums and prunes are still commercially important today.

What were some of the accomplishments of Dr. George Washington Carver?

Because of the work of Dr. George Washington Carver (1864–1943) in plant diseases, soil analysis, and crop management, many southern farmers who adopted his methods increased their crop yields and profits. Carver developed recipes using cowpeas, sweet potatoes, and peanuts. He eventually made 118 products from sweet potatoes, 325 from peanuts, and 75 from pecans. He promoted soil diversification and the adoption of peanuts, soybeans, and other soil-enriching crops. His other work included developing plastic material from soybeans, which Henry Ford (1863–1947) later used in part of his automobile. Carver also extracted dyes and paints from the Alabama red clay and worked with hybrid cotton. He was a widely talented man who became an almost mythical American folk hero.

What is the origin of land plants?

Many scientists believe that land plants evolved from green algae. Green algae, especially the charaophytes, share a number of biochemical and metabolic traits with plants. Both contain the same photosynthetic pigments—carotenes and xanthophylls as well as chlorophylls *a* and *b*. Cellulose is a major component of the cell walls of plants and algae, and both store their excess carbohydrates as starch. In addition, some aspects of cell division, particularly the formation of new cross-walls, only occurs in plants and certain charaophytes, such as species of the genera *Cerara* and *Colechaete*.

What was the basis of the earliest classification systems of plants?

The earliest classifications of plants were based on whether the plant was considered medicinal or was shown to have other uses. *De re Rustica* by Cato the Censor (234–149 B.C.E.) lists 125 plants and was one of the earliest catalogs of Roman plants. Gaius Plinius Secundus (23–79 C.E.), known as Pliny the Elder, wrote *Historia Naturalis*, which was published in the first century. The book was one of the earliest catalogs of significant plants in the ancient world, describing more than one thousand plants. Plant classification became more complicated as more and more plants were discovered. One of the earliest plant taxonomists was Italian botanist Caesalpinus (1519–1603). In 1583, he classified more than 1,500 plants according to various attributes, including leaf formation and the presence of seeds or fruit.

How do artificial systems of classification differ from natural systems of classification?

An artificial system of classification groups organisms based on arbitrary features rather than true evolutionary relationships. John Ray (1627–1705) was the first botanist to base plant classification on the presence of multiple similarities and features. His *Historia Plantarum Generalis*, published between 1686 and 1704, was a detailed classification of more than eighteen thousand plants. The book included a distinction between monocotyledon and dicotyledon flowering plants. French botanist J. P. de Tournefort (1656–1708) was the first to characterize genus as a taxonomic rank that falls between the ranks of family and species. De Tournefort's classification system included nine thousand species in seven hundred genera. Swedish naturalist Carolus Linnaeus published *Species Plantarum* in 1753. It organized plants into twenty-four classes based on reproductive features. The Linnaean system of binomial nomenclature remains the most widely used system for naming plants and animals. It is considered an artificial system since it often does not reflect natural relationships.

During the late eighteenth century, several natural systems of classification were proposed. A natural system of classification is more likely to reflect evolutionary relationships between species. French botanist Antoine-Laurent de Jussieu (1686–1758) published *Genera Plantarum*. The tome *Prodromus Systematis Naturalis Regni Vegetabilis* was started in 1824 by Swiss botanist Augustin Pyrame de Candolle (1778–1841) and completed fifty years later. Another *Genera Plantarum* was published between 1862 and 1883 by British botanists George Bentham (1800–1884) and Sir Joseph Dalton Hooker (1817–1911).

When was the first phylogenetic system of classification of plants proposed?

Phylogeny is the evolutionary history of a group of organisms. Charles Darwin's ideas on evolution began to influence systems of classification during the late nineteenth century. The first major phylogenetic system of plant classification was proposed around the close of the nineteenth century. *Die natürlichen Pflanzenfamilien* (*The Natural Plant Families*), one of the most complete phylogenetic systems of classification and still in use through the twentieth century, was published between 1887 and 1915 by German botanists Adolf Engler (1844–1930) and Karl Prantl (1849–1893). Their system recognizes about one hundred thousand species of plants, organized by their presumed evolutionary sequence.

English polymath Joseph Priestley discovered oxygen and that plants were responsible for converting carbon dioxide into oxygen molecules.

What were some of the earliest ideas about plant nutrition?

The ancient Greeks and Romans believed that plants derived their food from the soil. Belgian scientist Jan Baptista van Helmont (1577–1644) was the first to perform an experiment to test the hypothesis of the ancient Greeks and Romans. He planted a willow tree in a container of soil and fed it only water. At the end of five years, the weight of the willow tree had increased by 164 pounds (74.4 kilograms), while the weight of the soil had decreased by 2 ounces (57 grams). Van Helmont concluded that the plant had received all its nourishment from the water and none from the soil.

British theologian and scientist Joseph Priestley (1733–1804) demonstrated that air was "restored" by plants. In 1771, Priestley conducted an experiment in which he placed a lighted candle in a glass container and allowed it to burn until extinguished by lack of oxygen. He then put a plant into the same chamber and allowed it to grow for a month. Repeating the candle experiment a month later, he found that the candle would now burn. Priestley's experiments showed that plants release oxygen (O_2) and take in carbon dioxide (CO_2) produced by combustion. Dutch physician Jan Ingenhousz (1730–1799) confirmed Priestley's ideas, emphasizing that air is "restored" only by green plants in the presence of sunlight.

PLANT DIVERSITY

What are the four major groups of plants?

The four major groups of plants are nonvascular; seedless vascular; flowering, seed-bearing vascular; and nonflowering, seed-bearing vascular. Plants are divided into phyla based on whether they are vascular (containing vascular tissue consisting of cells joined into tubes that transport water and nutrients) or nonvascular. The phyla of vascular plants are then further divided into seedless plants and those that contain seeds. Plants with seeds are divided into flowering and nonflowering groups. Nonvascular plants have traditionally been called bryophytes. Because bryophytes lack a system for conducting water and nutrients, they are restricted in size and live in moist areas close to the ground. Examples of bryophytes are mosses, liverworts, and hornworts. Examples of seedless, vascular plants are ferns, horsetails, and club mosses. The conifers, which are cone-bearing, are seed-bearing, nonflowering vascular plants. The majority of plants are seed-bearing, flowering, vascular plants known as angiosperms.

What are the phyla of plants?

The following chart lists many of the phyla (major divisions) of plant species.

Phyla	Number of Species	Characteristics	Example
Bryophyta	12,000	Nonvascular	Mosses
Hepaticophyta	6,500	Nonvascular	Liverworts
Anthocerotophyta	100	Nonvascular	Hornworts
Psilophyta	6	Vascular, homosporous, no differentiation between root and shoot	Whisk ferns
Lycophyta	1,000	Vascular, homosporous or heterosporous	Club mosses
Sphenophyta	15	Vascular, homosporous	Horsetails
Pterophyta	12,000	Vascular, homosporous	Ferns
Cycadophyta	100	Vascular, heterosporous, seed-forming	Cycads (commonly known as "sago palms")
Ginkgophyta	1	Vascular, heterosporous, seed-forming, deciduous tree	Ginkgo
Gnetophyta	70	Vascular, heterosporous, seed-forming	Ephedra, shrubs, vines
Coniferophyta	550	Vascular, heterosporous, seed-forming	Conifers (pines, spruces, firs, yews, and redwoods)
Anthophyta	240,000	Vascular, heterosporous, seed-forming	Flowering plants

How are plants identified based on their growth patterns?

Herbaceous or nonwoody plants die at the end of each growing season. Woody plants add a new layer of wood each year.

What is the alternation of generations in plants?

All plants exhibit an alternation of generations between diploid sporophytes and haploid gametophytes. Sporophytes produce haploid spores as a result of meiosis. The spores grow into multicellular, haploid individuals known as gametophytes. Spores are the first cells of the gametophyte generation. Gametophytes produce gametes as a result of mitosis. Male and female gametes fuse to form a zygote, which grows into a sporophyte. The zygote is the first cell of the following sporophyte generation.

Who first demonstrated the alternation of generations?

In the mid-nineteenth century, German botanist Wilhelm Hofmeister (1824–1877) was the first to demonstrate the alternation of generations. Hofmeister studied mosses, ferns, and seed plants.

What are the main features of bryophytes?

Bryophytes are generally small, compact plants that rarely grow to more than 8 inches (20 centimeters) tall. They have parts that appear leaflike, stemlike, and rootlike and lack vascular tissue (xylem and phloem). Most species have rhizoids, a cuticle, a cellular jacket to retain moisture around sperm-producing and egg-producing structures, and large gametophytes that hold on to sporophytes. They require water to reproduce sexually. In nature, they are noted for their intense shades of green. Mosses, liverworts, and hornworts are bryophytes. They are known to inhabit almost all environments from hot, dry deserts to the coldest regions of the Antarctic but are most often found in moist environments. They are most noticeable when they grow in a dense mass.

What is the purpose of rhizoids?

Rhizoids are a characteristic feature of mosses, liverworts, and hornworts. Rhizoids are slender, usually colorless projections that consist of a single cell or a few cells. They serve to anchor mosses, liverworts, and hornworts to their substrate and absorb water.

Some types of liverwort, such as this Common Hepatica, sport beautiful, little, blue flowers.

What feature of liverworts hints to their name?

Liverworts were named during the Middle Ages, when herbalists followed the theoretical approach known as the Doctrine of Signatures. The core philosophy of this perspective was that if a plant part resembled a part of the human body, it would be useful in treating ailments of that organ or part. The thallus of thalloid liverworts resembles a lobed liver. Therefore, in line with the philosophy presented by the doctrine, the plant was used to treat liver ailments. The word "liver" was combined with "wort," which means herb, to form the name "liverwort."

Why are mosses and liverworts important?

Some mosses are decomposers that break down the substrata and release nutrients for the use of more complex plants. Mosses play an important role in controlling soil erosion. They perform this function by providing ground cover and absorbing water. Mosses are also bioindicators of air pollution. Under conditions of poor air quality, few mosses will exist. Peat is used as fuel to heat homes and generate electricity. Bryophytes are among the first organisms to invade areas that have been destroyed by a fire or volcanic eruption.

Liverworts provide food for animals. Due to their ability to retain moisture, they also assist in the decay of logs and aid in the disintegration of rocks into soil.

Are all plants commonly called "moss" true mosses?

Some plants with the common name "moss" are not true mosses or bryophytes. For example, Irish moss (*Chondrus crispus*) and related species are actually red algae. Iceland moss (*Cetraria islandica*) and reindeer moss (*Cladonia rangiferina*) are lichens. Club mosses (genus *Lycopodium*) are seedless, vascular plants, and Spanish moss (*Tillandsia usneoides*) is a flowering plant in the pineapple family.

What are the main features of vascular plants?

Vascular plants have leaves, roots, cuticles, stomata, specialized stems, tissues that conduct efficiently, and, in most cases, seeds. Their sporophytes are large, dominant, and nutritionally independent.

How many groups of seedless, vascular plants have been identified?

Seedless, vascular plants are categorized into four groups. These include: ferns of the phylum *Pterophyta*, which is the largest group; the whisk ferns of the family *Psilophyta*; the club mosses of the plant division *Lycophyta*; and the horsetails of the class *Arthrophyta*.

What are the distinguishing features of ferns?

Ferns have roots, stems, and leaves, but they do not have flowers or seeds. They usually reproduce sexually by tiny spores. The leaves of a fern are often called fronds. They display a variety of different shapes and divisions from a simple, undivided frond to those with many divisions, giving the familiar, lacy look of a fern. The new fronds that emerge are called fiddleheads. They are called fiddleheads since they are tightly coiled and resemble the top of a violin.

How do ferns reproduce?

Ferns have two generations—a diploid sporophyte generation and a haploid gametophyte generation. The leafy plant we recognize as a fern is the sporophyte generation. Spores are generally released in the summer. Once released, they must land on a moist, protected environment to germinate and grow into gametophytes. Male and female reproductive structures develop on the lower surface of different gametophytes, although in some cases, both male and female structures may develop on the same gametophyte. At sexual maturity, the male structures release sperm that move through the moist environment to fertilize the egg in the female structures. The fertilized egg develops into an embryo, which is the beginning of the diploid sporophyte generation.

What are gymnosperms, and which plants are included in this group?

Gymnosperms (from the Greek words *gymnos*, meaning "naked," and *sperma*, meaning "seed") produce seeds that are totally exposed or borne on the scales of cones. The four phyla of gymnosperms are: Coniferophyta, conifers including pine, spruce, hemlock, and fir; Ginkgophyta, consisting of one species, the ginkgo or maidenhair tree; Cycadophyta, the cycads or ornamental plants; and Gnetophyta, a collection of very unusual vines and trees.

What is the oldest genus of living trees?

The genus *Ginkgo*, commonly known as maidenhair trees, comprises the oldest living trees. This genus is native to China, where it has been cultivated for centuries. It has not been found in the wild, and it is likely that it would have become extinct had it not been cultivated. Fossils of two-hundred-million-year-old ginkgoes show that the modern-day ginkgo is nearly identical to its forerunner. As of the early twenty-first century, only one living species of ginkgo remains, *Ginkgo biloba*. The fleshy coverings of the seeds produced by females of the species *Ginkgo biloba* have a distinctly foul odor. Horticulturists prefer to cultivate the male plant from shoots to avoid the odor and mess created by the female tree.

What plant produces the largest seed cones?

The largest seed cones are produced by cycads. They may be up to 1 yard (1 meter) in length and weigh more than 3.3 pounds (1.5 kilograms).

The Ginkgo bilboa tree is easily recognized by its distinctively shaped leaves.

In the life cycle of a pine tree, is the sporophyte or gametophyte dominant?

In the evolution of seed plants, one of the key terrestrial adaptations in reproduction

Do pine trees keep their needles forever?

Pine needles occur in groups, called fascicles, of two to five needles. A few species have only one needle per fascicle, while others have as many as eight. Regardless of the number of needles, a fascicle forms a cylinder of short shoots that are surrounded at their base by small, scalelike leaves that usually fall off after one year of growth. The needle-bearing fascicles are also shed a few at a time, usually every two to ten years, so that any pine tree, while appearing evergreen, has a complete change of needles every two to four years or less.

was the increasing dominance of the sporophyte generation. The mature pine tree is a sporophyte.

What are the distinguishing characteristics of fir, pine, and spruce trees?

An easy way to identify the various trees is to gently reach out and "shake hands" with a branch. The needles of each tree are different: pine needles come in packages, spruce needles are sharp and single, and fir needles are flat and friendly. The following chart outlines some of the distinguishing features of selected species of pine, fir, and spruce.

Characteristics of Fir, Pine, and Spruce Trees

Species	Needles	Cones
Balsam fir	Needles are 1–1.5 inches (2.54–3.81 centimeters) long, flat, and arranged in pairs opposite each other.	Upright, cylindrical, and 2–4 inches (5–10 centimeters) long.
Douglas fir	Needles are 1–1.5 inches (2.54–3.81 centimeters) long, occur singularly, and are very soft.	Cone scales have bristles that stick out.
Scotch pine	Two needles in each bundle; needles are stiff, yellow green, and 1.5–3 inches (3.81–7.62 centimeters) long.	2–5 inches (5–12.7 centimeters) long.
White pine	Five needles in each bundle; needles are soft and 3–5 inches (7.62–12.7 centimeters) long.	4–8 inches (10–20.3 centimeters) long.
White spruce	Dark-green needles are rigid but not prickly; needles grow from all sides of the twig and are less than an inch (2.54 centimeters) long.	1–2.5 inches (2.54–6.35 centimeters) long and hang downward.

Species	Needles	Cones
Blue spruce	Needles are roughly 1 inch (2.54 centimeters) long, grow from all sides of the branch, are silvery blue in color, and are very stiff and prickly.	3.5 inches (8.89 centimeters) long.

How long does it take to produce a mature pine cone?

From the time young cones appear on the tree, it takes nearly three years for them to mature. The sporangia of a pine tree are located on scalelike sporophylls that are densely packed in structures called cones. Conifers, like all seed plants, are heterosporous, meaning that male and female gametophytes develop from spores produced by separate cones. Small pollen cones produce microspores that develop into the male gametophytes or pollen grains. Larger, ovulate cones make megaspores that develop into female gametophytes. Each tree usually has both types of cones. This three-year process culminates in the production of male and female gametophytes, brought together through pollination, and the formation of mature seeds from the fertilized ovules. The scales of ovulate cones then separate, and the seeds are scattered by the wind. A seed that lands on a habitable place germinates, its embryo emerging as a pine seedling.

Are giant redwood trees found only in California?

Although redwoods extend somewhat into southern Oregon, the vast majority of giant redwoods are found in California. The genus *Sequoia* includes two species, which are commonly known as the redwood and the big tree. Both can be seen in either Redwood National Park or Sequoia National Park. At the latter park, the most impressive tree is known as the General Sherman Tree. It is 272 feet (83 meters) tall and has a diameter of 32 feet (9.75 meters) and a circumference of 101 feet (30.8 meters). The weight of the tree is estimated to be more than 6,000 tons (5,443

The General Sherman sequoia tree in California's Sequoia National Park. Estimated to be over four thousand years old, this forest mammoth weighs more than six thousand tons.

metric tons). Other trees found in Sequoia National Park exceed 300 feet (91.4 meters) in height but are more slender. The General Sherman Tree is about four thousand years old, the oldest living thing next to the bristlecone pine. Approximately 150 million years ago, these giant trees were widespread across the Northern Hemisphere. While the size of these giant trees implies that they are composed of very strong wood, the opposite is true. The wood is useless as timber because it is brittle and shatters into splintery, irregular pieces when struck. Perhaps the weakness of the wood is why so many of these giant trees still survive and have not been harvested by the logging industry.

Which tree species from the United States have lived the longest?

Of the 850 different species of trees in the United States, the oldest species is the bristlecone pine, *Pinus longaeva*. This species grows in the deserts of Nevada and southern California, particularly in the White Mountains. Some of these trees are believed to be over 4,600 years old. The potential life span of these pines is estimated to be 5,500 years old, but the potential age of the bristlecone pine is very young when compared to the oldest surviving species in the world, the maidenhair tree (*Ginkgo biloba*) of China. This species of tree first appeared during the Jurassic period some two hundred million years ago. Also called icho, or the ginkyo (meaning "silver apricot"), this species has been cultivated in Japan since 1100 B.C.E.

Is hemlock poisonous?

Two species are known commonly as hemlock: *Conium maculatum* and *Tsuga canadensis*. *Conium maculatum* is a weedy plant—considered by many as the most dangerous plant in North America. All parts of the plant are poisonous. In ancient times, minimal doses of the plant were used to relieve pain with a great risk of poisoning. *Conium maculatum* was used to carry out the death sentence in ancient times. Greek philosopher Socrates (470–399 B.C.E.) was condemned to death by drinking a potion made from hemlock. It should not be confused with *Tsuga canadensis*, a member of the evergreen family. No part of of the *Tsuga canadensis* tree is poisonous.

Why are gymnosperms important in our daily lives?

Gymnosperms account for approximately 75 percent of the world's timber and a large amount of the wood pulp used to make paper. In North America, the white spruce, *Picea glauca*, is the main source of pulpwood used for newsprint and other paper. Other spruce wood is used to manufacture violins and similar string instruments because the wood produces a desired resonance. The Douglas fir, *Pseudotsuga menziesii*, provides more timber than any other North American tree species and produces some of the most desirable lumber in the world. The wood is strong and relatively free of knots. Uses for the wood include house framing, plywood production, structural beams, pulpwood, railroad ties, boxes, and crates. Since most naturally occurring areas of growth have been harvested, the Douglas fir is being grown in managed forests. The wood from the redwood *Sequoia sempervirens* is used for furniture, fences, posts, some construction, and has various garden uses.

How are gymnosperms used in the chemical industry?

Gymnosperms are important in making resin and turpentine. Resin, the sticky substance in the resin canals of conifers, is a combination of turpentine, a solvent, and a waxy substance called rosin. Turpentine is an excellent paint and varnish solvent but is also used to make deodorants, shaving lotions, medications, and limonene—a lemon flavoring used in the food industry. Resin has many uses; it is used by baseball pitchers to improve their grip on the ball and by batters to improve their grip on the bat; violinists apply resin to their bows to increase friction with the strings; dancers apply resin to their shoes to improve their grip on the stage.

What factors have contributed to the success of seed plants?

Seed plants do not require water for sperm to swim to an egg during reproduction. Pollen and seeds have allowed them to grow in almost all terrestrial habitats. The sperm of seed plants is carried to eggs in pollen grains by the wind or animal pollinators such as insects. Seeds are fertilized eggs that are protected by a seed coat until conditions are proper for germination and growth.

What are the two major groups of angiosperms, and what are the major differences between the groups?

Angiosperms—made up of the largest number of plant species (240,000)—are classified into two major groups, monocots and dicots. The description of monocots and dicots is based on the first leaves that appear on the plant embryo. Monocots have one seed leaf, while dicots have two seed leaves. Approximately 65,000 species of monocots and 175,000 species of dicots exist. Orchids, bamboo, palms, lilies, grains, and many grasses are examples of monocots. Dicots include most trees that are nonconiferous, shrubs, ornamental plants, and many food crops.

Among angiosperms, what is the most important family?

Angiosperms, commonly known as flowering plants, include the grass family. This family is of greater importance than any other family of flowering plants. The edible grains of cultivated grasses, known as cereals, are the basic foods of most civilizations. Wheat, rice, and corn are the most extensively grown of all food crops. Other important cereals are barley, sorghum, oats, millet, and rye.

What is the most widely cultivated cereal in the world?

Wheat is the most widely cultivated cereal in the world; the grain supplies a major percentage of the nutrients needed by the world's population. Wheat is one of the oldest domesticated plants, and it has been argued that it laid the foundation for Western civilization. Domesticated wheat had its origins in the Near East at least nine thousand years ago. Wheat grows best in temperate, grassland biomes that receive 12–36 inches (30–90 centimeters) of rain per year and have relatively cool temperatures. Some of the top wheat-producing countries are Argentina, Canada, China, India, Ukraine, and the United States.

Wheat is one of the first plants to be domesticated and is the single most cultivated food in the world, feeding billions of people.

Which part of the wheat plant (*Triticum aestivum*) is used to make flour?

Nearly two dozen species of wheat exist. The most important ones for commercial use are common wheat, *Triticum aestivum* (sometimes referred to as *Triticum vulgare*), and durum wheat (*Triticum durum*). These varieties of common wheat account for 90 percent of the wheat grown worldwide. Durum wheat accounts for 5–7 percent of the wheat grown, and all other species account for the remainder of the wheat grown. Wheat is a monocotyledon grass whose fruit, the grain or kernel, contains one seed. The endosperm and embryo of the wheat plant are surrounded by the pericarp, or fruit wall, and the remains of the seed coat. More than 80 percent of the volume of the wheat kernel is made up of the starchy endosperm. White flour is made by milling the starchy endosperm. Wheat bran constitutes approximately 14 percent of the wheat kernel and is found in the covering layers and the outermost layer of the kernel, called the aleurone layer. Wheat germ is the embryo of the wheat plant and represents approximately 3 percent of the wheat kernel.

What is the difference between spring wheat and winter wheat?

Wheat is an annual plant with several varieties. Spring wheat is a summer annual whose seeds are planted in the spring and harvested in the fall of the same year. Winter wheat is a winter annual whose seeds are planted in the fall, establishing an abundant root system, and grown until the weather becomes too cold for growth. The established root system ensures rapid growth in the spring of the following year. It is ready for harvest in early summer.

What are some of the uses of angiosperms?

Angiosperms produce lumber, ornamental plants, and a variety of foods. Some examples of economically important angiosperms are:

Common Family Name	Genus Name	Products and Uses
Gourd	*Cucurbitaceae*	Food (melons and squashes)
Grass	*Poaceae*	Cereals, forage, ornamentals
Lily	*Liliaceae*	Ornamentals and food (onions)

Common Family Name	Genus Name	Products and Uses
Maple	*Aceraceae*	Lumber and maple sugar
Mustard	*Brassicaceae*	Food (cabbage and broccoli)
Olive	*Oleaceae*	Lumber, oil, and food
Palm	*Arecaceae*	Food (coconut), fiber, oils, waxes, furniture
Rose	*Rosaceae*	Fruits (apple and cherry), ornamentals (roses)
Spurge	*Euphorbiaceae*	Rubber, medicinals (castor oil), food (cassava), ornamentals (poinsettia)

What are carnivorous plants?

Carnivorous plants are plants that attract, catch, and digest animal prey, absorbing the bodily juices of prey for the nutrient content. More than four hundred species of carnivorous plants exist. The species are classified according to the nature of their trapping mechanism. All carnivorous plants have traps made of modified leaves with various incentives or attractants, such as nectar or an enticing color, that can lure prey. Active traps display rapid motion in their capture of prey.

What are the three major types of traps in carnivorous plants?

Carnivorous plants have three major types of traps: active traps, semi-active traps, and passive traps. The Venus fly trap, *Dionaea muscipula*, and the bladderwort, *Utricularia vulgaris*, have active traps that imprison victims. Each leaf is a two-sided trap with trigger hairs on each side. When the trigger hairs are touched, the trap shuts tightly around the prey. Semi-active traps employ a two-stage trap in which the prey is caught in the trap's adhesive fluid. As prey struggles in the fluid, the plant is triggered to slowly tighten its grip. The sundew (*Drosera capensis*) and butterwort (*Pinguicula vulgaris*) have semi-active traps. Passive traps entice insects using nectar. The passive-trap leaf has evolved into a shape resembling a vase or pitcher. Once lured to the leaf, the prey falls into a reservoir of accumulated rainwater and drowns. An example of the passive trap is the pitcher plant (*Sarracenia purpurea*).

The sundew is one example of a carnivorous plant. It uses sticky, sweet bait on its trichomes (hairs) to entrap hungry insects.

What are succulents?

A group of more than thirty plant families, including the amaryllis, lily, and cactus families, form what is known as the succulents (from the Latin word *succulentis*, meaning "fleshy" or "juicy"). Most members of the group are resistant to droughts, as they are dry-weather plants. Even when they live in moist, rainy environments, these plants need very little water.

Why are native plants beneficial?

Native plants are plants that occur naturally in a particular region, ecosystem, or habitat without human introduction. In general, they were found in the area prior to the time when it was explored and settled by European immigrants. Since these plants evolved locally, they are adapted to the soil and climate (temperature and precipitation) of the area. Furthermore, they form a symbiotic relationship with the indigenous wildlife, e.g., insects and birds, since they provide food and shelter for the wildlife, while the native wildlife pollinate the plants and control pests. Native plants require less watering, less commercial fertilizer, fewer pesticides and insecticides, and preserve the native biodiversity.

What was the historical significance of hemp?

During the early years of colonial America, hemp (*Agave sisalana*) was as common as cotton is now. It was an easy crop to grow, requiring little water, no fertilizers, and no pesticides. The fabric looks and feels like linen. It was used for uniforms of soldiers, paper (the first two drafts of the Declaration of Independence were written on hemp paper), and an all-purpose fabric. Betsy Ross's (1752–1836) flag was made of red, white, and blue hemp.

What is the difference between poison ivy, oak, and sumac?

These North American woody plants grow in almost any habitat and are quite similar in appearance. Each variety of plant has three-leaf compounds that alternate berrylike fruits and rusty, brown stems. Poison ivy (*Rhus radicans*) grows like a vine rather than a shrub and can grow very high, covering tall, stationary items such as trees. The fruit of *Rhus radicans* is gray in color and is without "hair," and the leaves of the plant are slightly lobed. *Rhus toxicodendron*, commonly known as poison oak, usually grows as a shrub, but it can also climb. Its leaflets are lobed and resemble the leaves of oak trees, and its fruit is hairy. Poison sumac (*Rhus vernix*) grows only in acidic, wet swamps of North America. This shrub can grow as high as 12 feet (3.6 meters). The fruit it produces hangs in a cluster and ranges from gray to brown in color. Poison sumac has dark-green leaves that are sharply pointed, compound, and alternating; it also has inconspicuous flowers that are yellowish green. All parts of poison ivy, poison oak, and poison sumac can cause serious dermatitis.

PLANT STRUCTURE AND FUNCTION

What are the major parts of vascular plants?

Vascular plants consist of roots, shoots, and leaves. The root system penetrates the soil and is below ground. The shoot system consists of the stem and the leaves.

What is the difference between the root system and the shoot system of vascular plants?

The root system is the part of the plant below ground level. It consists of the roots that absorb water and the various ions necessary for plant nutrition. The root system anchors the plant into the ground. The shoot system is the part of the plant above ground level. It consists of the stem and the leaves. The stem provides the framework for the positioning of the leaves. The leaves are the sites of photosynthesis.

Does a relationship exist between the size of the root system and the size of the shoot system?

Growing plants maintain a balance between the size of the root system (the surface area available for the absorption of water and minerals) and the shoot system (the photosynthesizing surface). The total water- and mineral-absorbing surface area in young seedlings usually far exceeds the photosynthesizing surface area. As the plant ages, the root-to-shoot ratio decreases. Additionally, if the root system is damaged, reducing the water- and mineral-absorbing surface area, shoot growth is reduced by lack of water, minerals, and root-produced hormones. Similarly, reducing the size of the shoot system limits root growth by decreasing the availability of carbohydrates and shoot-produced hormones to the roots.

What are the specialized cells in plants?

All plant cells have several common features, such as chloroplasts, a cell wall, and a large vacuole. In addition, a number of specialized cells are found only in vascular plants. They include:

- *Parenchyma cells*—Parenchyma (from the Greek words *para*, meaning "beside," and *en* + *chein*, meaning "to pour in") cells are the most common cells found in leaves, stems, and roots. They are often spherical in shape with only primary cell walls. Parenchyma cells play a role in food storage, photosynthesis, and aerobic respiration. They are living cells once they reach maturity. Most nutrients in plants, such as corn and potatoes, are contained in starch-laden parenchyma cells. These cells comprise the photosynthetic tissue of a leaf, the flesh of fruit, and the storage tissue of roots and seeds.
- *Collenchyma cells*—Collenchyma (from the Greek word *kola*, meaning "glue") cells have thickened primary cell walls and lack secondary cell walls. They form strands or continuous cylinders just below the surfaces of stems or leaf stalks. The most common function of collenchyma cells is to provide support for parts of the plant that are still growing, such as the stem. Similar to parenchyma cells, collenchyma cells are living cells once they reach maturity.
- *Sclerenchyma cells*—Sclerenchyma (from the Greek word *skleros*, meaning "hard") cells have tough, rigid, thick secondary cell walls. These secondary cell walls are hardened with lignin, which is the main chemical component of wood. It makes the cell

walls more rigid. Sclerenchyma cells provide rigid support for the plant. Sclerenchyma cells are divided into two types—fiber and sclereid. Fiber cells are long, slender cells that usually form strands or bundles. Sclereid cells, sometimes called stone cells, occur singly or in groups and have various forms. They have a thick, very hard secondary cell wall. Most sclerenchyma cells are dead once they reach maturity.

- *Xylem*—Xylem (from the Greek word *xylos*, meaning "wood") is the main water-conducting tissue of plants and consists of dead, hollow, tubular cells arranged end to end. The water transported in xylem replaces what is lost via evaporation through stomata. The two types of water-conducting cells are tracheids and vessel elements. Water flows from the roots of a plant up through the shoot via pits in the secondary walls of the tracheids. Vessel elements have perforations in their end walls to allow the water to flow between cells.
- *Phloem*—The two kinds of cells in the food-conducting tissue of plants, the phloem (from the Greek word *phloios*, meaning "bark"), are sieve cells and sieve-tube members. Sieve cells are found in seedless vascular plants and gymnosperms, while sieve-tube members are found in angiosperms. Both types of cells are elongated, slender, tubelike cells arranged end to end with clusters of pores at each cell junction. Sugars (especially sucrose), other compounds, and some mineral ions move between adjacent food-conducting cells. Sieve-tube members have thin primary cell walls but lack secondary cell walls. They are alive once they reach maturity.
- *Epidermis*—Several types of specialized cells occur in the epidermis, including guard cells, trichomes, and root hairs. Flattened epidermal cells, one layer thick and coated by a thick layer of cuticle, cover all parts of the primary plant body.

What are meristems?

Meristems (from the Greek word *meristos*, meaning "divided") are unspecialized cells that divide and generate new cells and tissues. Apical meristems, found at the tips of all roots and stems, are responsible for a plant's primary growth. The vascular cambium and cork cambium are the meristems responsible for a plant's secondary growth.

What is a seed?

A seed is a mature, fertilized ovule. It consists of the seed embryo and the nutrient-rich tissue, called the endosperm. The embryo consists of a miniature root and shoot. Once

Do plants ever stop growing?

Unlike many organisms that stop growing when they reach maturity, plants continue to grow during their entire life span. Unlimited, prolonged plant growth is described as indeterminate. The apical meristem produces an unrestricted number of lateral organs indefinitely.

the seed is protected and enclosed in a seed coat, it ceases further development and becomes dormant.

What are the advantages of seed dormancy?

The time during which a seed is dormant (when growth and development do not occur) allows for the dispersal of seeds. The plant can send its seeds into new environments. Dormancy assures survival of the plant since germination does not occur until conditions are favorable for plant growth.

What conditions are necessary for seed germination?

Seeds remain dormant until the optimum conditions of temperature, oxygen, and moisture are available for germination and further development. In general, the best temperatures for seed germination in most plants are 77–86°F (25–30°C). In addition to these external factors, some seeds undergo a series of enzymatic and biochemical changes prior to germination.

What are heirloom seeds?

Heirloom seeds have become popular in recent years. However, they have no precise, standard definition. In general, seeds are either hybrid seeds or open-pollinated seeds.

This is the entrance to the Global Seed Vault located on Norway's Svalbard archipelago. Here, millions of seeds from a wide variety of crop and other plants are stored in the permafrost as a hedge against a global disaster. Thankfully, seeds can remain dormant for many years!

Hybrids are the result of a controlled process of breeding. Most hybrid plants are bred with selected traits (e.g., drought resistant, disease resistant, or pest resistant). The seeds of a hybrid plant are not self-sustaining. Growing seeds saved from a hybrid plant will produce a plant with the traits of the grandparent plants or earlier generations but not of the hybrid plant. In contrast, seeds from open-pollinated plants are plants that have been pollinated naturally by birds, bees, other insects, other animals, rain, or wind. Seeds saved from open-pollinated plants will produce plants that are similar to the parent plant. Heirloom seeds are open-pollinated seeds. Some experts add an age requirement of fifty years for seeds to be considered heirloom seeds. Others believe that heirloom seeds should also have a story associated with them—for example, a family experience of growing the plant successfully under difficult conditions, giving the seeds historical, cultural, or familiar significance.

How does asexual propagation differ from sexual propagation in plants?

Sexual propagation requires seeds and pollen to produce new plants. Seeds are formed when pollen (the male) is transferred to an egg (the female). In plants that self-pollinate, every seed will be genetically identical to the parent plant. When cross-pollination occurs, the resulting seed is the combination of the genetic material from two parent plants that result in a unique, new plant.

Asexual reproduction requires taking a part of a plant which is used to regenerate the plant into a new plant. The new plant is genetically identical to the parent plant. Stems, roots, or leaves may be used to propagate a plant asexually. Some of the techniques used for asexual plant propagation are cuttings, layering, division or separation, and grafting.

How are seedless grapes grown?

Since seedless grapes cannot reproduce in the manner that grapes usually do (i.e., dropping seeds), growers have to take cuttings from the plants, root them, and then plant the plant cuttings. Seedless grapes come from a naturally occurring mutation in which the hard seed casing fails to develop. Although the exact origin of seedless grapes is un-

Do seedless watermelons occur naturally?

Seedless watermelon was first introduced in 1988 after fifty years of research. A seedless watermelon plant requires pollen from a seeded watermelon plant. Farmers frequently plant seeded and seedless plants close together and depend on bees to pollinate the seedless plants. The white "seeds," also known as pods, found in seedless watermelons serve to hold a fertilized egg and embryo. Because a seedless melon is sterile and fertilization cannot take place, pods do not harden and become a black seed as occurs in seeded watermelons.

known, they might have been first cultivated thousands of years ago in present-day Iran or Afghanistan.

How does the shoot develop following germination?

Shoot development is classified based on whether the cotyledons (seed leaves) are carried above ground or remain below ground. Seed germination, during which the cotyledons are carried above ground, is called epigeous. The food stored in the cotyledons is digested, and the products are transported to the growing parts of the young seedling. When the seedling becomes established and is no longer dependent upon the stored food in the seed for nutrition, the cotyledons gradually decrease in size, wither, and fall off. Seed germination during which the cotyledons remain underground is called hypogeous. The seedling uses the stored food from the cotyledons for growth and then the cotyledons decompose. The cotyledons remain in the soil during the entire process.

What are the parts of a stem?

A stem has nodes and internodes. The nodes are the points where the leaves are attached to the stem. The internodes are the parts of the stem between the nodes.

What are the functions of stems?

The four main functions of stems are to 1) support leaves; 2) produce carbohydrates; 3) store materials, such as water and starch; and 4) transport water and solutes between roots and leaves. Stems provide the link between the water and dissolved nutrients of the soil and the leaves.

How does water move up a tree?

Water is carried up a tree through the xylem tissue in a process called transpiration. Constant evaporation from the leaf creates a flow of water from root to shoot. The roots of a tree absorb the vast majority of water that a tree needs. The properties of cohesion and adhesion allow the water to move up a tree regardless of its height. Cohesion allows the individual water molecules to stick together in one continuous stream. Adhesion permits the water molecules to adhere to the cellulose molecules in the walls of xylem cells. When the water reaches a leaf, water is evaporated, thus allowing additional water molecules to be drawn up through the tree.

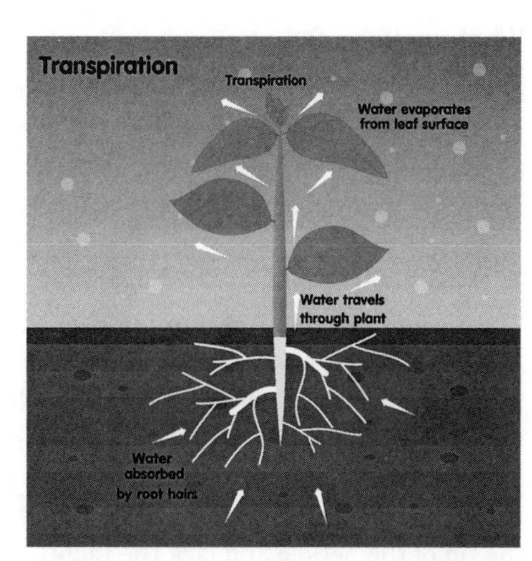

Water is circulated in a plant through the process of transpiration. As water evaporates through the stomates (pores) of the leaves, more water is pulled into the leaves by negative water pressure. This negative pressure goes all the way through to the roots, pulling water upwards. Transpiration is also important to keep the plant cool.

What famous artist first observed that the number of rings in the cross-section of a tree trunk indicates its age?

The painter Leonardo da Vinci noticed that the number of rings in the cross-section of a tree trunk corresponded to the age of the tree. He also saw that the year's dampness can be determined by the space between the tree's rings. The farther apart the rings, the more moisture that was contained in the ground around the tree.

How are tree rings used to date historical events?

The study of tree rings is known as dendrochronology. Every year, trees produce an annular ring composed of one wide, light ring and one narrow, dark ring. During spring and early summer, tree stem cells grow rapidly larger, thus producing the wide, light ring. In winter, growth is greatly reduced and cells are much smaller, producing the narrow, dark ring. In the coldest part of winter or the dry heat of summer, no cells are produced. Comparing pieces of dead trees of unknown age with the rings of living trees allows scientists to establish the date when the fragment was part of a living tree. This technique has been used to date the ancient pueblos throughout the southwestern United States. A subfield of dendrochronology is dendroclimatology. Scientists study the tree rings of very old trees to determine climatic conditions of the past. The effects of droughts, pollution, insect infestations, fires, volcanoes, and earthquakes are all visible in tree rings.

What are leaves?

Leaves are the main photosynthetic organ for plants. They are also organized to maximize sugar production while making sure little water loss occurs. Therefore, they are also important in gas exchange and water movement throughout the whole plant. Leaves are found in a variety of shapes, sizes, and arrangements.

What is the difference between simple leaves and compound leaves?

The blades of simple leaves are undivided, although they may have deep lobes. By contrast, the blades of compound leaves consist of clearly separated leaflets. Each leaflet usually has its own petiole, which is called a petiolule. Compound leaves are divided into two types: pinnately compound leaves and palmately compound leaves. The leaflets in pinnately compound leaves arise from either side of an axis called the rachis, which is an extension of the petiole. The leaflets in palmately compound leaves originate at the tip of the petiole and lack the rachis.

Which culinary herbs are from the leaves of plants?

Many of the herbs used to enhance flavors in foods are from the leaves of nonwoody plants. The following lists the common and scientific names of plants whose leaves are used in cooking:

- Basil (*Ocimum basilicum)*
- Bay leaves (*Laurus nobilis)*
- Dill (*Anethum graveolens)*
- Oregano (*Origanum vulgare)*
- Parsley (*Petroselinum crispum)*
- Peppermint (*Mentha piperita)*
- Rosemary (*Rosmarinus officinalis)*
- Sage (*Salvia officinalis)*
- Tarragon (*Artemesia dracunculus)*
- Thyme (*Thymus vulgaris)*

Why do tree leaves change color in the fall?

The carotenoids (pigments in the photosynthesizing cells), which are responsible for the fall colors, are present in the leaves during the growing season. However, the colors are eclipsed by the green chlorophyll. Toward the end of summer, when the chlorophyll production ceases due to declining daylight and a decrease in temperature, the other colors of the carotenoids (such as yellow, orange, red, or purple) become visible. Listed below are the autumn leaf colors of some common trees.

Tree	Color
Sugar maple, sumac	Flame red and orange
Red maple, dogwood, sassafras, scarlet oak	Dark red
Poplar, birch, tulip tree, willow	Yellow
Ash	Plum purple
Oak, beech, larch, elm, hickory, sycamore	Tan or brown
Locust	Stays green until leaves drop
Black walnut, butternut	Drops leaves before they turn color

What important organelle is found in the mesophyll layer of the leaf?

The mesophyll (from the Greek words *mesos*, meaning "middle," and *phylum*, meaning "leaf") of a leaf consists of masses of parenchyma cells that are packed with chloroplasts important for photosynthesis. The palisade parenchyma comprises columnar layers of

How many leaves are on a mature tree?

Leaves are one of the most conspicuous parts of a tree. A maple tree (genus *Acer*) with a trunk 3 feet (1 meter) wide has approximately one hundred thousand leaves. Oak (genus *Quercus*) trees have approximately seven hundred thousand leaves. Mature American elm (*Ulmus americana*) trees can produce more than five million leaves per season.

parenchyma cells found beneath the epidermis of many leaves. The spongy parenchyma is a mass of cells that are irregular in shape and often highly branched. Large, intercellular spaces in the spongy parenchyma function in gas exchange and the passage of water vapor from the leaves. These spaces are connected to the stomata.

What is photosynthesis, and why is it important?

Photosynthesis (from the Greek word *photo*, meaning "light," and synthesis, from the Greek word *syntithenai*, meaning "to put together") is the process by which plants use energy derived from light in order to make food molecules from carbon dioxide and water. Oxygen (O_2) is produced as a waste product of this process. Photosynthesis is a dual-staged process with multiple components. Ultimately, photosynthesis is the process that provides food for the entire world. Each year, more than 250 billion metric tons of sugar are created through photosynthesis. Photosynthesis is a source of food not only for plants but also all organisms that are not capable of internally producing their own food, including humans.

Who was the first person to formulate a chemical equation for photosynthesis in plants and bacteria?

Scientists believed that carbon, oxygen, and hydrogen were essential for plant nutrition and photosynthesis since the eighteenth century. Evidence of photosynthesis's two-stage process was first presented by F. F. Blackman (1866–1947) in 1905. Blackman had iden-

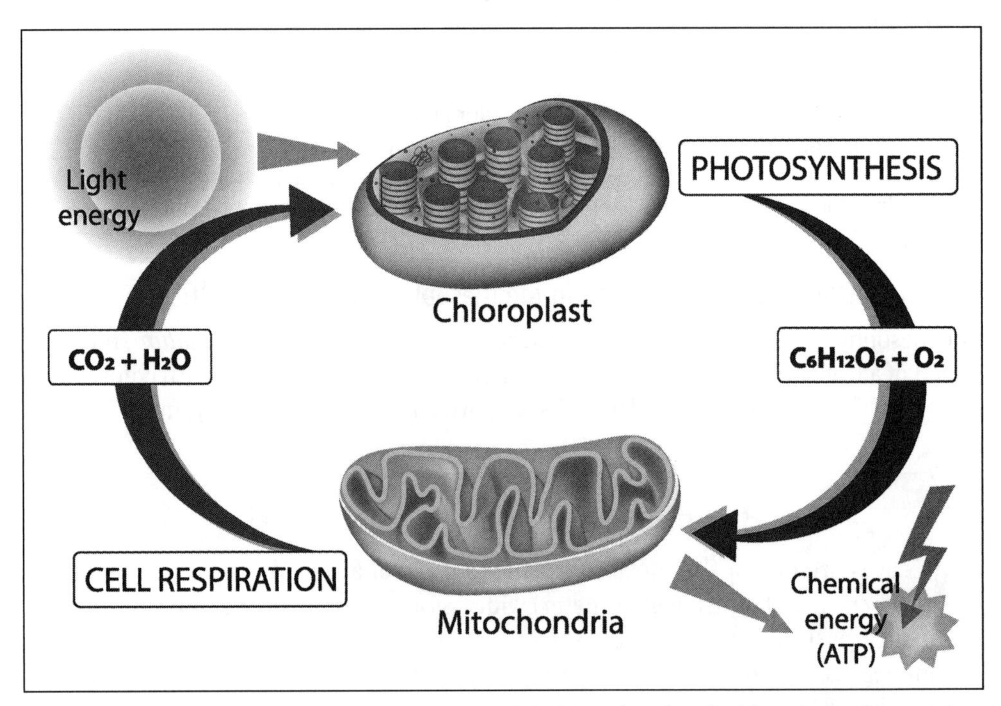

Photosynthesis is the process in which plants convert sunlight (energy), carbon dioxide, and water into sugar (energy) and oxygen.

In addition to chlorophyll, what are other pigments found in plants?

Plant pigments give all plants, including flowers, their color. In addition to chlorophyll, carotenoids and flavonoids are major plant pigments. As discussed above, chlorophyll is responsible for the green color of plant leaves. More than 750 carotenoids occur naturally, which are the source for the yellow, orange, and red colors of many plants. Many fruits and vegetables, including carrots, yams, papayas, cantaloupes, watermelons, spinach, kale, tomatoes, and oranges, contain carotenoids. Scientists have identified thousands of different flavonoids. Some of the major classes of flavonoids are anthocyanins, flavanols, chalcones, and aurones. These pigments give fruits, vegetables, and flowers their red, orange, blue, purple, and yellow colors.

tified that both a light-dependent stage and a light-independent stage occur during photosynthesis. Around 1930, Cornelis Bernardus van Niel (1923–1977), a Dutch-born American scientist, became the first person to propose that water rather than carbon dioxide was the source of the oxygen that resulted from photosynthesis. He formulated the general chemical equation of:

$$CO_2 + 2H_2A \longrightarrow (CH_2O) + 2A + H_2O$$

The letter A in the equation represents oxygen in the oxygen-evolving photosynthetic reactions of green plants and sulfur in the sulfur-producing photosynthetic reactions of purple bacteria.

In 1937, Robert Hill (1899–1991) discovered that chloroplasts are capable of producing oxygen in the absence of carbon dioxide only when the chloroplasts are illuminated and provided with an artificial electron acceptor. Hill's work supported van Niel's equation.

What is tropism?

Tropism is the movement of a plant in response to a stimulus. The categories include:

- Chemotropism—A response to chemicals by plants in which incurling of leaves may occur.
- Gravitropism—Formerly called geotropism, a response to gravity in which the plant moves in relation to gravity. Shoots of a plant are negatively geotropic (growing upward), while roots are positively geotropic (growing downward).
- Hydrotropism—A response to water or moisture in which roots grow toward the water source.
- Paraheliotropism—A response by the plant leaves to avoid exposure to the Sun.
- Phototropism—A response to light in which the plant may be positively phototropic (moving toward the light source) or negatively phototropic (moving away from the

light source). Main axes of shoots are usually positively phototropic, whereas roots are generally insensitive to light.

- Thermotropism—A response to temperature by plants.
- Thigmotropism or haptotropism—A response to touch by the climbing organs of a plant. For example, the plant's tendrils may curl around a support in a springlike manner.

What are the major classes of plant hormones?

The five major classes of plant hormones are auxins, gibberellins, cytokinins, ethylene, and abscisic acid.

Hormone	Principal Action	Where Produced or Found in Plant
Auxins	Elongate cells in seedlings, shoot tips, embryos, leaves	Shoot apical meristem
Gibberellins	Elongate and divide cells in seeds, roots, shoots, young leaves	Apical portions of roots and shoots
Cytokinins	Stimulate cell division (cytokinesis) in seeds, roots, young leaves, fruits	Roots
Ethylene	Hastens fruit ripening	Leaves, stems, young fruits
Abscisic acid	Inhibits growth; closes stomata	Mature leaves, fruits, root caps

When were the major classes of plant hormones identified, and who is associated with their identification?

- *Auxins*—Charles Darwin and his son, Francis (1845–1925), performed some of the first experiments on growth-regulating substances. They published their results in

What are some commercial uses of plant hormones?

Plant hormones are used in a variety of applications to control some aspect of plant development. Auxins are used in commercial herbicides as weed killers. Another use of auxins is to stimulate root formation. It is often referred to as the "rooting hormone" and applied to cuttings prior to planting. Some hormones are used to increase fruit production and prevent preharvest fruit drop. Gibberellins are sprayed on Thompson seedless grapes during the flowering stage to thin the flowers on each cluster, thus allowing the remaining flowers to spread out and develop larger fruit. Gibberellins are also used to enhance germination and stimulate the early emergence of seedlings in grapes, citrus fruits, apples, peaches, and cherries. When used on cucumber plants, gibberellins promote the formation of male flowers, which is useful in the production of hybrid seeds.

1881 in *The Power of Movement in Plants*. In 1926, Frits W. Went (1903–1990) isolated the chemical substance responsible for elongating cells in the tips of oat (genus *Avena*) seedlings. He named this substance auxin, from the Greek word *auxein*, meaning "to increase."

- *Gibberellins*—In 1926, Japanese scientist Eiichi Kurosawa discovered a substance produced by a fungus, *Gibberella fujikuroi*, that caused a disease ("foolish seedling disease") in rice (*Oryza sativa*) seedlings in which the seedlings would grow rapidly but appear sickly and then fall over. Japanese chemists Teijiro Yabuta (1888–1977) and Yusuke Sumiki (1901–1974) isolated the compound and named it gibberellin in 1938.
- *Cytokinins*—Johannes van Overbeek discovered a potent growth factor in coconut (*Cocos nucifera*) milk in 1941. In the 1950s, Folke Skoog (1908–2001) was able to produce a thousandfold purification of the growth factor but was unable to isolate it. Carlos O. Miller (1923–2012), Skoog, and their colleagues succeeded in isolating and identifying the chemical nature of the growth factor. They named the substance kinetin and the group of growth regulators to which it belonged cytokinins because of their involvement in cytokinesis or cell division.
- *Ethylene*—Even before the discovery of auxin in 1926, ethylene was known to have effects on plants. In ancient times, the Egyptians would use ethylene gas to ripen fruit. During the 1800s, shade trees along streets with lamps that burned ethylene, the illuminating gas, would become defoliated from leaking gas. In 1901, Dimitry Neljubov demonstrated that ethylene was the active component of illuminating gas.
- *Absicisic acid*—Philip F. Wareing (1914–1996) discovered large amounts of a growth inhibitor in the dormant buds of ash and potatoes that he called dormin. Several years later in the 1960s, Frederick T. Addicott (1912–2009) reported the discovery in leaves and fruits of a substance capable of accelerating absicission that he called abscisin. It was soon discovered that dormin and abscisin were identical chemically.

What ingredient did Steward add to the nutrient broth when culturing cells?

In addition to supplying the cultured cells with sugars, minerals, and vitamins, Steward added coconut milk. Coconut milk contains a substance that induces cell division. Subsequent research identified this substance as cytokinins, a group of plant hormones (growth regulators) that stimulate cell division. Once the cultured cells began dividing, they were transplanted on agar media, where they formed roots and shoots and developed into plants.

Who showed that plant cells were totipotent?

Totipotency is the ability of a cell to grow into a mature organism (plant) since all of the organism's genetic material is contained in the cell. In 1958, Frederick Campion Steward (1904–1993), a botanist at Cornell University, successfully regenerated an entire carrot plant from a tiny piece of phloem. Small pieces of tissue from carrots were grown in a nutrient broth. Cells that broke free from the fragments dedifferentiated, meaning that they reverted to unspecialized cells. However, as these unspecialized cells grew, they divided and redifferentiated back into specialized cell types. Eventually, cell division and redifferentiation produced entire new plants. Each unspecialized cell from the nutrient broth expressed its genetic potential to make all the other cell types in a plant.

Bees play an important role in pollinating plants, but they are not the only pollinators. Butterflies, moths, bats, birds, and a variety of insects and mammals also help pollinate plants.

Who was the first person to recognize the role of bees in pollination?

The role of bees in pollination was discovered by Joseph Gottlieb Kölreuter (1733–1806) in 1761. He was the first to realize that plant fertilization occurs with the help of pollen-carrying insects.

Why is pollination important?

Pollination is essential for the process of fertilization to produce offspring in plants. Effective pollination occurs when viable pollen is transferred to a plant's stigmas, ovule-bearing organs, or ovules (seed precursors). Since plants are immobile organisms, they generally required external agents to transport their pollen from where it is produced to where fertilization can occur. Some plants self-pollinate, that is, they have the ability to transfer their own pollen to their own stigmas. Most plants require cross-pollination. During cross-pollination, the pollen of one plant is moved by an agent to the stigma of a different plant. Cross-pollination allows for new genetic material to be introduced and therefore seems more advantageous.

What are some examples of pollinators?

The most common pollinators are animals, including bees, birds, butterflies, moths, other insects, bats, and some larger animals, such as the lemur. Wind and water are also capable of transferring pollen from one plant to another plant. Many times, flowers offer one or more "rewards" to attract these agents—sugary nectar, oil, solid food bodies, perfume, a place to sleep, or, sometimes, the pollen itself. Other times, the plant can "trap" the agent into transporting the pollen.

How does a plant's structure accommodate a pollinator?

Plant structure often accommodates the type of agent used by the plant for pollination. For example, plants such as grasses and conifers, whose pollen is carried by the wind, tend to have a simple structure lacking petals, with freely exposed and branched stigmas to catch airborne pollen and dangling anthers (pollen-producing parts) on long filaments. This type of anther allows the light, round pollen to be easily caught by the wind. These plants are found in areas such as prairies and mountains, where insect agents are rare. In contrast, semi-enclosed, nonsymmetrical, long-lived flowers such as irises, roses, and snapdragons have a "landing platform" and nectar in the flower base to accommodate insect agents such as the bee. The sticky, abundant pollen can easily become attached to the insect to be borne away to another flower.

What are the parts of a flower?

The basic parts of a flower are:

- *Sepal*—Found on the outside of the bud or on the underside of the open flower. It serves to protect the flower bud from drying out. Some sepals ward off predators by their spines or chemicals.
- *Petals*—Serve to attract pollinators and are usually dropped shortly after pollination occurs.
- *Nectar*—Contains varying amounts of sugar and proteins that can be secreted by any of the floral organs. It usually collects inside the flower cup near the base of the cup formed by the flower parts.
- *Stamen*—The male part of a flower, which consists of a filament and anther, where pollen is produced.
- *Pistil*—The female part of a flower, which consists of the stigma, style, and ovary, containing ovules. After fertilization, the ovules mature into seeds.

What is meant by an "imperfect" flower?

An imperfect flower is one that is unisexual, having either stamens (male parts) or pistils (female parts) but not both.

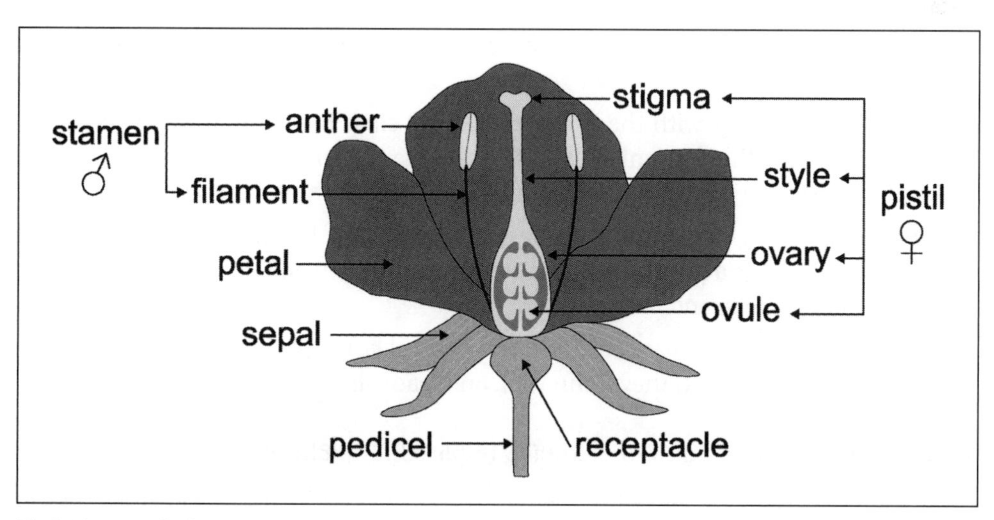

The basic parts of a flower are illustrated here.

Which parts of a tree are helpful in identifying the tree species?

All parts of a tree provide distinguishing features, which are used to identify the tree species.

- Leaf type and shape—Leaves may be needles, scales, or broadleaf in a variety of shapes. Some common leaf shapes are oval, long and narrow, round, triangular, or heart-shaped.
- Leaf bud shape and arrangement—Buds may have pointed or rounded tips. They may be arranged directly opposite each other on a branch or be placed staggered on alternate sides of the branch. Spiral buds, which whorl around the stem, are another variation.
- Bark—The color and texture of the bark of a tree may be used to identify a tree. Bark ranges in color from gray to various shades of brown to silver and white. Bark may be smooth, shiny, papery, scaly, or furrowed. Some bark changes as the tree matures and ages.
- Size, shape, and location—In a natural setting, different trees will have different shapes, for example, tall and narrow or with branches that spread out more fully. Some trees prefer a location closer to water, while others thrive in drier soils.
- Flowers—Flowers may be single blooms or in clusters, with variation in flower color.
- Fruit and seeds—Whether edible or for plant propagation, fruit and seeds vary in size, shape, color, and type of fruit.

SOIL

What are the different types of soil?

Soil is the weathered outer layer of Earth's crust. It is a mixture of tiny rock fragments and organic matter. Soils are categorized broadly into three types: clay, sandy, and loam.

Clay soils are heavy, with the particles sticking close together. Most plants have a hard time absorbing the nutrients in clay soil, and the soil tends to become waterlogged. Clay soils can be good for a few deep-rooted plants, such as mint, peas, and broad beans.

Sandy soils are light and have particles that do not stick together. Sandy soil is good for many alpine and arid plants, some herbs such as tarragon and thyme, and vegetables such as onions, carrots, and tomatoes.

Loam soils are a well-balanced mix of smaller and larger particles. They provide nutrients to plant roots easily, they drain well, and loam also retains water very well.

What are the essential nutrient elements required for plant growth?

Essential nutrients are chemical elements that are necessary for plant growth. An element is essential for plant growth when 1) it is required for a plant to complete its life

What do the numbers on a bag of fertilizer indicate?

The three numbers, such as 15–20–15, refer to the percentages by weight of macronutrients found in the fertilizer. The first number stands for nitrogen, the second for phosphorus, and the third for potassium. In order to determine the actual amount of each element in the fertilizer, multiply the percentage by the fertilizer's total weight in pounds. For example, a fifty-pound bag of 15–20–15 contains 7.5 pounds of nitrogen, 10 pounds of phosphorus, and 7.5 pounds of potassium. The remaining pounds are filler.

cycle (to produce viable seeds); 2) it is part of a molecule or component of the plant that is itself essential to the plant, such as the magnesium in the chlorophyll molecule; and 3) the plant displays symptoms of deficiency in the absence of the element. Essential nutrients are also referred to as essential minerals and essential inorganic nutrients.

What is the best soil pH for growing plants?

The value of the pH ("potential of hydrogen") of soil represents the hydrogen ion concentration in a soil sample. It expresses the relative alkalinity–acidity of the soil sample. The neutral point in the scale is seven. Nutrients such as phosphorous, calcium, potassium, and magnesium are most available to plants when the soil pH is between 6.0 and 7.5. Under highly acidic (low pH) conditions, these nutrients become insoluble and relatively unavailable for uptake by plants. However, some plants, such as rhododendrons, grow better in acidic soils. High soil pH can also decrease the availability of nutrients. If the soil is more alkaline than pH 8.0, phosphorous, iron, and many trace elements become insoluble and unavailable for plant uptake.

What is the composition of synthetic soil?

Synthetic soil is composed of a variety of organic and inorganic materials. Inorganic substances used include pumice, calcinated clay, cinders, vermiculite, perlite, and sand. Vermiculite and perlite are used for water retention and drainage. Organic materials used include wood residues, manure, sphagnum moss, plant residues, and peat. Sphagnum peat moss is also helpful for moisture retention and lowers the pH of the mixture. Lime may be added to offset the acidity of peat. Synthetic soil may also be referred to as growing medium, soil mixes, potting mixture, plant substrate, greenhouse soil, potting soil, and amended soil. Most synthetic soils are deficient in important mineral nutrients, which can be added during the mixing process or with water.

What is hydroponics?

This term refers to growing plants in some medium other than soil; the inorganic plant nutrients (such as potassium, sulfur, magnesium, and nitrogen) are continuously supplied to the plants in solution. Hydroponics is mostly used in areas containing little soil or unsuitable soil. Since it allows precise control of nutrient levels and oxygenation of the roots, it is often used to grow plants used for research purposes. Julius von Sachs (1832–1897), a researcher in plant nutrition, pioneered modern hydroponics. Research plants have been grown in solution culture since the mid-1800s. William Gericke (1882–1970), a scientist at the University of California, defined the word hydroponics in 1937. In the fifty years that hydroponics has been used on a commercial basis, it has been adapted to many situations. NASA will be using hydroponics in the space station for crop production and to recycle carbon dioxide into oxygen. Although successful for research, hydroponics has many limitations and may prove frustrating for the amateur gardener.

ZOOLOGY

INTRODUCTION AND HISTORICAL BACKGROUND

What are the main characteristics of animals?

Animals are a very diverse group of organisms, but all of them share a number of characteristics. Animals are multicellular eukaryotes that are heterotrophic, ingesting and digesting food inside the body. Animal cells lack the cell walls that provide support in the bodies of plants and fungi. Most animals have muscle systems and nervous systems, which are responsible for movement and rapid response to stimuli in their environment. In addition, most animals reproduce sexually, with the diploid stage dominating the life cycle. In most species, a large, nonmotile egg is fertilized by a small, flagellated sperm, thus forming a diploid zygote. The transformation of a zygote to an animal of a specific form depends on the controlled expression in the developing embryo of special regulatory genes.

Who is the "father of zoology"?

Aristotle is considered the "father of zoology." His contributions to zoology include vast quantities of information about the variety, structure, and behavior of animals; the analysis of the parts of living organisms; and the beginnings of the science of taxonomy.

Who is the "father of modern zoology"?

Conrad Gessner (1516–1565), a Swiss naturalist, is often credited as the "father of modern zoology" based on his three-volume *Historia Animalium*, which served as a standard reference work throughout Europe in the sixteenth and seventeenth centuries.

Who is considered the founder of experimental zoology?

Abraham Trembley (1710–1784), a Swiss scientist, is considered the founder of experimental zoology. Much of his research involved studying the regeneration of hydras.

How are animals classified?

Animals belong to the kingdom Animalia. Most biologists divide the kingdom into two subkingdoms: 1) Parazoa (from the Greek words *para*, meaning "alongside," and *zoa*, meaning "animal") and 2) Eumetazoa (from the Greek words *eu*, meaning "true"; *meta*, meaning "later"; and *zoa*, meaning "animal"). The only existing animals classified as Parazoa are the sponges (phylum Porifera). Sponges are very different from other animals and function much like colonial, unicellular protozoa, even though they are multicellular. Cells of sponges can be versatile and change form and function, and they are not organized into tissues and organs. They also lack symmetry. All other animals have true tissues, are symmetrical, and are classified as Eumetazoa.

How can animals be grouped according to body symmetry?

Symmetry refers to the arrangement of body structures in relation to the axis of the body. Most animals exhibit either radial or bilateral body symmetry. Animals such as jellyfish, sea anemones, and starfish have radial symmetry. In radial symmetry, the body has the general form of a wheel or cylinder, and similar structures are arranged as spokes from a central axis. The bodies of all other animals are marked by bilateral symmetry, a design in which the body has right and left halves that are mirror images of each other. A bilaterally symmetrical body plan has a top and a bottom, also known respectively as the dorsal and ventral portions of the body. It also has a front (or anterior) end and a back (or posterior) end.

ANIMAL CHARACTERISTICS

What is the difference between an invertebrate and a vertebrate?

Invertebrates are animals that lack a backbone. Almost all animals (99 percent) are invertebrates. Of the more than one million identified animals, only about 52,000 have a vertebral column; these are referred to as vertebrates. Most biologists believe that the millions of species that have yet to be discovered are mainly invertebrates.

What is the largest invertebrate?

The largest invertebrate is the giant squid, *Architeuthis dux*, which averages 30–53 feet (9–16 meters) in length including its tentacles. It may reach a length of 69 feet (21 meters). These animals have the largest eyes, up to 10 inches (25 centimeters) in diameter, in the animal kingdom. It is believed that they generally live on or near the ocean

A display at New York City's Museum of Natural History depicts a giant squid battling a sperm whale. These squid were the stuff of sailor legends. A video recording of living giant squid was not successfully taken until 2004!

bottom at a depth of 3,281 feet (1,000 meters), or slightly more than a half mile below the surface of the sea.

What is the heaviest invertebrate?

The giant clam, *Tridacna maxima*, is the heaviest invertebrate—it may weigh as much as 122 pounds (270 kilograms). The shells of this bivalve may be as long as 5 feet (1.5 meters).

Which group of invertebrates lives in both marine and freshwater and is one of the most important of all animals?

Copepods, tiny crustaceans, are the link between the photosynthetic life in the ocean or pond and the rest of the aquatic food web. They are primary consumers, grazing on algae in the waters of the oceans and ponds. These organisms, among the most abundant multicellular animals on Earth, are then consumed by a variety of small predators, which are eaten by larger predators, and so on. Virtually all animal life in the ocean depends on the copepods either directly or indirectly. Although humans do not eat copepods directly, our sources of food from the ocean would disappear without the copepods.

What are the largest and smallest vertebrates?

Largest Vertebrates	Name	Length and Weight
Sea mammal	Blue or sulphur-bottom whale (*Balaenoptera musculus*)	100–110 feet (30.5–33.5 meters) long; weighs 135–209 tons (122.4–189.6 tonnes)
Land mammal	African bush elephant (*Loxodonta africana*)	Bull is 10.5 feet (3.2 meters) tall at shoulder; weighs 5.25–6.2 tons (4.8–5.6 tonnes)
Bird	North African ostrich (*Struthio c. camelus*)	8–9 feet (2.4–2.7 meters) tall; weighs 345 pounds (156.5 kilograms)
Fish	Whale shark (*Rhincodon typus*)	41 feet (12.5 meters) long; weighs 16.5 tons (15 tonnes)
Reptile	Saltwater crocodile (*Crocodylus porosus*)	14–16 feet (4.3–4.9 meters) long; weighs 900–1,500 pounds (408–680 kilograms)
Amphibian	Chinese giant salamander (*Andrias davidianus*)	5.9 feet (1.8 meters) long; weighs 140 pounds (64 kilograms)

Smallest vertebrates	Name	Length and weight
Sea mammal	Commerson's dolphin (*Cephalorhynchus commersonii*)	Weighs 50–70 pounds (236.7–31.8 kilograms)
Land mammal	Bumblebee or Kitti's hog-nosed bat (*Craseonycteris thong longyai*) or the pygmy shrew (*Suncus erruscus*)	Bat is 1 inch (2.54 centimeters) long; weighs 0.062–0.07 ounces (1.6–2 grams); shrew is 1.5–2 inches (3.8–5 centimeters) long; weighs 0.052–0.09 ounces (1.5–2.6 grams)
Bird	Bee hummingbird (*Mellisuga helenea*)	2.25 inches (5.7 centimeters) long; weighs 0.056 ounces (1.6 grams)
Fish	Angler fish (male) (*Photocorynus spiniceps*)	0.24 inches (6.2 millimeters) long
Reptile	Dwarf gecko (*Spaerodactylus ariasae*)	0.63 inches (1.6 centimeters) long
Amphibian	(*Paedophryne amanuensis*) This tiny frog has no common name.	0.30 inches (0.76 centimeters) long

Do any one-eyed animals exist?

No mammals, birds, reptiles, or amphibians have only one eye. Some fish, such as the flounder, are thought of as having only one eye but actually have two eyes. One eye migrates to join the other on the left or right side. Only invertebrates have one eye. Two examples, both crustaceans, are water fleas and the cyclops, which lives in pond scum in stagnant bodies of water.

What is the longest-living animal species?

The animal with the longest recorded life span is the Aldabra Giant Tortoise (sometimes called Marion's tortoise; *Testudo sumeirei*) of the Seychelles Islands in the Indian Ocean. One tortoise lived to be more than 152 years. The quahog (*Mercenaria mercenaria*), or hard-shell clam, can survive for 150 years. The mammal with the longest recorded life is man, which can live more than 110 years. Among nonhuman species, the Asiatic elephant holds the record at seventy-eight years. The shortest-lived mammal is the pocket gopher, which lives for about twenty months.

The Aldabra tortoise can live to be over 150 years old. As far as anyone knows, that makes it the longest-living animal on the planet.

ANIMAL BEHAVIOR

Who were some prominent scientists to study animal behavior?

Aristotle wrote ten volumes on the *Natural History of Animals*. The Roman naturalist Pliny the Elder also extensively observed and recorded observations of organisms in his book *Natural History*. In more recent times, Charles Darwin recorded (in his journal) the behavior of the marine iguanas of the Galapagos Islands. Darwin also published a book, *The Expression of the Emotions of Man and Animals* (1872), in which he showed how natural selection would favor specialized behavioral patterns for survival. However, it was not until 1953, when Niko Tinbergen (1907–1988) documented and published his studies of gulls (*The Herring Gull's World*) and their begging techniques, that the field of ethology—the study of animal behavior—was established.

Who first won the Nobel Prize for work on animal behavior?

Although no specific Nobel Prize exists for the study of animal behavior, the first prize to be awarded for the scientific study of animal behavior was awarded jointly in 1973 in the category Physiology or Medicine to Konrad Lorenz (1903–1989), Karl von Frisch (1886–1982), and Niko Tinbergen. Each ethologist had a main research area for which he was well known: Lorenz for imprinting behavior of birds; von Frisch for the "dance" of the honeybees; and Tinbergen for aggressive behavior of stickleback fish.

What is sociobiology, and who is the "father" of sociobiology?

Sociobiology, considered a subdiscipline of behavioral ecology by some researchers, is the study of the social organization of a species. Sociobiology attempts to develop rules that explain the evolution of certain social systems. Edward O. Wilson (1929–) formulated the general biological principles that govern both social behavior and organization of all kinds of animals when he published *Sociobiology: The New Synthesis* in 1975. He found that social insects, such as bees and wasps, have a complex hierarchy, with strict rules governing who reproduces, who gets food, and who defends the colony. Wilson became interested in animal behavior as a child when he watched ant colonies. He is one of the most prolific writers in ecology, having written more than thirty books, won two Pulitzer Prizes, and discovered hundreds of new species. The application of sociobiology to humans initially generated controversy as to whether the same forces drive highly organized animal and human societies.

Besides humans, which animals are the most intelligent?

According to Edward O. Wilson, a behavioral biologist, the ten most intelligent animals are the following:

1. Chimpanzee (two species)
2. Gorilla
3. Orangutan
4. Baboon (seven species, including drill and mandrill)
5. Gibbon (seven species)
6. Monkey (many species, especially macaques, the patas, and the Celebes black ape)
7. Smaller-toothed whale (several species, especially killer whale)
8. Dolphin (many of the approximately eighty species)
9. Elephant (two species)
10. Pig

Among the invertebrates, which are the most intelligent?

Most specialists agree that cephalopods—octopi, squids, and nautili—are the most intelligent invertebrates. These animals can make associations among stimuli and have

Do dogs and cats have good memories?

Dogs do have long-term memories, especially for those whom they love. Cats have a memory for things that are important to their lives. Some cats seem to have extraordinary "memories" for finding places. Taken away from their homes, they seem able to remember where they live. The key to this "homing" ability could be a built-in celestial navigation, similar to that used by birds, or the cats' navigational ability could be attributed to the cats' sensitivity to Earth's magnetic fields. When magnets are attached to cats, their normal navigational skills are disrupted.

been used as models for studying learning and memory. Octopi can be trained to perform many tasks, including distinguishing between objects and opening jars to obtain food.

Who discovered the "dance of the bees"?

In 1943, Karl von Frisch published his study on the dance of the bees. It is a precise pattern of movements performed by returning forager (worker) honeybees in order to communicate the direction and distance of a food source to the other workers in the hive. The dance is performed on the vertical surface of the hive, and two kinds of dances have been recognized: the round dance (performed when food is nearby) and the waggle dance (done when food is farther away).

What were the contributions of Jane Goodall and Dian Fossey to ethology?

Jane Goodall (1934–) is a primate ethologist who is famous for her studies of chimpanzees in Tanzania. She began her career as a secretary in Nairobi, Kenya, for Louis B. Leakey (1903–1972) before beginning to study chimpanzees. Goodall's research showed that chimpanzees could make and use tools (a behavior previously attributed only to humans). She was also able to distinguish the individual personalities among the chimpanzees she studied. After a career spanning more than fifty years, Goodall continues her work through the Jane Goodall Institute. She focuses on protecting chimpanzees from extinction and environmental crises.

Dian Fossey (1932–1985) was an occupational therapist who, inspired by the writings of the naturalist George Schaller (1933–), decided to study the endangered mountain gorilla of Africa. She was trained in fieldwork by Jane Goodall and went on to watch and record the behavior of mountain gorillas in Zaire and Rwanda. She eventually obtained a Ph.D. in zoology from Cambridge University and in 1983 published a book on her studies, *Gorillas in the Mist*. In 1985, she was found murdered in her cabin in Rwanda; her death is still unsolved.

Famous for her studies of chimpanzees, Jane Goodall is a champion of wildlife conservation.

Who was Ivan Pavlov?

Ivan Pavlov (1849–1936) was a Russian physiologist who became famous for his experiments with dogs in which the animals performed a specific behavior upon being confronted with a certain stimulus. In these well-known investigations, minor surgery was performed on a dog so that its saliva could be measured. The dog was deprived of food, a bell was sounded, and meat powder was placed in the dog's mouth. The meat powder caused the hungry dog to salivate; this is an example of an unconditioned reflex. However, eventually, after many trials, the dog would salivate at the sound of the bell without meat powder being offered. This is an example of a conditioned reflex or classical Pavlovian conditioning. Although he never thought much of the then-fledgling science of psychology, Pavlov's work on conditioned reflexes has been far reaching from elementary education to adult training programs. Pavlov was awarded the 1904 Nobel Prize in Physiology or Medicine for his study of the physiology of digestion.

SPONGES AND COELENTERATES

What is the most primitive group of animals?

Sponges (phylum Porifera, from the Latin words *porus*, meaning "pore," and *fera*, meaning "bearing") represent the most primitive animals. These organisms are aggregates of specialized cells without true tissues or organs, with little differentiation and integration, and with no body symmetry. A sponge's body is perforated by holes that lead to an inner water chamber. Sponges pump water through those pores and expel it through a large opening at the top of the chamber. While water is passing through the body, nutrients are engulfed, oxygen is absorbed, and waste is eliminated. Sponges are distinctive in possessing choanocytes, special flagellated cells whose beating drives water through the body cavity and characterizes them as suspension feeders (also known as filter feeders).

What is the basic composition of a sponge?

A sponge is supported by a skeleton made of hard crystals called spicules, whose shape and composition are important features in taxonomy. Calcareous sponges have spicules of calcium carbonates, the material of marble and limestone. The silica spicules of the hexactinellid, or glass, sponges are formed into a delicate, glassy network. Demosponges have siliceous spicules and a network of fibrous protein, spongir, that is similar to collagen. The demosponges are the source of natural household sponges, which are made by soaking dead sponges in shallow water until all the cellular material has decayed, leaving the spongin network behind. However, most sponges sold now for household use are plastic and have nothing to do with real sponges.

How much water does an average sponge circulate in a day?

A sponge that is 4 inches (10 centimeters) tall and 0.4 inches (1 centimeter) in diameter pumps about 23 quarts (22.5 liters) of water through its body in one day. To obtain enough food to grow by 3 ounces (100 grams), a sponge must filter about 275 gallons (1,000 kilograms) of seawater.

Do more marine or freshwater sponges exist?

Approximately five thousand species of marine (saltwater) sponges and 150 species of freshwater sponges exist.

What accounts for the various colors of sponges?

Living sponges may be brightly colored—green, blue, yellow, orange, red, or purple—or they may be white or drab. The bright colors are due to bacteria or algae that live on or within the sponge.

What animals are members of the phylum Cnidaria?

Corals, jellyfish, sea anemones, and hydras are members of the phylum Cnidaria. The name Cnidaria (from the Greek word *knide*, meaning "nettle," and the Latin word *aria*, meaning "like" or "connected with") refers to the stinging structures that are characteristic of some of these animals. These organisms have a digestive cavity with only one opening to the outside; this opening is surrounded by a ring of tentacles used to capture food and defend against predators. Cells in the tentacles and outer body surface contain stinging, harpoonlike structures called nematocysts. Cnidarians are the first group in the animal hierarchy to have their cells organized into tissues.

What are some interesting features of jellyfish?

Jellyfish live close to the shores of most oceans and spend most of their time floating near the surface. Jellyfish have bell-shaped bodies that are between 95–96 percent water. They have a muscular ring around the margin of the bell that contracts rhythmically to propel them through the water. Jellyfish are carnivores, subduing their prey with stinging tentacles and drawing the paralyzed animal into the digestive cavity. Jellyfish, largely gelatinous, are opaque, so you can see through their bodies.

An orange Pacific sea nettle jellyfish is one of the largest of the jellyfish. It also has a painful sting.

What is the largest jellyfish?

The largest jellyfish is the genus *Cyanea*. It may be more than 6.5 feet (2 meters) in diameter and have tentacles of 98 feet (30 meters) long; it is among the largest invertebrates.

How does a nematocyst work?

A nematocyst is a specialized organelle found in all cnidarians. Each nematocyst features a coiled, threadlike tube lined with a series of barbed spines. The nematocyst is used to capture prey and may also be used for defense purposes. When it is triggered to discharge, the extremely high osmotic pressure within the nematocyst (140 atmospheres) causes water to rush into the capsule, increasing the hydrostatic pressure and expelling the thread with great force. The barb instantly penetrates the prey, stinging it with a highly toxic protein.

Are the stings of jellyfish and Portuguese man-of-war fatal to humans?

The stings of a jellyfish can be very painful and dangerous to humans, but they are generally not fatal. Most stings cause a painful, burning sensation that lasts for several hours. Welts and itchy skin rashes may also appear. Only the sting of the box jelly, or sea wasp (*Chironex fleckeri*), can result in death in humans. The box jelly is the only jellyfish for which a lifesaving, specific antidote exists.

What gives coral their colors?

Coral have a symbiotic relationship with zooxanthellae. Zooxanthellae are photosynthetic dinoflagellates (one-celled animals) that give coral their characteristic colors of pink, purple, and green. Coral that have expelled the zooxanthellae appear white.

How are coral reefs formed, and how fast are they built?

Coral reefs grow only in warm, shallow water. The calcium carbonate skeletons of dead corals serve as a framework upon which layers of successively younger animals attach themselves. Such accumulations, combined with rising water levels, slowly lead to the formation of reefs that can be hundreds of meters deep and long. The coral animal, or polyp, has a columnar form; its lower end is attached to the hard floor of the reef, while the upper end is free to extend into the water. A whole colony consists of thousands of individuals. Coral can be hard or soft, depending on the type of skeleton secreted. The polyps of hard corals deposit around themselves a solid skeleton of calcium carbonate (chalk), so most swimmers see only the skeleton of the coral; the animal is in a cuplike formation into which it withdraws during the daytime. The major reef builder in Florida and Caribbean waters, the *Montastrea annularis* (star coral), requires about one hundred years to form a reef just 3 feet (1 meter) high.

What is the origin of the name "hydra"?

Hydra are named after the multiheaded monster of Greek mythology that was able to grow two new heads for each head cut off. When a hydra is cut into several pieces, each piece is able to grow all the missing parts and become a complete, whole animal.

What are the characteristics of hydra?

Hydra, a well-known member of phylum Cnidaria, is a tiny (0.4 inches [1 centimeter] in length) organism found in freshwater ponds. It exists as a single polyp that sits on a basal disk that it uses to glide around. It can also move by somersaulting. It usually has six to ten tentacles, which it uses to capture food. Hydras reproduce both sexually and asexually through budding.

WORMS

What three groups are included in flatworms?

Flatworms belong to the phylum Platyhelminthes. They are flat, elongated, acoelomate animals that exhibit bilateral symmetry and have primitive organs. The members of the flatworms are 1) planarians; 2) flukes; and 3) tapeworms.

What are the most common tapeworm infections in humans?

Tapeworms, members of the class Cestoda, have long, flat bodies with a linear series of sets of reproductive organs. Each set or segment is called a proglottid.

Tapeworm	Means of Infection
Beef tapeworm (*Taenia saginata*)	Eating rare beef; most common of all tapeworms in humans
Pork tapeworm (*Taenia solium*)	Eating rare pork; less common than beef tapeworm
Fish tapeworm (*Diphyllobothrium latum*)	Eating rare or poorly cooked fish; fairly common in the Great Lakes region of the United States

How numerous are roundworms?

Roundworms, or nematodes, are members of the phylum Nematoda (from the Greek word *nematos*, meaning "thread") and are numerous in two respects: 1) the number of known and potential species and 2) the total number of these organisms in a habitat. Approximately twelve thousand species of nematodes have been named, but it has been estimated that if all species were known, the number would be closer to five hundred

What is unusual about the fish tapeworm?

The fish tapeworm is the largest cestode that infects humans. It can grow to a length of 66 feet (20 meters). By comparison, the beef tapeworm may only reach a length of 33 feet (10 meters).

thousand. Nematodes live in a variety of habitats ranging from the sea to soil; 6 cubic inches (100 cubic centimeters) of soil may contain several thousand nematodes. A square yard (0.85 square meters) of woodland or agricultural soil may contain several million of them. Good topsoil may contain billions of nematodes per acre.

What is the most famous roundworm?

One soil nematode, *Caenorhabditis elegans*, is widely cultured and has become a model research organism in developmental biology. The study of this animal was begun in 1963 by Sydney Brenner (1927–2019), who received the Nobel Prize in Physiology or Medicine in 2002. The species normally lives in soil but is easily grown in the laboratory in Petri dishes. It is only about .06 inches (1.5 millimeters) long, has a simple, transparent body consisting of only 959 cells, and grows from zygote to mature adult in only three and a half days. The genome (genetic material) of *Caenorhabditis elegans*, consisting of fourteen thousand genes, was the first animal genome to be completely mapped and sequenced. The small, transparent body of this nematode allows researchers to locate cells in which a specific, developmentally important gene is active. These cells show up as bright green spots in a photograph because they have been genetically engineered to produce a green fluorescent protein known as GFP. The complete "wiring diagram" of its nervous system is known, including all the neurons and connections between them. Much of the knowledge of nematode genetics and development gained from the study of *Caenorhabditis elegans* is transferable to the study of other animals.

What are the most common roundworm infections in humans in the United States?

Roundworm	Means of Infection
Hookworm (*Ancylostoma duodenale* and *Necator americanus*)	Contact with soil-based juveniles; common in southern states
Pinworm (*Enterobius vermicularis*)	Inhalation of dust that contains ova and by contamination with fingers; most common worm parasite in United States
Intestinal roundworm (*Ascaris lumbricoides*)	Ingestion of embryonated ova in contaminated food; common in rural Appalachia and southeastern states
Trichina worm (*Trichinella spiralis*)	Ingestion of infected meat; occurs occasionally in humans throughout North America

In what ways are earthworms beneficial?

Earthworms help maintain fertile soil. An earthworm literally eats its way through soil and decaying vegetation. As it moves about, the soil is turned, aerated, and enriched by nitrogenous wastes. Charles Darwin calculated that a single earthworm could eat its own weight in soil every day. Much of what is eaten is then excreted on Earth's surface in the form of "casts." The worms then rebury these casts with their burrowing process. In addition, Darwin claimed that 2.5 acres (1 hectare) of soil might contain 155,000 earthworms, which in one year would bring eighteen tons of soil to the surface and in twenty years might build a new layer 3 inches (11 centimeters) thick.

Scientists were stunned to discover creatures like these giant tube worms surviving at the bottom of the ocean with the help of thermal vents.

What are the major groups of segmented worms?

Members of the phylum Annelida, the segmented worms, have bilateral symmetry and a tubular body that may have one hundred to 175 ringlike segments. The three classes of segmented worms are 1) Polychaeta, the sandworms and tubeworms; 2) Oligochaeta, the earthworms; and 3) Hirudinea, the leeches.

What are the giant tube worms?

These worms were discovered near the hydrothermal (hot water) ocean vents in 1977 when the submersible *Alvin* was exploring the ocean floor of the Galapagos Ridge (located 1.5 miles [2.4 kilometers] below the Pacific Ocean surface and 200 miles [322 kilometers] from the Galapagos Islands). Growing to lengths of 5 feet (1.5 meters), *Riftia pachyptila Jones*, named after worm expert Meredith Jones (1926–1996) of the Smithsonian Museum of Natural History, lack both a mouth and a gut and are topped with feathery plumes composed of over two hundred thousand tiny tentacles. The phenomenal growth of these worms is due to their internal food source—symbiotic bacteria, over one hundred billion per ounce of tissue, that live within the worms' troposome tissues. To these troposome tissues, the tube worms transport absorbed oxygen from the water together with carbon dioxide and hydrogen sulfide. Utilizing this supply, the bacteria in turn produce carbohydrates and proteins that the worms need to thrive.

What is the largest leech?

Most leeches are between 0.75 and 2 inches (2–6 centimeters) in length, but some "medicinal" leeches reach 8 inches (20 centimeters) long. The giant of all leeches is the Amazonian *Haementeriam ghilanii* (from the Greek word *haimateros*, meaning "bloody"), which reaches 12 inches (30 centimeters) in length.

Why are leeches important in the field of medicine?

The medical leech, *Hirudo medicinalis*, is used to remove blood that has accumulated within tissues as a result of injury or disease. Leeches have also been applied to fingers or toes that have been surgically reattached to the body. The sucking by the leech unclogs small blood vessels, permitting blood to flow normally again through the body part. The leech releases hirudin, secreted by the salivary glands, which is an anticoag-

How long have leeches been used for medicinal purposes?

Leeches have been used in the practice of medicine since ancient times. During the 1800s, leeches were widely used for bloodletting because of the mistaken idea that body disorders and fevers were caused by an excess of blood. Leech collecting and culture were practiced on a commercial scale during this time. William Wadsworth's (1770–1850) poem "The Leech-Gatherer" was based on this use of leeches. In 2004, the Food and Drug Administration (FDA) approved the commercial marketing of leeches (*Hirudo medicinalis*) for medical purposes.

ulant that prevents blood from clotting and dissolves preexisting clots. Other salivary ingredients dilate blood vessels and act as an anesthetic. A medicinal leech can absorb as much as five to ten times its body weight in blood. Complete digestion of this blood takes a long time, and these leeches feed only once or twice a year in this manner.

MOLLUSKS AND ECHINODERMS

What are the major groups of mollusks?

Mollusks are categorized into four major groups: 1) chitons; 2) gastropods, which include snails, slugs, and nudibranches; 3) bivalves, which include clams, oysters, and mussels; and 4) cephalopods, which include squids and octopods. Although mollusks vary widely in external appearance, most share the following body plan: 1) a muscular foot, usually used for movement; 2) a visceral mass, containing most of the internal organs; and 3) a mantle, a fold of tissue that drapes over the visceral mass and secretes a shell (in organisms that have a shell).

What is the largest group of mollusks?

The gastropods (class Gastropoda), which include snails, slugs, and their relatives, is the largest and most diverse group of mollusks. It includes more than forty thousand different species and comprises the second largest group of related animals. Only the insects comprise a larger group. Most gastropods are marine animals, but many freshwater species exist. Garden snails and slugs have adapted to land.

What is the difference between snails and slugs?

Snail is a common name applied to the molluscan class Gastropoda and includes land snails, sea snails, and freshwater snails. A key characteristic of this group is that they have a coiled shell that is large enough for the animal to retract into completely. Gastropods that lack a shell, or have only an internal shell, are often commonly called slugs.

Both snails and slugs move by gliding along on a mucous-secreting muscle. The dried mucous becomes a silvery slime trail that remains after the snail or slug has moved across an area. Moisture is critical to the survival of snails and slugs. On sunny days, they hide in moist, shady areas in order to preserve moisture. Although snails are both pests and vectors of disease, they are also beneficial to humans as food. Their shells are used as decorative objects and in jewelry. Slugs are eaten by snakes, toads, and starlings. Some species are edible by humans.

How fast does a snail move?

Many snails move at a speed of less than 3 inches (8 centimeters) per minute. This means that if a snail did not stop to rest or eat, it could travel 16 feet (4.8 meters) per hour.

How many tentacles do cephalopods have?

Octopods have eight tentacles or arms, squids have ten tentacles, and the chambered nautilus has as many as ninety tentacles.

How are pearls created?

Cultured pearls are produced by both freshwater and marine mollusks. Most of the world's cultured pearls (known as freshwater pearls) are produced by freshwater mussels belonging to the family *Unionidae*. Most saltwater pearls are produced by three species of oysters belonging to the genus *Pinctada*, including *Pinctada imbricata*, *Pinctada maxima*, and *Pinctada margaritifera*.

Within the body of these mollusks is a curtainlike tissue called the mantle. Certain cells on the side of the mantle toward the shell secrete nacre, also known as mother-of-pearl, during a specific stage of the shell-building process. A pearl is the result of an oyster's reaction to a foreign body, such as a piece of sand or a parasite, within the oyster's shell. The oyster neutralizes the invader by secreting thin layers of nacre around the foreign body, eventually building it into a pearl. The thin layers are alternately composed of calcium carbonate, argonite, and conchiolin. Irritants intentionally placed within an oyster result in the production of what are called cultured pearls.

What are the major groups of echinoderms?

According to one common classification, six principal groups of echinoderms (phylum Echinodermata, from the Greek words *echina*, meaning "spiny," and *derma*, meaning "skin") exist:

1. Class Crinoidea (sea lilies and feather stars)
2. Class Asteroidea (sea stars, also called starfish)
3. Class Ophiuroidea (basket stars and brittle stars)
4. Class Eichinoidea (sea urchins and sand dollars)
5. Class Holothuroidea (sea cucumbers)

6. Class Concentricycloidea (sea daisies that live on waterlogged wood in the deep ocean, which were first discovered in 1986)

Do all starfish, or sea stars, have five arms?

Starfish, also called sea stars, are members of the class Asteroidea. Their bodies consist of a central disk. Five to more than twenty arms or rays radiate from the central disk.

Not all starfish (sea stars) have five arms. This crown-of-thorns sea star, for example, can have up to twenty-one arms.

Why are sea urchins frequently used to study embryonic development?

Sea urchins are a useful model system for studying many problems in early animal development. Historically, sea urchins were a key system in elucidating a variety of classic developmental problems, including the mechanisms of fertilization, egg activation, cleavage, gastrulation, and the regulation of differentiation in the early embryo. In addition, early studies of the molecular basis of early embryology were carried out in this system. Gametes can be obtained easily, sterility is not required, and the eggs and early embryos of many species are beautifully transparent. The early development of sea urchin embryos is also highly synchronous—that is, when a batch of eggs is fertilized, all of the resulting embryos typically develop on the same schedule, making possible biochemical and molecular studies of early embryos.

ARTHROPODS: CRUSTACEANS

Why are arthropods considered the most biologically successful phylum of animals?

Members of the phylum Arthropoda are characterized by jointed appendages and an exoskeleton of chitin. More than one million species of arthropods are currently known to science, and many biologists believe that millions more will be identified. Arthropods are the most biologically successful group of animals because they are the most diverse and live in a greater range of habitats than do the members of any other phylum of animals.

What are the major groups of arthropods?

Subphylum	Class	Examples
Chelicerata	Merostomata	Horseshoe crabs

Subphylum	Class	Examples
Chelicerata	Arachnida	Spiders, scorpions, ticks, mites
Crustacea	Malacostraca	Lobsters, crabs, shrimps, isopods
Crustacea	Copepoda	Copepods
Crustacea	Cirripedia	Barnacles
Unirania	Insecta	Grasshoppers, roaches, ants, bees, butterflies, flies, beetles
Unirania	Chilopoda	Centipedes
Unirania	Diplopoda	Millipedes

How large is the arthropod population?

Zoologists estimate that the arthropod population of the world, including crustaceans, spiders, and insects, numbers about a billion million (10^{18}) individuals. More than one million arthropod species have been described, with insects making up the vast majority of them. In fact, two out of every three organisms known on Earth are arthropods, and the phylum is represented in nearly all habitats of the biosphere. About 90 percent of all arthropods are insects, and about half of the named species of insects are beetles.

What are the only sessile crustaceans?

Barnacles are the only sessile (permanently attached to one location) crustaceans. They were described by nineteenth-century naturalist Louis Agassiz (1807–1873) as "nothing more than a little shrimplike animal standing on its head in a limestone house and licking food into its mouth." Accumulations of barnacles may become so great that the speed of a ship may be reduced by 30–40 percent, necessitating dry-docking the ship to remove the barnacles.

Which arthropods can potentially cause health problems in humans?

Arthropod	Effect on Human Health
Black widow spider (*Latrodectus mactans*)	Venomous bite
Brown recluse or violin spider (*Loxosceles reclusa*)	Venomous bite
Scorpion (*Centruroides exilicauda*)	Venomous bite
Chiggers (*Trombiculid mites*)	Dermatitis
Itch mite (*Sarcoptes scabiei*)	Scabies
Deer tick (*Ixodes dammini*)	Bite transmits Lyme disease
Dog tick, wood tick (*Dermacentor* species)	Bite transmits Rocky Mountain spotted fever
Mosquitoes	Bite transmits disease (West Nile virus, encephalitis, filarial worms

Arthropod	Effect on Human Health
Horseflies, deerflies	Female has painful bite
Houseflies	Many transmit bacteria and viruses
Fleas	Dermatitis
Bees, wasps, ants	Venomous stings (single stings not dangerous unless person is allergic)

ARTHROPODS: SPIDERS

Are spiders dangerous?

Most spiders are harmless organisms that, rather than being dangerous to humans, are actually allies in the continuing battle to control insects. Most venom produced by spiders to kill prey is usually harmless to humans. However, two spiders in the United States can produce severe or even fatal bites. They are the black widow spider (*Latrodectus mactans*) and the brown recluse spider (*Loxosceles reclusa*). Black widows are shiny black, with a bright red "hourglass" on the underside of the abdomen. The venom of the black widow is neurotoxic and affects the nervous system. About 4 out of 1,000 black widow bites have been reported as fatal. Brown recluse spiders have a violin-shaped strip on their back. The venom of the brown recluse is hemolytic and causes the death of tissues and skin surrounding the bite. Their bite can be mild to serious and sometimes fatal.

What first aid measures can be used for a bite by a black widow spider?

The black widow spider (*Latrodectus mactans*) is common throughout the United States. Age, body size, and degree of sensitivity determine the severity of symptoms following a bite. The initial symptom often includes a pinprick with a dull, numbing pain, followed by swelling. An ice cube may be placed over the bite to relieve pain. Between 10 and 40 minutes after the bite, severe abdominal pain and rigidity of stomach muscles develop. Muscle spasms in the extremities, ascending paralysis, and difficulty in swallowing and breathing may follow. The mortality rate is less than 1 percent, but anyone who has been bitten should see a doctor. Doctors may decide to give pain relievers and muscle relaxants. Spider bite victims who are elderly, infants, and those with allergies are most at risk and may require hospitalization or antivenin.

How long does it take the average spider to weave a complete web?

The average orb-weaver spider takes 30–60 minutes to completely spin its web. These species of spiders (order Araneae) use silk to capture their food in a variety of ways ranging from the simple trip wires used by large, bird-eating spiders to the complicated and beautiful webs spun by orb spiders. Some species produce funnel-shaped webs, and other communities of spiders build communal webs.

A completed web features several spokes leading from the initial structure. The number and nature of the spokes depend on the species. The spider replaces any damaged threads by gathering up the thread in front of it and producing a new one behind it. The orb web must be replaced every few days because it loses its stickiness (and its ability to entrap food).

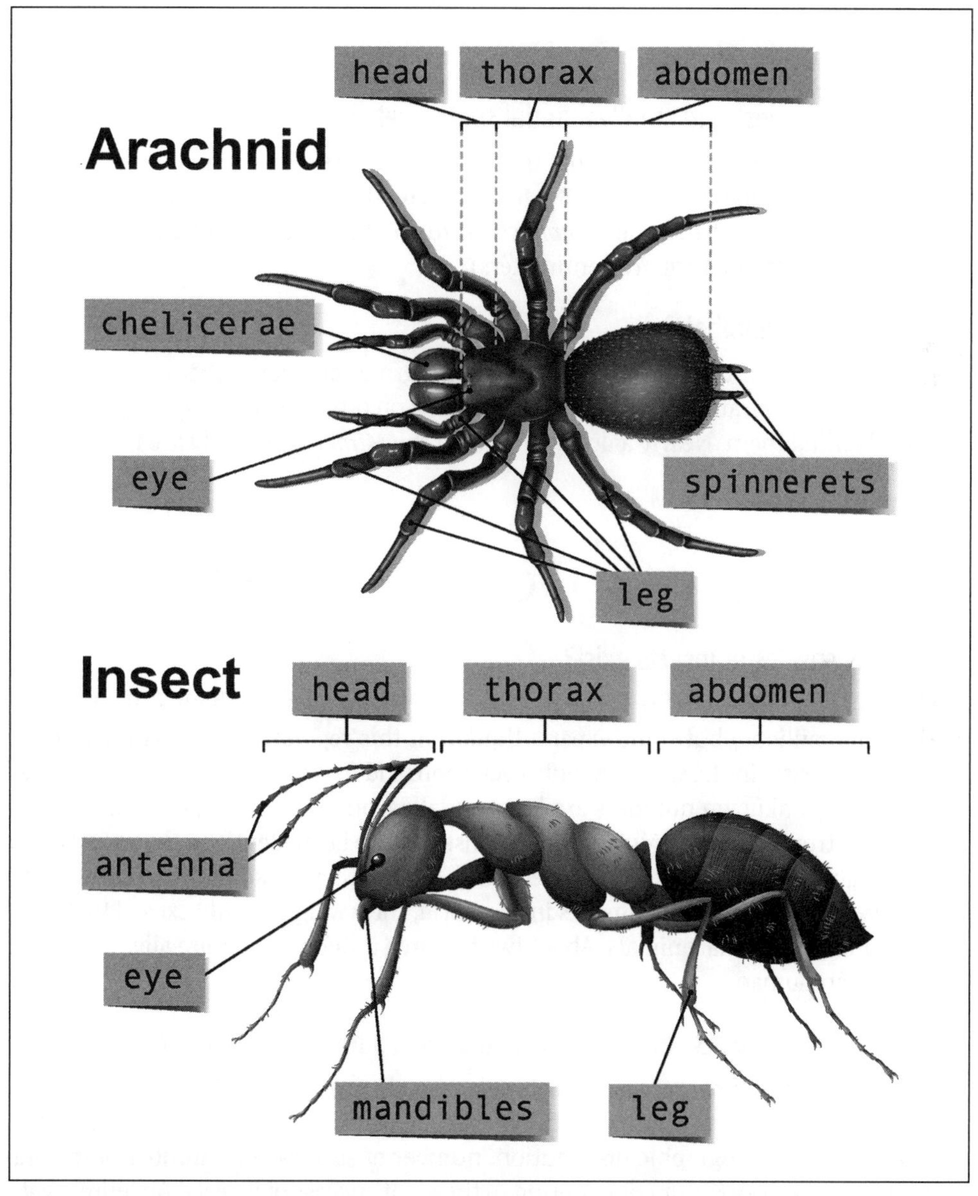

Using a spider and an ant as examples, one can see the fundamental difference between arachnids and insects in this illustration. The easiest comparison concerns how spiders have two main body parts (the head and thorax are combined into a cephalothorax) and insects have three distinct body parts.

Which is stronger—steel or the silk from a spider's web?

Spider silk is stronger. Well known for its strength and elasticity, the strongest spider silk has tensile strength second only to fused quartz fibers and five times greater than that of steel of equivalent weight. Tensile strength is the longitudinal stress that a substance can bear without tearing apart.

What are the largest and smallest aerial spider webs?

The largest aerial webs are spun by the tropical orb weavers of the genus *Nephila*, which produce webs that measure up to 18.9 feet (6 meters) in circumference. The smallest webs are produced by the species *Glyphesis cottonae*; their webs cover an area of about 0.75 square inches (4.84 square centimeters).

Are any spiders cannibals?

Yes, the members of the family *Mimetidae* are a group of spiders that are sometimes called cannibal or pirate spiders. These spiders feed on other spiders, creeping into their webs and killing them. Nearly a dozen species of the *Mimetidae* family are found in the United States.

ARTHROPODS: INSECTS

How many species of insects exist?

Estimates of the number of recognized insect species range from about 750,000 to upward of one million—but some experts think that this represents less than half of the number that exists in the world. About seven thousand new insect species are described each year, but unknown numbers are lost annually from the destruction of their habitats, mainly tropical forests. More species of insects have been identified than of all other groups of animals combined. What insects lack in size, they make up for in sheer numbers. If we could weigh all the insects in the world, their weight would exceed that of all the remaining terrestrial animals. About two hundred million insects are alive at any one time for each human.

Why do some biologists consider insects the most successful group of animals?

With almost one million described species (and perhaps millions more not yet discovered and identified), class Insecta is the most successful group of animals on Earth in terms of diversity, geographic distribution, number of species, and number of individuals. Flight is one key factor contributing to the great success of insects. An animal with the ability to fly can escape many predators, find food and mates, and disperse to new habitats much faster than an animal that must crawl about on the ground.

What is the largest group of insects that have been identified and classified?

The largest group of insects that has been identified and classified is the order Coleoptera, beetles and weevils, with some 350,000–400,000 species. Beetles are the most dominant form of life on Earth, as one of every five living species is a beetle.

What is a "bug," biologically speaking?

The biological meaning of the word "bug" is significantly more restrictive than in common usage. People often refer to all insects as "bugs," even using the word to include such organisms as bacteria and viruses as well as glitches in computer programs. In the strictest biological sense, a "bug" is a member of the order Hemiptera, also called true bugs. Members of Hemiptera include bedbugs, squash bugs, clinch bugs, stink bugs, and water striders.

What is the heaviest insect in the world?

The heaviest insect in the world is the goliath beetle (*Goliathus goliatus*) from Africa, which can weigh up to 3.5 ounces (100 grams).

What are the stages of insect metamorphosis?

Metamorphoses (marked structural changes in the growth processes) are divided into two types: complete and incomplete. In complete metamorphosis, the insect (such as the ant, moth, butterfly, termite, wasp, or beetle) goes through all the distinct stages of growth to reach adulthood. In incomplete metamorphosis, the insect (such as the grasshopper, cricket, or louse) does not go through all the stages of complete metamorphosis.

Complete Metamorphosis

- *Egg*—One egg is laid at a time or many (as much as ten thousand).
- *Larva*—What hatches from the eggs is called a larva. A larva can look like a worm.
- *Pupa*—After reaching its full growth, the larva hibernates, developing a shell or pupal case for protection. A few insects (e.g., the moth) spin a hard covering called a "cocoon." The resting insect is called a pupa (except the butterfly, which is called a chrysalis) and remains in the hibernation state for several weeks or months.
- *Adult*—During hibernation, the insect develops its adult body parts. When it has matured physically, the fully grown insect emerges from its case or cocoon.

Incomplete Metamorphosis

- *Egg*—One egg or many eggs are laid.
- *Early-stage nymph*—Hatched insect resembles an adult but smaller in size. However, those insects that would normally have wings have not yet developed them.
- *Late-stage nymph*—At this time, the skin begins to molt (shed), and the wings begin to bud.
- *Adult*—The insect is now fully grown.

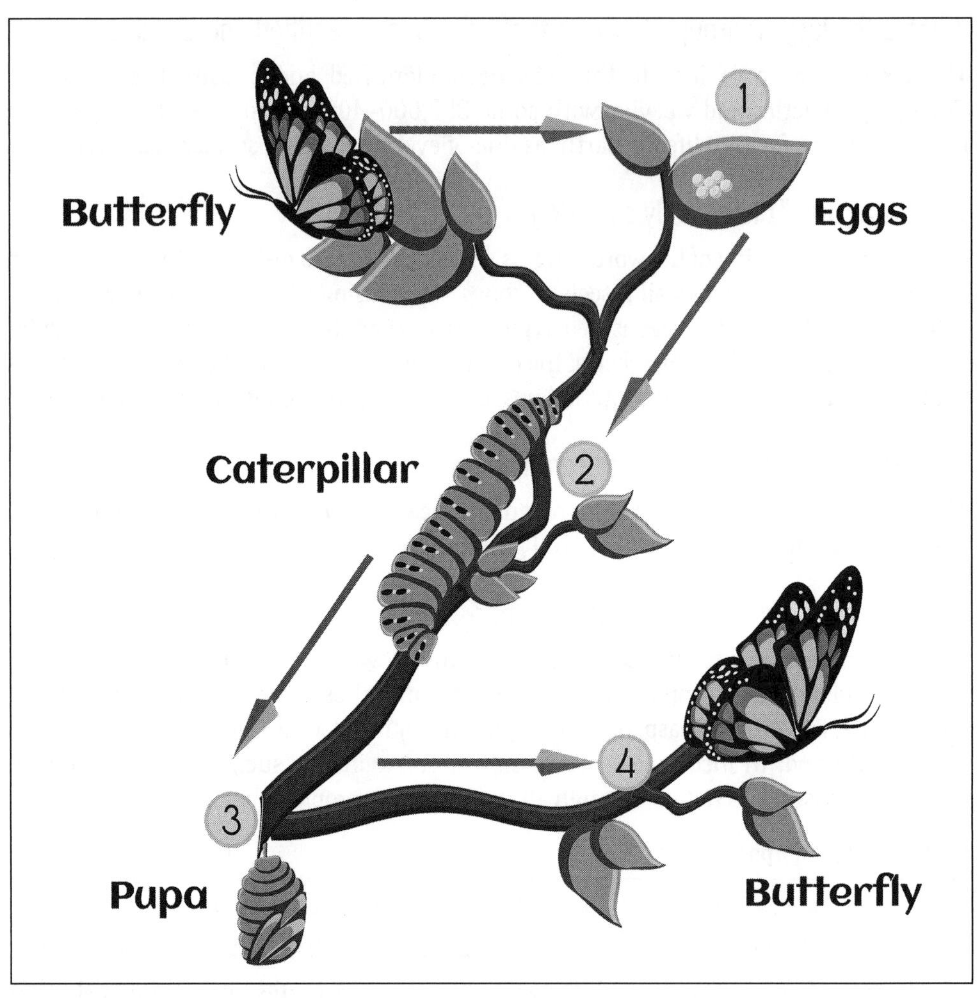

Butterfly lifecycles are a familiar example of a complete metamorphosis from egg to adult.

How do insects survive over the winter in cold-weather climates?

As cold-blooded animals, insects are particularly susceptible to the freezing temperatures of winter in cold-weather climates. While some species migrate to warmer climates during the winter, some remain in the same locale throughout the year. The insects that do not migrate have one stage of their life cycle that is resistant to freezing temperatures. In order to prepare for the winter, some insects will secrete natural antifreeze, such as glycerol, into their body fluids. Glycerol prevents the formation of ice crystals, which will fatally rupture the cells of a butterfly. Many species that overwinter in an area remain as chrysalises and complete their metamorphosis the following spring or summer. Some species spend the winter as caterpillars burrowing into leaf litter. Others remain as adults hibernating in holes or crevices of trees, logs, or under bark. A few species spend the winter as eggs hatching the following spring or summer.

Why are insects often found in amber?

People have long been infatuated with amber, the fossilized form of ancient tree resin, a semiprecious stone used for jewelry and mosaics. Amber from the Dominican Republic contains an average of one insect in every hundred pieces. Some pieces of amber contain thousands of insects—both whole insects and insect fragments. These insects were probably crawling or lodged on the outside of a tree about thirty million years ago and became trapped by a glob of sticky tree resin, which continued to ooze around the animal matter and eventually fossilized. Scientists are able to study these insects, many of which are extinct but may turn out to be missing links to modern-day species.

What are some beneficial insects?

Beneficial insects include bees, wasps, flies, butterflies, moths, and others that pollinate plants. Many fruits and vegetables depend on insect pollinators for the production of seeds. Insects are an important source of food for birds, fish, and many animals. In some countries, such insects as termites, caterpillars, ants, and bees are eaten as food by people. Products derived from insects include honey and beeswax, shellac, and silk. Some predators such as mantises, ladybugs or lady beetles, and lacewings feed on other harmful insects. Other helpful insects are parasites that live on or in the body of harmful insects. For example, some wasps lay their eggs in caterpillars that damage tomato plants.

How is the light in fireflies produced?

The light produced by fireflies (*Photinus pyroles*), or lightning bugs, is a kind of heatless light called bioluminescence. It is caused by a chemical reaction in which the substance luciferin undergoes oxidation when the enzyme luciferase is present. The flash is a photon of visible light that radiates when the oxidating chemicals produce a high-energy state, which then reverts back to their normal state. The flashing is controlled by the nervous system and takes place in special cells called photocytes. The nervous system, photocytes, and tracheal end organs control the flashing rate. The air temperature also seems to be correlated with the flashing rate. The higher the temperature, the shorter the interval between flashes—8 seconds at 65°F (18.3°C) and 4 seconds at 82°F (27.7°C). Scientists are uncertain as to why this flashing occurs. The rhythmic flashes could be a means of attracting prey or enabling mating fireflies to signal in heliographic codes (that differ from one species to another), or they could serve as a warning signal.

What is the most destructive insect in the world?

The most destructive insect is the desert locust (*Schistocera gregaria*), the locust of the Bible, whose habitat ranges from the dry and semi-arid regions of Africa and the Middle East through Pakistan and northern India. This short-horn grasshopper can eat its own weight in food a day, and during long migratory flights, a large swarm can consume 20,000 tons (18,144,000 kilograms) of grain and vegetation a day.

How does a butterfly differ from a moth?

Butterflies vs. Moths

Characteristic	Butterflies	Moths
Antennae	Knobbed	Unknobbed
Active time of day	Day	Night
Coloration	Bright	Dull
Resting wing position	Vertically above body	Horizontally beside body

While these guidelines generally hold true, exceptions do exist. Moths have hairy bodies, and most have tiny hooks or bristles linking the fore-wing to the hind-wing; butterflies do not have either characteristic.

Do butterflies migrate?

Many populations of butterflies migrate, sometimes long distances, to and from areas that are only suitable for part of the year. Migration allows these species to avoid unfavorable conditions, including weather, food shortage, or overpopulation. In some lepidopteran species, all individuals migrate, while in others, only some individuals migrate. The best-known lepidopteran migration is that of the Monarch butterfly. Each fall between August and October, this insect heads south. Those who summer east of the Rocky Mountains spend the winter in Mexico, while those who summer west of the Rockies winter in California. Unlike most other migratory insects, the Monarchs who return in the spring are the same ones who left the previous winter.

How many flowers need to be tapped for bees to gather enough nectar to produce one pound of honey?

Bees must gather 4 pounds (1.8 kilograms) of nectar, which requires the bees to tap about two million flowers, in order to produce 1 pound (454 grams) of honey. A bee

What is Colony Collapse Disorder?

Colony Collapse Disorder (CCD) is the decline in the honeybee population due to the death of the honeybees. CCD has no single, known cause. Possible causes are the use of pesticides, attacks on bee colonies by parasitic mites, establishment of Africanized honeybees ("killer bees"), viruses, fungi, and complications from poor diets due to lack of native plants and flowers that nourish bees. The decline of the honeybee population impacts our food sources.

colony may produce a harvest of 60–100 pounds (27–45 kilograms) of honey per year. A little more than one-third of the honey produced by the bees is retained in the hive to sustain the population.

How many bees are in a bee colony?

On average, a bee colony contains from fifty to seventy thousand bees. The members of each honeybee colony are a queen, worker bees, and drones. The queen is a fertile female bee who is the mother of all the bees in the colony. A queen may lay up to three thousand eggs per day. The worker bees are infertile females who perform all the work associated with the hive, including food preparation, guarding the hive, feeding the queen, drones, and brood, and heating and cooling the hive. Worker bees live an average of six to eight weeks in the summer. The most common cause of death of worker bees is wearing out their wings. During their lifetime, a worker bee will fly the equivalent of one and a half times the circumference of Earth gathering nectar to produce honey. The average worker gathers approximately one teaspoon of nectar to produce 1/12 of a teaspoon of honey during their life. Drones begin life as an unfertilized egg. Their task in the colony is to mate with a virgin queen. However, only one in a thousand drones have the opportunity to mate.

Which crops rely on honeybees for pollination?

Almonds, apples, cherries, pears, apricots, avocados, kiwi, melons, blueberries, cranberries, seed onions, and seed alfalfa are examples of crops that rely on honeybee pollination. The almond crop in the United States relies entirely on honeybees for pollination.

Beekeeping isn't just about honey. Beekeepers also raise the insects to pollinate crops such as almonds, melons, and alfalfa. There is increasing concern about Colony Collapse Disorder, which is killing bees at an alarming rate.

How do bees make honey?

The first step in the production of honey is collecting nectar secreted by flowering plants. After extracting the nectar from a flower, the honeybees store the nectar in their stomach, called a "crop," while they return to the hive. While in the crop, the bees secrete enzymes into the nectar, altering the chemical composition and pH of the nectar. Upon returning to the hive, the bees regurgitate the liquid nectar into the mouth of another worker bee. The regurgitation process is repeated as the nectar is passed from one worker bee to another until it is deposited into a honeycomb. Once the liquid is deposited into a honeycomb, the workers fan their wings in order to speed the evaporation of the excess liquid in the nectar. When the water content is reduced from 70 percent to 20 percent, the honeycomb is sealed with beeswax. The finished honey product is food for the bees during the winter. Honey is slightly acidic, with a pH between 3.5 and 4 (similar to orange juice), which discourages the growth of bacteria. For centuries, it has been used as a sweetener by humans.

What is beeswax?

Beeswax is secreted by four pairs of glands under the abdomen of worker bees and is similar to human earwax. While all insects secrete some protective wax to coat their wings, bees use beeswax to seal the honey in the honeycomb. After harvesting the honey, secondary uses of beeswax include candles, natural cosmetics, such as lip balm and other lotions, wood protectors and polish for furniture, and other lubricants.

What are migratory beekeepers?

A migratory beekeeper is a person who transports his or her bee colonies to different areas to produce better honey or to collect fees for pollinating such crops as fruit trees, almonds, and alfalfa. The beekeepers frequently travel north in the spring and summer to pollinate crops and then back south in the fall and winter to maintain the colonies in the warmer southern weather. Approximately one thousand migratory beekeepers operate in the United States, transporting approximately two million bee colonies a year.

What are "killer bees"?

Africanized honeybees—the term entomologists prefer rather than killer bees—are a hybrid originating in Brazil, where African honeybees were imported in 1956. The breeders, hoping to produce a bee better suited to producing more honey in the tropics, instead found that African bees soon hybridized with and mostly displaced the familiar European honeybees. Although they produce more honey, Africanized honeybees (*Apis mellifera scutellata*) also are more dangerous than European bees because they attack intruders in greater numbers. Since their introduction, they have been responsible for approximately one thousand human deaths. In addition to such safety issues, concern is growing regarding the effect of possible hybridization on the U.S. beekeeping industry.

In October 1990, the bees crossed the Mexican border into the United States; they reached Arizona in 1993. In 1996, six years after their arrival in the United States,

Africanized honeybees could be found in parts of Texas, Arizona, New Mexico, and California. Africanized honeybees are currently found in Nevada, Utah, Oklahoma, Arkansas, Louisiana, and Florida. Their migration northward has slowed partially because they are a tropical insect and cannot live in colder climates. Most believe they were introduced into Florida via cargo ships through commercial shipping ports. Experts have suggested two possible ways of limiting the spread of the Africanized honeybees. The first is drone flooding, a process by which large numbers of European drones are kept in areas where commercially reared European queen bees mate, thereby ensuring that only limited mating occurs between Africanized drones and European queens. The second method is frequent requeening, in which a beekeeper replaces a colony's queen with one of his or her own choosing. The beekeeper can then be assured that the queens are European and that they have already mated with European drones.

Do wasps have any useful role?

In addition to serving as pollinators of various plants, wasps are predators or parasites of various injurious pests. Wasps have been used to attack the worms and beetles that infect plants such as corn, cotton, soybeans, tomatoes, snap and lima beans, wheat, oats, and barley. Fig trees and ornamental ficus trees depend on specific wasp species for pollination.

How much weight can an ant carry?

Ants are the "superweight lifters" of the animal kingdom. They are strong in relation to their size and can carry objects ten to twenty times their own weight—some species can carry objects up to fifty times their own weight. Ants are able to carry these objects great distances and even climb trees while carrying them. This is comparable to a one-hundred-pound person picking up a small car, carrying it seven to eight miles on his back, and then climbing the tallest mountain while still carrying the car!

How are ants distinguished from termites?

Both insect orders—ants (order Hymenoptera) and termites (order Isoptera)—have segmented bodies with multijointed legs. Listed below are some differences.

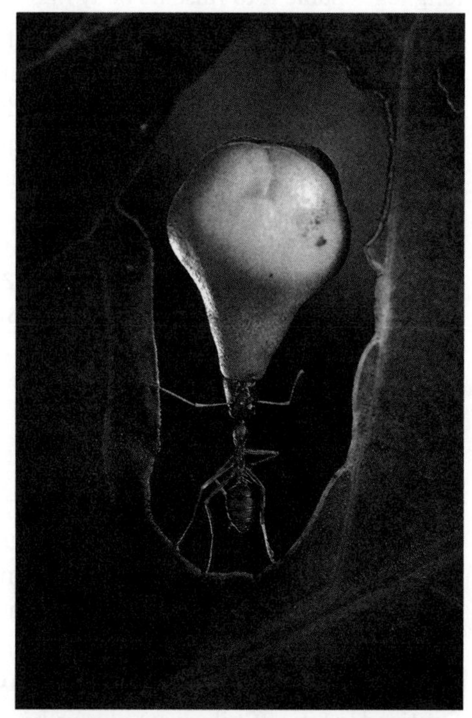

Ants can heft extraordinary weights compared to their body sizes, sometimes lifting or dragging food that weighs twenty times more than they do!

Characteristic	Ant	Termite
Wings	Two pairs with the front pair being much longer than the back pair	Two pairs of equal length
Antenna	Bends at right angle	Straight
Abdomen	Wasp-waist (pinched in)	No wasp-waist

Do termites have any natural predators?

Birds, ants, spiders, lizards, and dragonflies have been seen preying on young, winged termites when they emerge and fly from a home colony to establish new colonies. Termites are generally most vulnerable to predators when they emerge from their home colony. Chimpanzees are also known to use sticks as tools to forage for termites.

What is a daddy longlegs?

The name applies to two different kinds of invertebrates. The first is a harmless, nonbiting, long-legged arachnid. Also called a harvestman, it is often mistaken for a spider, but it lacks the segmented body shape that a spider has. Although it has the same number of legs (eight) as a spider, the harvestman's legs are far longer and thinner. These very long legs enable it to raise its body high enough to avoid ants or other small enemies.

Harvestmen are largely carnivorous, feeding on a variety of small invertebrates such as insects, spiders, and mites. They never spin webs as spiders do. They also eat some juicy plants and in captivity can be fed almost anything edible from bread and milk to meat. Harvestmen also need to drink frequently. The term "daddy longlegs" also is used for a cranefly—a thin-bodied insect with long, thin legs that has a snoutlike proboscis with which it sucks water and nectar.

How long have cockroaches been on Earth?

The earliest cockroach fossils are about 280 million years old. Cockroaches (order Dictyoptera) are nocturnal, scavenging insects that eat not only human food but book bindings, ink, and whitewash as well.

Do male mosquitoes bite humans?

No. Male mosquitoes live on plant juices, sugary saps, and liquids arising from decomposition. They do not have a biting mouth that can penetrate human skin as female mosquitoes do. In some species, the females, who lay as many as two hundred eggs, need blood to lay their eggs. These are the species that bite humans and other animals.

Do centipedes actually have one hundred legs and millipedes one thousand legs?

Centipedes (class Chilopoda) always have an uneven number of pairs of walking legs varying from fifteen to more than 171. The true centipedes (order Scolopendromorpha) have twenty-one or twenty-three pairs of legs. Common house centipedes (*Scutigera*

How do fleas jump so far?

The jumping power of fleas comes both from strong leg muscles and from pads of a rubberlike protein called resilin. The resilin is located above the flea's hind legs. To jump, the flea crouches, squeezing the resilin, then it relaxes certain muscles. Stored energy from the resilin works like a spring, launching the flea. A flea can jump well both vertically and horizontally. Some species can jump 150 times their own length. To match that record, a human would have to spring over the length of two and a quarter football fields—or the height of a one-hundred-story building—in a single bound. The common flea (*Pulex irritans*) has been known to jump 13 inches (33 centimeters) in length and 7.25 inches (18.4 centimeters) in height.

coleoptrato) have fifteen pairs of legs. Centipedes are all carnivorous and feed mainly on insects. Millipedes (class Diplopoda) have thirty or more pairs of legs. They are herbivores, feeding mainly on decaying vegetation.

CHORDATES AND VERTEBRATES

What are the major characteristics of all chordates?

All chordates share a notochord, dorsal nerve cord, and pharyngeal gill pouches. The notochord, a cartilaginous supporting rod, runs along the dorsal part of the body. It is always found in embryos, but in most vertebrates, it is replaced during development by a backbone of bony or cartilaginous vertebrae. The tubular dorsal nerve cord, dorsal to the notochord, is formed during development by an infolding of the ectoderm. In vertebrates, the nerve cord eventually becomes encased and thus protected by the backbone. The pharyngeal gill pouches appear during embryonic development on both sides of the throat region, the pharynx.

What are the three major groups of chordates?

The chordates are divided into three subphyla: Tunicata, Cephalochordata, and Vertebrata. Tunicates are like little, leathery bags that are either free living or attach to pilings, rocks, and seaweeds. They are also called sea squirts because a disturbed animal may contract and shoot streams of water from both of its siphons.

The subphylum Cephalochordata contains the amphioxus or lancelet (*Branchiostoma*), which looks like a small fish and has the three chordate features as an adult. Amphioxus also shows clear serial segmentation or metamerism (from the Greek words *meta*, meaning "between, among, after," and *meros*, meaning "part"). It is divided lengthwise into a series of muscle segments. Vertebrates, which comprise the third chordate subphylum, retain the same metamerism in internal structures.

What was the first group of vertebrates?

The first vertebrates were fish that appeared five hundred million years ago. They were agnathans (from the Greek words *a*, meaning "without," and *gnath*, meaning "jaw"), small, jawless fish up to about 8 inches (20 centimeters) long and also known as ostracoderms ("shell skin") because their bodies were covered with bony plates, most notably a head shield protecting the brain.

What are the main features shown by all vertebrates?

Animals in the subphylum Vertebrata are distinguished from other chordates by several features. Most prominent is the endoskeleton of bone or cartilage, centering around the vertebral column (spine or backbone). Composed of separate vertebrae (showing internal metamerism), a vertebral column combines flexibility with enough strength to support even a large body. Other vertebrate features include 1) complex dorsal kidneys; 2) a tail (lost via evolution in some groups) extending by the anus; 3) a closed circulatory system with a single, well-developed heart; 4) a brain at the anterior end of the spinal cord with ten or more pairs of cranial nerves; 5) a cranium (skull) protecting the brain; 6) paired sex organs in both males and females; and 7) two pairs of movable appendages—fins in the fish, which evolved into legs in land vertebrates.

What is the largest group of vertebrates?

The largest group of vertebrates is fish. They are a diverse group and include almost 25,000 species—almost as many as all the other groups of vertebrates combined. Most members of this group are osteichythes, or "bony fish," which include basses, trout, and salmon.

SHARKS AND FISH

What general characteristics do all fish have in common?

All fish have the following characteristics: 1) gills that extract oxygen from water; 2) an internal skeleton with a skin that surrounds the dorsal nerve cord; 3) single-loop blood circulation, in which the blood is pumped from the heart to the gills and then to the rest of the body before returning to the heart; and 4) nutritional deficiencies, particularly some amino acids that must be consumed and cannot be synthesized.

How is the age of a fish determined?

One way to determine the age of a fish is by its scales, which have growth rings just as trees do. Scales have concentric, bony ridges or "circuli," which reflect the growth patterns of the individual fish. The portion of the scale that is embedded in the skin contains clusters of these ridges (called "annuli"); each cluster marks one year's growth cycle.

How do fish swimming in a school change their direction simultaneously?

The movement, which confuses predators, happens because fish detect pressure changes in the water. The detection system, called the lateral line, is found along each side of the fish's body. Along the line are clusters of tiny hairs inside cups filled with a jellylike substance. If a fish becomes alarmed and turns sharply, it causes a pressure wave in the water around it. This wave pressure deforms the "jelly" in the lateral line of nearby fish. This moves the hairs that trigger nerves, and a signal is sent to the brain telling the fish to turn.

When and why do salmon migrate?

Most salmon species migrate during the fall, sometime between September through November. Salmon spend their early life in rivers and then swim out to sea, where they live their adult lives and gain most of their body mass. When they have matured, they migrate or "run up" freshwater rivers to spawn in what is called the salmon run. The salmon run is the time when salmon, which have migrated from the ocean, swim to the upper reaches of rivers, where they spawn on gravel beds. Usually, they return with uncanny precision to the natal river where they were hatched, even to the very spawning ground of their birth. It is thought that, when they are in the ocean, they use magnetoception to locate the general position of their natal river, and once close to the river, they use their sense of smell to home in on the river entrance and even their natal spawning ground. After spawning, all Pacific salmon and most Atlantic salmon die, and the salmon life cycle starts over again.

What are chondrichthyes?

Chondrichthyes are fish that have a cartilaginous skeleton rather than a bony skeleton; they include such organisms as sharks, skates, and rays.

How many kinds of sharks exist, and how many are dangerous?

The United Nations's Food and Agricultural Organization lists 354 species of sharks ranging in length from 6 inches (15 centimeters) to 49 feet (15 meters). While thirty-five species are known to have attacked humans at least once, only a dozen do so on a regular basis. The relatively rare great white shark (*Carcharodan car-*

Migrating upstream to their original places of hatching is an arduous task that many salmon do not survive, but they are compelled to spawn in their home rivers and streams.

charias) is the largest predatory fish. The largest specimen accurately measured was 20 feet, 4 inches (6.2 meters) long and weighed 5,000 pounds (2,270 kilograms).

This image of a great white shark about to bite a swimmer could terrify anyone, but the number of fatalities due to sharks is actually very low.

Is the whale shark a mammal or a fish?

The whale shark (*Rhincodon typus*) is a shark, not a whale. It is, therefore, a fish. This species' name merely indicates that it is the largest of all shark species (weighing 40,000 pounds [18,144 kilograms] or more and growing to lengths of 49 feet [15 meters] or more) and the largest fish species in the world. However, it is completely harmless to humans.

How many shark attacks occur per year?

The International Shark Attack File (ISAF) reported 130 cases of shark attacks worldwide during 2018 of which thirty-two were in the United States. The ISAF differentiates between provoked attacks and unprovoked attacks. Provoked attacks are those that occur when a human initiates contact with a shark—for example, after grabbing a shark. Unprovoked attacks are those that occur in the shark's natural habitat without any human provocation. The vast majority of shark attacks are unprovoked and occur near shore. Most unprovoked attacks involve surfers and water skiers.

Unprovoked Shark Attacks

Year	Attacks Worldwide	Fatal Attacks Worldwide	Attacks in United States	Fatal Attacks in United States
2013	77	10	48	1
2014	73	3	52	0
2015	98	6	59	1
2016	81	4	53	0
2017	155	5	53	0

Some advice to avoid shark attacks includes swimming in groups, avoiding the water during darkness or twilight when sharks are most active, using extra caution in murky waters, avoiding water when sharks are present, and avoiding wearing shiny jewelry since the reflected light resembles the sheen of fish scales. If attacked, it is recommended to be proactive by hitting the shark repeatedly on the nose.

How much electricity does an electric eel generate?

An electric eel (*Electrophorus electricus*) has current-producing organs made up of electric plates on both sides of the vertebral column running along almost its entire

Does a shark really smell blood from a great distance away in the water?

Sharks have an extremely sensitive sense of smell for blood in the water. In fact, a few molecules of blood can be enough to draw a shark. Studies have shown that 70 percent of a relatively small shark brain, by weight and volume, is devoted to smell. Also, strong evidence exists that most shark attacks on people are cases of mistaken identity or reflex action. Many such attacks occur where visibility is low, at night, or in turbid water. The smell of blood might stimulate a shark to try to feed on a person it mistakes for a sea turtle or sea lion, but the taste of suntan oil or neoprene rubber might lead to rejection. Most shark attacks are not fatal.

body length. The charge—on the average of 350 volts but as great as 550 volts—is released by the central nervous system. The shock consists of four to eight separate charges, which last only two- to three-thousandths of a second each. These shocks, used as a defense mechanism, can be repeated up to 150 times per hour without any visible fatigue to the eel. The most powerful electric eel, found in the rivers of Brazil, Colombia, Venezuela, and Peru, produces a shock of 400–650 volts.

AMPHIBIANS AND REPTILES

What group of animals was the first to make a partial transition from water to land?

Amphibians have made a partial transition to terrestrial life. The living amphibians include newts, salamanders, frogs, and toads. Although lungfish made a partial transition to living out of the water, amphibians were the first to struggle onto land and become adapted to a life of breathing air while not constantly surrounded by water.

What does the word "amphibian" mean?

The word "amphibian," from the Greek word *amphibia*, means "both lives" and refers to the animals' double life on land and in water. The usual life cycle of amphibians begins with eggs laid in water, which develop into aquatic larvae with external gills; in a development that recapitulates its evolution, the fishlike larvae develop lungs and limbs and become adults.

What are the major groups of amphibians?

The following chart illustrates the three major groups of amphibians:

Amphibian Groups

Examples	Order	Number of Living Species
Frogs and toads	Anura (Salientia)	3,450
Salamanders and newts	Caudata (Urodela)	360
Caecilians	Apoda (Gymnophiona)	160

What group of vertebrates were the first terrestrial vertebrates?

Reptiles were the first group of vertebrates that were truly terrestrial. A number of adaptations led to their successful terrestrial life.

What features of reptiles enabled them to become true land vertebrates?

Legs were arranged to support the body's weight more effectively than in amphibians, allowing reptile bodies to be larger and to run. Reptilian lungs were more developed with a greatly increased surface area for gas exchange than the saclike lungs of amphibians. The three-chambered heart of reptiles was more efficient than the three-chambered amphibian heart. In addition, the skin was covered with hard, dry scales to minimize water loss. However, the most important evolutionary adaptation was the amniotic egg, in which an embryo could survive and develop on land. The eggs were surrounded by a protective shell that prevented the developing embryo from drying out.

What is the difference between a reptile and an amphibian?

Reptiles are clad in scales, shields, or plates, and their toes have claws; amphibians have moist, glandular skins, and their toes lack claws. Reptile eggs have a thick, hard, or parchmentlike shell that protects the developing embryo from moisture loss, even on dry land. The eggs of amphibians lack this protective outer covering and are always laid in water or in damp places. Young reptiles are miniature replicas of their parents in general appearance if not always in coloration and pattern. Juvenile amphibians pass through a larval, usually aquatic, stage before they metamorphose (change in form and structure) into the adult form. Reptiles include alligators, crocodiles, turtles, and snakes. Amphibians include salamanders, toads, and frogs.

An eastern diamondback rattlesnake prepares to strike, displaying terrifying, venomous fangs.

What groups of reptiles are living today?

The three orders of reptiles that are alive today are 1) Chelonia, which includes turtles, terrapins, and tortoises; 2) Squamata, which includes lizards and snakes; and 3) Crocodilia, which includes crocodiles and alligators.

Which venomous snakes are native to the United States?

Rattlesnakes

Snake	Average Length
Eastern diamondback (*Crotalus adamateus*)	33–65 inches (84–165 centimeters)
Western diamondback (*Crotalus atrox*)	30–65 inches (76–419 centimeters)
Timber rattlesnake (*Crotalus horridus horridus*)	32–54 inches (81–137 centimeters)
Prairie rattlesnake (*Crotalus viridis viridis*)	32–46 inches (81–117 centimeters)
Great Basin rattlesnake (*Crotalus viridis lutosus*)	32–46 inches (81–117 centimeters)
Southern Pacific rattlesnake (*Crotalus viridis helleri*)	30–48 inches (76–122 centimeters)
Red diamond rattlesnake (*Crotalus ruber ruber*)	30–52 inches (76–132 centimeters)
Mojave rattlesnake (*Crotalus scutulatus*)	22–40 inches (56–102 centimeters)
Sidewinder (*Crotalus cerastes*)	18–30 inches (46–76 centimeters)

Moccasins

Snake	Average Length
Cottonmouth (*Agkistrodon piscivorus*)	30–50 inches (76–127 centimeters)
Copperhead (*Agkistrodon contortrix*)	24–36 inches (61–91 centimeters)
Cantil (*Agkistrodon bilineatus*)	30–42 inches (76–107 centimeters)

Coral snakes

Snake	Average Length
Eastern coral snake (*Micrurus fulvius*)	16–28 inches (41–71 centimeters)

Are tortoises and terrapins the same as turtles?

The terms "turtle," "tortoise," and "terrapin" are used for various members of the order Testudines (from the Latin word *testudo*, meaning "tortoise"). In North American usage, they are all correctly called turtles. The term "tortoise" is often used for land turtles. In British usage, the term "tortoise" is the inclusive term, and "turtle" is only applied to aquatic members of the order. Terrapins are small turtles usually belonging to either the Geoemydidae or Emydidae family.

BIRDS

What is the most successful and diverse group of terrestrial vertebrates?

Birds, members of the class Aves, are the most successful of all terrestrial vertebrates. Twenty-eight orders of living birds, with almost ten thousand species, are distributed over almost the entire Earth. The success of birds is basically due to the development of the feather.

Do all birds fly?

No. Among the flightless birds, the penguins and the ratites are the best known. Ratites include emus, kiwis, ostriches, rheas, and cassowaries. They are called ratite because they lack a keel on the breastbone. All of these birds have wings but lost their power to fly millions of years ago. Many birds that live isolated on oceanic islands (for example, the great auk) apparently became flightless in the absence of predators and the consequent gradual disuse of their wings for escape.

What bird has the biggest wingspan?

Three members of the albatross family, the wandering albatross (*Diomedea exculans*), the royal albatross (*Diomedea epomophora*), and the Amsterdam Island albatross (*Diomeda amsterdiamensis*), have the greatest wingspan of any bird species, with a spread of 8–11 feet (2.5–3.3 meters).

How fast does a hummingbird fly, and how far does the hummingbird migrate?

Hummingbirds fly at speeds up to 71 miles (80 kilometers) per hour. The longest migratory flight of a hummingbird documented to date is the flight of a rufous hummingbird from Ramsey Canyon, Arizona, to near Mount St. Helens, Washington, a distance of 1,414 miles (2,277 kilometers). Bird-banding studies are now in progress to verify that a few rufous hummingbirds do make a 11,000–11,500-mile (17,699–18,503-kilometer) journey along a super Great Basin high route, a circuit that could take a year to complete. Hummingbird studies, however, are difficult to complete because so few banded birds are recovered.

Why do birds migrate annually?

Migratory behavior in birds is inherited; however, birds will not migrate without certain physiological and environmental stimuli. In the late summer, the decrease in sunlight stimulates the pituitary gland and the adrenal gland of migrating birds, causing them to produce the hormones prolactin and corticosterone respectively. These hormones in turn cause the birds to accumulate large amounts of fat just under the skin, providing

What is the longest-living bird species?

Bird bands returned to ornithologists indicate that the wandering albatross is frequently listed as the longest-living bird in the wild. Two species of the wandering albatross live for at least forty years and may live for as long as eighty years in the wild. The albatross enjoys such a long life because it nests on remote islands in southern seas in and around Antarctica, far removed from most predators. In contrast to the longevity of the albatross, songbirds have an average life expectancy of only eight months in the wild.

them with enough energy for the long migratory flights. The hormones also cause the birds to become restless just prior to migration. The exact time of departure, however, is dictated not only by decreasing sunlight and hormonal changes but also by such conditions as the availability of food and the onset of cold weather.

The major wintering areas for North American migrating birds are the southern United States and Central America. Migrating ducks follow four major flyways south: the Atlantic flyway, the Mississippi flyway, the central flyway, and the Pacific flyway. Some bird experts propose that the birds return north to breed for several reasons: (1) birds return to nest because their young can get a huge insect supply there; (2) Earth's latitude in the summer in the Northern Hemisphere is higher, which means the longer the daylight is available to the parents to find food for their young; (3) less competition exists for food and nesting sites in the north; (4) fewer mammal predators are in the north for nesting birds (which are particularly vulnerable during the nesting stage); and (5) birds migrate south to escape the cold weather, and they return north when the weather improves and is more temperate.

Which bird migrates the greatest distance?

The arctic tern (*Sterna paradisaea*) migrates the longest distance of any bird. They breed from subarctic regions to the very limits of land in the arctic of North America and Eurasia. At the end of the northern summer, the arctic tern leaves the north on a migration of more than 11,000 miles (17,699 kilometers) to its southern home in Antarctica. A tern tagged in July on the arctic coast of Russia was recovered the next May near Fremantle, Australia, a record 14,000 miles (22,526 kilometers) away.

Why do geese fly in formation?

Aerodynamicists have suspected that long-distance migratory birds, such as geese and swans, adapt the "V" formation in order to reduce the amount of energy needed for such long flights. According to theoretical calculations, birds flying in a "V" formation can fly some 10 percent farther than a lone bird can. Formation flying lessens the drag (the air pressure that pushes against the wings). The effect is similar to flying in a thermal upcurrent, where less total lift power is needed. In addition, when flying, each bird creates behind it a small area of disturbed air. Any bird flying directly behind it would be caught in this turbulence. In the "V" formation of Canada geese, each bird flies not directly behind the other but to one side or above the bird in front.

Migratory birds like these Canada geese fly in formation to lessen drag, which allows them to fly up to 10 percent farther than if they were flying alone.

How do birds know their migratory path?

Birds are known to use a variety of things in combination as migratory aids. They include landmarks, the angle of the Sun, stars, odors, and even the magnetic field of Earth. While exhaustive studies of bird migration have been done, scientists still do not know exactly what they use for a map and a compass, much less exactly how they use them. Major research on the homing pigeon still has not answered the question of how a bird carried a hundred miles in a covered box immediately orients itself and heads accurately for home.

How does a homing pigeon find its way home?

Scientists currently have two hypotheses to explain the homing flight of pigeons. Neither has been proved to the satisfaction of all the experts. The first hypothesis involves an "odor map." This theory proposes that young pigeons learn how to return to their original point of departure by smelling different odors that reach their home in the winds from varying directions. They would, for example, learn that a certain odor is carried on winds blowing from the east. If a pigeon were transported eastward, the odor would tell it to fly westward to return home. The second hypothesis proposes that a bird may be able to extract its home's latitude and longitude from Earth's magnetic field. It may be proven in the future that neither theory explains the pigeon's navigational abilities or that some synthesis of the two theories is plausible.

How are birds related to dinosaurs?

Birds are essentially modified dinosaurs with feathers. Robert T. Bakker (1945–) and John H. Ostrom (1928–2005) did extensive research on the relationship between birds and dinosaurs in the 1970s and concluded that the bony structure of small dinosaurs was very similar to *Archaeopteryx*, the first animal classified as a bird, but that dinosaur fossils showed no evidence of feathers. They proposed that birds and dinosaurs evolved from the same source.

How do birds learn to sing the distinctive melody of their respective species?

The ability to learn the proper song appears to be influenced by both heredity and experience. Scientists have speculated that a bird is genetically programmed with the ability to recognize the song of its own species and with the tendency to learn its own song. As a bird begins to sing, it goes through a stage of practice (which closely resembles the babbling of human infants), through which it perfects the notes and structure of its distinctive song. In order to produce a perfect imitation, the bird must apparently hear the

Why is *Archaeopteryx* important?

rchaeopteryx is the first known bird. It had true feathers that provided insulation and allowed this animal to form scoops with its wings for catching prey.

song from an adult during its first months of life.

What are the natural predators of the penguin?

The leopard seal (*Hydrurga leptonyx*) is the principal predator of both the adult and juvenile penguin. The penguin may also be caught by a killer whale while swimming in open water. Eggs and chicks that are not properly guarded by adults are often devoured by skuas and sheathbills.

The word "bald" in bald eagle does not mean "lacking feathers"; rather, "bald" once meant someone with white hair, which makes sense when one looks at an example of the U.S. national emblem.

How do woodpeckers know where to peck to find food?

Woodpeckers have a very good sense of hearing. Several species of woodpeckers can actually hear carpenter ant larvae and can peck right into a nest of them. Woodpeckers also tap tree trunks to find the hollow places where insects might be living.

When was the bald eagle adopted as the national bird of the United States?

On June 20, 1782, the citizens of the newly independent United States of America adopted the bald or "American" eagle as their national emblem. At first, the heraldic artists depicted a bird that could have been a member of any of the larger species, but by 1902, the bird portrayed on the seal of the United States of America had assumed its proper white plumage on the head and tail. The choice of the bald eagle was not unanimous; Benjamin Franklin preferred the wild turkey. Oftentimes a tongue-in-cheek humorist, Franklin thought the turkey a wily but brave, intelligent, and prudent bird. He viewed the eagle, on the other hand, as having "a bad moral character" and "not getting his living honestly," preferring instead to steal fish from hardworking fishhawks. He also found the eagle a coward that readily flees from the irritating attacks of the much smaller kingbird.

MAMMALS

What are the distinguishing characteristics of mammals?

All mammals are warm-blooded vertebrates. They have lungs to breathe air. Two characteristics, common to all mammals that distinguish them from all other animals, are mammary glands and fur or hair. Mammary glands produce milk for the females to nourish their young. Although it may be scant, all mammals have some fur or hair covering their bodies. In addition, mammals have three middle ear bones.

What are some mammals that have pouches?

Marsupials (meaning "pouched" animals) differ from all other living mammals in their anatomical and physiological features of reproduction. Most female marsupials, including kangaroos, bandicoots, wombats, banded anteaters, koalas, opossums, wallabies, and tasmanian devils, possess an abdominal pouch (called a marsupium), in which their young are carried. In some small, terrestrial marsupials, however, the marsupium is not a true pouch but merely a fold of skin around the mammae (milk nipples).

The short gestation period in marsupials (in comparison to other similarly sized mammals) allows their young to be born in an "undeveloped" state. Consequently, these animals have been viewed as "primitive" or second-class mammals. However, some now see that the reproductive process of marsupials has an advantage over that of placental mammals. A female marsupial invests relatively few resources during the brief gestation period, more so during the lactation (nursing) period, when the young are in the marsupium. If the female marsupial loses its young, it can conceive again sooner than a placental mammal in a comparable situation.

What freshwater mammal is venomous?

The male duck-billed platypus (*Ornithorhynchus anatinus*) has venomous spurs located on its hind legs. When threatened, the animal will drive them into the skin of a potential enemy, inflicting a painful sting. The venom this action releases is relatively mild and generally not harmful to humans.

How does a bat catch flying insects in total darkness?

Bats use sound waves for communication and navigation. They emit supersonic radiation ranging from as low as 200 hertz to as high as 30,000 hertz. The sounds are emitted through the bat's nostrils or mouth and are aided by a complex flap structure to provide precise directivity to the radiation. Echo returns from the emissions allow a bat to pick out a tiny, flying insect some distance ahead. Highly sensitive ears and an ability to maneuver with great agility enable many bats to fly around in a darkened cave, catching insects without fear of collision.

What is the difference between porpoises and dolphins?

Marine dolphins (family *Delphinidae*) and porpoises (family *Phocoenidae*) together comprise about forty species. The chief differences between dolphins and porpoises occur in the snout and teeth. True dolphins have a beaklike snout and cone-shaped teeth. True porpoises have a rounded snout and flat or spade-shaped teeth.

How do the great whales compare in weight and length?

Whale	Average Weight (tons/kilograms)	Greatest Length (feet/meters)
Sperm	35/31,752	59/18
Blue	84/76,204	98.4/30

Which mammals lay eggs and suckle their young?

The duck-billed platypus (*Ornithorhynchus anatinus*), the short-nosed echidna or spiny anteater (*Tachyglossus aculeatus*), and the long-nosed echidna (*Zaglossus bruijni*), indigenous to Australia, Tasmania, and New Guinea, are the only three species of mammals that lay eggs (a nonmammalian feature) but suckle their young (a mammalian feature). These mammals (order Monotremata) resemble reptiles in that they lay rubbery, shell-covered eggs that are incubated and hatched outside the mother's body. In addition, they resemble reptiles in their digestive, reproductive, and excretory systems and in a number of anatomical details (eye structure, presence of certain skull bones, pectoral [shoulder] girdle, and rib and vertebral structures). They are, however, classed as mammals because they have fur and a four-chambered heart, nurse their young from gland milk, are warm-blooded, and have some mammalian skeletal features.

Whale	Average Weight (tons/kilograms)	Greatest Length (feet/meters)
Finback	50/45,360	82/25
Humpback	33/29,937	49.2/15
Right	50 (est.)/45,360 (est.)	55.7/17
Sei	17/15,422	49.2/15
Gray	20/18,144	39.3/12
Bowhead	50/45,360	59/18
Bryde's	17/15,422	49.2/15
Minke	10/9,072	29.5/9

What is the name of the seal-like animal in Florida?

The West Indian manatee (*Trichechus manatus*), in the winter, moves to more temperate parts of Florida, such as the warm headwaters of the Crystal and Homosassa rivers in central Florida or the tropical waters of southern Florida. When the air temperature rises to 50°F (10°C), it will wander back along the Gulf Coast and up the Atlantic Coast as far as Virginia. Long-range, offshore migrations to the coast of Guyana and South America have been documented. In 1893, when the population of manatees in Florida was reduced to several thousand, the state gave it legal protection from being hunted or commercially exploited. However, many animals continue to be killed or injured by the encroachment of humans. Entrapment in locks and dams, collisions with barges and powerboat propellers, and other man-made objects cause at least 30 percent of the manatee deaths, which total 125–130 annually.

Which member of the cat family lives in the desert?

The sand cat (*Felis margarita*) is the only member of the cat family tied directly to desert regions. Found in North Africa, the Arabian Peninsula, Turkmenistan, Uzbekistan, and

The slow, docile manatee of Florida is endangered due to loss of habitat and, often, because they are injured and killed by boat propellers in shallow waters.

western Pakistan, the sand cat has adapted to extremely arid desert areas. The padding on the soles of its feet is well suited to the loose, sandy soil, and it can live without drinking freestanding water. Having sandy or grayish-ochre, dense fur, its body length is 17.5–22 inches (45–57 centimeters). Mainly nocturnal (active at night), the cat feeds on rodents, hares, birds, and reptiles.

The Chinese desert cat (*Felis bieti*) does not live in the desert, as its name implies, but inhabits the steppe country and mountains. Likewise, the Asiatic desert cat (*Felis silvestris ornata*) inhabits the open plains of India, Pakistan, Iran, and Asiatic Russia.

What is the only American canine that can climb trees?

The gray fox (*Urocyon cinereoargenteus*) is the only American canine that can climb trees.

Which bear lives in a tropical rain forest?

The Malayan sun bear (*Ursus malayanus*) is one of the rarest animals in the tropical forests of Sumatra, the Malay Peninsula, Borneo, Myanmar, Thailand, and southern China. The smallest bear species, with a length of 3.3–4.6 feet (1–1.4 meters) and weighing 60–143 pounds (27–65 kilograms), it has a strong, stocky body. Against its black, short fur, it has a characteristic orange-yellow-colored crescent across its chest, which, according to legend, represents the rising sun. With powerful paws having long, curved claws to help it climb trees in the dense forests, it is an expert tree climber. The sun

What is the only four-horned animal in the world?

The four-horned antelope (*Tetracerus quadricornis*) is a native of central India. The males have two short horns usually 4 inches (10 centimeters) in length between their ears and an even shorter pair 1–2 inches (2.5–5 centimeters) long between the brow ridges over their eyes. Not all males have four horns, and in some, the second pair eventually falls off. The females have no horns at all.

bear tears at tree bark to expose insects, larvae, and the nests of bees and termites. Fruit, coconut palms, and small rodents are also part of its diet. Sleeping and sunbathing during the day, it is active at night. Unusually shy and retiring, cautious and intelligent, the sun bear is declining in population as its native forests are being destroyed.

What is the largest terrestrial mammal in North America?

The bison (*Bison bison*) is the largest terrestrial mammal in North America. It weighs 3,100 pounds (1,406 kilograms) and is 6 feet (1.8 meters) high.

Do camels store water in their humps?

The hump or humps do not store water since they are fat reservoirs. The ability to go long periods without drinking water, up to ten months if plenty of green vegetation and dew is around to feed on, results from a number of physiological adaptations. One major factor is that camels can lose up to 40 percent of their body weight with no ill effects. A camel can also withstand a variation of its body temperature by as much as 14 degrees. A camel can drink thirty gallons of water in 10 minutes and up to fifty gallons over several hours. A one-humped camel is called a dromedary or Arabian camel; a Bactrian camel has two humps and lives in the wild in the Gobi Desert. Today, the Bactrian is confined to Asia, while most of the Arabian camels are on African soil.

How many quills does a porcupine have?

For its defensive weapon, the average North American porcupine has about thirty thousand quills or specialized hairs, comparable in hardness and flexibility to slivers of celluloid and so sharply pointed that they can penetrate any hide. The quills that do the most damage are the short ones that stud the porcupine's muscular tail. With a few lashes, the porcupine can send a rain of quills that have tiny, scalelike barbs into the skin of its adversary. The quills work their way inward because of their barbs and the involuntary muscular action of the victim. Sometimes, the quills can work themselves out, but other times, the quills pierce vital organs, and the victim dies.

Slow-footed and stocky, porcupines spend much of their time in the trees using their formidable incisors to strip off bark and foliage for their food and supplement their

The easiest way to distinguish between an African and an Asian elephant is to look at the ears, which are much larger on an African elephant. They also differ in head shape, back shape, tusk length, and overall size.

diets with fruits and grasses. Porcupines have a ravenous appetite for salt; as herbivores (plant-eating animals), their diets have insufficient salt, so natural salt licks, animal bones left by carnivores (meat-eating animals), yellow pond lilies, and other items having a high salt content (including paints, plywood adhesives, and human clothing that bears traces of sweat) have a strong appeal to porcupines.

What is the difference between an African elephant and an Indian elephant?

The African elephant (*Loxodonta africana*) is the largest living land animal, weighing up to 8.25 tons (7,500 kilograms) and standing 10–13 feet (3–4 meters) at the shoulder. The Indian elephant (*Elephas maximus*) weighs about six tons (5,500 kilograms), with a shoulder height of 10 feet (3 meters). Other differences are:

African vs. Asian Elephant Traits

African Elephant	Indian Elephant
Larger ears	Smaller ears
Gestation period of about 670 days	Gestation period of about 610 days
Ear tops turn backward	Ear tops turn forward
Concave back	Convex back
Three toenails on hind feet	Four toenails on hind feet
Larger tusks	Smaller tusks
Two fingerlike lips at tips of trunk	One lip at tip of trunk
Single-domed head	Two-domed head

What is the chemical composition of a skunk's spray?

The chief odorous components of the spray have been identified as crotyl mercaptan, isopentyl mercaptan, and methyl crotyl disulfide in the ratio of 4:4:3. The liquid is an oily, pale-yellow, foul-smelling spray that can cause severe eye inflammation. This defensive weapon is discharged from two tiny nipples located just inside the skunk's anus—either as a fine spray or a short stream of rain-sized drops. Although the liquid's range is 6.5–10 feet (2–3 meters), its smell can be detected 1.5 miles (2.5 kilometers) downwind.

Why were Clydesdale horses used as war horses?

The Clydesdales were among a group of European horses referred to as the Great Horses, which were specifically bred to carry the massively armored knights of the Middle Ages. These animals had to be strong enough to carry a man wearing as much as 100 pounds (45 kilograms) of armor as well as up to 80 pounds (36 kilograms) of armor on their own bodies. However, the invention of the musket quickly ended the use of Clydesdales and other Great Horses on the battlefield as speed and maneuverability became more important than strength.

PETS AND DOMESTICATED ANIMALS

Are all dogs descended from wolves?

The best scientific evidence, based on behavior and genetics, indicates that all modern dogs are descended from an ancestral wolf species. The ancestral wolf was probably an Asian wolf.

How did so many different dog breeds develop?

Genetic variability is built into a species so it can adapt as conditions change. In wild form with the coyote and wolf, buffering systems exist so that all the genes they have are not necessarily expressed. In domestication, however, the system of genes that buffer differences from the norm are bred out. Dog breeders have selected away from the buffering system, so all possibilities can appear, and dogs can freely show all their genetic variability from dachshund to wolfhound. A tremendous choice exists in varieties like coat, leg proportions, and nose length, and without the buffering genes, breeders can select desired characteristics and, over time, develop a particular breed.

What is the oldest breed of dog?

Dogs are the oldest domestic animal, originating twelve to fourteen thousand years ago. They are believed to be descendants of wild canines, most likely wolves, which began to frequent human settlements where food was more readily available. The more aggressive canines were probably driven off or killed, while the less dangerous ones were kept to guard, hunt, and later herd other domesticated animals, such as sheep. Attempts at selectively breeding desirable traits likely began soon after.

The oldest purebred dog is believed to be the saluki. Sumerian rock carvings in Mesopotamia that date to about 7000 B.C.E. depict dogs bearing a striking resemblance to the saluki. The dogs are 23–28 inches (58–71 centimeters) tall, with a long, narrow head. The coat is smooth and silky and can be white, cream, fawn, gold, red, grizzle (bluish-gray) and tan, black and tan, or tricolor (white, black, and tan). The tail is long and feathered. The saluki has remarkable sight and tremendous speed, which makes it an excellent hunter.

The oldest American purebred dog is the American foxhound. It descends from a pack of foxhounds belonging Robert Brooke (1602–1655), who was colonial governor of Maryland in 1652. These dogs were crossed with other strains imported from England, Ireland, and France to develop the American foxhound. This dog stands 22–25 inches (56–63.5 centimeters) tall. It has a long, slightly domed head, with a straight, squared-out muzzle. The coat is of medium length and can be any color. They are used primarily for hunting.

What are the different classifications of dogs?

Dogs are divided into groups according to the purpose for which they have been bred.

Dog Classifications

Group	Purpose	Representative Breeds
Sporting dogs	Retrieving game birds and waterfowl	Cocker spaniel, English setter, English springer spaniel, golden retriever, Irish setter, Labrador retriever, pointer
Hounds	Hunting	Basenji, beagle, dachshund, foxhound, greyhound, saluki, Rhodesian ridgeback
Terriers	Hunting small animals such as rats and foxes	Airedale terrier, Bedlington terrier, bull terrier, fox terrier, miniature schnauzer, Scottish terrier, Skye terrier, West Highland white terrier
Toy dogs	Small companions or lap dogs	Chihuahua, Maltese, Pekingese, Pomeranian, pug, Shih Tzu, Yorkshire terrier
Herding dogs	Protect sheep and other livestock	Australian cattle dog, bouviers des Flandres, collie, German shepherd, Hungarian puli, Old English sheepdog, Welsh corgi

Group	Purpose	Representative Breeds
Working dogs	Herding, rescue, and sled dogs	Alaskan malamute, boxer, Doberman pinscher, great Dane, mastiff, St. Bernard, Siberian husky
Non-sporting dogs	No specific purpose, not toys	Boston terrier, bulldog, Dalmatian, Japanese akita, keeshond, Lhasa apso, poodle

Which breed is known as the wrinkled dog?

The shar-pei, or Chinese fighting dog, is covered with folds of loose skin. It stands 18–20 inches (46–51 centimeters) tall and weighs up to 50 pounds (22.5 kilograms). Its solid-colored coat can be black, red, fawn, or cream. The dog originated in Tibet or the northern provinces of China some two thousand years ago. The People's Republic of China put such a high tax on shar-peis, however, that few people could afford to keep them, and the dog was in danger of extinction, but a few specimens were smuggled out of China, and the breed has made a comeback in the United States, Canada, and the United Kingdom. Although bred as a fighting dog, the shar-pei is generally an amiable companion.

Why are Dalmatians "firehouse dogs"?

Before automobiles, coaches and carriages were often accompanied by dogs that kept horses company and guarded them from theft. Dalmatians were particularly well known for the strong bond they formed with horses, and firemen, who often owned the strongest and speediest horses in the area, kept the dogs at the station to deter horse thieves. Although fire engines have replaced horses, Dalmatians have remained a part of firehouse life both for the appeal of these beautiful dogs and for their nostalgic tie to the past.

The adorable shar-pei is notable for its wrinkly skin. The breed almost disappeared in China, but fans of shar-peis brought them back in the United States, Canada, and the United Kingdom.

Which breed is known as the voiceless dog?

The basenji dog does not bark. When happy, it will make an appealing sound described as something between a chortle and a yodel. It also snarls and growls on occasion. One of the oldest breeds of dogs and originating in central Africa, the basenji was often given as a present to the pharaohs of ancient Egypt. Following the decline of the Egyptian civilization, the basenji was still valued in central Africa for its hunting

What is the rarest breed of dog?

The Tahltan bear dog, of which only a few remain, is thought to be the rarest breed of dog. In danger of extinction, this breed was once used by the Tahltan Indians of western Canada to hunt bears, lynxes, and porcupines.

prowess and its silence. The dog was rediscovered by British explorers in the nineteenth century, although it was not widely bred until the 1940s.

The basenji is a small, lightly built dog with a flat skull and a long, rounded muzzle. It measures 16–17 inches (40–43 centimeters) in height at the shoulder and weighs 22–24 pounds (10–11 kilograms). The coat is short and silky in texture. The feet, chest, and tail tip are white; the rest of the coat is chestnut red, black, or black and tan.

What are "designer dogs"?

Designer dogs or crossbreed dogs are dogs that have been intentionally bred from two or more recognized dog breeds. Unlike mixed-breed dogs (also called mutts), which have unidentified parental lineage and the breed cannot be traced, designer breeds have parents with validated breeds. If a designer breed is traced back to the original parents with a full review of the pedigree, an established kennel club may register it as recognized as a new breed in its own right.

Which are some popular designer dog breeds?

Some of the most popular designer dog breeds are:

1. Cockapoo (Cocker Spaniel and Poodle)
2. Maltipoo (Maltese and Miniature Poodle)
3. Labradoodle (Labrador Retriever and Poodle)
4. Goldendoodle (Golden Retriever and Poodle)
5. Mal-shi (Maltese and Shih-Tzu)
6. Puggle (Pug and Beagle)
7. Schnoodle (Poodle and Schnauzer)
8. Yorkipoo (Yorkshire Terrier and Poodle)
9. Pomchi (Pomeranian and Chihuahua)
10. Cheeks (Chihuahua and Pekingese)

Which dogs are the easiest to train?

In a study of fifty-six popular dog breeds, the top breeds to train were Shetland sheepdogs, Shih Tzus, miniature toy and standard poodles, Bichons Frises, English Springer Spaniels, and Welsh Corgis.

What is the most recent method of tagging a dog?

A computer-age dog tag now exists. A microchip is implanted painlessly between the dog's shoulder blades. The semiconductor carries a ten-digit code, which can be read by a scanner. When the pet is found, the code can be phoned into a national database to locate the owner. The microchip can store the dog's license number and medical condition and the owner's address and phone number.

Which breeds of dogs are most often used as service dogs?

A service dog is a working dog that has specially trained skills that provide assistance and service in a variety of settings. The term "service dog" has different meanings in different parts of the world. In the United States, the term refers to any type of assistance dog specifically trained to help people who have disabilities, such as visual impairment, hearing impairments, mental disorders (such as post-traumatic stress disorder), seizures, mobility impairments, and diabetes. Labrador Retrievers, Golden Retrievers, Labrador/Golden Retriever crossbred dogs, and German Shepherd dogs are among the most common dog breeds working as service dogs today in the United States. Although dogs of almost any breed or mix of breeds may be capable of becoming a service dog, very few dogs have the requisite health and temperament qualities.

How is the age of a dog or cat computed in human years?

When a cat is one year old, it is about twenty years old in human years. Each additional year is multiplied by four. Another source counts the age of a cat slightly differently. At age one, a cat's age equals sixteen human years. At age two, a cat's age is twenty-four human years. Each additional year is multiplied by four. On average, pet cats live thirteen to fourteen years.

When a dog is one year old, it is about fifteen years old in human years. At age two, it is about twenty-four; after age two, each additional year is multiplied by four. On average, pet dogs live eight to fourteen years. The larger breeds (Great Dane, Irish Wolfhound, and St. Bernard) tend to live shorter, while the smaller breeds (Yorkshire Terrier, Toy Poodle, and Chihuahua) tend to live longer.

The Golden Retriever is one breed of dog that makes a great service dog as well as an excellent family dog and therapy dog.

What is the original breed of domestic cat in the United States?

The American shorthair is believed by some naturalists to be the original domestic cat in America. It is descended from cats brought to the New World from Europe by the early settlers. The cats readily adapted to their new environment. Selective breeding to enhance the best traits began early in the twentieth century.

The American shorthair is a very athletic cat with a lithe, powerful body, excellent for stalking and killing prey. Its legs are long, heavy, and muscular, ideal for leaping and coping with all kinds of terrain. The fur, in a wide variety of color and coat patterns, is thick enough to protect the animal from moisture and cold but short enough to resist matting and snagging.

Although this cat makes an excellent house pet and companion, it remains very self-sufficient. Its hunting instinct is so strong that it exercises the skill even when well provided with food. The American shorthair is the only true "working cat" in the United States.

What is a tabby cat?

"Tabby," the basic feline coat pattern, dates back to the time before cats were domesticated. The tabby's coat is an excellent form of camouflage. Each hair has two or three dark and light bands, with the tip always dark. The basic tabby pattern has four variations.

The mackerel (also called striped or tiger) tabby has a dark line running down the back from the head to the base of the tail, with several stripes branching down the sides. The legs have stripes, and the tail has even rings with a dark tip. The stomach has two rows of dark spots. Above the eyes is a mark shaped like an "M," and dark lines run back to the ears. Two dark, necklacelike bands appear on the chest.

The blotched, or classic, tabby markings seem to be the closest to those found in the wild. The markings on the head, legs, tail, and stomach are the same as the mackerel tabby. The major difference is that the blotched tabby has dark patches on the shoulder and side, rimmed by one or several lines.

The spotted tabby has uniformly shaped round or oval, dark spots all over the body and legs. The forehead has an "M" on it, and a narrow, dark line runs down the back.

The Abyssinian tabby has almost no dark markings on its body; they appear only on the forelegs, the flanks, and the tail. The hairs are banded except on the stomach, where they are light and unicolored.

What controls the formation of the color points on a Siamese cat?

The color points are due to the presence of a recessive gene, which operates at cooler temperatures, limiting the color to well-defined areas—the mask, ears, tail, lower legs, and paws—the places at the far reaches of the cardiovascular system of the cat.

Siamese cats come in four classic varieties. Seal-points have a pale fawn to cream-colored coat with seal-brown markings. Blue-points are bluish-white with slate-blue markings. Chocolate-points are ivory-colored with milk-chocolate-brown-colored markings. Lilac-points have a white coat and pinkish-gray markings. Some newer varieties also exist with red, cream, and tabby points.

The Siamese originated in Thailand (once called Siam) and arrived in England in the 1880s. They are medium-sized and have long, slender, lithe bodies, with long heads and long, tapering tails. Extroverted and affectionate, Siamese are known for their loud, distinctive voices, which are impossible to ignore.

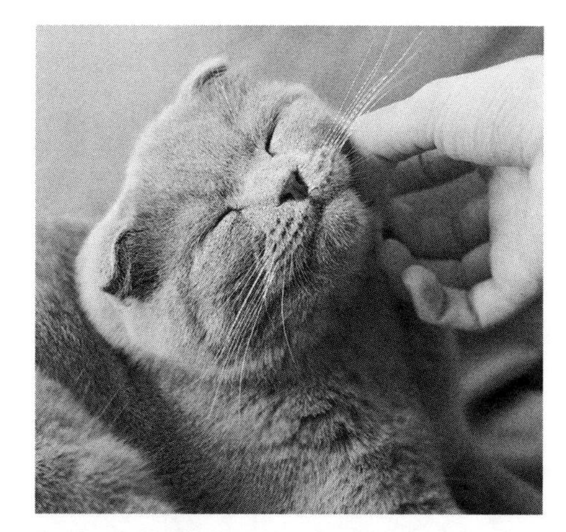

Purring is usually interpreted to mean a cat is happy, content, and well-fed. It might also be a signal that the cat is not a threat (at least, not while purring!)

Why and how do cats purr?

Experts cannot agree on how or why cats purr or from where the sound originates. Some think that the purr is produced by the vibration of blood in a large vein in the chest cavity. Where the vein passes through the diaphragm, the muscles around the vein contract, nipping the blood flow and setting up oscillations. These sounds are magnified by the air in the bronchial tubes and the windpipe. Others think that purring is the vibrations of membranes, called false vocal cords, located near the vocal cords. No one knows for sure why a cat purrs, but many people interpret the sound as one of contentment. Cats are also known to purr while in pain, such as while giving birth or dying, possibly as a way to soothe themselves.

Why do cats have whiskers?

The function of a cat's whiskers is not fully understood. They are thought to have something to do with the sense of touch. Removing them can disturb a cat for some time. Some people believe that the whiskers act as antennae in the dark, enabling the cat to identify things it cannot see. The whiskers may help the cat to pinpoint the direction from which an odor is coming. In addition, the cat is thought to point some of its whiskers downward to guide it when jumping or running over uneven terrain at night.

Which types of birds make the best pets?

Several birds make good house pets and have a reasonable life expectancy:

Bird	Life Expectancy in Years	Considerations
Finch	2–3	Easy care
Canary	8–10	Easy care; males sing
Budgerigar (parakeet)	8–15	Easy care
Cockatiel	15–20	Easy care; easy to train
Lovebird	15–20	Cute but not easy to care for or train
Amazon parrot	50–60	Good talkers but can be screamers
African grey parrot	50–60	Talkers; never scream

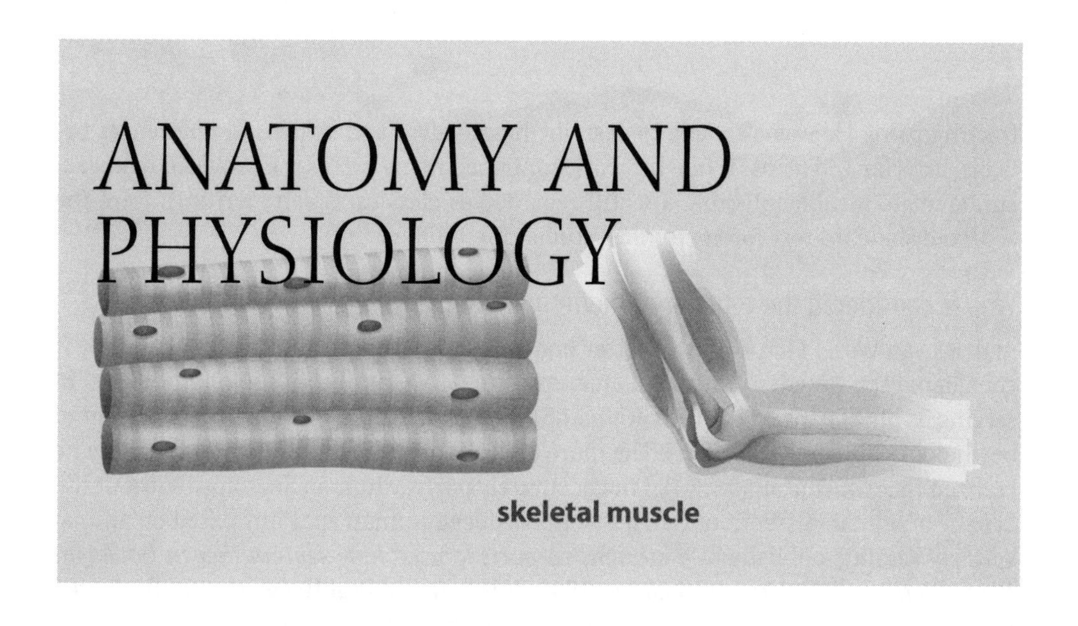

INTRODUCTION AND HISTORICAL BACKGROUND

Which scientific disciplines study the structure and function of organisms?

The disciplines of anatomy and physiology study the structure and function of organisms. Anatomy (from the Greek words *ana* and *temnein*, meaning "to cut up") is the study of the structure of an organism's cells, tissues, and organs, including their form and organization. Physiology (from the Greek *physis*, meaning "nature" or "origin") is the study of the function of the various parts of living organisms. Anatomy and physiology are usually studied together to achieve a complete understanding of an organism.

What were Aristotle's contributions to anatomy?

Aristotle wrote several works laying the foundations for comparative anatomy, taxonomy, and embryology. He investigated carefully all kinds of animals, including humans. His works on life sciences, *On Sense and Sensible Objects*, *On Memory and Recollection*, *On Sleep and Waking*, *On Dreams*, *On Divination by Dreams*, *On Length and Shortness of Life*, *On Youth and Age*, and *On Respiration*, are collectively called *Parva Naturalia*.

Whose work during the Roman era became the authority on anatomy?

Galen (130–200), a Greek physician, anatomist, and physiologist living during the time of the Roman Empire, was one of the most influential and authoritative authors on medical subjects. His writings include *On Anatomical Procedures*, *On the Usefulness of the Parts of the Body*, *On the Natural Faculties*, and hundreds of other treatises. Since human dissection was forbidden, Galen made most of his observations on different animals. He correctly described bones and muscles and observed muscles working in contracting pairs. He was also able to describe heart valves and structural differences between arteries and veins. While his work contained many errors, he provided many accurate anatomical details that are still regarded as classics. Galen's writings were the accepted standard text for anatomical studies for 1,400 years.

Who is considered the founder of modern anatomy?

Andries van Wesel (1514–1564), better known by his Latin name, Andreas Vesalius, is considered the founder of modern anatomy. Born to a family with a long tradition of serving as physicians to the Holy Roman Emperors, Vesalius was the first physician to begin doing his own dissections. The more dissections Vesalius performed, the more he realized that Galenic anatomy did not accurately portray human anatomy. Without the benefit of human dissection, Galen's works deduced human anatomy based on animal species. Vesalius published *De humani corporis fabrica libri septem* (*Seven Books on the Structure of the Human Body*) in 1543. It is believed that the work was illustrated by Flemish artist Jan Stephan van Calcar (1499–1546). It has endured not only as one of the most influential books on human anatomy but also for the beauty of the exquisite illustrations.

When was the term "physiology" first used?

The term "physiology" was first used by the Greeks as early as 600 B.C.E. to describe a philosophical inquiry into the nature of things. It was not until the sixteenth century that the term was used in reference to vital activities of healthy humans. During the nineteenth century, its usage was expanded to include the study of all living organisms using chemical, physical, and anatomical experimental methods.

French scientist Claude Bernard, considered the father of physiology, also originated the concept of homeostasis and was the first to propose the use of blind experiments to eliminate the problem of subjective biases.

Who is considered the founder of physiology?

As an experimenter, Claude Bernard (1813–1878) enriched physiology by his introduction of numerous new concepts into the field. The most famous of these concepts is that of the *milieu intérieur* or internal environment. The complex functions of the various organs are closely interrelated and are all directed to maintaining the constancy of internal conditions despite external changes. All cells exist in this aqueous (blood and lymph) internal environment, which bathes the cells and provides a medium for the elementary exchange of nutrients and waste material.

What is homeostasis?

Homeostasis (from the Greek words *homois*, meaning "same," and *stasis*, meaning "standing still") is the state of inner balance and stability maintained by an organism despite constant changes in the eternal environment. Nearly everything that occurs in the body helps to maintain homeostasis. The term was coined by Walter Bradford Cannon (1871–1945), who elaborated on Claude Bernard's concept of the *milieu intérieur* (internal environment). He referred to the body's ability to maintain a relative constancy in its internal environment.

What are the four levels of structural organization in animals?

Every animal has four levels of hierarchical organization: cell, tissue, organ, and organ system. Each level in the hierarchy is of increasing complexity, and all organ systems work together to form an organism. (For more information about the cell, see the chapter entitled "Biology.")

TISSUE

What are the four major types of tissue?

A tissue (from the Latin word *texere*, meaning "to weave") is a group of similar cells that perform a specific function. The four major types of tissue are epithelial, connective, muscle, and nerve. Each type of tissue performs different functions, is located in a different part of the body, and has certain distinguishing features. The table below explains these differences.

Characteristics of Tissues

Tissue	Function	Location	Distinguishing Features
Epithelial	Protection, secretion, absorption, excretion	Covers body surfaces, covers and lines internal organs, compose glands	Lacks blood vessels
Connective	Bind, support, protect, fill spaces, store fat, produce blood cells	Widely distributed throughout the body	Matrix between cells, good blood supply
Muscle	Movement	Attached to bones, in the walls of hollow internal organs, heart	Contractile
Nerve	Transmit impulses for coordination, regulation, integration, and sensory reception	Brain, spinal cord nerves	Cells connect to each other and other body parts

What are some examples of epithelial tissue?

Epithelial tissue, also called epithelium (from the Greek words *epi*, meaning "on," and *thele*, meaning "nipple") consists of densely packed cells. It covers every surface, both external and internal, of the body. It forms a barrier, allowing the passage of certain substances while impeding the passage of other substances. One example of epithelial tissue is the outer layer of the skin, called the epidermis. Other examples of epithelial tissue are the lining of the lungs and the inner surfaces of the esophagus, stomach, and intestines.

Is epithelial tissue replaced during an animal's lifetime?

Epithelial cells are constantly being replaced and regenerated during an animal's lifetime. The epidermis (outer layer of the skin) is renewed every two weeks, while the epithelial lining of the stomach is replaced every two to three days. The liver, a gland consisting of epithelial tissue, easily regenerates after portions are removed surgically.

What is a gland?

Glands are secretory cells or multicellular structures that are derived from epithelium and often stay connected to it. They are specialized for the synthesis, storage, and secretion of chemical substances. Glands are classified as either endocrine or exocrine glands. Exocrine glands have ducts to carry the secretions. Mucus, saliva, perspiration, earwax, oil, milk, and digestive enzymes are examples of exocrine secretions. Endocrine glands are ductless and release the secretions directly into the extracellular fluid.

What are the major types of connective tissue?

Connective tissue is categorized into six major types. They are loose connective tissue, adipose tissue, blood, collagen, cartilage, and bone. A unique characteristic of connective tissue is that the cells of connective tissue are spaced widely apart and are scattered through a nonliving, extracellular material called a matrix. The matrix is different in the different types of connective tissue and may be a liquid, jelly, or solid.

What is the matrix in blood?

Blood is a loose connective tissue whose matrix is a liquid called plasma. Blood consists of red blood cells (erythrocytes), white blood cells (leukocytes), and platelets (thrombocytes), which are tiny pieces of bone marrow cell. Plasma also contains water, salts, sugars, lipids, and amino acids. Blood is approximately 55 percent plasma and 45 percent formed elements.

How strong is bone?

Bone is one of the strongest materials found in nature. It is a rigid, connective tissue that has a matrix of collagen fibers embedded in calcium salts. Most of the skeletal system is comprised of bone, which provides support for muscle attachment and protects the internal organs. One cubic inch of bone can withstand loads of at least 19,000 pounds

(8,626 kilograms), which is approximately the weight of five standard-size pickup trucks. This is roughly four times the strength of concrete. Bone's resistance to load is equal to that of aluminum and light steel. Ounce for ounce, bone is actually stronger than steel and reinforced concrete since steel bars of a comparable size would weigh four or five times as much as bone.

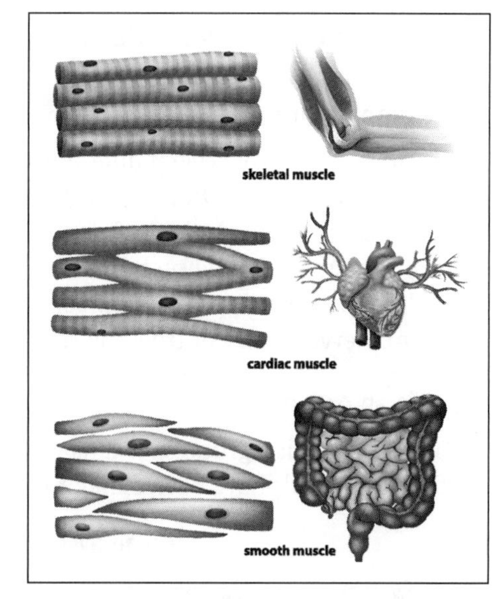

The three types of muscles.

What are the three types of muscle tissue?

Vertebrate animals have three types of muscle tissue: 1) smooth muscle; 2) skeletal muscle; and 3) cardiac muscle. Muscle tissue, consisting of bundles of long cells called muscle fibers, is specialized for contraction. It enables body movements as well as the movement of substances within the body. Smooth muscle and cardiac muscle contraction is involuntary since it occurs without intervention of the animal. Skeletal muscles, allowing an animal to move, lift, and utter sounds, is voluntary since the animal consciously contracts the muscle.

What type of cell is found in nerve tissue?

Neurons are specialized cells that produce and conduct "impulses," or nerve signals. Neurons consist of a cell body, which contains a nucleus and two types of cytoplasmic extensions: dendrites and axons. Dendrites are thin, highly branched extensions that receive signals. Axons are tubular extensions that transmit nerve impulses away from the cell body, often to another neuron. Nerve tissue also has supporting cells, called neuroglia or glial cells, which nourish the neurons, insulate the dendrites and axons, and promote quicker transmission of signals.

How many different types of neurons are found in nerve tissue?

Neurons are divided into three main types: 1) sensory neurons; 2) motor neurons; and 3) interneurons (also called association neurons). Sensory neurons conduct impulses from sensory organs (eyes, ears, and the surface of the skin) into the central nervous system. Motor neurons conduct impulses from the central nervous system to muscles or glands. Interneurons are neither sensory neurons nor motor neurons. They permit elaborate processing of information to generate complex behaviors. Interneurons comprise the majority of neurons in the central nervous system.

ORGANS AND ORGAN SYSTEMS

What is an organ?

An organ is a group of several different tissues working together as a unit to perform a specific function or functions. Each organ performs functions that none of the component tissues can perform alone. This cooperative interaction of different tissues is a basic feature of animals, including humans. The heart is an example of an organ. It consists of cardiac muscle wrapped in connective tissue. The heart chambers are lined with epithelium. Nerve tissue controls the rhythmic contractions of the cardiac muscles.

What is an organ system?

An organ system is a group of organs working together to perform a vital body function. Vertebrate animals have twelve major organ systems.

Organ Systems and Their Functions

Organ System	Components	Functions
Cardiovascular and circulatory	Heart, blood, and blood vessels	Transports blood throughout the body, supplying nutrients and carrying oxygen to the lungs and wastes to kidneys
Digestive	Mouth, esophagus, stomach, intestines, liver, and pancreas	Ingests food and breaks it down into smaller chemical units
Endocrine	Pituitary, adrenal, thyroid, and other ductless glands	Coordinates and regulates the activities of the body
Excretory	Kidneys, bladder, and urethra	Removes wastes from the bloodstream
Immune	Lymphocytes, macrophages, and antibodies	Removes foreign substances
Integumentary	Skin, hair, nails, and sweat glands	Protects the body
Lymphatic	Lymph nodes, lymphatic capillaries, lymphatic vessels, spleen, and thymus	Captures fluid and returns it to the cardiovascular system
Muscular	Skeletal muscle, cardiac muscle, and smooth muscle	Allows body movements
Nervous	Nerves, sense organs, brain, and spinal cord	Receives external stimuli, processes information, and directs activities
Reproductive	Testes, ovaries, and related organs	Carries out reproduction
Respiratory	Lungs, trachea, and other air passageways	Exchanges gases—captures oxygen (O_2) and disposes of carbon dioxide (CO_2)
Skeletal	Bones, cartilage, and ligaments	Protects the body and provides support for locomotion and movement

Which organ system is not essential to the survival of an individual organism?

The reproductive system is not essential to the survival of an individual organism. Each individual organism can survive without its reproductive systems; however, it is essential for the survival of the species.

CARDIOVASCULAR AND CIRCULATORY SYSTEM

What are the functions of the circulatory system?

The circulatory system provides a transport system between the heart, lungs, and tissue cells. The primary function is to supply nutrients to all the cells of an organism. It also transports wastes from the cells to waste-disposal organs, for example, lungs and kidneys. The circulatory system plays a vital role in maintaining homeostasis.

Are the cardiovascular system and the circulatory system the same thing or two different things?

The circulatory system is a general term referring to the heart, blood and blood vessels, and lymph and lymph vessels. The cardiovascular system usually refers to the heart (cardio) and blood vessels (vascular).

What are the differences between an open and a closed circulatory system?

In an open circulatory system, found in many invertebrates (e.g., spiders, crayfish, and grasshoppers), the blood is not always contained within the blood vessels. Periodically, the blood leaves the blood vessels to bathe the tissues with blood and then returns to the heart. No interstitial body fluid exists separate from the blood. A closed circulatory system, also called a cardiovascular system, is found in all vertebrate animals and many invertebrates; in a closed system, the blood never leaves the blood vessels.

Do all animals have red blood?

Some invertebrates, such as flatworms and cnidarians, possess a clear, watery tissue that contains some phagocytic cells, a little protein, and a mixture of salts similar to seawater in lieu of blood. Invertebrates with an open circulatory system have a fluid referred to as hemolymph (from the Greek word *haimo*, meaning "blood," and the Latin word *lympha*, meaning "water"). Invertebrates with a closed circulatory system have blood that is contained in blood vessels. Vertebrates have blood vessels and blood composed of plasma and formed elements.

The color of the blood is related to the compounds that transport oxygen. Hemoglobin, containing iron, is red and is found in all vertebrates and a few invertebrates. Annelids (segmented worms) have either a green pigment, chlorocruorin, or a red pigment, hemerythrin. Some crustaceans (arthropods having divided bodies and generally having gills) have a blue pigment, hemocyanin, in their blood.

What are the functions of blood?

The functions of blood can be divided into three general categories: transportation, regulation, and protection.

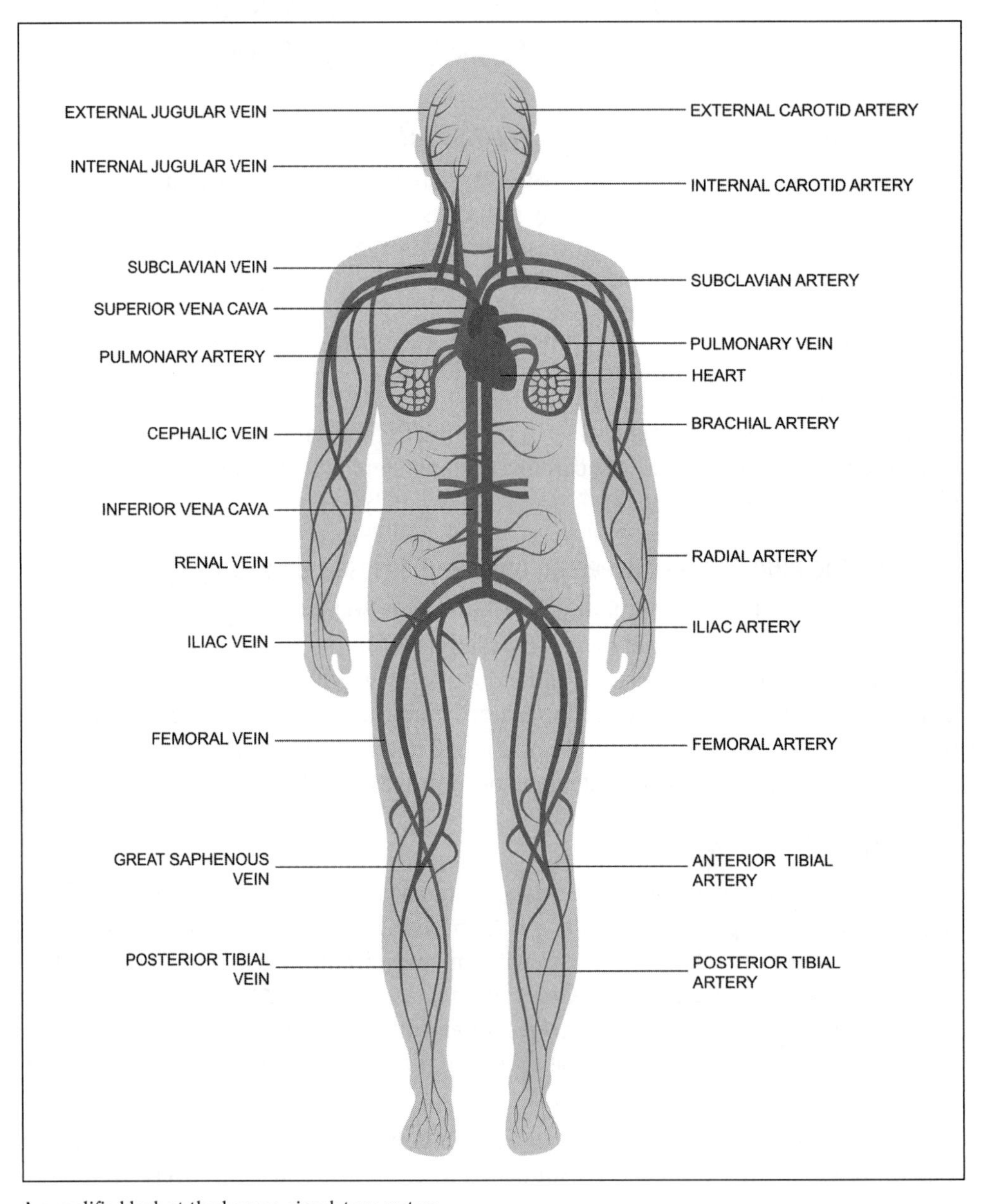

A symplified look at the human circulatory system.

Why does the blood seen under the skin in humans look blue?

The blood in human veins is oxygen-poor blood since the hemoglobin carried in the blood in the arteries has been distributed to various cells and tissues. The "blue blood" in the veins is a combination of light passing through the skin on the surface and the oxygen-poor blood, which is dark red versus the bright red of oxygen-rich blood.

Functions of Blood

Function	Examples
Transportation	Gases (oxygen and carbon dioxide), nutrients, metabolic waste
Regulation	Body temperature, normal pH, fluid volume/pressure
Protection	Against blood loss, against infection

What are blood groups?

More than twenty genetically determined blood group systems among humans are known today, but the ABO and Rh systems are the most important ones used to type blood for human blood transfusions. Different species of animals have varying numbers of blood groups.

Species	Number of Blood Groups
Human	20+
Pig	16
Cow	12
Chicken	11
Horse	9
Sheep	7
Dog	7
Rhesus monkey	6
Mink	5
Rabbit	5
Mouse	4
Rat	4
Cat	2

Who discovered the ABO system of typing blood?

Austrian physician Karl Landsteiner (1868–1943) discovered the ABO system of blood types in 1909. Landsteiner had investigated why blood transfused from one individual was sometimes successful and other times resulted in the death of the patient. He theorized that several different blood types must exist. A person with one type of blood will have antibodies to the antigens in the blood type they do not have. If a transfusion occurs between two individuals with different blood types, the red blood cells will clump together, blocking the blood vessels.

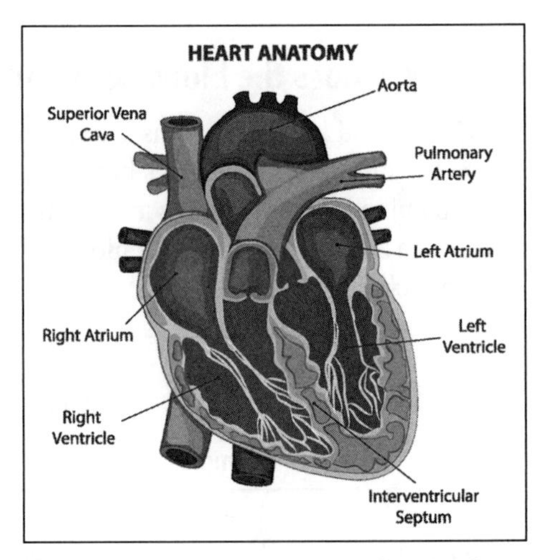

The human heart is a big muscle with four chambers for pumping blood. Blood goes into the right atrium, then to the right ventricle, which squeezes and pumps blood to the lungs. After the blood is oxygenated, it goes back to the heart via the pulmonary veins, enters the left atrium, then the left ventricle, which then pumps it out of the heart to the rest of the body.

What is the Rh factor?

In addition to the ABO system of blood types, blood types can also be grouped by the Rhesus factor, or Rh factor, an inherited blood characteristic. Discovered independently in 1939 by Philip Levine (1900–1987) and R. E. Stetson (1886–1967) and in 1940 by Karl Landsteiner and A. S. Weiner (1907–1976), the Rh system classifies blood as either having the Rh factor or lacking it. Pregnant women are carefully screened for the Rh factor. If a mother is found to be Rh-negative, the father is also screened. Parents with incompatible Rh factors can lead to potentially fatal blood problems in newborn infants. The condition can be treated with a series of blood transfusions.

How many miles of blood vessels are contained in the human body?

If they could be laid end to end, the blood vessels in the human body, including the arteries, arterioles, capillaries, venules, and veins, would span about 60,000 miles (96,500 kilometers). This would be enough to encircle Earth more than two times.

How much blood does the average-sized adult human have?

An adult man has 5.3–6.4 quarts, or 1.5 gallons (5–6 liters), of blood, while an adult woman has 4.5–5.3 quarts, or 1.2 gallons (4–5 liters). Differences are due to the sex of the individual, body size, fluid and electrolyte concentrations, and amount of body fat.

How hard does the human heart work?

The human heart squeezes out about 2 ounces (71 grams) of blood at every beat and pumps at least 2,500 gallons (9,450 liters) of blood daily. On the average, the adult heart beats seventy to seventy-five times a minute. The rate of the heartbeat is determined in part by the size of the organism. Generally, the smaller the size, the faster the heartbeat. Thus, women's hearts beat six to eight beats per minute faster than men's hearts do. At birth, the heart of a baby can beat as fast as 130 times per minute. The following chart compares the resting heart rate of several mammals:

Mammal	Resting Heart Rate (beats per minute)
Human	75
Horse	48
Cow	45–60
Dog	90–100
Cat	110–140
Rat	360
Mouse	498

Which of the major blood types are the most common in the United States?

The following table lists the major blood types and their rate of occurrence in the United States.

Distribution of Blood Types in the United States

Blood Type	Frequency (U.S.)
O+	37.4%
O–	6.6%
A+	35.7%
A–	6.3%
B+	8.5%
B–	1.5%
AB+	2.4%
AB–	0.6%

Who first demonstrated that blood circulates?

William Harvey (1578–1657) was the first person to demonstrate that blood circulates in the bodies of humans and other animals. Harvey's hypothesis was that the heart is a pump for the circulatory system, with blood flowing in a closed circuit. Harvey conducted his research on live organisms as well as dissection of dead organisms to demonstrate that when the heart pumps, blood flows into the aorta. He observed that when an

How does a giraffe overcome gravity in order for blood to circulate to its head?

The head of a giraffe may be as much as 6.5 feet (2 meters) above the level of its heart. In order for blood to reach the brain, the blood pressure of a giraffe is nearly twice that of humans. The arteries of a giraffe have exceptionally thick walls to withstand the high pressure. The veins have valves to facilitate the return of blood from the legs to the heart. The valves in the neck impede the backflow of blood to the head when a giraffe lowers its head to drink.

artery is slit, all the blood in the system empties. Finally, Harvey demonstrated that the valves in the veins serve to return blood to the heart.

DIGESTIVE SYSTEM

What are the steps of food processing for animals?

The first step is for animals to ingest food. The food is then broken down via the digestive process into molecules that the organism can absorb for energy. Once the food is digested, it is absorbed through the digestive tract to provide energy for the organism. The final step of food processing is elimination. During elimination, undigested material is passed out of the digestive tract.

What are the main organs of the digestive system and their functions?

The digestive system includes the mouth, alimentary canal or gastrovascular cavity, esophagus, stomach, small intestine, large intestine, and anus. The mouth is the opening through which food is ingested. In animals with a second opening for elimination, the digestive system contains an alimentary canal, a tube allowing for the passage of food from the mouth to the anus. The esophagus is another channel through which food passes on the way to the stomach. The stomach (or "crop" in certain species, such as birds) stores food and is the primary site of chemical digestion. Following digestion in the stomach where food is broken down with acid and enzymes, the food passes into the small intestine. Nutrients are absorbed via the intestine. Much shorter than the small intestine, though greater in diameter, is the large intestine, also called the colon. Here, the solid material remaining after digestion is compacted and then eliminated via the anus.

What are examples of animals with a gastrovascular cavity?

Many invertebrate animals, such as hydra and sea anemones, have a gastrovascular cavity that serves as the site of digestive activities. Once the food is caught, it is moved to the animals' gastrovascular cavity, where digestion occurs. The organism's cells are in contact with the gastrovascular cavity, so it is not necessary for other organs to transport the food for absorption. Elimination of waste matter also occurs in the gastrovascular cavity.

How do continuous feeders differ from discontinuous feeders?

Continuous feeders, also known as filter feeders, are aquatic animals that constantly feed by having water filled with food particles (e.g., small plankton or fish) entering through the mouth. Continuous feeders do not need a storage area, such as a stomach, for food. Discontinuous feeders must hunt for food on a regular basis. They need a storage area to house food until it is digested.

Are the teeth of an animal different based on its diet?

The teeth of an animal reflect the type of diet the animal eats.

- Herbivores, animals that eat only plant matter, have sharp incisors to bite off blades of grass and other plant matter. They also have a system of flat premolars and molars for grinding and crushing grasses and plant matter.
- Carnivores, animals that eat other animals, have pointed incisors and enlarged canine teeth to tear off pieces of meat. Their premolars and molars are jagged to aid in chewing flesh.
- Omnivores, animals that eat both plants and other animals, have nonspecialized teeth to accommodate a diet of both plant material and animals.

Do other mammals lose baby teeth the same way that humans do?

Most mammals have two sets of teeth during their lives. They are born without teeth and remain toothless while they are nourished by their mother's milk. They develop deciduous teeth around the time they are weaned. The deciduous teeth are sometimes called baby teeth or milk teeth. As they mature, the deciduous set of teeth are lost and replaced by a permanent set of teeth.

Why does a cow have four stomachs?

Cows are ruminants. Ruminants eat rapidly and do not chew much of their food completely before they swallow it. They have four stomachs to process their low-quality diet of grass. The four sections are 1) the rumen; 2) the reticulum; 3) the omasum; and 4) the abomasum. The liquid part of their food enters the first chamber, called the reticulum, while the solid part of their food enters the rumen, where it softens. Bacteria in the rumen initially break the food material down as a first step in digestion. Ruminants later regurgitate the partially liquefied plant parts into the mouth, where they continue to munch it in a process known as "chewing their cud." Cows chew their cud about six to eight times per day, spending a total of 5–7 hours in rumination. The chewed cud goes directly into the other chambers of the stomach, where various microorganisms assist in further digestion. Herbivores have a longer small intestine to allow maximum time

What is unusual about the teeth of sharks?

Sharks were among the first vertebrates to develop teeth. The teeth are not set into the jaw but rather sit atop it. They are not firmly anchored and are easily lost. The teeth are arranged in six to twenty rows, with the ones in front doing the biting and cutting. Behind these teeth, others grow. When a tooth breaks or is worn down, a replacement moves forward. One shark may eventually develop and use more than twenty thousand teeth in a lifetime.

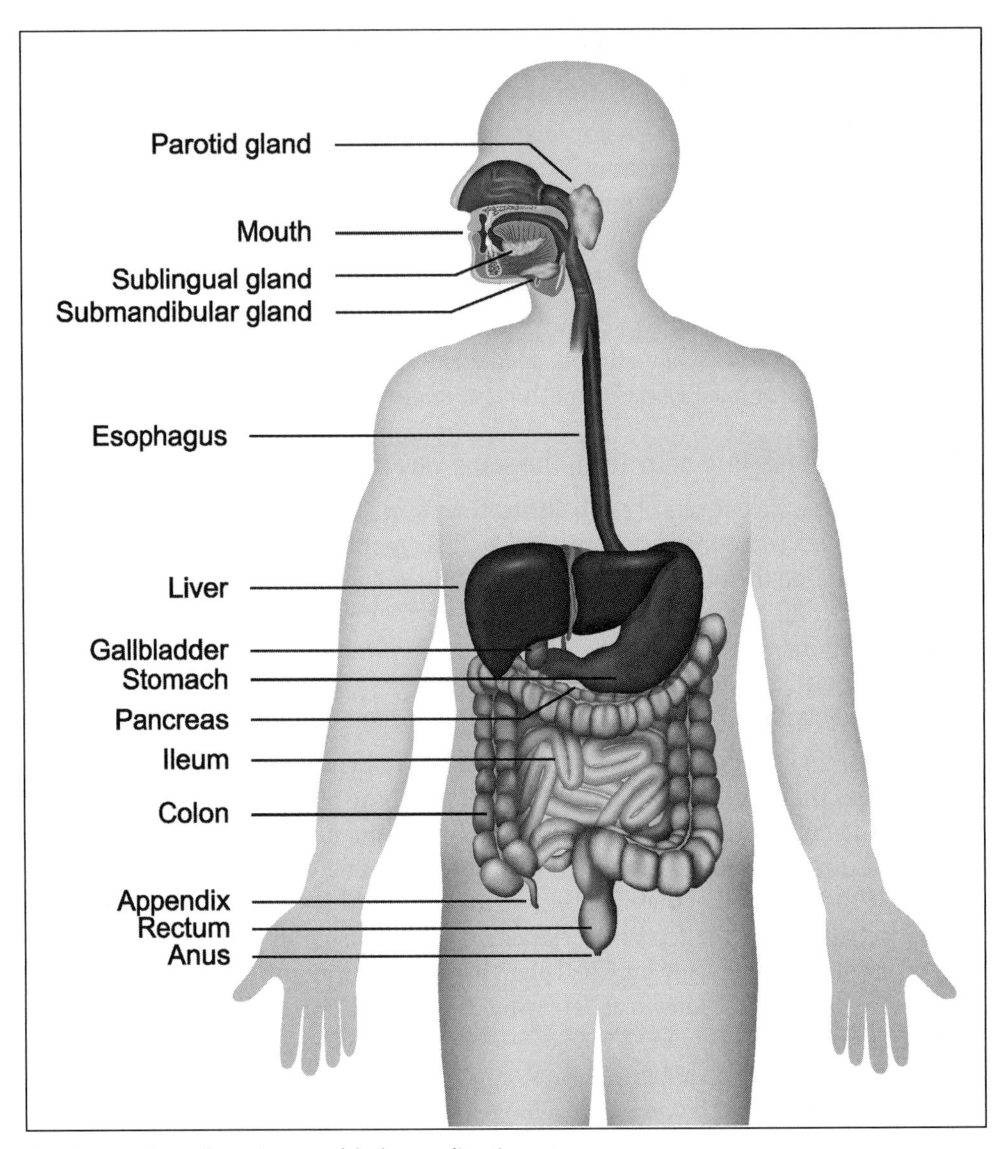

This diagram shows the main parts of the human digestive system.

for the absorption of nutrients. Other ruminants are bison, buffalo, goats, antelopes, sheep, deer, and giraffe.

How long does it take food to digest in humans?

The entire digestive tract in humans, from the mouth to the anus, is approximately 30 feet (9 meters) long. The stomach holds a little under 2 quarts (1.9 liters) of semidigested food that stays in the stomach for 3–5 hours. The stomach slowly releases food to the rest of the digestive tract. Fifteen hours or more after the first bite started down the alimentary canal, the final residue of the food is passed along to the rectum and is excreted through the anus as feces.

ENDOCRINE SYSTEM

What is the function of the endocrine system?

The endocrine system is the main chemical-regulating system of an organism. Hormones, chemicals made and secreted by endocrine glands or neurosecretory cells, are the main messengers of the endocrine system. Hormones are transported in the blood to all parts of the body and interact with target cells (cells that contain hormone receptors). They regulate metabolic rate, growth, maturation, and reproduction.

What are hormones?

Hormones are chemical messengers that are secreted by the endocrine glands into the blood. They produce a specific effect on the activity of cells that are remotely located from their point of origin. Hormones are transported via the bloodstream to reach specific cells, called target cells, in other tissues. Target cells have special receptors on their outer membranes that allow the individual hormones to bind to the cell. The hormones and receptors fit together much like a lock and key.

Who discovered the first animal hormone?

British physiologists William Bayliss (1860–1924) and Ernest Starling (1866–1927) discovered secretin in 1902. They used the term "hormone" (from the Greek word *horman*, meaning "to set in motion") to describe the chemical substance they had discovered that stimulated an organ at a distance from the chemical's site of origin. Their famous experiment using anesthetized dogs demonstrated that diluted hydrochloric acid, mixed with partially digested food, activated a chemical substance in the duodenum (the upper part of the small intestine). This activated substance (secretin) was released into the bloodstream and came into contact with the cells of the pancreas. In the pancreas, it stimulated secretion of digestive juice into the intestine through the pancreatic duct.

What are some vertebrate endocrine glands and their hormones?

Vertebrates have ten major endocrine glands as shown in this table.

Endocrine Gland	Hormone	Target Tissue	Principal Function
Posterior pituitary	Antidiuretic hormone (ADH)	Kidneys	Stimulates water reabsorption by kidneys
Posterior pituitary	Oxytocin	Uterus, mammary glands	Stimulates uterine contractions and milk ejection

Endocrine Gland	Hormone	Target Tissue	Principal Function
Anterior pituitary	Growth hormone (GH)	General	Stimulates growth, especially cell division and bone growth
Anterior pituitary	Adrenocorticotropic hormone (ACTH)	Adrenal cortex	Stimulates adrenal cortex
Anterior pituitary	Thyroid-stimulating hormone (TSH)	Thyroid gland	Stimulates thyroid
Anterior pituitary	Luteinizing hormone (LH)	Gonads	Stimulates ovaries and testes
Anterior pituitary	Follicle-stimulating hormone (FSH)	Gonads	Controls egg and sperm production
Anterior pituitary	Prolactin (PRL) glands	Mammary	Stimulates milk production
Anterior pituitary	Melanocyte-stimulating hormone (MSH)	Skin	Regulates skin color in reptiles and amphibians; unknown function in humans
Thyroid	Thyroxine (thyroid hormone)	Liver	Stimulates and maintains metabolic processes to maintain normal growth and development
Thyroid	Calcitonin	Bone	Lowers blood calcium level
Parathyroid	Parathyroid hormone (PTH)	Bone, kidneys, digestive tract	Raises blood calcium level
Adrenal medulla	Epinephrine (adrenaline) and norepinephrine (noradrenaline)	Skeletal muscle, cardiac muscle, blood vessels	Initiates stress responses; raises heart rate, blood pressure, metabolic rates; constricts certain blood vessels
Adrenal cortex	Aldosterone tubules	Kidney	Stimulates kidneys to reabsorb sodium and excrete potassium
Adrenal cortex	Cortisol	General	Increases blood glucose
Pancreas	Insulin	Liver	Lowers blood glucose level; stimulates formation and storage of glycogen
Pancreas	Glucagon	Liver, adipose tissue	Raises blood glucose level
Ovary	Estrogens	General; female reproductive structures	Stimulates development of secondary sex characteristics in females and uterine lining
Ovary	Progesterone	Uterus, breasts	Promotes growth of uterine lining; stimulates breast development

Endocrine Gland	Hormone	Target Tissue	Principal Function
Testes	Estadriol; androgens (testosterone)	General; Male reproductive structures	Stimulates development of male sex organs and spermatogenesis
Pineal gland	Melatonin	Gonads, pigment cells	Involved in daily and seasonal rhythmic activities (circadian cycles); influences pigmentation in some species

What are the "fight-or-flight" hormones?

Epinephrine and norephinephrine are released by the adrenal glands in times of stress. The familiar feelings of a pounding, racing heart, increased respiration, elevated blood pressure, and goose bumps on the skin are responses to stressful circumstances.

EXCRETORY SYSTEM

What are the functions of the excretory system?

The excretory system is responsible for removing waste products from an organism. It also plays a vital role in regulating the water and salt balance in the organism.

Do invertebrates have specialized organs as part of their excretory system?

Many animals, such as sponges, jellyfish, tapeworms, and other invertebrates, do not have distinct excretory organs. Rather, they rid their bodies of waste through diffusion. Larger, more complex animals require specialized, often tubular, organs to rid their bodies of waste. For example, flatworms such as planarians have tubules that collect wastes and expel them to the outside via pores. Segmented worms such as earthworms have nephridia (tubules with a ciliated opening) in each segment. Fluid from the body cavity is propelled through the nephridia. Wastes are expelled through a pore to the outside, while certain substances are reabsorbed. Insects have a unique excretory system that consists of Malpighian tubules. Waste products enter the Malpighian tubules from the body cavity. Water and other useful substances are reabsorbed, while uric acid passes out of the body.

What are the specialized organs in the excretory system of vertebrates?

Vertebrate animals have kidneys to dispose of the metabolic waste in their bodies. In humans, urine is manufactured in the kidneys. The human excretory system also includes the bladder for storage of urine and ureters to transport the urine from the kidneys to the bladder. Urine is transported from the bladder to the outside of the body through the urethra.

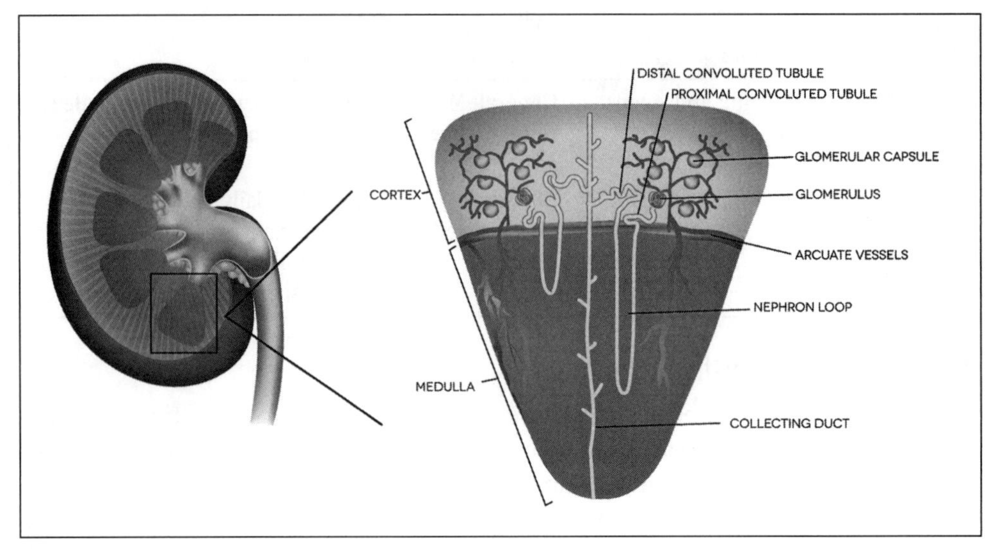

The anatomy of a human nephron, the functional working unit of a kidney.

Where are nephrons located?

Nephrons are the functional working unit of the human kidney. Blood is filtered in the nephrons and toxic wastes are removed, while water and necessary nutrients are reabsorbed into the body. Each kidney contains approximately one million nephrons. Each nephron produces a minute quantity of urine.

What is the difference between the waste products of ammonia, urea, and uric acid?

Ammonia, urea, and uric acid are nitrogenous waste products. They are the result of the breakdown of various molecules, including nucleic acids and amino acids. Since it is highly toxic, excretion of pure ammonia is possible only for aquatic animals because ammonia is very soluble in water. Urea and uric acid are excreted by terrestrial animals. Urea is approximately one hundred thousand times less toxic than ammonia, so it may be stored in the body and eliminated with relatively little water loss. Uric acid requires very little water for disposal and is often excreted as a paste or dry powder. Guano, the solid white droppings of seabirds and bats, is an example of uric acid excretion.

Type of Nitrogenous Waste	Animals	Animal Habitat
Ammonia	Aquatic invertebrates, bony fish, amphibian larvae	Water
Urea	Adult amphibians, mammals, sharks	Land; water (sharks)
Uric acid	Insects, birds, reptiles	Land

How much urea does a human produce each day?

Humans produce approximately 0.75 ounces (21 grams) of urea each day. In addition to urea, human urine contains water and other chemical compounds and ions.

Which types of fish do not drink water?

Freshwater fish never drink water separate from ingesting food. These fish are prone to gain water since their body fluids are hypotonic (containing a lesser concentration of salts) to the surrounding water. They imbibe water through their gills to maintain the correct balance of salts in their bodies and excrete large quantities of diluted urine daily. It is estimated that freshwater fish eliminate a quantity of urine equal to one-third of their body weight each day.

Cartilaginous fish (e.g., sharks and rays) also do not need to drink water to maintain the balance of water (osmotic balance) in their bodies. They reabsorb the waste product urea, creating and maintaining a blood urea concentration that is one hundred times higher than that of mammals. Their kidneys and gills thus do not have to remove large quantities of salts from their bodies.

Which animals drink seawater?

Marine bony fish (e.g., tuna, flounder, and halibut) drink seawater almost constantly to replace water lost by osmosis and through their gills. It is estimated that they drink an amount equal to 1 percent of their body weight each hour, an amount comparable to a human drinking 1.5 pints or nearly 3 cups (700 milliliters) of water every hour around the clock. The gills eliminate most of the excess salts obtained by drinking large quantities of seawater. These fish excrete small quantities of urine that is isotonic to their body fluids.

Marine birds also drink seawater. They have salt glands near their eyes which excrete the excess quantity of salty liquid. The salty liquid dribbles down their beaks.

Marine reptiles, such as crocodiles, turtles, and sea snakes, drink seawater. They also have salt glands near their eyes or noses to eliminate excess salt.

Marine mammals, such as whales, primarily depend on the water content of the food they eat for hydration rather than on drinking seawater. Whales eat mostly shrimp-like krill and fish, which have a high percentage of water in the body tissue. If these animals drank seawater, they would need a mechanism to eliminate the high salt content from their bodies. Some marine mammals, such as certain species of seals and sea lions, drink seawater on occasion. Studies have shown that the urine of these species contains up to two and a half times more salt than seawater does.

IMMUNE AND LYMPHATIC SYSTEMS

How does the immune system work?

The immune system fights invading organisms that have penetrated the body's general defenses. There are two main components of the immune system: white blood cells and antibodies circulating in the blood. The antigen-antibody reaction forms the basis for this immunity. When an antigen (antibody generator)—a harmful bacterium, virus, fungus, parasite, or other foreign substance—invades the body, a specific antibody is generated to attack the antigen. The antibody is produced by B lymphocytes (B cells) in the spleen or lymph nodes. An antibody may either destroy the antigen directly or it may "label" it so that a white blood cell (called a macrophage, or scavenger cell) can engulf the foreign intruder. After exposure to an antigen, a later exposure to the same antigen will produce a faster immune system reaction. The necessary antibodies will be produced more rapidly and in larger amounts.

What are the primary functions of the lymphatic system?

The lymphatic system is responsible for maintaining proper fluid balance in tissues and blood, in addition to its role defending the body against disease-causing agents. The primary functions of the lymphatic system are: 1) to collect the interstitial fluid that consists of excess water and proteins and return it to the blood; 2) to transport lipids and other nutrients that are unable to enter the bloodstream directly; and 3) to protect the body from foreign cells and microorganisms.

What are the three major components of the lymphatic system?

The lymphatic system consists of the lymphatic vessels, lymph, and lymphoid organs. Together these components form a network that collects and drains most of the fluid that seeps from the bloodstream and accumulates in the space between cells.

How do T cells differ from B lymphocytes?

Lymphocytes are one variety of white blood cells and are part of the body"s immune system. T cells, responsible for dealing with most viruses, for handling some bacteria and fungi, and for cancer surveillance, are one of the two main classes of lymphocytes. T lym-

How does immunization provide protection from disease?

Artificial immunization, through vaccination, induces the antigen-antibody reaction to protect an animal from certain diseases. The animal is exposed to a safe dose of antigen to produce effective antibodies as well as a "readiness" for any future attacks of the harmful antigen.

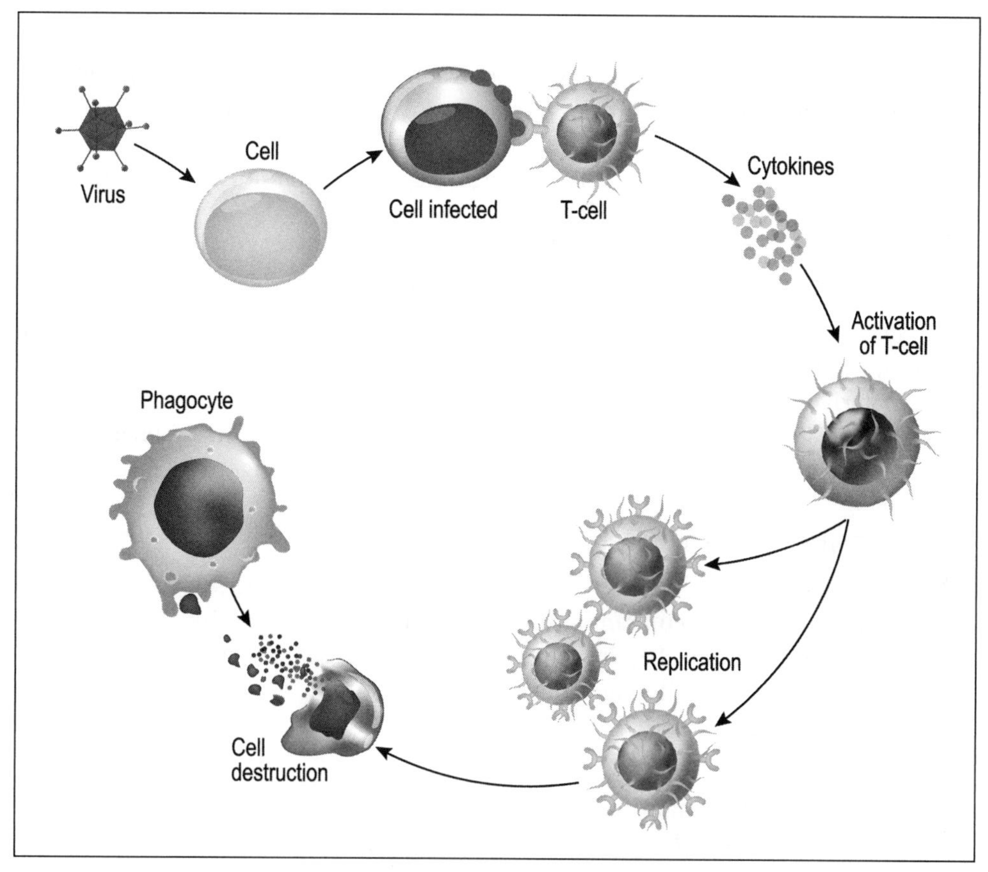

Activation of T cells to destroy invading or dangerous cells, viruses, or bacteria.

phocytes, or T cells, compose about 60 to 80 percent of the lymphocytes circulating in the blood. They have been "programmed" in the thymus to perform particular functions.

Killer T cells are sensitized to multiply when they come into contact with antigens (foreign proteins) on abnormal body cells (cells that have been invaded by viruses, cells in transplanted tissue, or tumor cells). These killer T cells attach themselves to the abnormal cells and release lymphokines (a type of cytokine) to destroy them.

Helper T cells assist killer cells in their activities and control other aspects of the immune response. When B lymphocytes, which compose approximately 10 to 15 percent of total lymphocytes, contact the antigens on abnormal cells, the lymphocytes enlarge and divide to become plasma cells. Then the plasma cells secrete vast numbers of immunoglobulins or antibodies into the blood, which attach themselves to the surfaces of the abnormal cells, to begin a process that will lead to the destruction of the invaders.

NK (natural killer) cells account for the remaining 5 to 10 percent of the circulating lymphocytes. They attack foreign cells, normal cells infected with viruses, and cancer cells that appear in normal tissues.

Which other cells play a role in the immune system?

White blood cells, including macrophages, neutrophils, eosinophils, basophils, and mast cells, all have active roles in the immune system. For example, macrophages, in particular, are important phagocytic cells that destroy many pathogens. Mast cells are specialized cells of connective tissue. They release heparin, histamine, leukotrienes, and prostaglandins to stimulate the inflammatory response.

Why is phagocytosis so important to the human body?

Not only does phagocytosis allow bodies to remove potentially deadly invaders, it is also important in the maintenance of healthy tissues. Without this mechanism, nonfunctional materials would accumulate and interfere with the body's ability to function. A good example of the importance of phagocytosis is provided by the macrophages of the human spleen and liver, which dispose of more than ten billion aged blood cells daily.

What are nonspecific defenses?

Nonspecific defenses do not differentiate between various invaders. Barriers such as skin, hide, the mucous membrane lining the respiratory and digestive tracts, phagocytic white blood cells, and chemicals are nonspecific defenses. The nonspecific defenses are the first to respond to a foreign substance in the body.

What is specific resistance?

Specific resistance, or immunity, is the production of specific types of cells or specific antibodies to destroy a particular antigen.

What is an antigen?

An antigen is a substance that triggers the immune response causing the body to form and produce specific antibodies.

What is an antibody?

An antibody is a protein produced by B cells in response to an antigen. Antibodies are able to neutralize the antigens that provoke their production.

What are disorders of the immune system?

Allergies, autoimmune diseases, and immunodeficiency diseases are different kinds of disorders of the immune system. Allergies are abnormal sensitivities to a substance that is harmless to many other people. Common allergens include pollen, certain foods, cosmetics, medications, fungal spores, and insect venom.

Autoimmune diseases are diseases in which the immune system rejects the body's own molecules. Insulin-dependent diabetes, rheumatoid arthritis, systemic lupus erythematosus, and rheumatic fever are autoimmune diseases.

Do animals suffer from allergies?

Veterinarians report that dogs and cats suffer from allergies. They may be allergic to food, insect bites, dust, household chemicals, or pollen. Instead of having runny noses and watery eyes, animals experience itchy skin conditions, difficulty in breathing, or disruptions in the digestive tract.

The immune system fails in immunodeficiency diseases and disorders. It either fails to develop normally or the immune response is blocked.

How many classes of antibodies have been identified?

There are five classes of antibodies, known as immunoglobulins (Igs). The following lists the known classes:

Class	Description
IgG	Accounts for 80% of all antibodies in the blood; found in blood, lymph, and the intestines; the only antibody that crosses the placenta from mother to fetus; provides resistance against many viruses, bacteria, and bacterial toxins
IgA	Accounts for 10% to 15% of all antibodies in the blood; found mostly in secretions such as sweat, tears, saliva, and mucus; attacks pathogens before they enter internal tissues; levels reduce under stress-lowering resistance
IgM	Accounts for 5% to 10% of all antibodies in the blood; found in blood and lymph; first antibody secreted after exposure to an antigen; includes the anti-A and anti-B antibodies of ABO blood, which bind to A and B antigens during incompatible blood transfusions
IgD	Accounts for 0.2% of all antibodies in the blood; found in blood, lymph, and the surfaces of B cells; plays a role in the activation of B cells
IgE	Accounts for less than 0.1% of all antibodies in the blood; found on the surfaces of mast cells and basophils; stimulates cells to release histamine and other chemicals that accelerate inflammation; plays a role in allergic reactions

INTEGUMENTARY SYSTEM

What is the main function of the integumentary system?

The integumentary (from the Latin word *integere*, meaning "to cover") system provides a protective barrier between the internal organs and the changing environment external to the body. Skin, hair, glands, nails, feathers, and scales are organs of the integumentary system.

What are the various layers of human skin?

Skin is a tissue membrane that consists of layers of epithelial and connective tissues. The outer layer of the skin's epithelial tissue is the epidermis, and the inner layer of connective tissue is the dermis. The epidermis layer is replaced continually as new cells, produced in the stratum basale, mature and are pushed to the surface by the newer cells beneath; the entire epidermis is replaced in about twenty-seven days. The dermis, the lower layer, contains nerve endings, sweat glands, hair follicles, and blood vessels. The upper portion of the dermis has small, fingerlike projections called "papillae," which extend into the upper layer. The patterns of ridges and grooves visible on the skin of the soles, palms, and fingertips are formed from the tops of the dermal papillae.

Do animals other than humans have fingerprints?

It is known that gorillas and other primates have fingerprints. Of special interest, however, is that our closest relative, the chimpanzee, does not. Koala bears also have fingerprints. Researchers in Australia have determined that the fingerprints of koala bears closely resemble those of human fingerprints in size, shape, and pattern.

How fast do fingernails grow?

Healthy nails grow about 0.12 inches (3 millimeters) each month or 1.4 inches (3.5 centimeters) each year. It takes approximately three months for a whole fingernail to be re-

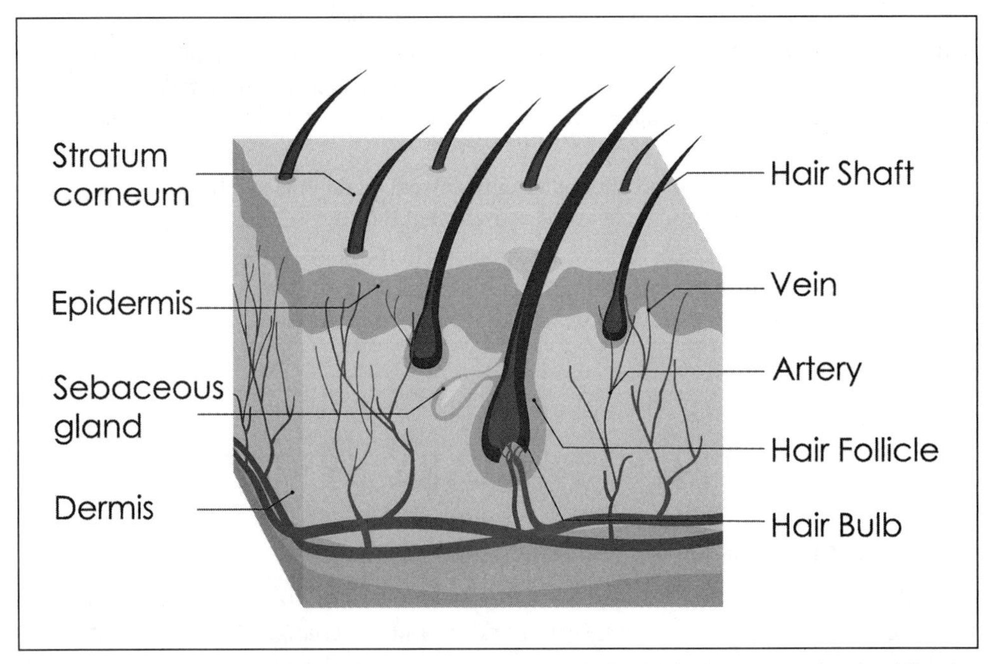

Hair begins under the skin at the hair follicle, which must be supplied with blood to maintain health, while sebaceous glands keep the hair from drying out.

placed. The middle fingernail grows the fastest because the longer the finger, the faster its nail growth. The thumbnail grows the slowest. Toenails grow more slowly than fingernails.

How many hairs are on the human body?

The average human body has approximately five million hairs. The amount of hair on the head varies from one individual to another. An average person has about one hundred thousand hairs on their scalp.

What are the functions of bird feathers?

Bird feathers have many purposes in addition to being essential for flight. Feathers provide insulation, waterproofing, and color for display and camouflage. The fuzzy down feathers of newly hatched water chicks keep them warm when they swim in cold water. Birds can arrange their feathers to trap air to help retain their body heat. On cold days, when birds "fluff their feathers," they are trapping extra air to keep warm. Another way birds control their temperature is by tucking their heads and feet into their feathers to stay warm or exposing them to keep cool. The interlocking pattern of feathers, with waterproof tips, allow birds to stay dry. Finally, the color of bird feathers is important for display in attracting a mate or camouflage.

What accounts for the different colors of bird feathers?

The vivid color of feathers is of two kinds: pigmentary and structural. Red, orange, and yellow feathers are colored by pigments called lipochromes deposited in the feather barbules as they are formed. Black, brown, and gray colors are from another pigment, melanin. Blue feathers depend not on pigment but on the scattering of shorter wavelengths of light by particles within the feather. These are structural feathers. Green colors are almost always a combination of yellow pigment and blue feather structure. Another kind of structural color is the beautiful iridescent color of many birds, which ranges from red, orange, copper, and gold to green, blue, and violet. Iridescent color is based on interference that causes light waves to reinforce, weaken, or eliminate each other. Iridescent colors may change with the angle of view.

Do all mammals have hair or fur?

Yes, hair or fur is a distinguishing feature of all mammals. Hair, composed of keratinized filament, provides protection and insulation. Land mammals that live in cold environments, such as polar bears, rely on a thick layer of fur (the collective term for hair) to trap a layer of air close to their skin to stay warm. Sea otters, which spend much of the time in the water, may have as much as one million hairs per square inch of skin to help maintain their body temperature. In warmer climates, large mammals are able to more easily maintain their body temperature and have fewer and shorter hairs.

How does amphibian skin differ from reptilian skin?

Amphibians have moist, glandular skins and their toes lack claws, while reptiles are clad in scales, shields, or plates and their toes have claws. Amphibians must live in moist environments in order to avoid drying out. Reptiles have dry skin that prevents water loss. The scales develop as surface cells fill with keratin—the same protein that forms human fingernails and bird feathers.

MUSCULAR SYSTEM

Who discovered how muscles work?

Hugh Huxley (1924–2013) and Andrew Huxley (1917–2012) (the scientists were unrelated) researched theories regarding muscle contraction. Hugh Huxley was initially a nuclear physicist who entered the field of biology at the end of World War II. He used both X-ray diffraction and electron microscopy to study muscle contraction. Andrew Huxley was a muscle biochemist who obtained data similar to Hugh's, indicating that the contractile proteins thought to be present in muscles are not contractile at all but rather slide past each other to shorten a muscle. This theory is called the sliding filament theory of muscle contraction.

How do muscle cells work?

Muscle cells—whether the skeletal muscles in the arms or legs, the smooth muscles that line the digestive tract and other organs, or the cardiac muscle cells in the heart—work by contracting. Skeletal muscle cells are comprised of thousands of contracting units known as sarcomeres. The proteins actin (thin filament) and myosin (thick filament) are the major components of the sarcomere. These units perform work by moving structures closer together through space. Sarcomeres in the skeletal muscles pull parts of the body through space relative to each other (e.g., walking or swinging your arms). To visualize how a sarcomere works:

- Interlace the fingers of your two hands with your palms facing toward you (represents actin, myosin); fingertips touching
- Push your fingers together so that the overall length from one thumb to the other is decreased (sarcomere length decreases); allow fingers to slide past each other without bending
- Any object attached to either thumb would be pulled through space as your fingers move together (sliding filament theory)

What are the four major characteristics of skeletal muscle?

The four major functional characteristics of skeletal muscle are:

1. Contractility, the ability to shorten, which causes movement of the structures to which the muscles are attached

2. Excitability, the ability to respond or contract in response to chemical and/or electrical signals
3. Extensibility, the capacity to stretch to the normal resting length after contracting
4. Elasticity, the ability to return to the original resting length after a muscle has been stretched

What is the difference between the origin and the insertion of a muscle?

The skeleton is a complex set of levers that can be pulled in many different directions by contracting or relaxing skeletal muscles. Most muscles extend from one bone to another and cross at least one joint. One end of a skeletal muscle, the origin, attaches to a bone that remains relatively stationary when the muscle contracts. The other end of the muscle, the insertion, attaches to another bone, which will undergo the greatest movement when the muscle contracts. When a muscle contracts, its insertion is pulled toward its origin. The origin is generally closer to the midline of the body, and the insertion is further away.

Do all muscle cells work the same way?

Although all muscles work by contracting, not all muscle types have sarcomeres, the muscle contraction units. Cardiac muscle cells have sarcomeres but use different support structures during contraction than those found in skeletal muscles. Smooth muscle cells do not use sarcomeres at all.

What sources do muscle cells use for energy?

Muscle cells use a variety of energy sources to power their contractions. For quick energy, the cells utilize their stores of ATP and creatine phosphate, which is another phosphate-containing compound. These stored molecules are usually depleted within the first twenty seconds of activity. The cells then switch to other sources, most notably glycogen, a carbohydrate that is made of glucose molecules strung together in long-branching chains.

Are smooth muscle contractions the same as skeletal muscle contractions?

Smooth and skeletal muscle contractions have both similarities and differences. Both types of muscles include reactions involving actin and myosin, both are triggered by membrane impulses and an increase in intracellular calcium ions, and both use energy from ATP. One difference between smooth and skeletal muscle contractions is that smooth muscle is slower to contract and to relax than skeletal muscle. Smooth muscle can maintain a forceful contraction longer with a set amount of ATP. In addition, smooth muscle fibers can change length without changing tautness (as when the stomach is full), while this does not occur in skeletal muscles.

Which are the largest and smallest muscles in the human body?

The largest muscle is the gluteus maximus (buttock muscle), which moves the thigh bone away from the body and straightens out the hip joint. It is also one of the stronger

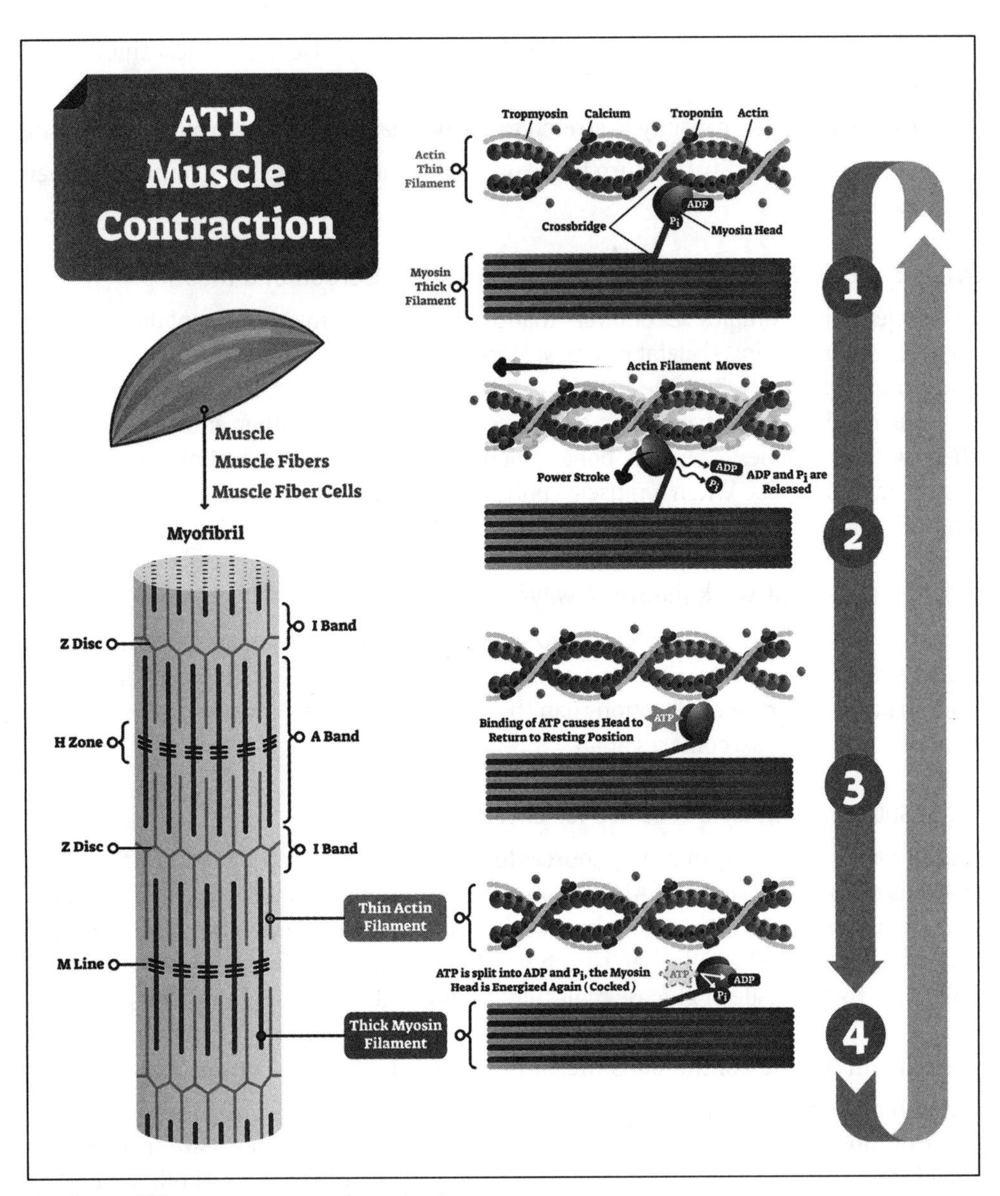

Muscles use ATP as an energy source for contraction.

muscles in the body. The smallest muscle is the stapedius in the middle ear. It is thinner than a thread and 0.05 inches (0.127 centimeters) in length. It activates the stirrup that sends vibrations from the eardrum to the inner ear.

What is the longest muscle in the human body?

The longest muscle is the sartorius, which runs from the waist to the knee. Its purpose is to flex the hip and knee.

Do humans have dark and white muscles similar to those of a chicken?

A chicken has white wing meat and dark leg meat, and humans are much the same in having dark leg muscles and white arm muscles. These differences in color are due to their uses and what demands are placed on the limbs. Dark muscle is specialized for endurance, and its color comes from a rich blood supply and high myoglobin content. Endurance in dark muscle is at the expense of speed. Your legs can carry you all day, but they can't move with the speed of a magician's hand. White muscle specializes in very fast contractions and movements, such as wildly clapping hands or swinging a tennis racquet. White muscle tires quickly because it is less supplied with blood.

NERVOUS SYSTEM

What is the nervous system?

The nervous system is an intricately organized, interconnected system of nerve cells that relays messages to and from the brain and spinal cord of an organism in vertebrates. It receives sensory input, processes the input, and then sends messages to the tissues and organs for an appropriate response. In vertebrates, the nervous system has two parts: 1) the central nervous system, consisting of the brain and spinal cord, and 2) the peripheral system, consisting of nerves that carry signals to and from the central nervous system.

How does the nervous system of invertebrates differ from that of vertebrates?

The least complex nervous system is the nerve net of cnidarians, such as hydras. The nerve net is a network of neurons located throughout the radially symmetric body. The neurons are in contact with one another and with muscle fibers within epidermal cells. These animals lack a head and brain. Invertebrates that display bilateral symmetry, such as planarians, annelids, and arthropods, have a brain (a concentration of neurons at the anterior or head end) and one or more nerve cords and the presence of a central nervous system. Vertebrates have a central nervous system and a peripheral nervous system.

How is the peripheral nervous system organized in vertebrates?

The peripheral nervous system has two divisions: the sensory division and the motor division. The sensory division has two sets of neurons. One set (from the eyes, ears, and other external sense organs) brings in information about the outside environment, while the other set supplies the central nervous system with information about the body itself, such as the acidity of the blood. The motor division includes the somatic nervous system and the autonomic nervous system. The somatic nervous system carries signals to skeletal muscles and skin, mostly in response to external stimuli. It controls voluntary actions. The neurons of the autonomic nervous system are involuntary. This latter system is further divided into the sympathetic and parasympathetic divisions. The sympathetic division prepares the body for intense activities. It is responsible for the "fight or flight" response. The parasympathetic division, or "housekeeper system," is involved in all responses associated with a relaxed state, such as digestion.

How is the vertebrate brain organized?

The vertebrate brain is divided into three regions: the hindbrain, the midbrain, and the forebrain. The size of each region of the brain varies from species to species. The hindbrain may be considered an extension of the spinal cord. Hence, it is often described as the most primitive portion of the brain. The primary function of the hindbrain is to coordinate motor reflexes. The midbrain is responsible for processing visual information. The forebrain is the center for processing sensory information in fish, amphibians, reptiles, birds, and mammals.

How is sensory information transmitted to the central nervous system?

Sensory information is transmitted to the central nervous system through a process that includes stimulation, transduction, and transmission. A physical stimulus (e.g., light or sound pressure) is converted into nerve cell electrical activity in a process called transduction. The electrical activity is then transmitted as action potentials to the central nervous system.

What are the main types of receptors?

Receptor cells are cells that receive stimuli. Each type of receptor responds to a particular stimulus. The five main types of receptors are pain receptors, thermoreceptors, mechanoreceptors, chemoreceptors, and electromagnetic receptors.

- *Pain receptors* are probably found in all animals. However, it is difficult to understand nonhuman perception of pain. Pain often indicates danger, and the animal or individual retreats to safety.
- *Thermoreceptors* in the skin are sensitive to changes in temperature. Thermoreceptors in the brain monitor the temperature of the blood to maintain proper body temperature.
- *Mechanoreceptors* are sensitive to touch and pressure, sound waves, and gravity. The sense of hearing relies on mechanoreceptors.
- *Chemoreceptors* are responsible for taste and smell.
- *Electromagnetic receptors* are sensitive to energy of various wavelengths including electricity, magnetism, and light. The most common types of electromagnetic receptors are photoreceptors that detect light and control vision.

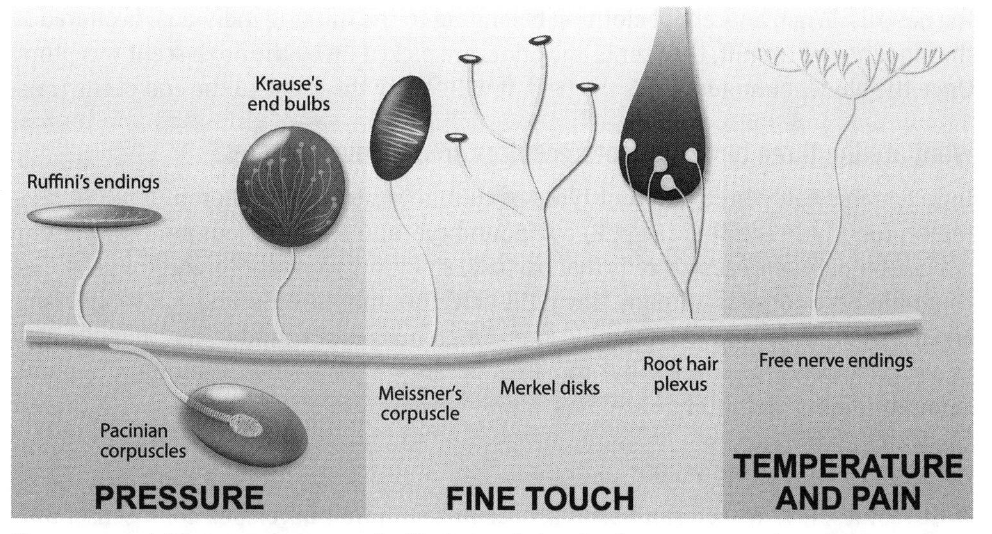

There are several different nerve receptors in skin that are designed to detect pressure, pain, and temperature.

How do animals and people identify smells?

The sense of smell allows animals and humans as well as other organisms to identify food, mates, and predators. This sense also provides sensory pleasure (e.g., flowers) and warnings of danger (e.g., chemical dangers). Specialized receptor cells in the nose have proteins that bind chemical odorants and cause the receptor cells to send electrical signals to the olfactory bulb of the brain. Cells in the olfactory bulb relay this information to olfactory areas of the forebrain to generate perception of smells.

Which insect has the best sense of smell?

Giant male silk moths (*Bombyx mori*) may have the best sense of smell in the world. Their antennae are covered with about 65,000 tiny bristles. Most of the bristles are chemoreceptors. The moths can smell a female's perfume nearly 7 miles (11 kilometers) away.

Which breed of dog may be trained to work as "trackers" for search and rescue operations?

Bloodhounds are very adept at working as "trackers" for search and rescue operations. In fact, the results of using a bloodhound for tracking are so accurate that the results are accepted as evidence in court. Bloodhounds have a large nose with an ultrasensitive set of scent membranes that gives them the ability to distinguish smells far greater than humans can do. It is estimated that they have approximately 230 million olfactory cells (also called "scent receptors"). This is forty times greater than the number in a human's nose. As an additional asset, a bloodhound's long, drooping ears drag on the ground as they hold their nose close to the ground, collecting odors and sweeping the odors toward the nostrils. When an item of clothing belonging to the missing individual is offered to the bloodhound to sniff, the scents and odors are picked up by the dog's scent receptors. Once the bloodhound identifies the trail, it will follow the scent to the end of the trail.

What are the three types of photoreceptors among invertebrates?

Invertebrates have three different types of eyes, represented by different types of photoreceptors. They are 1) eye cup; 2) compound eye; and 3) single-lens eye. The eye cup is a cluster of photoreceptor cells that partially shield adjacent photoreceptor cells. The compound eye consists of many tiny light detectors (photoreceptors). Crayfish, crabs, and nearly all insects have compound eyes. Single-lens eyes, found in cephalopods such as squids and octopi, are similar to cameras. They have a small opening, the pupil, through which light enters.

Do animals have color vision?

Most reptiles, fish, insects, and birds appear to have a well-developed color sense. Butterflies have highly developed sensory capabilities. They have the widest visual spectrum of any animal and are able to see from the red end of the spectrum all the way to the near-ultraviolet end. They are able to distinguish colors that humans are unable to distinguish. Most mammals are color-blind. Apes and monkeys have the ability to tell colors apart. Dogs and cats can distinguish shades of black, white, and gray. Dogs may be able to also distinguish various shades of blue, while cats may be able to distinguish various shades of greens and blues.

What is the difference in the functions of the rods and cones found in the eyes?

Rods and cones contain photoreceptors that convert light first to chemical energy and then into electrical energy for transmission to the vision centers of the brain via the optic nerve. Rods are specialized for vision in dim light; they cannot detect color, but they are the first receptors to detect movement and register shapes. Each human eye has about 125 million rods. Cones provide acute vision, functioning best in bright daylight. They allow us to see colors and fine detail. Cones are divided into three different types, which absorb wavelengths in the short (blue), middle (green), and long (red) ranges. Each eye has about seven million cones.

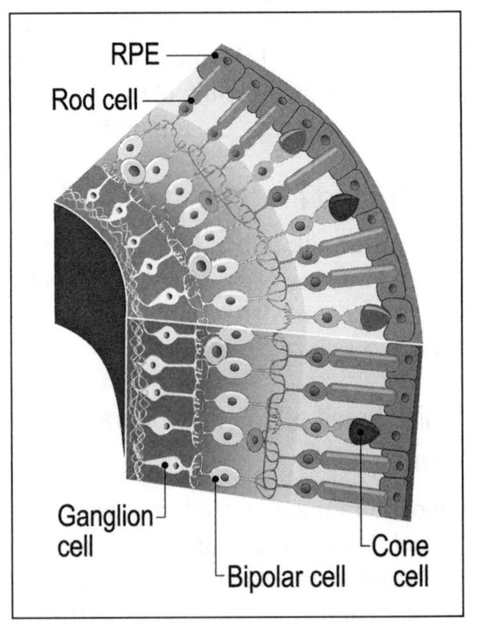

Rod cells in the eye are adept at detecting things in dim lighting, but they don't detect color; cone cells detect color well, but they don't work well at night. Animals that are active at night logically have more rod than cone cells.

How long does it take for a person to adapt to dim light?

Rods give us vision in dim light but not in color and not with sharp detail. They are hundreds of times more sensitive to light than cones, letting us detect shape and movement in dim light. This type of photoreceptor takes about 15 minutes to fully adapt to very dim light.

Do nocturnal animals have the same eyes and vision as diurnal animals?

Diurnal animals are active during the day, while nocturnal animals are active in the darkness of night. Nocturnal vertebrate animals have the same basic eye structure as diurnal animals, including a pupil that controls how much light enters the eye, a lens to focus the image onto the retina, and nerves that relay the information to the brain. However, the eyes of nocturnal animals are very large. When wide open, the pupils of owls and cats will appear to cover the entire front of the eye. Their eyes are often tubular rather than spherical, with a lens positioned close to the retina to allow as much light as possible to register on the retina. In addition, many nocturnal animals have a mirrorlike layer, called the tapetum, behind the retina. Light passes through the retina and is then reflected off the tapetum, giving the retinal cells another chance to absorb the rays of light. The retinas of nocturnal animals have many rods but few cones.

Why do cats' eyes shine in the dark?

A cat's eyes contain a special, light-conserving mechanism called the *tapetum lucidum*, which reflects any light not absorbed as it passes through the retina of each eye. The retina gets a second chance (so to speak) to receive the light, aiding the cat's vision even more. In dim light, when the pupils of the cat's eyes are opened the widest, this glowing or shining effect occurs when light hits them at certain angles. The *tapetum lucidum*, located behind the retina, is a membrane composed of fifteen layers of special, glittering cells that all together act as a mirror. The color of the glow is usually greenish or golden, but the eyes of the Siamese cat reflect a luminous ruby red.

Can animals hear different sound frequencies than humans?

The frequency of a sound is the pitch. Frequency is expressed in Hertz (Hz). Sounds are classified as infrasounds (below the human range of hearing), sonic range (within the range of human hearing), and ultrasound (above the range of human hearing).

Do owls move their eyes?

Owls do not have eye movements like humans. Their eyes are held in place by bones. In order to compensate for fixed eyeballs, they have very flexible necks that allow them to turn their heads about 270 degrees in either direction. They are also capable of moving their heads up and down about 90 degrees without moving their shoulders.

Animal	Frequency Range Heard (Hz)
Dog	15–50,000
Human	20–20,000
Cat	60–65,000
Dolphin	150–150,000
Bat	1,000–120,000

How sensitive is the hearing of birds?

In most species of birds, the most important sense after sight is hearing. Birds' ears are close to their bodies and covered with feathers. However, the feathers covering the ears do not have barbules, which would obstruct sound. Ears of different heights allow the bird to locate a sound. Nocturnal raptors, such as the great horned owl, have a very well-developed sense of hearing in order to be able to capture their prey in total darkness.

REPRODUCTIVE SYSTEM

What is the function of the reproductive system?

The function of the reproductive system is to produce new offspring. The reproductive system is essential for the survival of a species.

How does asexual reproduction differ from sexual reproduction?

Asexual reproduction produces offspring with the exact genetic material of the parent. Only one individual is needed to produce offspring via asexual reproduction. Sexual reproduction produces offspring by the fusion of two gametes (haploid cells) to form one zygote (diploid cell). The male gamete is the sperm, and the female gamete is the egg.

Some lizards can shed their tails at will to distract or run away from a predator; the tail can then regenerate. Neat trick!

What are some methods of asexual reproduction?

Budding, fission, and fragmentation are methods of asexual reproduction.

- In budding, a new individual begins as an outgrowth, or bud, of the parent. Eventually, the bud detaches from the parent and develops into a new individual. Budding is common among sponges and coelenterates such as hydras and anemones.

- Fission is the division of one individual into two or more individuals of almost equal size. Each new individual develops into a mature adult. Some corals reproduce by dividing longitudinally into two smaller but complete individuals.
- Fragmentation is the breaking of the parent into several pieces. It is accompanied by regeneration, when each piece develops into a mature individual. Sea stars are well known for reproducing by fragmentation and regeneration.

Can animals regenerate parts of their bodies?

Regeneration frequently occurs among primitive invertebrates. However, the ability for an animal to regenerate a part of its body lessens as the animal species becomes more complex. For example, a planarian (flatworm) can split symmetrically, with the two sides turning into clones of one other. In higher invertebrates, regeneration occurs in echinoderms, such as starfish, and arthropods, such as insects and crustaceans. Starfish are known for their ability to develop into complete individuals from one cut-off arm. Regeneration of appendages (limbs, wings, and antennae) occurs in insects such as cockroaches, fruit flies, and locusts and in crustaceans such as lobsters, crabs, and crayfish. For example, regeneration of a crayfish's missing claw occurs at its next molt (shedding of its hard cuticle exterior shell/skin in order to grow and the subsequent hardening of a new cuticle exterior). Sometimes, the regenerated claw does not achieve the same size as the missing claw. However, after every molt (a process that occurs two to three times a year), the regenerated claw grows and will eventually become nearly as large as the original claw.

Do any vertebrate animals have the ability to regenerate parts of their bodies?

Some amphibians and reptiles have the ability to replace a lost leg or tail. Salamanders, newts, frogs, and toads are able to regenerate limbs. Some species regenerate fully formed limbs, while others regenerate cartilage spikes rather than full limbs. The ability of lizards to self-amputate their tails is known as autotomy. It is often used as a way to escape from a predator. The vertebrae in lizard tails have fracture planes. Fracture planes are preformed breaks in the bone, along which the tails separate easily. When a tail is lost by autotomy, the tail tissues separate along fracture planes, minimizing bone and muscle tissue damage. When a new tail is grown, it is usually cartilaginous tissue with a spinal cord. It may have differences in coloring and markings than the original tail.

How does external fertilization differ from internal fertilization?

External fertilization is common among aquatic animals, including fish, amphibians, and aquatic invertebrates. Following an elaborate ritual of mating behavior to synchronize the release of eggs and sperm, both males and females deposit their gametes in the water at approximately the same time in close proximity to each other. The water protects the sperm and eggs from drying out. Fertilization occurs when the sperm reach the eggs. Internal fertilization requires that sperm be deposited in or close to the female reproductive tract. It is most common among terrestrial animals that either lay a shelled egg, such as reptiles and birds, or when the embryo develops for a period of time within the female body.

Which aquatic animals reproduce through internal fertilization?

Certain sharks, skates, and rays reproduce through internal fertilization. The pelvic fins are specialized to pass sperm to the female. In most of these species, the embryos develop internally and are born alive.

Which animal has the longest gestation period?

The animal with the longest gestation period is the alpine black salamander. This amphibian can have a gestation period of up to thirty-eight months at altitudes above 4,600 feet (1,402 meters) in the Swiss Alps. It bears two fully metamorphosed young salamanders following the long gestation period.

How many eggs does a spider lay?

The number of eggs varies according to the species. Some larger spiders lay more than two thousand eggs, but many tiny spiders lay one or two and perhaps no more than a dozen during their lifetime. Spiders of average size probably lay a hundred or so. Most spiders lay all their eggs at one time and enclose them in a single egg sac; others lay eggs over a period of time and enclose them in a number of egg sacs.

How many eggs are produced by sea urchins?

The number of eggs produced by sea urchins is enormous. It has been estimated that a female of the genus *Arbacia* contains about eight million eggs. In the much-larger genus *Echinus*, the number reaches twenty million.

How can you tell male and female lobsters apart?

The differences between male and female lobsters can only be seen when they are turned on their backs. In the male lobster, the two swimmerets (forked appendages used for swimming) nearest the carapace (the solid shell) are hard, sharp, and bony; in the female, the same swimmerets are soft and feathery. The female also has a receptacle that appears as a shield wedged between the third pair of walking legs. During mating, the male deposits sperm into this receptacle, where it remains for as long as several months until the female uses it to fertilize her eggs as they are laid.

What is a mermaid's purse?

Mermaid's purses are the protective cases in which the eggs of dogfish, skates, and rays are released into the environment. The rectangular purse is leathery and has long tendrils streaming from each corner. The tendrils anchor the case to seaweed or rocks, where the case is protected during the six to nine months it takes for the embryos to hatch. Empty cases often wash up on beaches.

How is the gender of alligator embryos determined?

The gender of an alligator is determined by the temperature at which the eggs are incubated. High temperatures of 90–93°F (32–34°C) result in males; low temperatures of 82–86°F (28–30°C) yield females. This determination takes place during the second or third week of the two-month incubation. Further temperature fluctuations before or after this time do not alter the gender of the young. The heat from the decaying matter on top of the nest incubates the eggs.

Which birds lay the largest and smallest eggs?

Generally speaking, the larger the bird, the larger the egg. The largest egg produced by any living bird is that of the North African ostrich (*Struthio camelus*). The average size of the egg is 6–8 inches (15–20.5 centimeters) in length and 4–6 inches (5–15 centimeters) in diameter. The smallest mature egg, measuring less than 0.39 inches (1 centimeter) in length, is that of the vervain hummingbird (*Mellisuga minima*) of Jamaica.

What is unusual about the way the emperor penguin's eggs are incubated?

Each female emperor penguin (*Aptenodytes forsteri*) lays one large egg. Initially, both sexes share in incubating the egg by carrying it on his or her feet and covering it with a fold of skin. After a few days of passing the egg back and forth, the female leaves to feed in the open water of the Arctic Ocean. Balancing their eggs on their feet, the male penguins shuffle about the rookery, periodically huddling together for warmth during blizzards and frigid weather. If an egg is inadvertently orphaned, a male with no egg will quickly adopt it. Two months after the female's departure, the chick hatches. The male feeds it with a milky substance he regurgitates until the female returns. Now padded with blubber, the females take over feeding the chicks with fish they have stored in their crops. The females do not return to their mate, however, but wander from male to male until one allows her to take his chick. It is then the male's turn to feed in open water and restore the fat layer they lost while incubating.

Which mammals have the shortest and longest gestation periods?

Gestation is the period of time between fertilization and birth. The shortest gestation period known among mammals is twelve to thirteen days, shared by three marsupials: the American or Virginian opossum (*Didelphis marsupialis*); the rare water opossum or yapok (*Chironectes minimus*) of central and northern South America; and the eastern native cat (*Dasyurus viverrinus*) of Australia. The young of each of these marsupials are born while still immature and complete their development in the pouch of their mother. While twelve to thirteen days is the average, the gestation period is sometimes as short as eight days. The longest gestation period for a mammal is that of the African elephant (*Loxodonta africana*), with an average of 660 days and a maximum of 760 days.

How does reproduction in marsupials differ from reproduction in other mammals?

Marsupials (meaning "pouched" animals) differ from all other living mammals in their anatomical and physiological features of reproduction. Most female marsupials, including kangaroos, bandicoots, wombats, banded anteaters, koalas, opossums, wallabies, and Tasmanian devils, possess an abdominal pouch (called a marsupium), in which their young are carried. In some small, terrestrial marsupials, however, the marsupium is not a true pouch but merely a fold of skin around the mammae (milk nipples).

Marsupials such as this wallaby give birth to very underdeveloped young, which then continue to mature within the mother's pouch.

The short gestation period in marsupials (in comparison to other similarly sized mammals) allows their young to be born in an "undeveloped" state. Consequently, these animals have been viewed as "primitive" or second-class mammals. However, some now see that the reproductive process of marsupials has an advantage over that of placental mammals. A female marsupial invests relatively few resources during the brief gestation period, more so during the lactation (nursing period), when the young are in the marsupium. If the female marsupial loses its young, it can conceive again sooner than a placental mammal in a comparable situation.

What are the organs of the male and female reproductive systems in humans?

The male reproductive organs and structures are the testes, a duct system that includes the epididymis and the vas deferens; the accessory glands, including the seminal vesicles and prostate gland; and the penis. The testes are the male gonads. They produce the male reproductive cells called sperm.

The female reproductive system organs include the ovaries, the uterine tubes, the uterus, the vagina, the external organs called the vulva, and the mammary glands. The paired ovaries are the female gonads. They produce the female gametes, called ova, and secrete the female sex hormones.

At what age does sperm production begin in males?

Sperm production begins with the onset of puberty, usually between ages eleven to fourteen in boys. It continues throughout the life of an adult male. It is estimated that during his lifetime, a normal male will produce as many as three hundred million sperm per day.

What is the female reproductive cycle?

The female reproductive cycle is a general term to describe both the ovarian cycle and the uterine cycle as well as the hormonal cycles that regulate them. The ovarian cycle is the monthly series of events that occur in the ovaries related to the maturation of an oocyte. The menstrual cycle is the monthly series of changes that occur in the uterus as it awaits a fertilized ovum.

RESPIRATORY SYSTEM

What is respiration?

Respiration is the exchange of gases (oxygen and carbon dioxide) between an animal and its environment. The process of respiration (gas exchange) has three phases: 1) breathing, when an animal inhales oxygen and exhales carbon dioxide; 2) transport of gases via the blood (circulatory system) to the body's tissues; and 3) at the cellular level, when the cells take in oxygen from the blood and in return add carbon dioxide to the blood.

Where does respiration take place?

Different types of animals have different respiratory organs for gas exchange. Four types of respiratory organs are 1) skin; 2) gills; 3) tracheae; and 4) lungs. Many invertebrates and some vertebrate animals, including amphibians, breathe through their skin. Many of the animals that breathe through their skin (a process known as cutaneous respiration) are small, long, and flattened—for example, earthworms and flatworms. All animals that rely on their skin for respiration live in moist, damp places in order to keep their body surfaces moist. Capillaries, small blood vessels, bring blood rich in carbon dioxide and deficient in oxygen to the skin's surface, where gaseous exchange takes place via diffusion.

Gills may be external extensions of the body surface, such as those found in aquatic insect larvae and some aquatic amphibians. Diffusion of oxygen occurs across the gill surface into capillaries, while carbon dioxide diffuses out of the capillaries into the environment. Fish and some other marine animals have internal gills. Water enters the animals through the mouth, then flows over the gills in a steady stream and out through gill slits. Although some animals with gills spend part of the time on land, they all must spend some time in moist, wet environments for the gills to function.

Where does respiration take place in insects?

Insects have a system of internal tubes, called tracheae, that lead from the outside world to internal regions of the body via spiracles. Gaseous exchange takes place in the tracheae. Some insects rely on muscles to pump the air in and out of the tracheae, while in others, the process is a passive exchange of gases.

Which group of animals have lungs for respiration?

Lungs are internal structures found in most terrestrial animals where gas exchange occurs. The lungs are lined with moist epithelium to avoid their becoming desiccated. Some animals, including lungfish, amphibians, reptiles, birds, and mammals, have special muscles to help move air in and out of the lungs. Some animals have lungs connected to the outside surface with special openings and do not require special muscles to move air in and out of the lungs.

What is the respiratory rate of various animals?

The respiratory rate for various animals is:

Animal	Breaths Per Minute
Diamondback snake	4
Horse	10
Human	12
Dog	18
Pigeon	25–30
Cow	30
Giraffe	32
Shark	40
Trout	77
Mouse	163

How much air does a person breathe in a lifetime?

Each year, an adult human may take between four and ten million breaths. The volume of air in each breath ranges from 500 milliliters for breaths at rest and 3,500 to 4,800 milliliters for each breath during strenuous exercise. During his or her life, the average person will breathe about 75 million gallons (284 million liters) of air. Per minute, the human body needs 2 gallons (7.5 liters) of air when lying down, 4 gallons (15 liters) when sitting, 6 gallons (23 liters) when walking, and 12 gallons (45 liters) or more when running.

How do air-breathing mammals such as whales and seals dive underwater for extended periods of time?

Seals and whales are able to dive underwater for extended periods of time because they are able to store oxygen. While humans store 36 percent of their oxygen in their lungs and 51 percent in their blood, seals store only approximately 5 percent of their oxygen in their lungs and 70 percent in their blood. They also store more oxygen in the muscle tissue—25 percent, compared with only 13 percent in human muscle tissue. While underwater, these

Whales such as these humpbacks can stay submerged in the ocean for many minutes because they store oxygen so efficiently in their blood.

mammals' heart rates and oxygen consumption rates decrease, allowing some species to remain underwater for up to 20 minutes at a time.

How long do marine mammals remain underwater?

Below are the maximum depths and the longest durations of time underwater by various aquatic mammals:

Mammal	Maximum Depth Feet/Meters	Maximum Time Underwater
Porpoise	984/300	6 minutes
Fin whale	1,148/350	20 minutes
Bottlenose whale	1,476/450	120 minutes
Weddell seal	1,968/600	70 minutes
Sperm whale	>6,562/ >2,000	75–90 minutes

What are the organs of the respiratory system in humans?

The respiratory system is divided into the upper respiratory system and the lower respiratory system. The upper respiratory system includes the nose, nasal cavity, and sinuses. The lower respiratory system includes the larynx, trachea, bronchi, bronchioles, and alveoli.

What is the essential relationship between the heart and lungs?

The teamwork of the heart and lungs ensures that the body has a constant supply of oxygen for metabolic activities and that the major waste product of metabolism, carbon dioxide, is continuously removed. This occurs through the pulmonary circulation, with the heart supplying blood that has moved through the body to the lungs. The lungs connect to the heart through blood vessels. The pulmonary artery delivers deoxygenated blood to the lungs from the right ventricle, and the pulmonary vein delivers oxygenated blood to the left atrium of the heart.

How does the body introduce oxygen to the blood, and where does this happen?

Blood entering the right side of the heart (right auricle or atrium) contains carbon dioxide, a waste product of the body. The blood travels to the right ventricle, which pushes it through the pulmonary artery to the lungs. In the lungs, the carbon dioxide is removed, and oxygen is added to the blood. The blood then travels through the pulmonary vein, carrying the fresh oxygen to the left side of the heart, first to the left auricle, where it goes through a one-way valve into the left ventricle, which must push the oxygenated blood to all portions of the body (except the lungs) through a network of arteries and capillaries. The left ventricle must contract with six times the force of the right ventricle, so its muscle wall is twice as thick as the right.

What are the two phases of breathing?

Breathing, or ventilation, is the process of moving air into and out of the lungs. The two phases are 1) inspiration, or inhalation, and 2) expiration, or exhalation. Inspiration is the movement of air into the lungs, while expiration is the movement of air out of the lungs. The respiratory cycle consists of one inspiration followed by one expiration. The volume of air that enters or leaves during a single respiratory cycle is called the tidal volume. Tidal volume is typically 500 milliliters, meaning that 500 milliliters of air enters during inspiration and the same amount leaves during expiration.

SKELETAL SYSTEM

What is the function of the skeletal system?

The skeletal system is a multifunctional system. The skeletal system provides support, allows an animal to move, and protects the internal organs and soft parts of an animal's body.

What are the three main types of skeletal systems?

The three main types of skeletal systems are hydrostatic skeleton, exoskeleton, and endoskeleton. A hydrostatic skeleton consists of fluid under pressure. This type of skeletal system is most common in soft, flexible animals such as hydras, planarians, and earthworms and other segmented worms. Hydras and planarians have a fluid-filled gastrovascular cavity. The body cavity, or coelom, of an earthworm is also filled with fluid.

Many aquatic and certain terrestrial animals have an exoskeleton. The exoskeleton is rigid and hard. Mollusks have an exoskeleton made of calcium carbonate. It grows with the animal during its entire lifetime. Another type of exoskeleton common among insects and arthropods is made from chitin. Chitin is a strong, flexible nitrogenous polysaccharide. While it provides excellent protection and allows for a large variety of movements, it does not grow with the animal. When an animal outgrows its skeleton, it must shed its skeleton and replace it with a larger one in a process known as molting.

An endoskeleton consists of bone and cartilage and grows with the animal throughout its life. It stores calcium salts and blood cells and consists of hard or leathery supporting elements situated among the soft tissues of an animal. Although most common among vertebrates, certain invertebrates such as sponges, sea stars, sea urchins, and other echinoderms have an endoskeleton of hard plates beneath their skin. This type of skeletal system allows for a wider range of movement than do the other two.

Which group of animals have a skeleton that is entirely made of cartilage?

The endoskeleton of sharks consists entirely of cartilage.

What are the upper and lower shells of a turtle called?

The turtle (order Testudines) uses its shell as a protective device. The upper shell is called the dorsal carapace, and the lower shell is called the ventral plastron. The shell's sections are referred to as the scutes. The carapace and the plastron are joined at the sides.

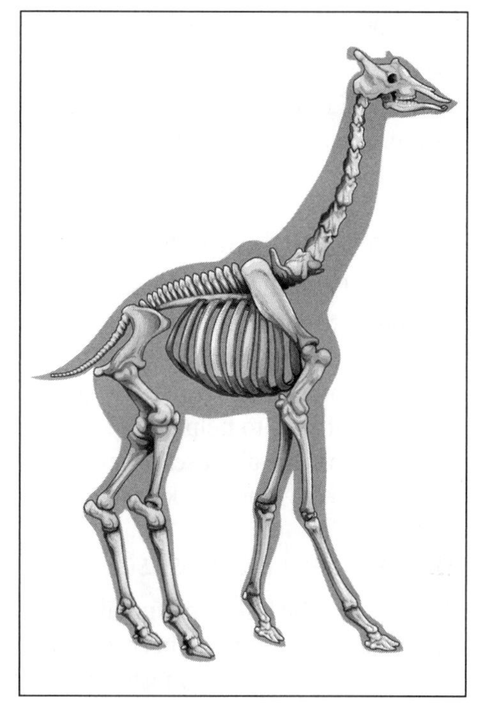

Humans have seven neck bones, so you might be surprised that giraffes *also* have just seven neck bones. They are just much bigger than a human's.

How many vertebrae are in the neck of a giraffe?

A giraffe neck has seven vertebrae, the same as other mammals, but the vertebrae are greatly elongated.

How many bones are in the human body?

Babies are born with about 300 to 350 bones, but many of these fuse together between birth and maturity to produce an average adult total of 206. Bone counts vary according to the method used to count them because a structure may be treated as either multiple bones or as a single bone with multiple parts.

Location	Number
Skull	22
Ears (pair)	6
Vertebrae	26
Sternum	3
Throat	1
Pectoral girdle	4
Arms (pair)	60
Hip bones	2
Legs (pair)	58
Total	206

What is the smallest bone in the human body?

The stapes (stirrup) in the middle ear is the smallest bone in the body. It measures 1.02–1.34 inches (2.6–3.4 centimeters) and weighs 0.00071–0.0015 ounces (0.002–0.004 grams).

Which two cervical vertebrae allow the head to move?

The first two cervical vertebrae, C1 and C2, allow the head to move. The first cervical vertebra, the C1 or atlas, articulates with the occipital bone of the skull and makes it possible for a person to nod his or her head. The second cervical vertebra, C2, known as the axis, forms a pivot point for the atlas to move the skull in a side-to-side rotation.

What are the problems an animal must overcome to move?

In contrast to other organisms, animals are able to move. The two forces an animal overcomes to move are gravity and friction. Aquatic animals do not have much difficulty overcoming gravity since they are buoyant in water. However, because water is dense, the problem of resistance (friction) is greater for these animals. Many of them have sleek shapes to help them swim. Terrestrial animals tend to have fewer problems with friction since air poses fewer problems of resistance than does water. However, terrestrial animals must work harder to overcome gravity.

Which mammal is the fastest?

The cheetah, the fastest mammal, can accelerate from 0 to 45 miles (64 kilometers) per hour in 2 seconds; it has been timed at speeds of 70 miles (112 kilometers) per hour over short distances. In most chases, cheetahs average around 40 miles (63 kilometers) per hour.

Which animals can run faster than a human?

Humans can run very short distances at almost 28 miles (45 kilometers) per hour maximum. Most of the speeds given in the table below are for distances of 0.25 miles (0.4 kilometers).

Animal	Maximum Speed (mph)	Maximum Speed (kph)
Pronghorn antelope	61	98.1
Wildebeest	50	80.5
Lion	50	80.5
Thomson's gazelle	50	80.5
Quarter horse	47.5	76.4
Elk	45	72.4
Cape hunting dog	45	72.4
Coyote	43	69.2
Gray fox	42	67.6
Hyena	40	64.4
Zebra	40	64.4
Mongolian wild ass	40	64.4
Greyhound	39.4	63.3
Whippet	35.5	57.1
Rabbit (domestic)	35	56.3
Mule deer	35	56.3

Animal	Maximum Speed (mph)	Maximum Speed (kph)
Jackal	35	56.3
Reindeer	32	51.3
Giraffe	32	51.3
White-tailed deer	30	48.3
Wart hog	30	48.3
Grizzly bear	30	48.3
Cat (domestic)	30	48.3
Human	27.9	44.9

Which is the fastest-swimming whale?

The orca or killer whale (*Orcinus orca*) is the fastest-swimming whale. In fact, it is the fastest-swimming marine mammal, with speeds that reach 31 miles (50 kilometers) per hour.

At what speeds do fish swim?

The maximum swimming speed of a fish is somewhat determined by the shape of its body and tail and by its internal temperature. The Indo-Pacific sailfish (*Istiophorus platypterus*) is considered to be the fastest fish species, at least for short distances, swimming at greater than 60 miles (95 kilometers) per hour. The speed of other fish is listed below:

Fish	Speed (mph/kph)
Yellowfin tuna (*Thunnus albacares*)	46.35 mph/74.5 kph
Wahoe (*Acanthocybium solandri*)	47.88 mph/77 kph

The sailfish—catching it is a popular challenge for skilled fishermen—is considered the fastest swimmer. It can speed through the water at 60 mph (95 kph).

Fish	Speed (mph/kph)
Flying fish	40+ mph/64+ kph
Trout	15 mph/24 kph
Blenny	5 mph/8 kph

What is the fastest snake on land?

The black mamba (*Dendroaspis polylepis*), a deadly, poisonous African snake that can grow up to 13 feet (4 meters) in length, has been recorded reaching a speed of 7 miles (11 kilometers) per hour. A particularly aggressive snake, it chases animals at high speeds, holding the front of its body above the ground.

How fast can a crocodile run on land?

In smaller crocodiles, the running gait can change into a bounding gallop that can achieve speeds of 2–10 miles (3–17 kilometers) per hour.

How fast do birds fly?

Different species of birds fly at different speeds. The following table lists the flight speeds of some birds:

Bird	Speed (mph)	Speed (kph)
Peregrine falcon	168–217	270.3–349.1
Swift	105.6	169.9
Hummingbird	71	114
Merganser	65	104.6
Golden plover	50–70	80.5–112.6
Mallard	40.6	65.3
Wandering albatross	33.6	54.1
Carrion crow	31.3	50.4
Herring gull	22.3–24.6	35.9–39.6
House sparrow	17.9–31.3	28.8–50.4
Woodcock	5	8

How fast does a hummingbird's wings move?

Hummingbirds are the only family of birds that can truly hover in still air for any length of time. They do so in order to hang in front of a flower while they perform the delicate task of inserting their slim, sharp bills into the flower's depths to drink nectar. Their thin wings are not contoured into the shape of aerofoils and do not generate lift in this way. Their paddle-shaped wings are, in effect, hands that swivel at the shoulder. They beat them in such a way that the tip of each wing follows the line of a figure eight lying on its side. The wing moves forward and downward into the front loop of the eight, creating lift. As it begins to come up and goes back, the wing twists through 180 degrees so that once again, it creates a downward thrust. The hummingbird's method of flying does have a major limitation. The smaller a wing, the faster it has to beat in order to produce sufficient downward thrust. An average-sized hummingbird beats its wings 25 times a second. The bee hummingbird, native to Cuba, is only 2 inches (5 centimeters) long and beats its wings at an astonishing two hundred times a second.

Do any mammals fly?

Bats (order Chiroptera, with 986 species) are the only truly flying mammals, although several gliding mammals are referred to as "flying" (such as the flying squirrel and flying lemur). The "wings" of bats consist of double membranes of skin stretching from the sides of the body to the hind legs and tail and are actually skin extensions of the back and belly. The wing membranes are supported by the elongated fingers of the forelimbs (or arms).

INTRODUCTION AND HISTORICAL BACKGROUND

What is ecology?

Ecology is the scientific study of how organisms interact with their environments. The environment includes both the physical and chemical conditions in which they live, such as the availability of water and temperature changes, as well as the relationships between organisms of the same species or different species. The term "ecology," from the Greek words *oikos*, meaning "house," and *logos*, meaning "logic or knowledge," was first introduced by Ernst Haeckel in 1866.

What is the Gaia hypothesis?

British scientists James Lovelock (1919–) and Lynn Margulis (1938–2011) proposed the Gaia hypothesis (named for Gaia, the Earth goddess of ancient Greece) in the 1970s. According to the theory, all living and nonliving organisms on Earth form a single unity that is self-regulated by the organisms themselves. Therefore, the whole planet can be considered a huge, single organism. Evidence for this theory is the stability of atmospheric conditions over eons. Many scientists regard it as a useful analogy but a difficult theory to test scientifically.

How is Henry David Thoreau associated with the environment?

Henry David Thoreau (1817–1862) was a writer and naturalist from New England. His most familiar work, *Walden*, describes the time he spent in a cabin near Walden Pond in Massachusetts. He is also known for being one of the first to write and lecture on the topic of forest succession. His work, along with that of John Muir (1838–1914) and others, has served to inspire those others to understand the natural world and provide for its conservation.

Who is considered the founder of modern conservation?

American naturalist John Muir is the father of conservation and the founder of the Sierra Club. He fought for the preservation of the Sierra Nevada in California and the creation of Yosemite National Park. He directed most of the Sierra Club's conservation efforts and was a lobbyist for the Antiquities Act, which prohibited the removal or destruction of structures of historic significance from federal lands. Another prominent influence was George Perkins Marsh (1801–1882), a Vermont lawyer and scholar. His book *Man and Nature* emphasized the mistakes of past civilizations that resulted in the destruction of natural resources. As the conservation movement swept through the country in the last three decades of the nineteenth century, a number of prominent citizens joined the efforts to conserve natural resources and to preserve wilderness areas. Writer John Burroughs (1837–1921), forester Gifford Pinchot (1865–1946), botanist Charles Sprague Sargent (1841–1927), and editor Robert Underwood Johnson (1857–1937) were early advocates of conservation.

Who was the first person to predict that the rate of human population growth would outpace the ability of the land to produce food?

Thomas Malthus in 1798 attempted to inform people that the human population, like any other population, had the potential to increase exponentially. Malthusian ideas were not well received, as he predicted that the rate of population growth would exceed the ability of the land to produce food. His work was later used by Charles Darwin in his explanation of the theory of natural selection.

Scottish American John Muir was a naturalist, glaciologist, author, and environmentalist who helped preserve what is now Yosemite National Park and much of the surrounding Sierra Nevada wilderness.

What is the ecological footprint?

Mathis Wackernagel (1962–) and William Rees (1943–) developed the concept of the ecological footprint in 1990. The ecological footprint is a measurement of how fast humans consume resources and generate waste compared to how fast nature can absorb the waste and generate new resources. Since the late 1970s, humanity has been in ecological overshoot; that is, the annual demands on nature exceed what Earth can generate in a year. It currently takes Earth one year and six months to regenerate what is used in one year.

What are the ecological resources and assets measured for the ecological footprint?

The ecological assets that a given population, for example, a city, state, or nation, uses to produce the natural resources that it consumes include plant-based food and fiber products, livestock and fish products, timber and other forest products, space for urban infrastructure, and absorption of waste, especially carbon emissions. The land includes crop land, forest and fishing grounds, and grazing lands.

Which countries reach their overshoot date the earliest and latest in the year?

Overshoot day is the date in the year when a country's demand for resources exceeds its capacity to regenerate the resources it consumes.

Country	Overshoot Day in 2018
Qatar	February 9
Luxembourg	February 19
United Arab Emirates	March 4
Mongolia	March 6
Bahrain	March 12
United States	March 15
Jamaica	December 13
Niger	December 16
Kyrgyzstan	December 17
Morocco	December 17
Vietnam	December 21

What is a carbon footprint?

A carbon footprint measures the total amount of greenhouse gases produced or generated and fossil fuels consumed minus activities that reduce greenhouse gas outputs. A carbon footprint may be calculated for an individual person or a group of individuals, such as a country.

What was the environmental significance of Rachel Carson's *Silent Spring*?

In the book *Silent Spring*, published in 1962, Rachel Carson (1907–1964) exposed the dangers of pesticides, particularly DDT, to the reproduction of species that prey upon the insects for whom the pesticide was intended. *Silent Spring* raised the public awareness and is considered a pivotal point at the beginning of the environmental movement.

Who started Earth Day?

The first Earth Day, April 22, 1970, was coordinated by Denis Hayes (1944–) at the request of Gaylord Nelson (1916–2005), U.S. senator from Wisconsin. Nelson is sometimes called the father of Earth Day. His main objective was to organize a nationwide public demonstration so large that it would get the attention of politicians and force the environmen-

President Richard Nixon (far left) created the EPA in 1970. He is seen here watching the first EPA director, William Ruckelshaus, get sworn in by U.S. Supreme Court chief justice Warren Burger while Mrs. Jill Ruckelshaus watches.

tal issue into the political dialogue of the nation. Important official actions that began soon after the celebration of the first Earth Day were the establishment of the Environmental Protection Agency (EPA); the creation of the President's Council on Environmental Quality; and the passage of the Clean Air Act, establishing national air quality standards. In 1995, Nelson received the Presidential Medal of Freedom for his contributions to the environmental protection movement. Earth Day continues to be celebrated each spring.

When was the Environmental Protection Agency created, and what does it do?

In 1970, President Richard M. Nixon (1913–1994) signed an executive order that created the Environmental Protection Agency (EPA) as an independent agency of the U.S. government. The creation of a federal agency by executive order rather than by an act of the legislative branch is somewhat uncommon. The EPA was established in response to public concern about unhealthy air, polluted rivers and groundwater, unsafe drinking water, endangered species, and hazardous waste disposal. The mission of the EPA is to protect human health and the environment. Responsibilities of the EPA include environmental research, monitoring, and enforcement of legislation regulating environmental activities. The EPA also manages the cleanup of toxic chemical sites as part of a program known as Superfund.

ECOSYSTEMS

What is an ecosystem?

An ecosystem is a community of species interacting with one another in a specific area and the physical aspects of the environment, such as soil, moisture, nutrients, and climatic factors.

What is biodiversity?

Biodiversity or biological diversity refers to the breadth of species represented within ecosystems or even on Earth as a whole. It includes genetic variability within a species, diversity of populations of a species, diversity of species within a natural community, or the wide array of natural communities and ecosystems throughout the world. Some scientists estimate that between fifteen and one hundred million species may be present throughout the world. Biodiversity is threatened at the present time more than at any other time in history. In the time since the North American continent was settled, as many as five hundred plant and animal species have disappeared. Some recent examples of threats to biodiversity in the United States include the following: 50 percent of the United States no longer supports its original vegetation; in the Plains, 99 percent of the original prairies are gone; and across the United States, we destroy one hundred thousand acres of wetlands each year.

What is a population?

A population is a group composed of all members of the same species that live in a specific geographical area at a particular time. An example of a population might include all the gray squirrels that live in a certain urban park. The areas occupied by a population could include the small area (measured in square millimeters) occupied by bacteria in a rotting apple to the vast areas of ocean (square kilometers) that include the territory of migrating sperm whales. Population ecology is the branch of ecology that studies the structure and changes within a population. Studies of specific populations will indicate the dynamics of the population in terms of active, ongoing growth, declining growth, or stability.

What is a community?

The term used by ecologists to describe a group of populations of different species living in the same place at the same time is "community." For example, when an ecologist studies not just the sparrows living in your backyard but also the insects the sparrows feed to their young and the plants that those insects eat, the researcher may draw conclusions about the community. Changes that occur in one species or one place in the community affect other members of the community.

What is a climax community?

Terrestrial communities of organisms move through a series of stages from bare earth or rock to forests of mature trees. This last stage is described as the "climax" because it is thought that, if left undisturbed, communities can remain in this stage in perpetuity. However, more recent studies suggest that climax may be only one part of a continuous cycle of successional stages in these communities.

What is ecological succession?

Ecological succession is the process in which one or more species in a community is replaced by other species. The most abundant organisms within an ecosystem are those that will thrive in the ecosystem's set of environmental conditions. When the environmental conditions change, the first set of species may no longer be optimal for the dominant species. The process of new species becoming established as the dominant species is ecological succession. Sometimes, the environmental conditions change due to the impact of the species on their own environments. For example, an ecosystem dominated by sun-loving plants may be replaced with plants and animals that thrive in the shade as the plants grow and produce more shade. Ecological succession may take place more rapidly following forest fires, windstorms, or human activities that alter the environment.

What are some of the physical and chemical properties that define a habitat?

A habitat is an environment where an organism exists. Physical properties that describe a habitat include the climate, such as the amount of rainfall/precipitation and temperature, topographic position, soil texture, soil moisture, and vegetation. Chemical properties that describe the habitat include soil acidity and the concentration of nutrients and toxins.

What is a niche?

A niche is the functional role that an organism, population, or species plays in an ecosystem. It describes features such as whether the species is active at night or during the day, what it eats, where it lives, and other aspects of day-to-day life, including its interactions with other organisms. From one environment (or community) to another, the niche may vary depending on how much competition the species faces.

Can two populations occupy the same niche?

According to G. F. Gause (1910–1988), an ecologist, two species that are in direct competition for the same resource cannot coexist if that resource is limited in some way. Limiting factors are based on the law of supply and demand. Those factors (resources) whose supply is less than demand can influence the distribution of species within a community. Examples of limiting factors include soil, minerals, temperature extremes, and water availability. The work of Gause and others predicts that under such conditions, one population will drive the other to extinction in that local area.

What is zero population growth?

Zero population growth, or ZPG, is the estimation of the birth rate necessary to maintain the size of the human population at its current level. As of now, the rate is estimated at 2.1, which means that each set of existing parents would need to have (on average) slightly more than two children during their lifetime. The extra 0.1 allows for infant mortality.

What is a keystone species?

A keystone species is a species that is crucial or essential to the ecosystem's community structure. Originally, a keystone species was always thought to be the top predator, such as the gray wolf. Scientists have found that wolf population sizes influence populations of both their prey and other species in the environment. However, a more recent viewpoint recognizes that less conspicuous species are also very important, as all species are interconnected in a biological community. Other examples of keystone species include the sea star, *Pisaster*, found along the coast of Washington State, and the black-tailed prairie dog of the prairie ecosystem. The sea star feeds on mussels and prevents the mussels from crowding out other species. The prairie dog is a critical source of food for larger predators; its burrowing loosens the soil, and its burrows act as a home for other creatures.

Keystone species like these prairie dogs are animals that are important to maintaining the health of a biome such as a forest or prairie. They help keep animal and plant populations in balance.

What is the difference between a native species and a non-native species?

Native species are those species that normally live and thrive in a particular ecosystem. Other species that migrate into, or are deliberately or accidentally introduced into, an ecosystem are called non-native species, also referred to as invasive, alien, and exotic species.

Why is it important to plant native species in landscapes?

Local native species form an ecosystem that functions without intervention. Native plants provide food and shelter for the other native species in an ecosystem—birds, insects, large animals, and other plants. Native plants are also adapted for the local climate, especially temperature, the amount of sunshine, and precipitation.

What is a bioinvader?

A bioinvader or invader species is an organism usually introduced into an ecosystem accidentally. These bioinvaders are non-native plants, animals, or other organisms and often overwhelm the native species. Some examples of bioinvaders are:

- *Kudzu*— a vine that was first introduced in the 1930s by the U.S. Soil Conservation Service for a good purpose—to control erosion. Kudzu now grows uncontrolled in the southeastern United States, pulling down powerlines and killing trees.
- *Purple loosestrife (Lythrum salicaria)*—a perennial weedy plant that was introduced to North America in the early 1800s most likely as a contaminant in European ship ballast. It was used as a medicinal herb. Once introduced, it became established along the East Coast and to Canada, spreading inland with the construction of waterways, drainage systems, canals, and highways. It outcompetes native species and is very difficult to eradicate.

- *Asian Long-Horned Beetle (Anoplophora glabripennis)*—a destructive, wood-boring pest that was first discovered in New York in 1996, spreading to the Midwest. It is believed that it arrived in the United States in cargo from Asia. It has the potential to destroy millions of acres of hardwood trees, including maple, birch, elm, willow, ash, and poplar. Currently, infested trees cannot be saved. The best protection is to check trees for signs of infestation (round holes made by the beetle) and remove infested trees to stop the spread.
- *Emerald Ash Borer (Agrilus planipennis)*—an invasive insect from Asia that kills ash trees. It arrived accidentally in cargo from Asia. First detected in the United States in 2002 in southeastern Michigan, the Emerald Ash Borer spread to thirty states. Infested trees lose most of their canopy within two years and die within three to four years. It is important to not move firewood from one location to another location since the larvae can survive hidden in the bark of firewood. Inspect trees regularly for infestation. Some insecticides are helpful, and some parasites are being investigated for biological control of the pest.
- *Brown Marmorated Stink Bug (Halyomorpha halys)*—native to East Asia. It possibly arrived in shipping material in the late 1990s and was first confirmed in the United States in 2001 in Allentown, Pennsylvania. It has spread to forty-four states and four provinces in Canada. In addition to damaging fruit crops, such as apples, peaches, grapes, and pears; vegetable crops, such as green beans, asparagus, and corn; and ornamental trees and shrubs, the stink bugs infest homes in affected areas. While they do not cause structural damage, they bring a foul odor into the homes.

What has been the impact of invasive species on ecosystems?

Some people tend to think of non-native species as threatening. In fact, most introduced and domesticated plant species, such as food crops and flowers, and animals, such as chickens, cattle, and fish, from around the world are beneficial to us. However, some non-native species can compete with and reduce a community's native species, causing unintended and unexpected consequences. In 1957, for example, Brazil imported wild African honeybees to help increase honey production. Instead, the bees displaced domestic honeybees and reduced the honey supply. Non-native species can spread rapidly if they find a new location with favorable conditions. In their new niches, these species often do not face the predators and diseases they face in their native niches, or they may be able to outcompete some native species in their new locations.

What has been the impact of zebra mussels on North American waterways?

Zebra mussels (*Dreissena polymorpha*) are black-and-white, striped, bivalve mollusks. They are hard-shelled species that adhere to hard surfaces with byssal threads. They were probably introduced to North America in 1985 or 1986 via a discharge of a foreign ship's ballast water into Lake St. Clair. They have spread throughout the Great Lakes, the Mississippi River, and as far east as the Hudson River. High densities of zebra mussels have been found in the intakes, pipes, and heat exchangers of waterways throughout the world. They can clog the water intakes of power plants, industrial sites, and public drinking water systems; foul boat hulls and engine-cooling water systems; and disrupt aquatic ecosystems. Water-processing facilities must be cleaned manually to rid the systems of the mussels. Zebra mussels are a threat to surface water resources because they reproduce quickly, have free-swimming larvae and rapid growth, lack competitors for space or food, and have no predators.

Zebra mussels are an out-of-control, invasive species from Europe that arrived in North America in the ballast water of ships.

Were killer algae ever discovered in the United States?

Caulerpa taxifolia, "killer algae," was introduced to the Mediterranean Sea in the mid-1980s when the Oceanographic Museum in Monaco dumped the bright green seaweed into the sea while cleaning the aquarium tanks. *Caulerpa taxifolia* now covers 32,000 acres of the coasts of France, Spain, Italy, and Croatia, devastating the Mediterranean ecosystem. The species continues to invade the Mediterranean today and appears to be unstoppable. It reproduces asexually through fragmentation, with fragments as small as 1 centimeter giving rise to viable plants. These algae smother other algal species, seagrasses, and sessile invertebrate communities. As an invasive species, they outcompete with other species for food and light. Native fish that consume *Caulerpa taxifolia* accumulate toxins, making the fish unsuitable for human consumption.

Caulerpa taxifolia appeared in southern California in 2000. It is believed that they were introduced by a marine aquarium owner dumping the contents of a fish tank into a storm water system. Steps were taken to control the spread of the species in southern California, and by 2006, the infestation was eradicated.

Who introduced the gypsy moth into the United States?

In 1869, Professor Leopold Trouvelot (1827–1895) brought gypsy moth egg masses from France to Medford, Massachusetts. His intention was to breed the gypsy moth (*Porthetria dispar*) with the silkworm to overcome a wilt disease of the silkworm. He placed the egg masses on a window ledge, and evidently, the wind blew them away. About ten years later, these caterpillars were numerous on trees in that vicinity, and in twenty years, trees in eastern Massachusetts were being defoliated. In 1911, a contaminated plant shipment from Holland also introduced the gypsy moth to Massachusetts.

Gypsy moths are now found throughout the entire northeastern United States and portions of Virginia, North Carolina, Ohio, and Michigan.

The gypsy moth lays its eggs on the leaves of oaks, birches, maples, and other hardwood trees. When the yellow, hairy caterpillars hatch from the eggs, they devour the leaves in such quantities that the tree becomes temporarily defoliated. Sometimes, this causes the tree to die. The caterpillars grow from 0.5 inches (3 millimeters) to about 2 inches (5.1 centimeters) before they spin a pupa, in which they will metamorphose into adult moths.

About forty-five kinds of birds, squirrels, chipmunks, and white-footed mice are natural predators of the gypsy moth. Among the thirteen imported natural enemies of the moth, two flies, *Compislura concinnata* (a tachnid fly) and *Sturnia scutellata*, parasitize the caterpillar. Other parasites and various wasps have also been tried as controls as well as spraying and male sterilization.

What is an indicator species?

Species that provide early warnings of damage to a community or an ecosystem are called indicator species. Birds are excellent biological indicators because they are found almost everywhere and are affected quickly by environmental changes such as the loss or fragmentation of their habitats and the introduction of chemical pesticides.

Why are amphibians an important group of indicator species?

Amphibians (frogs, toads, and salamanders) live part of their lives in water and part on land. Populations of some amphibians, frequently referred to as indicator species, are declining throughout the world. Amphibians were the first vertebrates on Earth and have been better at adapting to environmental changes through evolution than many other species. Many species of amphibians are having difficulty adapting to some of the rapid environmental changes that have taken place in the air and water and on the land from human activities in the past few decades. Since 1980, populations of hundreds of the world's almost six thousand amphibian species have been vanishing or declining in almost every part of the world, even in protected wildlife reserves and parks. In a 2008 assessment by the International Union for the Conservation of Nature and Natural Resources (IUCN), about 12 percent of all known amphibian species (and more than 80 percent of those in the Caribbean) are threatened with extinction, and populations of another 43 percent of the species are declining. Frogs are especially sensitive and vulnerable to environmental disruptions at various points in

Frogs and other amphibians are good indicator species because of their vulnerability to water pollution and UV radiation. The decline of amphibian species has been strong evidence that our environment is not healthy.

Who are the members of FrogWatch USA?

FrogWatch USA is a citizen science program under the auspices of the Association of Zoos and Aquariums that monitors frogs and toads from February through August. Individuals or groups are trained to listen to frog and toad calls (sounds) during the evenings and submit their observations to a national online database. These data are valuable for establishing programs to conserve frogs and toads and the wetlands they inhabit.

their life cycle. As juveniles (tadpoles), frogs live in water and eat plants; as adults, they live mostly on land and eat insects that can expose them to pesticides. The eggs of frogs have no protective shells to block harmful UV radiation or pollution. As adults, they take in water and air through their thin, permeable skins, which can readily absorb pollutants from the water, air, or soil. Also, they have no hair, feathers, or scales to protect them. No single cause has been identified to explain these amphibian declines, but factors ranging from prolonged drought and increases in UV radiation to parasites and habitat loss are probably responsible for most of the decline and disappearances among amphibian species.

How do members of a community interact with each other?

Members of a community interact in a number of different ways. Some form predator–prey relationships. Other members of a community form symbiotic relationships in which two or more species live together in a prolonged physical association. The three types of symbiosis or symbiotic relationships are 1) parasitism; 2) mutualism; and 3) commensalism.

- *Predator–prey* relationships occur when one species feeds on another species. One example of a predator–prey relationship is the coyote (predator) and elk (prey). The predator–prey relationship is important in maintaining populations of species.
- *Parasitism* is a relationship where one species (the parasite) benefits while the host species (the prey) is harmed. Parasites are usually smaller than their host organism. An example of a parasitic relationship is a tapeworm that lives in the intestines of a larger animal, absorbing nutrients from the host.
- *Mutualism* is a type of relationship between two species in which both members benefit. An example of a mutualistic relationship is between bean plants, such as soybeans, peas, peanuts, and the nitrogen-fixing bacteria *Rhizobium*. The plants provide a food source for the bacteria. The bacteria releases nitrogen into the soil, increasing the fertility of the soil and enhancing the crop.
- *Commensalism* is an interaction between two species in which one benefits, while the other is neither harmed nor helped. Few examples of commensalism exist since rarely does a relationship neither harm nor benefit the host species. One known ex-

ample is between clownfish and sea anemones. The clownfish have the ability to live among the stinging tentacles of the sea anemones eating the scraps of food from the sea anemones without being harmed by the stinging tentacles. Sea anemones have no known harm or benefit.

What is a trophic level?

A trophic level represents a step in the dynamics of energy flow through an ecosystem. The first trophic level is made up of the producers. The second level comprises those who consume the producers, also known as the primary consumers. Consumer levels are numbered according to their reliance on producers as a main source of energy. Primary consumers are those that rely heavily on producers, while secondary and tertiary (and even quaternary) consumers exploit other consumers as their preferred energy sources.

An ecological pyramid is one way to describe the distribution of energy, biomass, or individuals among the different levels of ecosystem structure. Since a limited amount of energy is available to each level, these trophic pyramids rarely rise above a third or fourth level of structure. Usually, only a few individuals are at the top of the pyramid. In 1942, Raymond L. Lindeman (1915–1942) was one of the first ecologists to refer to the "trophic dynamics" of ecosystems.

What are producers and consumers?

"Producers" and "consumers" are terms used to describe the different roles played by species within ecosystems. Producers are those who "fix" energy—meaning they take energy from one source, such as the Sun, and convert it into a form (their biomass) that makes it accessible to the consumers within the system.

What is bioaccumulation, and how is it different from biomagnification?

Some compounds are neither recycled by decomposers nor released into the atmosphere like energy. Instead, they remain in the ecosystem in virtually unchanged form as they are passed from one organism to another by predation. If a larger fish consumes

Why do ecosystems need decomposers?

While energy flows through ecosystems in only one direction, entering at the producer level and exiting as heat and the transfer of energy (as biomass) to consumers, chemical compounds can be reused over and over again. In a well-functioning ecosystem, some organisms make their living (their niche) by breaking down structures and recycling the compounds. These organisms are known as decomposers. Without these organisms, the chemicals used to build a tree would remain locked in the tree biomass for eternity instead of being returned to the soil after the tree's death. From this soil will spring new growth, beginning the cycle once again.

five smaller ones every day for several years, some of the compounds in the flesh of those little fish will be transferred to the larger fish as it builds and repairs its own structures. Over time, the larger fish will accumulate many units of such compounds. An example of these compounds is the pesticide DDT. The toxic effects of DDT may not be apparent in the small concentrations found in the little fish, but the accumulation over time in the larger fish will allow the effects to be magnified. This may become even more apparent as the chemicals move up the trophic pyramid to the top predators like the birds (or humans) that eat those larger fish. To describe this phenomenon, ecologists use the terms biomagnification and bioaccumulation in recognition of the disproportional effect these toxins have on the upper levels of the ecological pyramid.

How does a food chain differ from a food web?

A food chain refers to the transfer of energy from producers through herbivores through carnivores in a community. An example of a food chain would be little fish eating plankton and those little fish in turn being eaten by bigger fish. Food chains may overlap since many organisms eat more than one type of food. German zoologist Karl Semper (1832–1893) introduced the concept of the food chain in 1891.

The term food web is broader, as it includes interconnected food chains within a specific ecosystem. Many animals feed on different foods rather than exclusively on one single species of prey or one type of plant. Animals that use a variety of food sources have a greater chance of survival than those with a single food source. Complex food webs provide greater stability to a living community. A food web describes a nutritional portrait of the ecosystem. In 1927, Charles Elton (1868–1945) was one of the first scientists to diagram a food web, in his case a description of feeding relationships on Bear Island in the Arctic.

What is limnology?

Limnology is the study of freshwater ecosystems—especially lakes, ponds, and streams. These ecosystems are more fragile than marine environments since they are subject to greater extremes in temperature. The study of limnology includes the chemistry, physics, and biology of these bodies of water. F. A. Forel (1848–1931), a Swiss professor, has been called the father of limnology.

What is a wetland?

A wetland is an area that is covered by water for at least part of the year and has characteristic soils and water-tolerant plants. Examples of wetlands include swamps, marshes, bogs, and estuaries.

Type of Wetland	Typical Features
Swamp	Tree species such as willow, cypress, and mangrove
Marsh	Grasses such as cattails, reeds, and wild rice
Bog	Floating vegetation including mosses and cranberries
Estuary	Specially adapted flora and fauna such as crustaceans, grasses, and certain types of fish

A marsh (left) is a wetland with lots of wild grasses such as reeds and cattails; a bog (center) is distinguished by floating vegetation, sometimes including cranberries as seen here; and a swamp (right) is characterized by having various trees such as cypress, mangrove, and willows.

How many acres of wetlands have been lost in the United States?

The United States lost approximately one hundred million acres of wetland areas between colonial times and the 1970s. The most recent Wetlands Status and Trends report (2004–2009), published in 2011 by the U.S. Fish and Wildlife Service, estimates that the United States had 110.1 million acres (44.6 million hectares) of wetlands at that time. Wetlands composed 5.5 percent of the surface area of the United States. Ninety-five percent of all wetlands were freshwater, and the remaining 5 percent were in saltwater (marine) systems. In the five-year period from 2004 to 2009, the total wetland area declined by an estimated 62,300 acres (25,200 hectares) or an average annual loss of 13,800 acres (5,590 hectares), which is considered statistically insignificant. While freshwater wetlands increased in area slightly, marine areas saw a slight decrease. The reasons for wetland area increase or decrease reflect economic conditions, land use trends, changing regulation, and enforcement and climatic changes.

What is the ecological importance of estuaries?

Estuaries, the area where streams and rivers flow into the sea, provide water filtration and habitat protection. Depending on the geology and climate of an area, estuaries will have salt marshes, mangrove forests, or barrier beaches and islands. The many marsh grasses, such as cattails, serve as biological filters for the water flowing to the sea. They filter pollutants, such as herbicides, pesticides, and heavy metals, out of the water as well as excess sediments, such as silt and sand, and nutrients. Estuaries also stabilize shorelines and protect coastal areas. Natural estuaries are able to absorb much of the excess water from flooding and storm surge. In addition, estuaries provide an important habitat for many species of animals. Birds, such as mallards, stop in estuaries during their migrations. Many marine species of fish, including the American shad, Atlantic menhaden, and striped bass, spawn in the brackish waters of estuaries.

However, estuaries are also popular locations for human habitation and businesses. Contamination from shipping, household pollutants, and power plants are carried to the sea by rivers and streams and threaten the ecological health of many estuaries.

What are some major economic services provided by freshwater and marine ecosystems?

Marine and freshwater ecosystems provide a variety of economic services.

Marine Ecosystem	Freshwater Ecosystem
Food (human, animal, and pet food)	Food (human)
Pharmaceuticals	Drinking water
Oil, natural gas, and minerals	Irrigation water
Harbors and transportation	Transportation
Coastal habitats for humans	Hydroelectricity
Recreation	Recreation

ECOLOGICAL CYCLES

What is a biogeochemical cycle?

The elements that organisms need most (carbon, nitrogen, phosphorus, and sulfur) cycle through the physical environment, the organism, and then back to the environment. Each element has a distinctive cycle that depends on the physical and chemical properties of the element. Examples of biogeochemical cycles include the carbon and nitrogen cycles, both of which have a prominent gaseous phase. Examples of biogeochemical cycles with a prominent geologic phase include phosphorus and sulfur, where a large portion of the element may be stored in ocean sediments. Examples of cycles with a prominent atmospheric phase include carbon and nitrogen. These biogeochemical cycles involve biological, geologic, and chemical interactions.

What is the hydrologic cycle?

The hydrologic cycle takes place in the hydrosphere, which is the region containing all the water in the atmosphere and Earth's surface. It involves five phases: condensation,

What is the source of rain and other precipitation?

Solar energy drives winds that evaporate water from the surface of the oceans. The water vapor cools as it rises and then falls to the ground as rain, snow, or some other form of precipitation. Rain is part of the hydrologic cycle, which describes dynamic changes in the aquatic environment.

infiltration, runoff, evaporation, and precipitation. Rain, and other precipitation, is part of the hydrologic cycle.

What is the carbon cycle?

To survive, every organism must have access to carbon atoms. Carbon makes up about 49 percent of the dry weight of organisms. The carbon cycle includes movement of carbon from the gaseous phase (carbon dioxide in the atmosphere) to solid phase (carbon-containing compounds in living organisms) and then back to the atmosphere via decomposers. The atmosphere is the largest reservoir of carbon, containing 32 percent carbon dioxide. Biological processes on land shuttle carbon between atmospheric and terrestrial compartments, with photosynthesis removing CO_2 from the atmosphere and cell respiration returning CO_2 to the atmosphere.

How do plants obtain nitrogen?

Nitrogen is crucial to all organisms because it is an integral element of proteins and nucleic acids. The primary way that plants obtain nitrogen compounds is via the nitrogen cycle, which is a series of reactions involving several different types of bacteria, including nitrogen-fixing bacteria and denitrifying bacteria. During nitrogen fixation, symbiotic bacteria, which live in association with the roots of legumes, are able through a series of enzymatic reactions to make nitrogen available for plants. Plants must use nitrogen in its fixed form, such as ammonia, urea, or the nitrate ion, since the molecular nitrogen in Earth's atmosphere is very stable and does not easily combine with other elements.

What is the sulfur cycle?

The sulfur cycle provides a mineral nutrient that is essential for forming proteins and many cell organelles. Sulfur is taken up from the soil by fungi, bacteria, protists, and plants. From them, it passes on to animals. Unlike many other nutrients, sulfur is in short supply on land although overabundant in the oceans. Sulfur is made available in usable forms by bacteria. The sulfur cycle is rounded out when some of the sulfur compounds produced by oceanic organisms are transported by winds over the land, where they fall in rain and nourish plant life. Then they wash down into the seas again to continue the cycle.

What is the phosphorus cycle?

Phosphorus is essential for all cell membranes, genes, teeth, bones, and many enzymes. Phosphorus is abundant on the planet, but only certain compounds, phosphates, which are in limited supply, can be taken up by plants. Plants, aided by bacteria and mycorrhizal fungi, provide phosphorus. Phosphates reach the decomposers either directly when bacteria, protists, fungi, or plants die or indirectly after plants are eaten by animals; thus, it is returned to the nutrient cycle through the soil.

BIOMES

What is biogeography?

Biogeography is the study of the distribution, both current and past, of individual species in specific environments. One of the first biogeographers was Carolus Linnaeus, a Swedish botanist who studied the distribution of plants. Biogeography specifically addresses the questions of evolution, extinction, and dispersal of organisms in specific ecosystems.

What are the three major climate zones of Earth?

Earth shows a great diversity of species and habitats, places where these species can live. Some species live in water-covered habitats in aquatic life zones such as rivers, lakes, and oceans. Other species live on land or terrestrial habitats such as deserts, grasslands, and forests. Earth has three major climate zones: 1) tropical, where the climate is generally warm throughout the year; 2) temperate, where the climate is not extreme and typically changes through four different annual seasons; and 3) polar, where it is fiercely cold during the winter months and cool to cold during the summer months.

What is a biome?

A biome is a plant and animal ecosystem that covers a large geographical area. Complex interactions of climate, geology, soil types, water resources, and latitude all determine the kinds of plants and animals that thrive in different places. Fourteen major ecologi-

What are the two most important factors that determine the nature and location of biomes?

Temperature and precipitation are the most important determinants in biome distribution on land. If the general temperature range and precipitation level are known, the kind of biological community, that is, biome, can be predicted to occur there in the absence of human disturbance.

cal zones, called "biomes," exist over five major climatic regions and eight zoogeographical regions. Important land biomes include tundra, coniferous forests, deciduous forests, grasslands, savannas, deserts, chaparral, and tropical rain forests.

What are the general characteristics of biomes?

A biome is one of the world's most prominent ecosystems characterized by both vegetation and organisms particularly adapted to that environment.

Biome	Temperature	Precipitation	Vegetation	Animals
Arctic tundra	–40 to 18°C (–40 to 64°F)	Dry season, wet season	Shrubs, grasses, lichens, mosses	Birds, insects, mammals
Deciduous forest	Warm summers, cold winters	Low, distributed throughout year	Trees, shrubs, herbs, lichens, mosses	Mammals, birds, insects, reptiles
Desert	Hottest; great daily range	Driest, <10 inches (25 centimeters) of rain per year	Trees, shrubs, succulents, forbs	Birds, small mammals, reptiles
Taiga or coniferous forest	Cold winters, cool summers	Moderate	Evergreens, tamarack	Birds, mammals
Tropical rain forest	Hot	Wet season, short dry season	Trees, vines, epiphytes, fungi	Small mammals, birds, insects
Tropical savannah	Hot	Wet season, dry season	Tall grasses, shrubs, trees	Large mammals, birds, reptiles
Temperate grassland	Warm summers, cold winters	Seasonal drought, occasional fires	Tall grasses	Large mammals, birds, reptiles

Which biome occupies the largest portion of Earth?

The largest land biome is the taiga, which covers about 29 percent of the world's forest area. The taiga or boreal (northern) forest includes the enormous expanse of coniferous trees that extend to the tundra. This biome is a model of species uniformity, as only a few types of trees, including spruce, fir, and pine, are the ecological dominants, but they are present in great numbers. The taiga's vegetation supports large mammals such as moose, bears, and caribou as well as sables, minks, beavers, arctic hares, and other small, fur-bearing animals. The largest portions are located in Russia and Canada.

What are the major types of deserts?

The desert biome is characterized by a combination of low rainfall and varying average temperatures. Precipitation is rare and unpredictable in all types of deserts. The average precipitation is often 12 inches (30 centimeters) of precipitation (rain or snow) per year. The average temperature will depend on the location of the desert but can be very hot or very cold.

- The subtropical deserts, such as the Sahara and the Namib deserts of Africa, are hot and dry most of the year. In North America, the Chihuahuan, Sonoran, Mojave, and Great Basin deserts are hot and dry deserts. They have few plants except for small trees and ground-hugging shrubs. The land surface is hard and windblown and strewn with rocks and some sand. Most animals are nocturnal—becoming active at night when temperatures are cooler. Animals include insects, arachnids, reptiles, birds, and small mammals.

Taigas such as this one in Sakhalin Island, Russia, make up nearly a third of the planet's forested areas.

- The temperate deserts, also called the semiarid or cold-winter deserts, such as the Sonoran Desert in southeastern California, southwestern Arizona, and northwestern Mexico, have daytime temperatures that are high in summer and low in winter. Usually, more precipitation occurs in temperate deserts than in hot and dry deserts. The sparse vegetation consists of widely dispersed, drought-resistant shrubs and cacti or other succulents adapted to the lack of water and temperature variations. Animals may be either nocturnal (active at night) or diurnal (active during the day). Mammals, birds, insects, and reptiles live in the semiarid deserts.
- Cold deserts occur in the polar regions, including the Antarctic, Greenland, and the Nearctic realm. Winters are cold, summers are warm or hot, and precipitation, often in the form of snow, is low. Plants are usually deciduous with spiny leaves. Most of the animals are mammals due to the very cold winters.
- Coastal deserts are found in areas that are generally cool to moderately warm. The best-known coastal desert is the Atacama Desert in Chile. The average summer temperatures are cooler than the hot and dry deserts. Animals include mammals, birds, reptiles, and insects.

What are the three major types of grasslands?

Grasslands occur primarily in the interiors of continents in areas that are too moist for deserts to form and too dry for forests to grow. The three types result from combinations of low average precipitation and varying average temperatures. Grasslands persist because of a combination of seasonal drought, grazing by large herbivores, and occasional fires—all of which keep shrubs and trees from growing in large numbers. The three main types of grasslands are tropical, temperate, and cold (arctic tundra). One type of tropical grassland, called a savanna, contains widely scattered clumps of trees such as acacia, which are covered with thorns to keep some herbivores away. This biome usually has warm temperatures year-round with alternating dry and wet seasons. Tropical savannas in East Africa are home to gazelles, zebras, giraffes, and antelopes as well as predators such as lions, hyenas, and humans. Savanna plants, like those frequently found in deserts, are adapted to survive drought and extreme heat.

In a temperate grassland, winters can be bitterly cold, summers are hot and dry, and annual precipitation is sparse and falls unevenly throughout the year. Most of the above-ground parts of the grasses die and decompose each year and contribute to the deep, fertile topsoil. This topsoil is held in place by a thickly intertwined system of grass roots. However, if plowed, the topsoil could be blown away by high winds found in these biomes.

Cold grasslands, or arctic tundra, lie south of the arctic polar ice cap. During most of the year, these treeless plains are bitterly cold, swept by frigid winds, covered with ice and snow, and receive scant precipitation, which falls primarily as snow. Under the snow is a thick, spongy mat of grasses, mosses, lichens, and dwarf shrubs. Animals such as the arctic wolf, arctic fox, and musk oxen survive the intense winter cold through adaptations such as thick coats of fur and feathers, as in the case of the snowy owl.

How has the grasslands biome in the United States changed over the last two hundred years?

When Europeans first visited the American grasslands, they saw a territory that had essentially no trees but instead a sea of grasses that could grow up to ten feet tall in the wetter tallgrass prairie. Prairie soil was some of the most fertile soil on Earth. Almost all of the tallgrass and mixed-grass prairie in the United States were eventually plowed under for cultivation, while the short-grass prairie is now used for crops and cattle grazing. Prior to the coming of Europeans, twenty-two million acres of Illinois were prairie.

This restored grassland at the Morton Arboretum in Lisle, Illinois, shows what grasslands looked like in North America centuries ago before European settlers farmed and settled the land.

Today, that figure stands at approximately two thousand acres. From the former prairie, however, the United States feeds itself and some other areas of the world. Illinois, Kansas, and Nebraska are still grasslands, but the grasses that grow there now are not the native big bluestem or needlegrass but corn and wheat.

What are the three major types of forests?

Forests are lands that are dominated by trees. The three major types of forests are tropical, temperate, and cold (northern coniferous or boreal). These three types result from combinations of varying precipitation levels and varying average temperatures. Tropical rain forests have year-round, uniformly warm temperatures, high humidity, and almost daily heavy rainfall. Tropical rain forests are dominated by broadleaf evergreen plants, which keep most of their leaves year-round. The tops of the trees form a dense canopy that blocks most light from reaching the forest floor. Although tropical rain forests cover only about 2 percent of Earth's land surface, ecologists estimate that they contain at least 50 percent of Earth's known terrestrial plant and animal species. For example, a single tree in these forests may support several thousand different insect species, and other tropical rain forest plants are a source of chemicals used as blueprints for synthesizing many of the world's prescription drugs.

The second type of forest is the temperate deciduous forest. On a global basis, temperate forests have been degraded by various human activities, especially logging and urban expansion, more than any other terrestrial biome. However, within one to two hundred years, forests of this type that have been cleared by logging can return through secondary ecological succession.

Evergreen coniferous forests are also called boreal forests or taigas. These cold forests are found just south of the arctic tundra in northern regions around Earth. In this subarctic climate, winters are long and extremely cold, and winter sunlight is available only 6–8 hours per day. Most boreal forests are dominated by a few species of coniferous evergreen trees such as spruce, fir, cedar, hemlock, and pine that keep most of their leaves (or needles) year-round. Plant diversity is low because few species can survive the winters when soil moisture is frozen. Boreal forests or taigas contain a variety of wildlife. Year-round mammals include bears, wolves, moose, lynxes, and many burrowing rodent species. Caribou spend the winter in taigas and the summer in arctic tundras. Coastal coniferous forests or temperate rain forests are found in scattered coastal temperate areas with ample rainfall or moisture from the dense ocean fog. Thick stands of these forests with large conifers such as Sitka spruce, Douglas fir, and redwoods once dominated undisturbed areas of these biomes along the coast of North America from Canada to northern California in the United States.

What percentage of Earth's surface is forest area?

Prior to industrialization, the world's total forest area was estimated at 5.9 billion hectares. Currently, the world's total forest area is just over four billion hectares or 31 percent of the total land area. More than half of the world's forest areas are in the five most forest-rich countries: the Russian Federation, Brazil, Canada, the United States, and China. However, in 2017, scientists studying high-resolution satellite imagery discovered an additional 378 million hectares of forest area. Much of this newly found forest area is in the world's drylands. The additional amount of forest area increases total global forest area by 9 percent. It is nearly two-thirds the size of the Amazon rain forest.

What percentage of Earth's surface is tropical rain forest?

Rain forests account for approximately 7 percent of Earth's surface, or about three million square miles (7.7 million square kilometers).

What is the importance of the rain forest?

Rain forests play an important role in the ecological well-being of the planet. The large groups of plants found in rain forests help control levels of carbon dioxide in the atmosphere. It is estimated that the Amazon rain forest contains 90–140 billion metric tons of carbon, which helps to stabilize the climate globally. The vegetation and animal species in the rain forest are the source for a large variety of products. Half of all medicines prescribed worldwide are originally derived from wild products, and the U.S. National Cancer Institute has identified more than two thousand tropical rain forest plants with the potential to fight cancer. Rubber, timber, gums, resins and waxes, pesticides, lubricants, nuts and fruits, flavorings and dyestuffs, steroids, latexes, essential and edible oils, and bamboo are among the products that would be drastically affected by the depletion of the tropical forests. In addition, rain forests greatly influence patterns of rain deposition in tropical areas; smaller rain forests mean less rain.

Where is the largest rain forest located?

The Amazon River basin is the world's largest continuous tropical rain forest. It covers about 2.7 million square miles (6.9 million square kilometers) spanning eight countries—Brazil, Bolivia, Peru, Ecuador, Colombia, Venezuela, Guyana, Suriname—and the region of French Guiana. It is approximately the same size as the forty-eight contiguous states in the United States. The average rainfall in the Amazon River basin, encompassing the rain forest, is 80 inches (200 centimeters) annually. Interestingly, despite all the rain and thick forest growth, the soil in the Amazon region is quite sterile and not well suited to farming.

Rain forests are extremely diverse biologically. They help control carbon dioxide levels and are a resource for many products, including medicines. They need to be managed wisely and not bulldozed for farming.

How rapidly is deforestation occurring?

Agriculture, excessive logging, and fires are major causes of deforestation. Afforestation and the natural expansion of forests help to decrease the rate of deforestation. The rate of deforestation has decreased from 16 million hectares (61,776 square miles) per year during the 1990s to 13 million hectares (50,193 square miles) per year in the decade 2000–2010 (according to preliminary data for the decade). The net change in forest area for 2000–2010 is estimated at –5.2 million hectares (20,007 square miles) per year (an area about the size of Costa Rica), down from –8.3 million hectares (32,046 square miles) per year in the 1990s.

Who owns most of the forests in the United Sates?

The most recent (2013 published in 2015) National Woodland Owner Survey from the U.S. Forest Service found that 441 million acres, or more than half of the woods and forests in the United States, are owned and managed by private owners. Private owners may be families and individuals, corporate owners, or other private owners. Families and individuals account for 95 percent of the privately owned woods and forests. Corporate ownerships account for 4 percent of the owners. The remaining 1 percent are "other" owners, including Native American tribes, nongovernmental conservation and natural resource organizations, and other (nonfamily) unincorporated partnerships. Reasons for ownership include natural beauty or scenery, wildlife habitat, privacy, family lands, and investment. Forest- and woodland-management plans will ensure the sustainability of forests in the future.

POLLUTION AND WASTES

What is pollution?

The term "pollution" comes from the Latin word *pollutus*, meaning "something that has been made foul, unclean, or dirty." Pollution is often defined as unwanted or detrimental changes in a natural system. Most pollution is caused by human actions that alter the environment, although some naturally occurring changes, such as smoke damage caused by natural forest fires, are also a source of pollution. Many different substances are considered pollutants, although it is only when harmful ecological changes are caused by a substance that it acquires the status of a pollutant. The extent of damage caused by a substance is judged by humans. The perspective of different groups may have conflicting opinions as to whether the environment has been altered in a negative way. Pollution affects water, air, and land. Many extend the concept of pollution to soil pollution, light pollution, and noise pollution.

How are hazardous waste materials classified?

Hazardous waste materials are classified into four types—corrosive, ignitable, reactive, and toxic.

- *Corrosive* materials can wear away or destroy a substance. Most acids are corrosive and can destroy metal, burn skin, and give off vapors that burn the eyes.

How does the Environmental Protection Agency assess risk?

The Environmental Protection Agency (EPA) defines risk as the chance of harmful effects to humans or to ecological systems as a result of exposure to an environmental stressor. Stressors are any physical, chemical, or biological entity that can cause an adverse reaction. Three factors are taken into consideration when assessing risk: 1) how much of the chemical (stressor) is present in a given environment, such as in the air, water, or soil; 2) how much contact a person or ecological receptor, such as wildlife, has with the contaminated source; and 3) the inherent toxicity of the chemical. Risk assessments for both human and ecological health involve determining who or what is at risk, the nature of the environmental hazard (chemicals, physical, biological, or radiation), the source of and exposure to the environmental hazard, and the effects and nature of the hazard. The information compiled in a risk assessment is used to determine how to protect humans and the environment from various environmental stressors and contaminants.

- *Ignitable* materials can burst into flames easily. These materials pose a fire hazard and can irritate the skin, eyes, and lungs. Gasoline, paint, and furniture polish are ignitable.
- *Reactive* materials can explode or create poisonous gas when combined with other chemicals. Combining chlorine bleach and ammonia, for example, creates a poisonous gas.
- *Toxic* materials or substances can poison humans and other life. They can cause illness or death if swallowed or absorbed through the skin. Pesticides and household cleaning agents are toxic.

What is the Toxic Release Inventory?

The Toxic Release Inventory (TRI) is a government-mandated, publicly available compilation of information on the release of nearly 650 individual toxic chemicals by manufacturing facilities in the United States. These chemicals are used in the production of the products our industrial society consumes, such as pharmaceuticals, automobiles, clothing, and electronics. The chemicals on the list have been identified to cause significant adverse health effects, including cancer and other chronic conditions or significant adverse environmental effects. The law requires manufacturers to state the amounts of chemicals they release directly to air, land, or water or that they transfer to off-site facilities that treat or dispose of wastes. The U.S. Environmental Protection Agency compiles these reports into an annual inventory and makes the information available in a computerized database available for analysis.

In 2016 (report published in 2018), 21,605 facilities managed the release and disposal of 27.8 billion pounds (12.6 billion kilograms) of chemicals. Forty-four percent was recycled, 11 percent was used for energy recovery, 32 percent was treated, and 13 percent was disposed of or otherwise released. The majority of these releases (66 percent), 3.44 billion pounds (1.54 billion kilograms), were disposed on-site to land (including landfills and underground injection). Eighteen percent of what was remaining was released into the air, 11 percent went to off-site disposal or releases, and 6 percent was released into water.

Metal mining operations are a huge source of pollution. Forty-four percent of toxic chemical releases come from such industrial activities.

Which industries release the most toxic chemicals?

The metal mining industry released the greatest quantity of toxic chemicals for the year 2016, accounting for 44 percent of total chemical releases, primarily in the form of land disposal.

Industry	Total Releases	Percent of Total
Metal mining	1.52 billion pounds	44
Chemicals	49 million pounds	14
Electric utilities	36 million pounds	10
Primary metals	33 million pounds	10
Paper	160 million pounds	5
Hazardous waste management	140 million pounds	4
Food	120 million pounds	4
All others	310 million pounds	9

What environmental events can affect toxins in the environment?

Natural disasters and other environmental events are major sources of toxins in the environment, including those on land and in the air and water. Geological disasters include avalanches and landslides, earthquakes, sinkholes, and volcanic eruptions. Hydrological disasters include floods and tsunamis. Meteorological disasters include hurricanes and other cyclonic storms, tornadoes, blizzards, and related storms. The materials involved may range from biological and chemical pollutants to toxic gases and industrial chemicals.

Persistent Organic Pollutant (POP)	Use
Aldrin	Insecticide
Chlordane	Insecticide
DDT (dichlorodiphenyl-trichloroethane)	Insecticide
Dieldrin	Insecticide

What chemicals were initially banned by the Stockholm Convention on Persistent Organic Pollutants?

POPs is an acronym for Persistent Organic Pollutants. The Stockholm Convention on Persistent Organic Pollutants banned a dozen chemicals known as the "dirty dozen." All of these chemicals possess toxic properties, resist degradation, and are transported across long distances via air, water, and migratory species. The ban was adopted in 2001 and entered into force in 2004. Since POPs persist in the environment for a long time, they bioaccumulate in humans and animals and biomagnify in food chains. Health effects of these compounds include certain cancers, birth defects, and greater susceptibility to disease.

Persistent Organic Pollutant (POP)	Use
Endrin	Rodentcide and insecticide
Heptachlor	Fungicide
Hexachlorobenzene	Insecticide and fire retardant
Mirex™	Insecticide
Toxaphene™	Insecticide
PCBs (polychlorinated biphenyls)	Industrial chemical
Dioxins	By-product of certain manufacturing processes
Furans (dibenzofurans)	By-product of certain manufacturing processes

An additional sixteen POPs have been added to the list in subsequent years.

Persistent Organic Pollutant (POP)	Use
Alpha hexachlorocylohexane	Insecticide
Beta hexachlorocyclohexane	Insecticide
Chlordecone	Pesticide
Decabromodiphynyl ether (commercial mixture, c-DecaBDE)	Additive flame retardant (still acceptable for use in vehicles, aircraft, textile, additives in plastic housings, etc., polyurethane form for buildings)
Hexabromobiphenyl	Industrial chemical used as a flame retardant
Hexabromocyclododecane	Flame retardant additive (still acceptable as expanded polystyrene and extruded polystyrene in buildings in accordance with provisions)
Hexabromodiphenyl ether and heptabromodiphenyl ether (commercial octabromodiphenyl ether)	Industrial chemical used as a flame retardant
Hexachlorobutadiene	Solvent for other chlorine-containing compounds
Lindane	Insecticide (still acceptable for use as a pharmaceutical for control of head lice and scabies as second-line treatment)

Pentachlorobenzene	Fungicide, used in PCB products and a flame retardant
Pentachlorophenol and its salts and esters	Herbicide, insecticide, fungicide, algaecide, disinfectant, and an ingredient in antifouling paint (still acceptable for use in utility poles)
Perfluorooctane sulfonic acid and its salts and perfluorooctane sulfonyl fluoride	Industrial chemical used for electric and electronic parts, fire-fighting foam, photo imaging, hydraulic fluids, and textiles
Polychlorinated naphthalenes	Insulating coatings for electrical wires, wood preservatives, rubber and plastic additives, capacitor dielectrics, and lubricants (acceptable use for production of polyfluorinated naphthalenes)
Short-chain chlorinated paraffins (SCCPs)	Plasticizer in rubber, paints, adhesives, flame retardants for plastics as well as an extreme pressure lubricant in metal working fluids (still acceptable for use as additives in transmission belts, rubber conveyor belts, leather, lubricant additives, tubes for outdoor decoration bulbs, paints, adhesives, metal processing, and plasticizers)
Technical endosulfan and its related isomers	Insecticide (still acceptable for use in crop-pest complexes in accordance with provisions)
Tetrabromodiphenyl ether and pentabromodiphenyl ether (commercial pentabromodiphenyl ether)	Industrial chemical

How did DDT affect the environment?

Although DDT was synthesized as early as 1874 by Othmar Zeidler (1859–1911), it was Swiss chemist Paul Müller (1899–1965) who recognized its insecticidal properties in 1939. He was awarded the 1948 Nobel Prize in Physiology or Medicine for his development of dichloro-diphenyl-trichloro-ethene, or DDT. Unlike the arsenic-based compounds then in use, DDT was effective in killing insects and seemed not to harm plants and animals. In the following twenty years, it proved to be effective in controlling disease-carrying insects (mosquitoes that carry malaria and yellow fever and lice that carry typhus) and in killing many plant crop destroyers. Publication of Rachel Carson's *Silent Spring* in 1962 alerted scientists to the detrimental effects of DDT. Increasingly, DDT-resistant insect species and the accumulative hazardous effects of DDT on plant and animal life cycles led to its disuse in many countries during the 1970s.

What are PCBs?

Polychlorinated biphenyls (PCBs) are a group of chemicals that were widely used before 1970 in the electrical industry as coolants for transformers and in capacitors and other electrical devices. They caused environmental problems because they do not break down and can spread through the water, soil, and air. They have been linked by some scientists to cancer and reproductive disorders and have been shown to cause liver function abnormalities. Government action has resulted in the control of the use, disposal, and production of PCBs in nearly all areas of the world, including the United States.

Author Rachel Carson's seminal 1962 work *Silent Spring* warned of the devastating effects of DDT and other poisons on the environment.

What problems may be encountered when polyvinyl chloride (PVC) plastics are burned?

Chlorinated plastics, such as PVC, contribute to the formation of hydrochloric acid gases. They also may be a part of a mix of substances containing chlorine that form a precursor to dioxin in the burning process. Polystyrene, polyethylene, and polyethylene terephthalate (PET) do not produce these pollutants.

What is Superfund?

The discovery of toxic waste dumps, such as Love Canal in New York and Times Beach in Missouri, prompted the U.S. Congress to develop a program to clean up abandoned hazardous waste sites. In 1980, the U.S. Congress passed the Comprehensive Environmental Response, Compensation, and Liability Act, commonly known as the Superfund program. This law (along with amendments in 1986 and 1990) established the $16.3-billion Superfund program financed jointly by federal and state governments and by special taxes on chemical and petrochemical industries (which provide 86 percent of the funding). The purpose of the Superfund program is to identify and clean up abandoned hazardous waste dump sites and leaking underground tanks that threaten human health and the environment. To keep taxpayers from footing most of the bill, cleanups are based on the polluter-pays principle. The EPA is charged with locating dangerous dump sites, finding the potentially liable culprits, ordering them to pay for the entire cleanup, and suing them if they don't. When the EPA can find no responsible party, it draws money out of the Superfund for cleanup.

The Superfund Task Force was established in 2017. The Task Force established five recommendations:

1. Expediting cleanup and remediation
2. Reinvigorating responsible-party cleanup and reuse
3. Encouraging private investment
4. Promoting redevelopment and community revitalization
5. Engaging partners and stakeholders

Which states have the most hazardous waste (Superfund) sites?

As of 2018, 1,338 hazardous waste sites and fifty-three proposed sites were on the National Priorities List. All of the fifty states except North Dakota contain hazardous waste sites. The ten states with the most hazardous waste sites are:

- New Jersey: 114 sites
- California: 98 sites
- Pennsylvania: 92 sites
- New York: 85 sites
- Michigan: 65 sites
- Texas: 55 sites
- Florida: 53 sites
- Washington: 48 sites
- Illinois: 45 sites
- Indiana: 40 sites

How much garbage does the average American generate?

According to the Environmental Protection Agency, nearly 262 million tons of municipal waste was generated in 2015. This is equivalent to 4.48 pounds (2 kilograms) per person per day or approximately 1,635 pounds (742 kilograms) per person per year.

What are the components of municipal solid waste?

Municipal solid waste consists of things we commonly use and then throw away such as paper and packaging, food scraps, yard waste, tires, and large household items including old sofas, appliances, and computers. It does not include industrial, hazardous, or construction waste. The distribution of the municipal solid waste generated in 2015 is illustrated below:

Waste Product	Weight (millions of tons)	Percentage of Total
Paper and paperboard	68.05	25.9
Yard wastes	34.72	13.3
Food scraps	39.73	15.1
Plastics	34.50	13.1
Metals	24.00	9.1
Rubber, leather, and textiles	24.51	9.3
Wood	16.30	6.2
Glass	11.47	4.4
Other materials	5.16	3.6

How is municipal solid waste managed?

More than half of the municipal solid waste generated in the United States is discarded in landfills. The balance is either recovered through recycling programs or combusted with energy recovery. The following chart shows the management of municipal solid waste in the United States for 2015.

Where Does U.S. Garbage Go?

Method of Disposal	Amount (millions of tons)	Percentage of Total
Discarded in landfills	137.7	52.5
Recycling	67.8	25.8
Composting	23.4	8.9
Combustion with energy recovery	33.5	12.8
Total	262.4	100.0

A change has occurred in the management of municipal solid waste over several decades. While the generation of municipal solid waste has increased, the amount going to landfills has decreased, and the amount being recycled, composted, or combusted with energy recovery has increased.

Activity	1960	1970	1980	1990	2000	2010	2015
Generation of municipal solid waste	88.1	121.1	151.6	208.3	243	251.1	262.4
Recycling	5.6	8.0	14.5	29.0	53.0	65.3	67.8
Composting	Negligible	Negligible	Negligible	4.2	16.5	20.2	23.4
Combustion with energy recovery	0.0	0.5	2.8	29.8	33.7	29.3	33.5
Landfilling and other	82.5	112.6	134.3	145.3	140.3	136.3	137.7

*Units are millions of tons

- In 1960, 94 percent of the total amount of municipal solid waste generated was disposed in landfills.
- In 2015, slightly under 53 percent of the total amount of municipal solid waste generated was disposed in landfills.
- In 1960, the recycling rate was just over 6 percent of the total amount of municipal solid waste generated that was disposed in landfills.
- In 2015, the recycling rate increased to over 34 percent of the total amount of municipal solid waste generated that was disposed in landfills.

The United States has a poor record of recycling and instead dumps over half its solid waste in landfills.

How has disposal of solid waste to landfill facilities changed?

Since large amounts of land are available in the United States, landfilling has been an essential component of waste management for several decades. In areas where land is less available, combustion has a more significant role in waste management. However, combustion requires proper air-emission-control equipment to minimize the impact on air pollution. In 1960, 94 percent of all garbage was sent to landfills. During the following decades, although the amount of municipal solid waste increased, the amount going to landfills decreased. During the years 1990–2012, the total amount of waste going to landfills decreased from 145.3 million tons, 69.8 percent of the total municipal solid waste generated, to 135 million tons, representing 53.8 percent of the total municipal solid waste generated.

Year	Percentage of Municipal Solid Waste Being Disposed in Landfills
1960	93.6
1970	93.1
1980	88.6
1990	69.8
2000	57.6
2010	54.3
2015	52.5

What is sustainable materials management?

Sustainable materials management (SMM) is an approach to conserve resources, reduce waste, and minimize adverse environmental impacts from materials. This approach focuses on the entire life cycle of materials and products from the time of production through their use, reuse, recycling, and final disposal. It includes an objective of reducing toxic chemicals and environmental impacts throughout the material's life cycle with an ultimate goal of assuring sufficient resources for today's needs and those of the future. Products designed following an SMM philosophy will reduce environmental impacts beginning with the raw material acquisition for the product, using fewer, less toxic, and more durable materials. The product will be designed so at the end of its useful life, it can be easily disassembled, and the materials can be recycled or disposed of in an environmentally friendly manner.

WATER POLLUTION

What is eutrophication?

Eutrophication is a process in which the supply of plant nutrients in a lake or pond is increased. In time, the result of natural eutrophication may be dry land where water once flowed, caused by plant overgrowth. Natural fertilizers, washed from the soil, result in an accelerated growth of plants, producing overcrowding. As the plants die off, the dead and decaying vegetation depletes the lake's oxygen supply, causing fish to die. The accumulated dead plant and animal material eventually changes a deep lake to a shallow one, then to a swamp, and finally, it becomes dry land. While the process of eutrophication is a natural one, it has been accelerated enormously by human activities. Fertilizers from farms, sewage, industrial wastes, and some detergents all contribute to the problem.

What does it mean when a lake is brown or blue?

When a lake is brown, it usually indicates that eutrophication is occurring. This process refers to the premature "aging" of a lake, when nutrients are added to the water, usually due to runoff, which may be either agricultural or industrial in origin. Due to this rich supply of nutrients, blue-green algae begin to take over the green algae in the lake, and food webs within the lake are disturbed, leading to an eventual loss of fish. When a lake is blue, this usually means that the lake has been damaged by acid precipitation. The gradual drop in pH caused by exposure to acid rain causes disruption of the food webs, eventually killing most organisms. The end result is clear water, which reflects the low productivity of the lake.

Where do algal blooms occur?

Algal blooms occur in freshwater, brackish water (a mixture of salt and freshwater often found in estuaries where freshwater meets salt water), and marine (salt) water. They occur in the Great Lakes, other inland lakes, and in coastal waters of every state. Algal blooms may be green, blue, brown, yellow, orange, or red in color. Sometimes, they look like foam, scum, or mats on the surface of the water. An unpleasant odor of decomposing plants may be associated with some algal blooms.

Human activities such as farming and fertilizer runoff can lead to algal blooms, which can be harmful to fish and other aquatic life as well as humans.

Why do algal blooms occur?

Algal blooms are an overpopulation of algae or cyanobacteria (blue-green algae). The cause of algal blooms is still being investigated by researchers, but climatic conditions and human activities are indicated as factors contributing to algal blooms. Many algae flourish when wind and water conditions, such as water temperature, are favorable. More algal blooms form when the sea surface temperature increases. Human activities often provide the nutrients, especially phosphorus and nitrogen, necessary to feed algae. The source of the excess nutrients is often from fertilizers in agricultural uses and home lawns. These pollutants enter streams, rivers, lakes, and other waterways as runoff and flow into the larger bodies of water. Once the algae population in the water has an excess of nutrients, their population grows, leading to algal blooms. The thick masses of cyanobacteria absorb so much light that the water temperature increases. The algae at the base of the aquatic food webs prefer cooler temperatures and cannot compete successfully with rapidly growing cyanobacteria in the warmer temperatures.

How are algal blooms used as ecological indicators?

Algae are an important component of biological monitoring programs, especially for evaluating water quality. They are suited to water quality assessment because of their nutrient needs, rapid reproduction rate, and very short life cycle. Algae are valuable indicators of ecosystem conditions because they respond quickly both in species composition and densities to a wide range of water conditions due to changes in water chemistry. For example, increases in water acidity due to acid-forming chemicals that influence pond or lake pH levels as well as heavy metals discharged from industrial areas affect the composition of various groups of algae that are able to tolerate these conditions.

Are all algal blooms harmful?

Not all algal blooms are harmful. Many are nuisance blooms that discolor water, have a foul odor, and/or cause the water or fish to taste bad, but they are not harmful. Nuisance blooms are not dangerous and do not produce toxins that harm people or animals.

A small percentage of algae cause harmful algal blooms (HABs). The organisms that cause HABs produce toxins that are harmful to people, fish, shellfish, marine mammals, and birds. Swimming or playing in water contaminated with HABs, drinking contaminated water, or breathing the toxins (poisons) may cause illness or even death. Additionally, eating contaminated fish or shellfish may cause illness in people.

Algal blooms, both nuisance and HABs, can impact the environment by depleting oxygen in the water. Often, economic consequences of algal blooms occur through the loss of tourism when beaches and other recreational areas are closed. Fisheries closed due to HABs can lose millions of dollars in revenue, while many workers are without work.

What are the health effects of harmful algal blooms (HABs)?

The health effects of HABs vary with the different organisms that cause the HAB. Each different organism produces a different toxin, producing different health effects.

Organism	Water Type	Toxin	Target Tissue	Health Effects
Alexandrium sp.	Salt	Saxitoxins	Nerves and muscles	Paralytic shellfish poisoning, paralysis, death

Organism	Water Type	Toxin	Target Tissue	Health Effects
Karenia brevis	Salt	Brevetoxins	Nervous system Respiratory system	Gastrointestinal illness, muscle cramps, seizures, paralysis, respiratory problems, especially for asthmatics
Pseudo-nitzschia	Salt	Domoic acid	Nervous system	Amnesiac shellfish poisoning, vomiting, diarrhea, confusion, seizures, permanent short-term memory loss, death
Microcystis	Fresh	Microcystin	Liver	Gastrointestinal illness, liver damage

What causes a red tide?

A red tide is a popular term for a harmful algal bloom. It occurs when a population explosion occurs among toxic red dinoflagellates (members of the genera *Gymnodidium* and *Gonyaulax*, both protistans that have an unusual cellular plate or armor). The algal bloom may tint the water orange, red, or brown. A red tide caused by the organism *Karenia brevis* has occurred along Florida's Gulf Coast every summer for decades. During the early fall of 2018, the same red tide was detected along the east coast of Florida. The Florida red tide closes beaches and affects the fishing and shellfish industries.

What members of Cnidaria are economically important?

Reef-building corals are among the most important members of Cnidaria. Coral reefs are among the most productive of all ecosystems. They are large formations of calcium carbonate (limestone) in tropical seas laid down by living organisms over thousands of years. Fish and other animals associated with reefs provide an important source of food for humans, and reefs serve as tourist attractions. Many terrestrial organisms also benefit from coral reefs, which form and maintain the foundation of thousands of islands. By providing a barrier against waves, reefs also protect shorelines against storms and erosion.

How is coral bleaching related to changes in the environment?

Although corals can capture prey, many tropical species are dependent on photosynthetic algae (zooxanthellae) for nutrition. These algae live within the cells that line the digestive cavity of the coral. The symbiotic relationship between coral and zooxanthellae is mutually beneficial. The algae provide the coral with oxygen and carbon and nitrogen compounds. The coral supplies the algae with ammonia (waste product), from which the algae make nitrogenous compounds for both partners. Coral bleaching is the stress-induced loss of zooxanthellae that live in coral cells. In coral bleaching, the algae lose their pigmentation or are expelled from coral cells. Without the algae, coral become malnourished and die. The causes of coral bleaching are not completely understood, but it is believed that environmental factors are involved. Pollution, invasive bacteria such as *Vibrio*, salinity changes, temperature changes, and high concentrations of ultraviolet radiation (associated with the destruction of the ozone layer) all contribute to coral bleaching.

Coral bleaching results when the coral animals die, leaving only the calcium behind. Warming oceans have been blamed for the increased bleaching in coral reefs.

Why is marine debris an environmental concern?

Marine debris is defined by the National Oceanic and Atmospheric Administration as "any persistent solid material that is manufactured or processed and directly or indirectly, intentionally or unintentionally, disposed of or abandoned into the marine environment or the Great Lakes." Marine debris can hurt or kill animals. Some types of marine debris, such as fishing nets, can trap and entangle animals, leading to injury or death. Animals may ingest debris. For example, loggerhead sea turtles have been known to eat plastic bags, mistaking the bags for jellyfish. Many debris items, such as plastics, do not degrade in the ocean environment. Chemicals may accumulate in marine animals, eventually contaminating the food supply.

What are the sources of marine debris?

Marine debris comes from ocean-based sources and land-based sources. Ocean-based sources include fishing vessels, offshore oil and gas platforms, and cargo ships and other vessels. Examples of ocean-based debris are fishing gear (nets, traps, and lines) from fishing vessels; hard hats, fifty-five-gallon storage drums, and plastic pipes from offshore platforms; and cargo ranging from sneakers to televisions to lumber from cargo vessels.

Land-based sources of debris include items blown, swept, or washed out to sea. Examples of these items are industrial waste, litter, and other products that are dumped into waterways.

What is nonpoint source water pollution?

Nonpoint source (NPS) water pollution comes from many different sources. A major source of NPS is from rainfall and snowmelt. As the water from the precipitation flows into storm drains, it picks up pollutants and carries them to streams, rivers, lakes, wetlands, and oceans. Some runoff from precipitation flows directly into bodies of water. Examples of items that accumulate in runoff are street litter, such as cigarette butts; medical waste, such as syringes; and food and beverage packaging containers, including plastic cups and straws.

What are microplastics?

Microplastics are tiny pieces of plastic that are less than 5 millimeters (0.20 inches) in length or about the size of a sesame seed or piece of rice. Microplastics are either from large, plastic debris, such as plastic water bottles, that break into smaller and smaller pieces or very tiny pieces of manufactured polyethylene plastic. These tiny pieces of plastic, known as microbeads, were until recently added as exfoliants to health and beauty products, such as toothpaste. The microbeads were able to pass through water filtration systems and would eventually end up in the oceans and Great Lakes. The Microbead-Free Waters Act of 2015 banned the use of microbeads in rinse-off cosmetics.

Where is the Pacific Garbage Patch located?

The Pacific Garbage Patch is a nickname for an area in the North Pacific Ocean between Hawaii and California with a large concentration of marine debris. It is not a floating island of trash but rather a place where debris accumulates and is dispersed throughout the water column from the surface to the bottom of the ocean floor. Microplastics constitute the bulk of the debris in the Pacific Garbage Patch, although it also includes some larger pieces of debris. The debris accumulates due to rotating ocean currents called gyres. The North Pacific Subtropical Gyre is comprised of four major ocean currents and rotates in a clockwise direction. The area is constantly in motion, changing size and shape, due to the currents and winds. It is estimated to be 7–9 million square miles (18–23 million square kilometers) in size—approximately equivalent to three times the area of the continental United States. Although large in size, the "garbage patch" is barely visible due to the size of the microplastics. All marine debris may harm or kill wildlife, damage habitats, or be hazardous to ocean vessels and navigation. Researchers are continuing to investigate the environmental impact of the Pacific Garbage Patch.

How can individuals help collect data on marine debris?

The Marine Debris Tracker is a mobile application (app) that originated in 2010 as a partnership and joint initiative between the NOAA Marine Debris Program and the Southeast Atlantic Marine Debris Initiative with the University of Georgia's College of Engineering. Anyone who has downloaded the app on their phone (Apple or Android) can collect standardized data on marine debris. Data collected includes the type of debris from a list of common debris items (plastic, fishing gear, processed lumber, rubber, cloth, glass, and metal) and the location, which is recorded through GPS. Over 958,000 items from forty-six countries have been logged. The data may be analyzed to monitor cleanup and prevention efforts.

What were the top ten items collected during the International Coastal Cleanup?

The International Coastal Cleanup is an annual event sponsored by the Ocean Conservancy to remove trash from beaches and waterways around the world. During the 2018 cleanup event, 789,138 volunteers collected 20,824,689 items weighing 20,471,242 pounds (9,285,600 kilograms) of trash along 18,935 miles (30,472 kilometers) of coast worldwide. The top ten items collected were:

Workers with the NOAA Marine Debris Program are shown here clearing rubbish out of the Pacific Ocean. Several enormous garbage patches are floating in our oceans these days, and they are killing wildlife.

International Coastal Cleanup

Type of Garbage	Number Collected
Cigarette butts	2,412,151
Food wrappers	1,739,743
Plastic beverage bottles	1,569,135
Plastic bottle caps	1,091,107
Plastic grocery bags	757,523
Other plastic bags	746,211
Straws, stirrers	643,562
Plastic takeout/away containers	632,874
Plastic lids	624,878
Foam takeout/away containers	580,570

In the United States, 209,643 volunteers collected 5,860,996 items weighing 3,743,118 pounds (1,697,851 kilograms) of trash along 12,051 miles (19,392 kilometers) of coast. The top ten items collected from U.S. coasts were:

U.S. Coastal Cleanup

Type of Garbage	Number Collected
Cigarette butts	842,837
Food wrappers	345,241
Plastic bottle caps	286,678
Plastic beverage bottles	242,534
Beverage cans	168,885
Straws, stirrers	144,464
Glass beverage bottles	111,682
Plastic grocery bags	96,815
Metal bottle caps	93,917
Other plastic/foam packaging	92,209

What are the sources of oil pollution in the oceans?

Most oil pollution results from minor spillage from tankers and accidental discharges of oil when oil tankers are loaded and unloaded. Other sources of oil pollution include improper disposal of used motor oil, oil leaks from motor vehicles, routine ship maintenance, leaks in pipelines that transport oil, and accidents at storage facilities and refineries.

Oil Source	Percentage of Total Oil Spillage
Natural seepage	46
Discharges from consumption of oils (operational discharges from ships and discharges from land-based sources)	37
Accidental spills from ships	12
Extraction of oil	3

Have any species recovered following the *Exxon Valdez* oil spill?

The *Exxon Valdez* went aground on Bligh Reef in Alaska on March 24, 1989, spilling nearly 11 million gallons (nearly 305,000 barrels) of crude oil into the Prince William Sound. The oil spread in the Prince William Sound, polluting many miles of coastline and killing or injuring many plants and animals. It was estimated that 250,000 seabirds, 2,800 sea otters, three hundred harbor seals, 250 bald eagles, up to twenty-two killer

Cleanup of the massive 1989 *Exxon Valdez* oil spill along 1,300 miles (2,100 kilometers) of formerly pristine Alaskan coastline was an expensive, time-consuming process. Many of the mammals, fish, and birds that lived there have still not recovered.

whales, and billions of salmon and herring eggs died from the spill. Following the initial response and cleanup has been continued monitoring of the biology, chemistry, and geomorphology of the area. In a report published in 2014, twenty-five years after the accident, many species were listed as being recovered or recovering. The species that were still not recovered were herring, killer whale pod AT1, and pigeon guillemots.

- *Recovered species and habitats* include the rocky intertidal, bald eagle, river otter, common murre, sockeye salmon, pink salmon, cormorant, harbor seal, dolly varden, common loon, subtidal communities, rockfish, cutthroat trout, sea otter, and harlequin duck.
- *Recovering species and habitats* include clams, black oystercatcher, mussels, killer whale pod AB, Barrow's goldeneye, intertidal communities, designated wilderness, and sediments.

How do oil spills affect the environment?

Different types of oil may affect the environment differently. Four basic types of oil are summarized in the chart below.

Type of Oil	Example of Oil	Impacts on Environment
Very light oils	Jet fuels, gasoline	High concentration of toxic compounds Localized severe impact to water column and intertidal resources
Remediation:	Highly volatile; evaporating in one to two days No cleanup possible	
Light oils	Diesel, No. 2 fuel oil, light crudes	Moderate concentrations of toxic compounds Potential for long-term contamination of intertidal resources
Remediation:	Moderately volatile; will leave residue after a few days Cleanup can be very effective	
Medium oils	Most crude oils	Oil contamination of intertidal areas can be severe and long-term Oil impact to waterfowl and fur-bearing mammals can be severe
Remediation:	About one-third will evaporate within 24 hours Cleanup most effective if conducted quickly	
Heavy oils	Heavy crude oils, No. 6 fuel oil, Bunker C	Severe impacts to waterfowl and fur-bearing mammals (coating and ingestion) Heavy contamination of intertidal areas likely Long-term contamination of sediments possible

Type of Oil	Example of Oil	Impacts on Environment
Remediation:	Little or no evaporation or dissolution Weathers very slowly Shoreline cleanup difficult under all conditions	

Ingestion or inhalation of oil may poison animals or cause irritation. When the feathers of birds or fur of mammals, such as sea otters, become covered in oil, their ability to maintain body temperature is reduced, often leading to death due to hypothermia.

How do bacteria assist in cleaning up an oil spill?

Bioremediation is the degradation, decomposition, or stabilization of pollutants by microorganisms such as bacteria, fungi, and cyanobacteria. Oxygen and organisms are injected into contaminated soil and/or water (e.g., oil spills). The microorganisms feed on and eliminate the pollutants. When the pollutants are gone, the organisms die.

The BP Deepwater Horizon oil spill in the Gulf of Mexico in 2010 is the worst oil spill in U.S. history, releasing 4.1 million barrels of oil. Communities of bacteria grew exponentially into bacterial blooms following the oil spill. Various species of microbes consumed a vast range of chemicals and compounds in the spill.

AIR POLLUTION

What is air pollution?

Air pollution is the contamination of Earth's atmosphere at levels high enough to harm humans, other organisms, or other materials. The major air pollutants are particulate matter, ozone, carbon monoxide, nitrogen oxides, sulfur dioxide, and lead. Primary air pollutants, including carbon monoxide, nitrogen oxides, sulfur dioxide, and particulate matter, enter the atmosphere directly. Secondary air pollutants are harmful chemicals that form from other substances released into the atmosphere. Ozone and sulfur trioxide are examples of secondary air pollutants.

What are the sources of air pollution?

Most air pollutants originate from human-made sources, such as forms of transportation (cars, trucks, and buses), and industrial processes, such as refineries, iron and steel mills, paper mills, and chemical plants. Another significant source of air pollution is fuel combustion as a result of burning fossil fuels.

What are the health effects of the major air pollutants?

Common health effects of exposure to even low levels of air pollutants are irritated eyes and inflammation of the respiratory tract. Some evidence exists that exposure to air pollutants suppresses the immune system, increasing susceptibility to infections.

Pollutant	Source	Health Effects
Particulate matter	Industries, electric power plants, motor vehicles, construction, agriculture	Aggravates respiratory illnesses; long-term exposure may cause increased incidence of chronic conditions such as bronchitis; suppresses immune system; heavy metals and organic chemicals may cause cancer
Nitrogen oxides	Motor vehicles, industries, heavily fertilized farmland	Irritate respiratory tract; aggravate asthma and chronic bronchitis
Sulfur oxides	Electric power plants and other industries	Irritate respiratory tract; long-term exposure may cause increased incidence of chronic conditions such as bronchitis; suppress immune system; heavy metals and organic chemicals may cause cancer
Carbon monoxide	Motor vehicles, industries, fireplace	Reduces blood's ability to transport oxygen; low levels cause headaches and fatigue; higher levels cause mental impairment or death
Ozone	Formed in atmosphere	Irritates eyes and respiratory tract; produces chest discomfort; aggravates asthma and chronic bronchitis

What is the Air Quality Index (AQI)?

The Air Quality Index (AQI) is an index for reporting daily air quality. The AQI value indicates how clean or polluted the air is in a given location and whether it will cause any health effects. The color associated with a value serves as an easy way for people to determine whether the air quality in their community is reaching unhealthy levels. The U.S. Environmental Protection Agency calculates the AQI for five major air pollutants regulated by the Clean Air Act. The five major pollutants are ground-level ozone, particle pollution (also known as particulate matter), carbon monoxide, sulfur dioxide, and nitrogen dioxide. Corresponding national air quality standards have been established by the EPA. The AQI is divided into six categories:

Air Quality Index (AQI) Values	Color	Level of Health Concern	Health Precautions
0–50	Green	Good	Air quality is considered satisfactory, and air pollution poses little or no risk.
51–100	Yellow	Moderate	Air quality is acceptable; however, moderate health concerns may exist for small numbers of the population who are extremely sensitive to air pollution.
101–150	Orange	Unhealthy for sensitive groups	Members of sensitive groups may experience health effects. The general public is not likely to be affected.

Air Quality Index (AQI) Values	Color	Level of Health Concern	Health Precautions
151–200	Red	Unhealthy	Everyone may begin to experience health effects, with members of sensitive groups experiencing more serious health effects.
201–300	Purple	Very unhealthy	Health warnings of emergency conditions. The entire population is more likely to be affected.
301–500	Maroon	Hazardous	Health alert; everyone may experience more serious health effects.

The website *https://AirNow.gov* provides current information on the air quality for any place in the United States. It also includes an interactive map and archival information for air quality in the past.

What are the components of smog?

Smog, the most widespread pollutant in the United States, is a photochemical reaction resulting in ground-level ozone. Ozone, an odorless, tasteless gas in the presence of light, can initiate a chain of chemical reactions. Ozone is a desirable gas in the stratospheric layer of the atmosphere, but it can be hazardous to one's health when found near Earth's surface in the troposphere. The hydrocarbons, hydrocarbon derivations, and nitric oxides emitted from such sources as automobiles are the raw materials for photochemical reactions. In the presence of oxygen and sunlight, the nitric oxides combine with organic compounds, such as the hydrocarbons from unburned gasoline, to produce a whitish haze, sometimes tinged with a yellow-brown color. In this process, a large number of new hydrocarbons and oxyhydrocarbons are produced. These secondary hydrocarbon products may comprise as much as 95 percent of the total organics in a severe smog episode.

Air pollution is so bad in China—thanks to the reliance on coal for energy—that most Chinese people wear masks when they go outside in the thickened air. Other cities, such as Mexico City, Los Angeles, and others, have serious air pollution problems, too.

What were the goals of the Montréal Protocol?

The Montréal Protocol was signed in 1987 by members of the United Nations to phase out and reduce production of chlorofluorocarbons (CFCs) and other ozone-depleting substances (ODS) by 50 percent by 1998. Amendments and revisions to the Montréal Protocol were made in 1990,

What compounds cause ozone depletion in the stratosphere?

Research in the 1970s linked chlorofluorocarbons (CFCs), such as freon, to the depletion of the ozone layer. In 1978, the use of CFC propellants in spray cans was banned in the United States. In 1987, the Montréal Protocol was signed, and the signatory nations committed themselves to a reduction in the use of CFCs and other ozone-depleting substances.

1992, 1997, 1999, and 2016 to include other chemicals and accelerate the phase-out of certain chemicals. Production of CFCs, carbon tetrachloride, and methyl chlororform was phased out by 1996 in the United States and other highly developed countries. The United States and other developed countries are scheduled to complete the phase-out of hydrochlorofluorocarbons by 2020.

What is the Kyoto Protocol?

The Kyoto Protocol was adopted at a meeting in Kyoto, Japan, in December 1997 and entered into force in 2005. Participating countries committed to reduce national emissions over the period 2008–2012 to an average of 5 percent against the 1990 levels. The protocol covers these greenhouse gases: carbon dioxide, methane, nitrous oxide, hydrofluorocarbons, perfluorocarbons, and sulphur hexafluoride. The United States withdrew from the Kyoto Protocol in 2001.

In December 2012, the Doha Amendment to the Kyoto Protocol was adopted. This amendment provides for a second commitment period from January 1, 2013, to December 31, 2020, for countries to reduce their greenhouse gas emissions by at least 18 percent below 1990 levels. As of 2018, 117 countries had accepted the Doha Amendment, although it has not been accepted by the United States.

What is the main goal of the Paris Agreement?

The main goal of the Paris Agreement is to keep global temperature rise in the twenty-first century well below 2 degrees Celsius above pre-Industrial Revolution levels. Furthermore, an effort should be made to limit temperature increase to 1.5 degrees Celsius. The agreement was reached in 2015 and ratified in 2016. Initially, the United States ratified the agreement but has since indicated that the country will not maintain its commitment to the Paris Agreement.

What are some of the accomplishments achieved in reducing air pollution since the Clean Air Act was passed in 1970?

Since the passage of the Clean Air Act in 1970, the economy has grown dramatically due to advances in technology, while the levels of air pollution have decreased dramatically. One of the goals of the EPA was to set national air-quality standards for six common air pollutants—carbon monoxide, lead, nitrogen dioxide, ozone, particulate matter, and sulfur dioxide. Since the passage of the Clean Air Act in 1970 through 2017, the amount of these six pollutants in the air has decreased by an average of 73 percent, while the gross domestic product grew by 262 percent. Other accomplishments are:

- Large industrial sources (chemical plants, petroleum refineries, and paper mills) emit 1.5 million tons (1,361 metric tons) less of toxic air pollutants than in 1990
- New cars and trucks are 99 percent cleaner
- New locomotives are 90 percent cleaner than preregulation locomotives
- Sulfur in gasoline has been reduced by 90 percent, while in diesel fuel, it has been reduced by 99 percent
- Power plants have cut emissions that cause acid rain
- Fewer serious health effects in the American population caused by air pollution
- The use of ozone-depleting substances, such as CFCs and halons, has been phased out

What is acid rain?

Acid deposition is the fallout of acidic compounds in solid, liquid, or gaseous forms. Wet deposition occurs as precipitation, while dry deposition is the fallout of particulate matter. Acid rain is the best-known form of acid deposition. The term "acid rain" was

Acidified rain—the result of air pollution—can destroy entire ecosystems. It can also eat away at stone, architecture, and important archeological sites.

coined by British chemist Robert Angus Smith (1817–1884) who, in 1872, published *Air & Rain: The Beginnings of a Chemical Climatology*. Since then, "acid rain" has become an increasingly used term for rain, snow, sleet, or other precipitation that has been polluted by acids such as sulfuric and nitric acids. When gasoline, coal, or oil are burned, their waste products of sulfur dioxide and nitrogen dioxide combine in complex chemical reactions with water vapor in clouds to form acids. Acid rain may be either wet deposition or dry deposition from the atmosphere. Wet deposition refers to acid rain, fog, and snow. Dry deposition occurs in dry weather when the chemicals are incorporated into dust and/or smoke. It falls and sticks to the ground, buildings, cars, and trees. Dry deposited particles are washed off in rain, leading to increased runoff. The runoff makes the mixture more acidic. In the course of the hydrological cycle, about half of the acidity in the atmosphere falls back to Earth through dry deposition. Natural emissions of sulfur and nitrogen compounds, combined with those in air pollution, have resulted in severe ecological damage to lakes, streams, and forests. Acid rain also effects the composition of building materials, including metals (for example, bronze used in statues), marble, and limestone.

How acidic is acid rain?

Acidity or alkalinity is measured by a scale known as the pH (potential for hydrogen) scale. It runs from 0 to 14. Since it is logarithmic, a change in one unit equals a tenfold increase or decrease. Therefore, a solution at pH 2 is ten times more acidic than one at pH 3 and one hundred times as acidic as a solution at pH 4. Zero is extremely acid, 7 is neutral, and 14 is very alkaline. Any rain measuring below 5 is considered acid rain. Normal rain and snow containing dissolved carbon dioxide (a weak acid) measure about pH 5.6. Most acid rain in the United States has a pH of 4.2–4.4. The area centered around Lake Erie and Lake Ontario has acid rain with the highest pH in North America.

Which pollutants lead to indoor air pollution?

Indoor air pollution results from conditions in modern, high-energy efficiency buildings, which have reduced outside air exchange or have inadequate ventilation, chemical contamination, and microbial contamination. Indoor air pollution, also known as "sick building syndrome," can produce various symptoms, such as headache, nausea, and eye, nose, and throat irritation. In addition, houses are affected by indoor air pollution emanating from consumer and building products and from tobacco smoke. Below are listed some pollutants found in houses:

Pollutant	Sources	Effects
Asbestos	Old or damaged insulation; fireproofing; acoustical tiles	Many years later, chest and abdominal cancers and lung diseases
Biological pollutants	Bacteria; mold and mildew; viruses; animal dander and cat saliva; mites; cockroaches; pollen	Eye, nose, and throat irritation; shortness of breath; dizziness; lethargy; fever; digestive problems; asthma; influenza and other infectious diseases

Pollutant	Sources	Effects
Carbon monoxide	Unvented kerosene and gas heaters; leaking chimneys and furnaces; wood stoves and fireplaces; gas stoves; automobile exhaust from attached garages; tobacco smoke	At low levels, fatigue; at higher levels, impaired vision and coordination; headaches; dizziness; confusion; nausea; fatal at very high concentrations
Formaldehyde	Plywood, wall paneling, particle board, and fiberboard; foam insulation; fire and tobacco smoke; textiles and glues	Eye, nose, and throat irritations; wheezing and coughing; fatigue; skin rash; severe allergic reactions; may cause cancer
Lead	Automobile exhaust; sanding or burning of lead paint; soldering	Impaired mental and physical development in children; decreased coordination and mental abilities; kidneys, nervous system, and red blood cell damage
Mercury	Some latex paints	Vapors can cause kidney damage; long-term exposure can cause brain damage
Nitrogen dioxide	Kerosene heaters and unvented gas stoves and heaters; tobacco smoke	Eye, nose, and throat irritation; may impair lung function and increase respiratory infections in young children
Organic gases	Paints, paint strippers, solvents, and wood preservatives; aerosol sprays; cleansers and disinfectants; moth repellents; air fresheners; stored fuels; hobby supplies; dry-cleaned clothing	Eye, nose, and throat irritation; headaches; loss of coordination; nausea; damage to liver, kidneys, and nervous system; some organics cause cancer in animals and are suspected of causing cancer in humans
Pesticides	Products used to kill household pests and products used on lawns or gardens that drift or are tracked inside the house	Irritation to eye, nose, and throat; damage to nervous system and kidneys; cancer
Radon	Earth and rock beneath the home; well water; building materials	No immediate symptoms; estimated to cause about 10 percent of lung cancer deaths; smokers at higher risk

Why is exposure to asbestos a health hazard?

Asbestos is a general name given to six different minerals that occur naturally in rock and soil. Asbestos fibers were used in building materials between 1900 and the early 1970s as insulation for walls and pipes, as fireproofing for walls and fireplaces, in soundproofing and acoustic ceiling tiles, as a strengthener for vinyl flooring and joint compounds, and as a paint texturizer. Asbestos poses a health hazard only if the tiny fibers are released into the air, but this can happen with any normal fraying or cracking. Asbestos removal aggravates this normal process and multiplies the danger level—it should only be handled by a contractor trained in handling asbestos. Once released, the particles can hang suspended in the air for more than 20 hours.

Exposure to asbestos has long been known to cause asbestosis. This is a chronic, restrictive lung disease caused by the inhalation of tiny mineral asbestos fibers that scar lung tissues. Asbestos has also been linked with cancers of the larynx, pharynx, oral cavity, pancreas, kidneys, ovaries, and gastrointestinal tract. The American Lung Association reports that prolonged exposure doubles the likelihood that a smoker will develop lung cancer. It takes cancer fifteen to thirty years to develop from asbestos. Mesothelioma is a rare cancer affecting the surface lining of the pleura (lung) or peritoneum (abdomen) that generally spreads rapidly over large surfaces of either the thoracic or abdominal cavities. Current treatment methods include surgery, radiation, and chemotherapy, although mesothelioma continues to be difficult to control.

A worker wears protective gear from head to toe before removing asbestos from an old building. Asbestos fibers can be breathed in and cause lung disease.

How does the release of volatile organic compounds (VOCs) differ outdoors and indoors?

Volatile organic compounds (VOCs) are found both outdoors and indoors. They are regulated by the EPA for outdoor use to prevent the formation of ground-level ozone, a component of photochemical smog. Indoors, the main concern regarding VOCs is the adverse health effects on individuals exposed to them. These health effects include dizziness; headaches; nausea; irritation to the eyes, respiratory tract, and skin; and cancer. The EPA began to regulate the manufacture of paint and other coatings in the late 1990s to improve indoor air quality. Manufacturers developed new paints and coatings with low or zero VOC content. The carpet industry began to manufacture carpets and adhesives with very low emissions of VOCs.

What causes formaldehyde contamination in homes?

Formaldehyde contamination is related to the widespread construction use of wood products bonded with urea-formaldehyde resins and products containing formaldehyde. Major formaldehyde sources include subflooring of particle board; wall paneling made from hardwood plywood or particle board; and cabinets and furniture made from particle board, medium-density fiberboard, hardwood plywood, or solid wood. Urea-formaldehyde foam insulation (UFFI) has received the most media notoriety and regulatory attention. Formaldehyde is also used in drapes, upholstery, carpeting, wallpaper adhesives, milk cartons, car bodies, household disinfectants, permanent-press clothing, and paper towels. In particular, mobile homes seem to have higher formaldehyde levels than other houses. The release of formaldehyde into the air by these products (called outgassing) can develop poisoning symptoms in humans. The EPA classifies formaldehyde as a potential human carcinogen (cancer-causing agent).

Why is radon a health hazard?

Radon is a colorless, odorless, tasteless, radioactive, gaseous element produced by the decay of radium. It has three naturally occurring isotopes found in many natural materials, such as soil, rocks, well water, and building materials. Because the gas is continually released into the air, it makes up the largest source of radiation that humans receive. A National Academy of Sciences (NAS) report noted that radon was the second leading cause of lung cancer. It has been estimated that it may cause as much as 12 percent, or about 15,000–22,000 cases, of lung cancer deaths annually. Smokers seem to be at a higher risk than nonsmokers. The U.S. Environmental Protection Agency (EPA) recommends that in radon testing, the level should not be more than 4 picocuries per liter. The estimated national average is 1.5 picocuries per liter. Because EPA's "safe level" is equivalent to two hundred chest X-rays per year, some experts believe that lower levels are appropriate. The American Society of Heating, Refrigeration, and Air-Conditioning Engineers (ASHRAE) recommends 2 picocuries/liter. The EPA estimates that nationally, 8–12 percent of all houses are above the 4 picocuries/liter limit, whereas in another survey in 1987, it was estimated that 21 percent of homes were above this level.

What are greenhouse gases?

Scientists recognize carbon dioxide, methane, chlorofluorocarbons, nitrous oxide, and water vapor as significant greenhouse gases. Greenhouse gases account for less than 1 percent of Earth's atmosphere. These gases trap heat in Earth's atmosphere, preventing the heat from escaping back into space. Human activities, such as burning fossil fuels for gasoline in automobiles, electricity, and heat, account for the dramatic increase in the amount of greenhouse gas emissions in the United States in the last 150 years.

Emissions of Greenhouse Gases in the United States (Million Metric Tons of Gas)

Gas	1990	1995	2000	2005	2010	2013	2014	2015	2016
Carbon dioxide	5,121	5,439	6,001	6,132	5,701	5,520	5,569	5,421	5,311
Methane	780	767	709	689	694	663	664	665	657
Nitrous oxide	355	372	359	358	367	363	361	380	369
Fluorinated Gases*	100	117	148	142	161	164	169	172	173
Total	6,356	6,696	7,217	7,320	6,923	6,709	6,763	6,638	6,511

*Hydrofluorocarbons (HFCs), perfluorocarbons (PFCs), sulfur hexafluoride (SF_6), and nitrogen trifluoride (NF_3)

Which greenhouse gas has no natural sources?

Fluorinated gases have no natural sources. Emissions of fluorinated gases only come from human-related activities. The greatest source of emissions of fluorinated gases is as a substitute for ozone-depleting substances. Hydrofluorocarbons (HFCs) are used as refrigerants, replacing chlorofluorocarbons (CFCs) and hydrochlorofluorocarbons (HCFCs), which were banned by the Montréal Protocol. The newest amendment to the Montréal Protocol requires the phasing out of HFCs. Other uses of HFCs are aerosol propellants, foam-blowing agents, solvents, fire retardants, manufacture of semiconductors, production of aluminum, and transmission and distribution of electricity.

Emissions of carbon dioxide, methane, and nitrous oxide all have natural sources in addition to human-related activities.

What is the greenhouse effect?

The greenhouse effect is a warming near Earth's surface that results when Earth's atmosphere traps the Sun's heat. The atmosphere acts much like the glass walls and roof

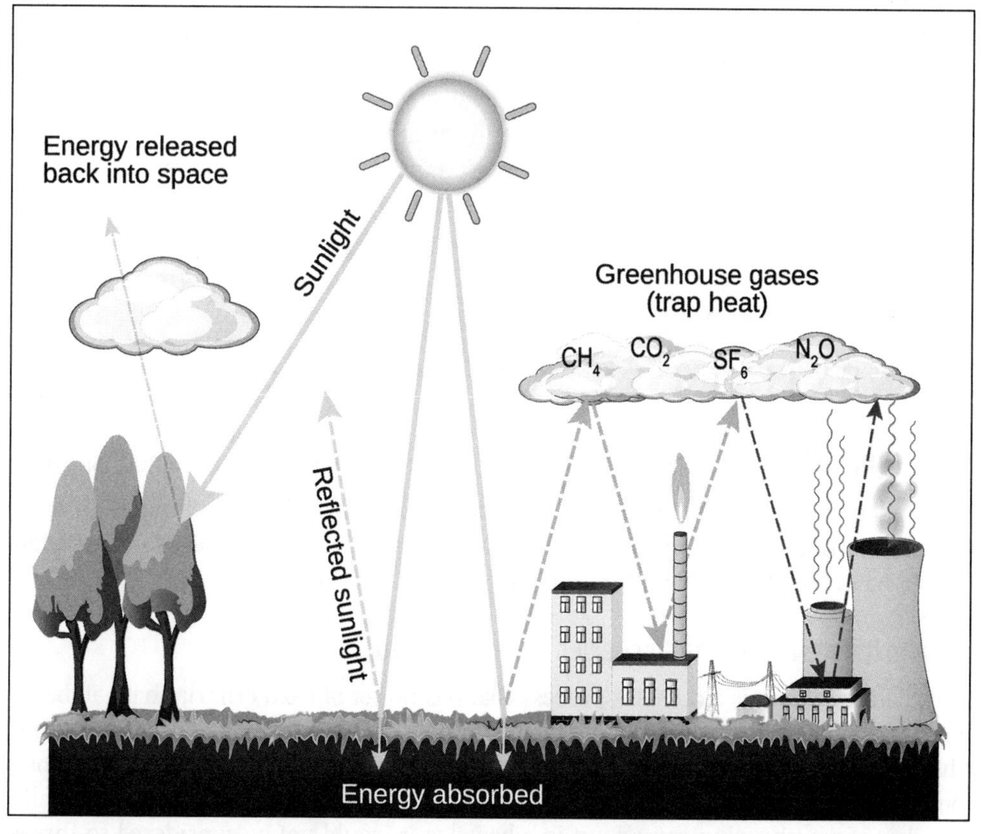

Certain gases can prevent heat from the sun from escaping back into space; the result is a greenhouse effect that leads to a warming of the atmosphere.

of a greenhouse. The effect was described by John Tyndall (1820–1893) in 1861. It was given the greenhouse analogy much later, in 1896, by Swedish chemist Svante Arrhenius (1859–1927). The greenhouse effect is what makes Earth habitable. Without the presence of water vapor, carbon dioxide, and other gases in the atmosphere, too much heat would escape, and Earth would be too cold to sustain life. Carbon dioxide, methane, nitrous oxide, and other greenhouse gases absorb the infrared radiation rising from Earth and hold this heat in the atmosphere instead of reflecting it back into space.

In the twentieth century, the increased buildup of carbon dioxide, caused by the burning of fossil fuels, has been a matter of concern. Some controversy exists concerning whether the increase noted in Earth's average temperature is due to the increased amount of carbon dioxide and other gases or is due to other causes. Volcanic activity, destruction of the rain forests, use of aerosols, and increased agricultural activity may also be contributing factors.

How do chlorofluorocarbons affect Earth's ozone layer?

Chlorofluorocarbons (CFCs) are hydrocarbons, such as freon, in which part or all of the hydrogen atoms have been replaced by fluorine atoms. These can be liquids or gases, are nonflammable and heat stable, and are used as refrigerants, aerosol propellants, and solvents. When released into the air, they slowly rise into Earth's upper atmosphere, where they are broken apart by ultraviolet rays from the Sun. Some of the resultant molecular fragments react with the ozone in the atmosphere, reducing the amount of ozone. The CFC molecules' chlorine atoms act as catalysts in a complex set of reactions that convert two molecules of ozone into three molecules of ordinary oxygen.

ENDANGERED AND EXTINCT PLANTS AND ANIMALS

What is the difference between an "endangered" species and a "threatened" species?

An "endangered" species is one that is in danger of extinction throughout all or a significant portion of its range. A "threatened" species is one that is likely to become endangered in the foreseeable future due to declining numbers.

How is it determined that a species is "endangered"?

This determination is a complex process that has no set of fixed criteria that can be applied consistently to all species. The known number of living members in a species is not the sole factor. A species with a million members known to be alive but living in only one small area could be considered endangered, whereas another species having a smaller number of members but spread out in a broad area would not be considered so threatened. Reproduction data—the frequency of reproduction, the average number of offspring born, the survival rate, etc.—enter into such determinations. In the United States, the director of the U.S. Fish and Wildlife Service (within the Department of the Interior) determines which species are to be considered endangered based on research and field data from specialists, biologists, botanists, and naturalists.

This photograph of an Eastern puma was taken in 1986, which is before the species was declared officially extinct in 2018.

According to the Endangered Species Act of 1973, a species can be listed if it is threatened by any of the following:

1. The present or threatened destruction, modification, or curtailment of its habitat or range.
2. Utilization for commercial, sporting, scientific, or educational purposes at levels that detrimentally affect it.
3. Disease or predation.
4. Absence of regulatory mechanisms adequate to prevent the decline of a species or degradation of its habitat.
5. Other natural or man-made factors affecting its continued existence.

If the species is so threatened, the director then determines the "critical habitat," that is, the species' inhabitation areas that contain the essential physical or biological features necessary for the species' preservation. The critical habitat can include non-habitation areas, which are deemed necessary for the protection of the species.

Which species have become extinct since the Endangered Species Act was passed in 1973?

Eleven species have been delisted due to extinction since the Endangered Species Act was passed.

First Listed	Date Delisted	Species Name
10/13/1970	1/15/1982	Tecopa pupfish (*Cyprinodon nevadensis calidae*)
3/11/1967	9/2/1983	Longjaw cisco (*Coregonus alpenae*)
3/11/1967	9/2/1083	Blue pike (*Stizostedion vitreum glaucum*)
3/11/1967	10/12/1983	Santa Barbara song sparrow (*Melospiza melodia graminea*)
6/14/1976	1/9/1984	Sampson's pearlymussel (*Epioblasma sampsoni*)
4/30/1980	12/4/1987	Amistad gambusia (*Gambisoa amistadensis*)
3/11/1967	12/12/1990	Dusky seaside sparrow (*Ammodramus maritimus nigrescens*)
8/27/1984	2/23/2004	Guam broadbill (*Myiagra freycineti*)
12/8/1977	2/23/2004	Mariana mallard (*Anas oustaleti*)
4/10/1979	10/28/2008	Caribbean monk seal (*Monachus tropicalis*)
6/4/1973	1/23/2018	Eastern puma (=cougar) (*Puma [= Felis] concolor couguar*)

Which species have been removed from the endangered species list because they have recovered?

Fifty-five species have been removed from the endangered species list because they have recovered.

First Listed	Date Delisted	Species Name
6/2/1970	2/4/1985	Brown pelican (U.S. Atlantic Coast, Florida, Alabama) (*Pelecanus occidentalis*)
6/2/1970	9/12/1985	Palau ground dove (*Gallicolumba canifrons*)
6/2/1970	9/12/1985	Palau fantail flycatcher (*Rhipidura lepida*)
6/2/1970	9/12/1985	Palau owl (*Pyroglaux podargina*)
3/11/1967	6/04/1987	American alligator (*Alligator mississippiensis*)
4/26/1978	9/14/1989	Rydberg milk-vetch (*Astragalus perianus*)
6/2/1970	06/16/1994	Gray whale (except where listed) (*Eschrichtius robustus*)
6/2/1970	10/5/1994	Arctic peregrine falcon (*Falco peregrinus tundrius*)
12/30/1974	3/9/1995	Eastern gray kangaroo (*Macropus giganteus*)
12/30/1974	3/9/1995	Red kangaroo (*Macropus rufus*)
12/30/1974	3/9/1995	Western gray kangaroo (*Macropus fuliginosus*)
6/2/1970	8/25/1999	American peregrine falcon (*Falco peregrinus anatum*)
3/11/1967	3/20/2001	Aleutian Canada goose (*Branta canadensis leucopareia*)
9/17/1980	8/27/2002	Robbins' cinquefoil (*Potentilla robbinsiana*)
3/11/1967	7/24/2003	Columbian white-tailed deer (Douglas County, Oregon) (*Odocoileus virginianus leucurus*)
6/2/1970	9/21/2004	Tinian monarch (old world flycatcher) (*Monarcha takatsukasae*)
5/22/1997	8/18/2005	Eggert's sunflower (*Helianthus eggertii*)
3/11/1967	8/8/2007	Bald eagle (lower 48 states) (*Haliaeetus leucocephalus*)
7/1/1985	9/25/2008	Virginia northern flying squirrel (*Glaucomys sabrinus fuscus*)
7/19/1990	10/7/2008	Hoover's woolly-star (*Eriastrum hooveri*)
3/28/2008	4/2/2009	Gray wolf (Northern Rocky Mountain DPS) (*Canis lupus*)
6/2/1970	12/17/2009	Brown pelican (except U.S. Atlantic Coast) (*Pelecanus occidentalis*)
9/5/1985	2/18/2011	Maguire daisy (*Erigeron maguirei*)
7/5/1979	9/2/2011	Tennessee purple coneflower (*Echinacea tennesseensis*)
8/30/1990	9/15/2011	Lake Erie water snake (*Nerodia sipedon insularum*)
9/3/1986	11/28/2011	Concho water snake (*Nerodia paucimaculata*)
6/2/1970	5/23/2012	Morelet's crocodile (*Crocodylus moreletii*)
7/31/1985	3/4/2013	Virginia northern flying Squirrel (*Glaucomys sabrinus fuscus*)
4/17/1989	6/14/2013	Magazine Mountain shagreen (*Inflectarius magazinensis*)
4/5/1990	12/4/2013	Steller sea lion (*Eumetopias jubatus*)
9/12/1977	4/1/2014	Island night lizard (*Xantusia riversiana*)
10/18/1993	3/23/2015	Oregon chub (*Oregonichthys crameri*)
3/11/1967	12/16/2015	Delmarva Peninsula fox squirrel (*Sciurus niger cinereus*)
6/11/1985	1/7/2016	Modoc sucker (*Catostomus microps*)
8/7/1984	2/11/2016	Johnston's frankenia (*Frankenia johnstonii*)
1/7/1992	3/11/2016	Louisiana black bear (*Ursus americanus luteolus*)
3/5/2004	9/12/2016	San Miguel Island fox (*Urocyon littoralis littoralis*)

First Listed	Date Delisted	Species Name
3/5/2004	9/12/2016	Santa Cruz Island fox (*Urocyon littoralis santacruzae*)
3/5/2004	9/12/2016	Santa Rosa Island fox (*Urocyon littoralis santarosae*)
4/7/1988	10/11/2016	White-haired goldenrod (*Solidago albopilosa*)
6/2/1970	10/11/2016	Humpback whale (*Megaptera novaeangliae*) • Southeastern Pacific
6/2/1970	10/11/2016	Humpback whale (*Megaptera novaeangliae*) • West Indies
6/2/1970	10/11/2016	Humpback whale (*Megaptera novaeangliae*) • Hawaii
6/2/1970	10/11/2016	Humpback whale (*Megaptera novaeangliae*) • Brazil
6/2/1970	10/11/2016	Humpback whale (*Megaptera novaeangliae*) • Gabon/Southwest Africa
6/2/1970	10/11/2016	Humpback whale (*Megaptera novaeangliae*) • West Australia
6/2/1970	10/11/2016	Humpback whale (*Megaptera novaeangliae*) • East Australia
6/2/1970	10/11/2016	Humpback whale (*Megaptera novaeangliae*) • Oceania
3/9/1978	5/1/2017	Gray wolf (*Canis lupus*) • Northern Rocky Mountains
12/2/1970	5/5/2017	Scarlet-chested parakeet (*Neophema splendida*)
6/2/1970	5/5/2017	Turquoise parakeet (*Neophema pulchella*)
5/27/1978	3/29/2018	Eureka Valley evening-primrose (*Oenothera avita ssp. eurekensis*)
10/6/1987	5/16/2018	Black-capped vireo (*Vireo atricapilla*)
9/30/1988	5/18/2018	Lesser long-nosed bat (*Leptonycteris curasoae yerbabuenae*)
10/14/1998	7/2/2018	Hidden Lake bluecurls (*Trichostema austromontanum ssp. compactum*)
10/20/1990	11/19/2018	Deseret milkvetch (*Astragalus desereticus*)

What are some milestones since the Endangered Species Act of 1973 was enacted?

- 1976—First invertebrate species (21 clams and 7 insects) gain federal protection.
- 1977—First plant species (the San Clemente Island Indian paintbrush, San Clemente Island larkspur, San Clemente Island broom, and San Clemente Island bushmallow) are listed as endangered.
- 1979—As a "last ditch effort," plans to capture the last remaining California condors to form a captive breeding flock to save the species from extinction are announced.

The Santa Cruz Island fox is one of a number of success stories of animals brought back from the edge of extinction thanks to conservation efforts.

- 1985—The brown pelican is the first species delisted (removed) from the Federal List of Endangered and Threatened Wildlife and Plants because of recovery.
- 1987—The American alligator is delisted following recovery.
- 1987—The red wolf is reintroduced back into the wild at Alligator River National Wildlife Refuge in North Carolina.
- 1987—The last dusky seaside sparrow dies in captivity. The species is extinct.
- 1991—The California condor is reintroduced back into the wild at Hopper Mountain National Wildlife Refuge in southern California.
- 1994—The Eastern North Pacific population of the gray whale is delisted following recovery.
- 1994—The Arctic peregrine falcon is delisted following recovery.
- 1995—The gray wolf is reintroduced to Yellowstone National Park and Idaho ending a 70-year absence.
- 1999—The American peregrine falcon is delisted following recovery.
- 2001—The Aleutian Canada goose is delisted following recovery.
- 2004—The California condor reproduces in the wild for the first time in seventeen years.
- 2007—The bald eagle recovers and is removed from protection under the Endangered Species Act.
- 2008—The polar bear is listed as threatened due to ice flows melting in the Arctic. This is the first species listed due to the impact of climate change.
- 2009—Brown pelicans are removed from protection under the Endangered Species Act since they are found across Florida and the Gulf and Pacific coasts as well as in the Caribbean and Latin America.
- 2011—The thirtieth anniversary of the rediscovery of the black-footed ferret and the twentieth anniversary of the successful return to the wild of the black-footed ferret.
- 2011—First successful captive breeding of the Ozark hellbender at the St. Louis Zoo.
- 2012—The gray wolf is delisted following recovery.

Why is the International Union for Conservation of Nature's Red List an important tool?

The International Union for Conservation of Nature's (IUCN) Red List was established in 1964. The Red List collects information on the status of animal, plant, and fungi species. Some of the information collected to assess the status of a species are population size, habitat and ecology, range, use and/or trade, threats, and conservation actions. To date, more than 93,000 species have been assessed. The goal is to assess 160,000 species by 2020. Not every species on the Red List is threatened or endangered. The Red

List is referred to as a "barometer of life," as it measures pressures on a species, which guide decisions for conservation actions to help prevent extinction and understand the status of the world's biodiversity. It provides valuable information to identify habitats that need to be protected and the reasons for the threats, such as climate change.

How many of the species assessed on the Red List are threatened with extinction?

More than 26,000 individual species, or 27 percent of all assessed species, are threatened with extinction.

Species	Percent of Assessed Species Threatened with Extinction
Amphibians	41
Reef-building corals	33
Sharks and rays	31
Selected crustaceans	27
Mammals	25
Birds	13
Cyads	63
Conifers	34

What is the rate of species extinction in the tropical rain forests?

Biologists estimate that tropical rain forests contain approximately half of Earth's animal and plant species. These forests contain 155,000 of the 250,000 known plant species and innumerable insect and animal species. Nearly one hundred species become extinct each day. This is equivalent to four species per hour. At the current rates, 5–10 percent of tropical rain forests will become extinct every decade. However, new species are also being discovered every year. Therefore, the rate of species extinction fluctuates.

Theories abound as to why dinosaurs became extinct at the end of the Cretaceous period, and even the rate of extinction (whether it was sudden and catastrophic or a slow decline) has been debated.

Why did dinosaurs become extinct?

Many theories exist as to why dinosaurs disappeared from Earth about sixty-five million years ago. Scientists debate whether dinosaurs became extinct gradually or all at once. The gradualists believe that the dinosaur population steadily declined at the end of the Cretaceous period. Numerous reasons have been proposed for this. Some claim that the dinosaurs' extinction was caused by biological changes that made them less competitive with other organisms, especially the mammals that were just beginning to appear. Overpopulation has been argued, as has the theory that mammals ate too many dinosaur eggs for the animals to reproduce themselves. Others believe that disease—everything from rickets to constipation—wiped them out. Changes in climate, continental drift, volcanic eruptions, and shifts in Earth's axis, orbit, and/or magnetic field have also been held responsible.

The catastrophists argue that a single disastrous event caused the extinction not only of the dinosaurs but also of a large number of other species that coexisted with them. In 1980, American physicist Luis Alvarez (1911–1988) and his geologist son, Walter Alvarez (1940–), proposed that a large comet or meteoroid struck Earth sixty-five million years ago. They pointed out that a high concentration of the element iridium is in the sediments at the boundary between the Cretaceous and Tertiary periods. Iridium is rare on Earth, so the only source of such a large amount of it had to be outer space. This iridium anomaly has since been discovered at over fifty sites around the world. In 1990, tiny, glass fragments, which could have been caused by the extreme heat of an impact, were identified in Haiti. A 110-mile- (177-kilometer-) wide crater in the Yucatan Peninsula, long covered by sediments, has been dated to 64.98 million years ago, making it a leading candidate for the site of this impact.

A hit by a large, extraterrestrial object, perhaps as much as 6 miles (9.3 kilometers) wide, would have had a catastrophic effect upon the world's climate. Huge amounts of dust and debris would have been thrown into the atmosphere, reducing the amount of sunlight reaching the surface. Heat from the blast may also have caused large forest fires, which would have added smoke and ash to the air. Lack of sunlight would kill off plants and have a dominolike effect on other organisms in the food chain, including the dinosaurs.

It is possible that the reason for the dinosaurs' extinction may have been a combination of both theories. The dinosaurs may have been gradually declining for whatever reason. The impact of a large object from space merely delivered the coup de grâce.

The fact that dinosaurs became extinct has been cited as proof of their inferiority and that they were evolutionary failures. However, these animals flourished for 150 million years. By comparison, the earliest ancestors of humanity appeared only about three million years ago. Humans have a long way to go before they can claim the same sort of success as the dinosaurs.

SUSTAINABILITY AND CONSERVATION

What is sustainability?

The most broad definition of sustainability is the practice of preserving Earth's resources for future generations. Sustainability encompasses ecologic, economic, and social issues. It includes the use and care of resources, changing patterns of consumption, and control of population growth, which impacts the total need for resources.

Is sustainability a new concept?

The concepts and philosophy of sustainability were a result of the environmental concerns of the 1960s and 1970s. In 1983, the United Nations (UN) established the World Commission on Environment and Development (WCED), also known as the Brundtland Commission, named after Gro Harlem Brundtland (1939–), the chair of the commission. In 1987, the WCED issued its report, "Our Common Future" (also known as the Brundtland Report). This report defined sustainable development as "development that meets the needs of the present without compromising the ability of future generations to meet their own needs." The elements of economic growth, social inclusion, and environmental protection must be balanced. In an effort to do so, in 2015 countries adopted the 2030 Agenda for Sustainable Development and its seventeen Sustainable Development Goals (SDGs). The SDGs include:

1. No poverty
2. Zero hunger
3. Good health and well-being
4. Quality education
5. Gender equality
6. Clean water and sanitation
7. Affordable and clean energy
8. Decent work and economic growth
9. Industry, innovation, and infrastructure
10. Reduced inequalities
11. Sustainable cities and communities
12. Responsible consumption and production
13. Climate action
14. Life below water
15. Life on land
16. Peace, justice, and strong institutions
17. Partnerships for the goals

What are the three "Rs" of sustainability?

Reduce, reuse, and recycle are the three "Rs" of sustainability. Reduction and reuse are considered the most effective ways to save natural resources, protect the environment, and save money. The benefits of reducing and reusing include:

- Prevents pollution by reducing the need to harvest new raw materials
- Saves energy
- Reduces greenhouse gas emissions

- Helps sustain the environment for future generations
- Reduces the amount of waste for recycling or disposal in landfills

How does recycling play a role in creating a sustainable environment?

Recycling is the process of collecting and processing materials and turning them into new products instead of throwing them away as trash. The benefits of recycling include:

- Reduces the amount of waste sent to landfills and incinerators
- Conserves natural resources (timber, water, and minerals)
- Prevents pollution by reducing the need to collect new materials
- Saves energy

What is a green product?

Green products are environmentally safe products that contain no chlorofluorocarbons, are degradable (can decompose), and are made from recycled materials. "Deep-green" products are those from small suppliers who build their identities around their claimed environmental virtues. "Greened-up" products come from the industry giants and are environmentally improved versions of established brands.

Is it possible to buy green electronics?

Electronic manufacturers are designing products that are more environmentally friendly. Consumers should look for products that:

- Contain fewer toxic constituents
- Use recycled materials in the new product
- Are designed for easy upgrading and disassembly
- Are energy efficient
- Use minimal packaging
- Have leasing or takeback options for reuse or recycling

What is a "green" building?

Green building, also called sustainable building, is the practice of creating structures and using processes that are responsible environmentally and resource-efficient throughout a building's life cycle from siting to design, construction, operation, maintenance, renovation,

What is eCycling?

ECycling is the reuse and recycling of electronic equipment. Donating used electronic equipment for use by others is the environmentally preferred alternative to discarding used electronics. If an electronic device cannot be reused, it should be eCycled.

and deconstruction. Green buildings are designed to reduce the impact of the built environment on human health and the natural environment by using resources, especially energy and water, efficiently; protecting occupant health; improving indoor environmental quality; and reducing waste, pollution, and environmental degradation. Green building first became a goal of the building industry during the 1990s.

Recycling is not just good for the planet; it can make good economic sense, too, by reusing materials, saving energy, and reducing the need for costly landfills.

What are green roofs?

Green roofs are roofs constructed with a vegetative layer grown on the rooftop. Benefits of a green roof are reduced energy use since green roofs absorb heat and act as insulators for buildings, reducing the energy needed for heating and cooling, and slowing stormwater runoff. The overall aesthetic value of a green roof can improve the quality of life.

Green walls are vegetated wall surfaces. The three major types of green walls are green façades, living walls, and retaining living walls. Green façades usually have vines and climbing plants that grow on supporting structures. Living walls have a greater diversity of plant species on them than green façades. They are either freestanding or attached to a structural wall. Green retaining living walls have the additional objective of stabilizing a slope to protect against erosion.

How does permaculture support the ideals of sustainability?

Permaculture is often thought of as a gardening technique, but it is more than that. The term was originally a contraction of the terms "permanent" and "agriculture" but has taken on the greater meaning of "permanent culture." Permaculture is described as a way of life and philosophy. At its core is the concept of designing all human systems so they integrate harmoniously with ecology. Systems designed based on the philosophy of permaculture have three core concepts: 1) caring for Earth; 2) caring for people; and 3) giving away the surplus, which will ensure success for the first two concepts.

What was the first U.S. national park?

On March 1, 1872, an act of the U.S. Congress signed by Ulysses S. Grant established Yellowstone National Park as the first national park. The action inspired a worldwide national park movement.

When was the U.S. National Park Service established?

The National Park Service was created by an act signed by President Woodrow Wilson on August 25, 1916. Its responsibility was to administer national parks and monuments.

The mission of the National Park Service is to preserve the natural and cultural resources for the enjoyment, education, and inspiration of this and future generations.

What is the largest national park in the United States?

The Wrangell St. Elias National Park in Alaska, encompassing 13.2 million acres, is the largest national park in the United States. It is the same size as Yellowstone National Park, Yosemite National Park, and the country of Switzerland combined. The boreal forest ecosystem, consisting of spruce, aspen, and balsam poplar with some muskeg and tussocks, is the predominant ecosystem in the park.

What causes most wildfires in the United States?

Nearly 90 percent of all wildland fires in the United States are caused by humans. Fires are caused by campfires left unattended, the burning of debris, discarding cigarettes in a negligent manner, and acts of arson. In 2017, 63,546 human-caused wildfires occurred in the United States, burning 4.8 million acres (1.9 million hectares) of land. Lightning strikes caused nearly eight thousand wildfires, burning 5.2 million acres (2.1 million hectares) of land.

Why are the number of forest fires increasing?

Some speculate as to the reason for the increase in wildfires. One reason is climate change. Warmer temperatures and more frequent periods of drought are conditions that permit wildfires to strengthen. Additionally, land-use practices, including the accumulation of biomass, presents fuel for wildfires.

How does ecological restoration achieve success?

Ecological restoration is the process of assisting the recovery of an ecosystem that has been degraded, damaged, or destroyed. In the natural world, ecosystems may be dam-

In what ways can forest fires be good for the environment?

Wildfires are critical to maintaining the integrity of forest and grassland ecosystems. Forest and grass fires, usually started by lightning, act as an ecologically renewing force by creating necessary conditions for plant germination and continued healthy growth to occur. The primary goal of fire management is to simulate the revitalizing aspects of natural fire cycles. Fire management also attempts to prevent large, catastrophic wildfires from occurring by removing accumulated debris from forests. Seen throughout the American West every summer, these extremely intense fires are caused primarily by decades of fire suppression, which has allowed heavy fuels—accumlated debris—to build up. Ironically, by attempting to prevent natural fires, humans have only increased their prevalence.

aged or destroyed due to natural events, such as forest fires or severe storms. The damaged or destroyed ecosystem will begin to reestablish itself with the community of native species. For example, certain plant germination and continued healthy growth of the forest is a result of wildfires. Ecosystems that have been degraded, damaged, or destroyed by human activities, such as the introduction of non-native species, overfishing or -hunting, land clearing and logging, coastal erosion, and mining, benefit from human-assisted restoration to encourage the process of ecological succession and recovery. Practitioners of ecological restoration do not carry out the work of ecosystem recovery but create the conditions for the ecosystem to revive itself. Some ecological restoration practices include removing non-native species to allow native species to flourish, controlling water to prevent erosion, and reintroducing wildlife. Success is achieved when the ecosystem is self-sustaining with the complete biodiversity of species interacting with each other. Ecological restoration is not a substitute for conservation practices that maintain an ecosystem's structure and function.

When was the first zoo in the United States established?

The Philadelphia Zoological Garden, chartered in 1859, was the first zoo in the United States. The zoo was delayed by the Civil War, financial difficulties, and restrictions on transporting wild animals. It opened in 1874 on 33 acres (13 hectares), and 282 animals were exhibited.

How do zoos play a role in conservation?

A primary goal of zoos accredited by the Association of Zoos and Aquariums (AZA) is the protection of wildlife and wild habitats. In addition to providing the best-quality care to

The Philadelphia Zoo is the largest zoological garden in the United States, having been opened in 1874. It currently houses nearly 1,300 animals, including such endangered species as the western lowland gorilla, Amur leopard, Chinese alligator, Bali mynah bird, Mhorr gazelle, and several lemur species.

animals in their care, zoos and aquariums also play an important role in educating zoo visitors and school students about wild animals, their habitats, and the importance of conservation and preservation.

How does the Species Survival Plan help sustain species of animals?

The Species Survival Plan program is a cooperatively managed program of the Association of Zoos and Aquariums. The plan oversees the population management of select species within the member institutions of the organization. These programs strive to maximize genetic diversity through breeding plans of a species to ensure the sustainability of a healthy, stable population of animals for the long-term future.

APPLIED SCIENCE AND TECHNOLOGY

INTRODUCTION AND HISTORICAL BACKGROUND

What is the difference between science and technology?

Science and technology are related disciplines but have different goals. The basic goal of science is to acquire a fundamental knowledge of the natural world. Outcomes of scientific research are the theorems, laws, and equations that explain the natural world. It is often described as a pure science. Technology is the quest to solve problems in the natural world with the ultimate goal of improving humankind's control of their environment. Technology is, therefore, often described as applied science: applying the laws of science to specific problems. The distinction between science and technology blurs since many times, researchers investigating a scientific problem will discover a practical application for the knowledge they acquire.

How do scientists and engineers differ?

Scientists and engineers both work to solve complex problems and seek answers to how things work in nature. Scientists focus on exploring and experimenting to make new discoveries, which are presented at conferences and published in peer-reviewed journals. Engineers apply the newest discoveries in science to develop, design, create, and build new products, new devices, and new systems, which is the technology used by society. Oftentimes, scientists and engineers collaborate to attain the final goal.

Why has STEM become important in education?

The fields of science, technology, engineering, and math (STEM) are essential for educating the future generations of students. STEM-education programs integrate these four separate disciplines into a cohesive program with real-world applications. These fields form the basis for the skills necessary for problem solving in the twenty-first century. Students with a strong background in STEM develop critical thinking and problem-solving skills in the context of creativity and collaboration, which are necessary for innovation. Many believe that STEM is important for the United States to remain a leader in advancing science and technology.

What percentage of the STEM workforce is comprised of women?

The STEM workforce is crucial for generating new ideas, receiving and commercializing patents, and providing flexibility and critical thinking in the modern economy. According to statistics compiled by the U.S. Department of Commerce:

- Overall, nearly as many women as men hold undergraduate degrees, but women make up only about 30 percent of all STEM degree holders.
- Women filled 47 percent of all jobs in the United States but held only 24 percent of STEM jobs.
- Women with STEM degrees are less likely to work in STEM occupations than men with STEM degrees. Women are more likely to work in the fields of education or health care.

What are the main branches of engineering?

Traditionally, the four main branches of engineering are civil, mechanical, electrical, and chemical engineering. Other areas of engineering, many of which are subspecialties of the main branches of engineering, are aerospace, agricultural, biomedical, computer, environmental, fire protection, geotechnical, nuclear, petroleum, sanitary, and transportation.

- *Civil engineers* design, construct, operate, and maintain construction projects, including roads, buildings, tunnels, airports, bridges, dams, and water supply and treatment systems.
- *Mechanical engineering* is one of the most diverse branches of engineering. The focus is to design, develop, build, and test mechanical and thermal devices, including tools, engines, and machines. Some of the areas that mechanical engineers tackle problems in are automotive, aerospace, biotechnology, electronics, environmental control, and manufacturing.
- *Electric engineers* design, develop, test, and supervise the manufacturing of electrical equipment, including electric motors, radar and navigation systems, communications systems, and power-generation equipment. Related to electric engineers are electronics engineers, who design and develop electronic equipment such as broadcast and communications systems, including global positioning systems (GPS) and portable music players.
- *Chemical engineers* focus on the production or use of chemicals, fuel, drugs, food, and other products. Chemical engineering encompasses applications from the basic sciences of chemistry, physics, biology, mathematics, and computing.

What are the major time periods of technology?

The history of technology is comprised of five major time periods. They are:

1. The Ancient World: 8000 B.C.E.–300 C.E. and Middle Ages: through 1599
2. The Age of Scientific Revolution: 1600–1790
3. The Industrial Revolution: 1791–1890
4. The Electrical Age: 1891–1934
5. The Atomic and Electronic Age: 1935–present

What were some of the technical advances made in each of the five major time periods of technology?

Many technical advances became possible due to the discoveries and achievements of researchers engaged in basic science research. Technical progress in the Ancient World (8000 B.C.E.–300 C.E.) and Middle Ages (up until 1599) was made in developing tools, materials, fixtures, and methods and procedures implemented by farmers, animal herders, cooks, tailors, and builders. Aqueducts, plumbing, the wheel, and sailing vessels are examples of technical advances in the earliest period of technology.

During the Age of Scientific Revolution (1600–1790), most achievements were attributed to a relatively small number of individuals. Key instruments were invented that enabled an examination of the natural universe or to measure natural phenomena. Some of these instruments were the telescope, the microscope, the mercury barometer and thermometer, the spring balance, and the pendulum clock.

The principles that governed machines were thoroughly examined during the years of the Industrial Revolution (1791–1890). Newly discovered principles were put in place, making it possible to build better machines and reduce principles to mathematical formulas. The use of interchangeable parts, steam engines, the electric telegraph, transcontinental railroad, the phonograph, lightweight cameras, the typewriter, and news of the first automobiles were all accomplishments of the Industrial Revolution.

The most important invention of the Electrical Age (1891–1934) was the automobile, which enjoyed widespread use by 1910. Other consumer items that were introduced and changed the everyday life of the average person during these years were safety razors, thermos bottles, electric blankets, cellophane, rayon, X-rays, and the zipper.

In the Atomic and Electronic Age (1935–present), we are constantly witnessing a multitude of achievements and advances in science and engineering. A major advancement was the discovery of nuclear power and its use for both generating power and weaponry. Some examples of modern technological advances include turbojet engines, computers, genetic engineering, and the Internet.

What were some of the greatest engineering achievements of the twentieth century?

The National Academy of Engineering published a list of the twenty greatest achievements of the twentieth century in 2000. The achievements were selected based on their impact on society's quality of life. The absence of any one achievement would greatly change our daily lives—from health benefits to how we travel, work, and spend our leisure time.

1. Electrification
2. Automobile
3. Airplane
4. Water supply and distribution
5. Electronics
6. Radio and television
7. Mechanization of agriculture
8. Computers
9. Telephone
10. Air conditioning and refrigeration
11. Highways
12. Spacecraft and space travel
13. Internet
14. Imaging
15. Household appliances
16. Health technologies and devices
17. Petroleum and petrochemical technologies
18. Laser and fiber optics
19. Nuclear technologies
20. High-performance materials

What is biomimicry, and how is it related to sustainability?

Biomimicry means looking at nature as a model to finding solutions to everyday problems. For example, by understanding how a gecko hangs on to surfaces, we have learned how to create specialized adhesives. By looking at how lotus plants repel dirt, we have created paint that can be cleaned with water. Biomimicry uses nature as a model, and the solutions are often more sustainable because elements in the natural world have evolved over time to sustain their place on the planet. Ecological sustainability is how biological systems remain diverse and productive indefinitely. Healthy wetlands and forests are examples of sustainable biological systems. Nature has already solved many of the challenges we face in the world today. Nature is, by its own virtue, sustainable. It has found sustainable solutions over millennia for almost every problem.

What is nanotechnology?

Nanotechnology is a relatively new field of science that aims to understand matter of dimensions between 1 and 100 nanometers. Nanomaterials may be engineered or occur

How large is a nanometer?

A nanometer equals one-billionth of a meter. A sheet of paper is about 100,000 nanometers thick. As a comparison, a single-walled carbon nanotube, measuring 1 nanometer in diameter, is one hundred thousand times smaller than a single strand of human hair, which measures 100 micrometers in diameter.

in nature. Some of the different types of nanomaterials, named for their individual shape and dimensions, are nanoparticles, nanotubes, and nanofilms. Nanoparticles are bits of material where all the dimensions are nanosized. Nanotubes are long, cylindrical strings of molecules whose diameter is nanosized. Nanofilms have a thickness that is nanosized, but the other dimensions may be larger. Researchers are developing ways to apply nanotechnology to a wide variety of fields, including transportation, sports, electronics, and medicine. Specific applications of nanotechnology include fabrics with added insulation but without additional bulk. Other fabrics are treated with coatings to make them stainproof. Nanorobots are being used in medicine to help diagnose and treat health problems. In the field of electronics, nanotechnology could shrink the size of many electronic products. Researchers in the food industry are investigating using nanotechnology to enhance the flavor of food. They are also searching for ways to introduce antibacterial nanostructures into food packaging.

COMPUTERS

What is an algorithm?

An algorithm is a set of clearly defined rules and instructions for the solution of a problem. It is not necessarily applied only in computers but can be a step-by-step procedure for solving any particular kind of problem. A nearly four-thousand-year-old Babylonian banking calculation inscribed on a tablet is an algorithm, as is a computer program that consists of step-by-step procedures for solving a problem.

The term is derived from the name of Muhammad ibn Mūsā al-Khwārizmī, a Baghdad mathematician who introduced Hindu numerals (including 0) and decimal calculation to the West. When his treatise was translated into Latin in the twelfth century, the art of computation with Arabic (Hindu) numerals became known as algorism.

Who invented the computer?

In 1823, British visionary Charles Babbage (1792–1871) persuaded the British government to finance an "analytical engine." This would have been a machine that could undertake any kind of calculation. It would have been driven by steam, but the most important innovation was that the entire program of operations was stored on a punched tape. Babbage's machine was not completed in his lifetime because the technology available to him was not sufficient to support his design. However, in 1991, a team led by Doron Swade (1946–) at London's Science Museum built the "analytical engine" (sometimes called the "difference engine") based on Babbage's work. Measuring 10 feet (3 meters) wide by 6.5 feet (2 meters) tall, it weighed three tons and could calculate equations down to 31 digits. The feat proved that Babbage was way ahead of his time, even though the device was impractical because one had to turn a crank hundreds of times in order to generate a single calculation. Modern computers use electrons, which travel at nearly the speed of light.

English mathematician and mechanical engineer Charles Babbage is credited with conceiving the digital programmable computer.

Based on the concepts of British mathematician Alan M. Turing (1912–1954), the earliest programmable electronic computer was the 1,500-valve Colossus, formulated by Max Newman (1897–1985), built by T. H. Flowers (1905–1998), and used by the British government in 1943 to crack the German codes generated by the cipher machine Enigma.

What is the Turing Award?

The Turing Award, considered the Nobel Prize in computing, is awarded annually by the Association for Computing Machinery to an individual who has made a lasting contribution of major technical importance in the computer field. The award, named for British mathematician Alan M. Turing, was first presented in 1966. Google, Inc., provides the financial support for the the award, which was raised to $1,000,000 in 2014. Prior to 2014, the award was $250,000. Recent winners of the Turing Award include:

A genius mathematician and cryptanalyst, Alan Turing is considered the father of theoretical computer science.

Turing Award Recipients, 2010–2018

Year	Award Recipient
2010	Leslie Gabriel Valiant (1949–)
2011	Judea Pearl (1936–)
2012	Silvio Micali (1954–) and Shafi Goldwasser (1959–)
2013	Leslie Lamport (1941–)
2014	Michael Stonebraker (1943–)
2015	Diffie Whitfield (1944–) and Martin Hellman (1945–)
2016	Tim Berners-Lee (1955–)
2017	John L. Hennessy (1952–) and David Patterson (1947–)
2018	Yoshua Bengio (1964–), Geoffrey E. Hinton (1947–), and Yann LeCun (1960–)

What was MANIAC?

MANIAC (mathematical analyzer, numerator, integrator, and computer) was built at the Los Alamos Scientific Laboratory under the direction of Nicholas C. Metropolis (1915–

What was the Turk?

The Turk was the name for a famous chess-playing automaton. An automaton, such as a robot, is a mechanical figure constructed to act as if it moves by its own power. On a dare in 1770, a civil servant in the Vienna Imperial Court named Wolfgang von Kempelen (1734–1804) created a chess-playing machine. This mustached, man-sized figure, carved from wood, wore a turban, trousers, and robe and sat behind a desk. In one hand, it held a long Turkish pipe, implying that it had just finished a pregame smoke, and its innards were filled with gears, pulleys, and cams. The machine seemed to be a keen chess player and dumbfounded onlookers by defeating all the best human chess players. It was a farce, however; its moves were surreptitiously made by a man hiding inside.

The Turk, so dubbed because of the outfit similar to traditional Turkish garb, is regarded as a forerunner to the Industrial Revolution because it created a commotion over devices that could complete complex tasks. Historians argue that it inspired people to invent other early devices such as the power loom and the telephone, and it even was a precursor to concepts such as artificial intelligence and computerization. Today, however, computer chess games are so sophisticated that they can defeat even the world's best chess masters. In May 1997, the Deep Blue chess computer defeated World Champion Garry Kasparov (1963–). Deep Blue was a 32-node IBM RS/6000 SP high-performance computer that used Power Two Super Chip processors (P2SC). Each node had a single microchannel card containing eight dedicated VLSI chess processors for a total of 256 processors working in tandem, allowing Deep Blue to calculate one to two hundred billion chess moves within 3 minutes.

1999) between 1948 and 1952. It was one of several different copies of the high-speed computer built by John von Neumann (1903–1957) for the Institute for Advanced Studies (IAS). It was constructed primarily for use in the development of atomic energy applications, specifically the hydrogen bomb.

A reconstruction of the Turk automaton people believed could play chess.

It originated with the work on ENIAC (electronic numerical integrator and computer), the first fully operational, large-scale, electronic digital computer. ENIAC was built at the Moore School of Electrical Engineering at the University of Pennsylvania between 1943 and 1946. Its builders, John Presper Eckert Jr. (1919–1995) and John William Mauchly (1907–1980), virtually launched the modern era of the computer with ENIAC.

What are the features of modern computer games?

Modern computer games run as real-time simulations, in which players must interact continuously with the game situation. Computer games include state-of-the-art animation and graphics with multiplayer design. Furthermore, some social networks serve as platforms for games that involve cooperation.

What are the components of a computer?

Computers have two major components: the hardware and the software. Hardware consists of all the physical devices needed to actually build and operate a computer. Examples of computer hardware are the central processing unit (CPU), hard drive, memory, modems, and external devices, such as the keyboard, monitor, printers, scanners, and other devices that can be physically touched. Software is an integral part of a computer and consists of the various computer programs that allow the user to interact with it and specify the tasks that the computer performs. Without software, a computer is merely a collection of circuits and metal in a box unable to perform even the most basic functions.

What is the difference between a bit and a byte?

A byte, a common unit of computer storage, holds the equivalent of a single character, such as a letter ("A"), a number ("2"), a symbol ("$"), a decimal point, or a space. It is usually equivalent to eight data bits and one parity bit. A bit (a binary digit), the smallest unit of information in a digital computer, is equivalent to a single "0" or "1." The parity bit is used to check for errors in the bits making up the byte. Eight data bits per byte is the most common size used by computer manufacturers.

What are some of the units for computer storage space?

Computer storage space for hard drives and other storage media, for example, flash drives, is calculated in base 2 using binary format with a byte as the basic unit. The most common units of computer storage are:

Computer Storage Space Units

Unit	Equivalent
Kilobyte (KB)	1,024 bytes
Megabyte (MB)	1,024 kilobytes or 1,048,576 bytes
Gigabyte (GB)	1,024 megabytes or 1,073,741,824 bytes
Terabyte (TB)	1,024 gigabytes or 1,099,511,627,776 bytes
Petabyte (PB)	1,024 terabytes or 1,125,899,906,842,624 bytes
Exabyte (EB)	1,024 petabytes
Zettabyte (ZB)	1,024 exabytes
Yottabyte (YB)	1,024 zettabytes

What is a silicon chip?

A silicon chip is an almost pure piece of silicon, usually less than 1 centimeter square and about 0.5 millimeters thick. It contains hundreds of thousands of miniaturized electronic circuit components, mainly transistors, packed and interconnected in layers beneath the surface. These components can perform control, logic, and memory functions. A grid of thin, metallic strips is located on the surface of the chip; these wires are used for electrical connections to other devices. The silicon chip was developed independently by two researchers, Jack Kilby (1923–2005) of Texas Instruments in 1958 and Robert Noyce (1927–1990) of Fairchild Semiconductor in 1959. Kilby was awarded the Nobel Prize in Physics in 2000 for his discovery of the silicon chip. Kilby and Noyce received the first Charles Stark Draper Prize in Engineering in 1989 for the invention and development of the semiconductor microchip. The Draper Prize is considered the preeminent award in engineering, comparable to the Nobel Prize.

While silicon chips are essential to almost all computer operations today, a myriad of other devices depend on them as well, including mobile devices, microwave ovens, and automobiles.

What is Moore's Law?

Gordon Moore (1929–), cofounder of Intel®, a top microchip manufacturer, observed in 1965 that the number of transistors per microchip—and hence a chip's processing power—would double about every year and a half. The press dubbed this Moore's Law. Despite claims that this ever-increasing trend cannot perpetuate, history has shown that microchip advances are, indeed, keeping pace with Moore's prediction.

What are the sizes of silicon chips?

Small silicon chips may be no more than 1/16 of an inch square by 1/30 of an inch thick and hold up to tens of thousands of transistors. Large chips, the size of a postage stamp, can contain hundreds of millions of transistors.

Chip Size	Number of Transistors per Chip
SSI—small-scale integration	< 100
MSI—medium-scale integration	100–3,000
LSI—large-scale integration	3,000–100,000
VLSI—very large-scale integration	100,000–1,000,000
ULSI—ultralarge-scale integration	> 1,000,000

Are any devices being developed to replace silicon chips?

When transistors were introduced in 1948, they demanded less power than fragile, high-temperature vacuum tubes; allowed electronic equipment to become smaller, faster, and more dependable; and generated less heat. These developments made computers much more economical and accessible; they also made portable radios practical. However, the smaller components were harder to wire together, and hand wiring was both expensive and error-prone.

In the early 1960s, circuits on silicon chips allowed manufacturers to build increased power, speed, and memory storage into smaller packages, which required less electricity to operate and generated even less heat. While through most of the 1970s, manufacturers could count on doubling the components on a chip every year without increasing the size of the chip, the size limitations of silicon chips are becoming more restrictive. Though components continue to grow smaller, the same rate of shrinking cannot be maintained.

Researchers are investigating different materials to use in making circuit chips. For example, gallium arsenide and germanium are possibilities for newer chips. In addition, researchers are investigating newer designs for transistors so they can fit more into a smaller area in order to increase performance and power.

How is the speed of a CPU measured?

The central processing unit (CPU) of a computer is where almost all computing takes place in all computers including mainframes, desktops, laptops, and servers. The CPU of almost every computer is contained on a single chip. Separate from the "real-time clock" that keeps track of the time of day, the CPU clock sets the tempo for the processor and measures the transmission speed of electronic devices. The clock is used to synchronize data pulses between sender and receiver. A 1-megahertz clock manipulates a set number of bits one million times per second. In general, the higher the clock speed, the quicker the data is processed. However, newer versions of software often require quicker computers just to maintain their overall processing speed.

What is a hard drive of a computer?

Hard disks, formerly called hard disk drives and more recently just hard drives, were invented in the 1950s. They are storage devices in desktop computers, laptops, servers, and mainframes. Hard disks use a magnetic recording surface to record, access, and erase data in much the same way that magnetic tape records, plays, and erases sound or images. A read/write head, suspended over a spinning disk, is directed by the central processing unit (CPU) to the sector where the requested data is stored or where the data is to be recorded. A hard disk uses rigid, aluminum disks coated with iron oxide to store data. Data are stored in files, which are named collections of bytes. The bytes could be anything from the ASCII codes for the characters of a text file to instructions for a software application to the records of a database to the pixel colors for an image. Hard drive size ranges from several hundred gigabytes to more than one terabyte.

A hard disk rotates from 5,400 to 7,200 revolutions per minute (rpm) and is constantly spinning (except in laptops, which conserve battery life by spinning the hard disk only when in use). An ultrafast hard disk has a separate read/write head over each track on the disk so that no time is lost in positioning the head over the desired track; accessing the desired sector takes only milliseconds, the time it takes for the disk to spin to the sector.

Hard drive performance is measured by data rate and seek time. Data rate is the number of bytes per second that the hard drive can deliver to the CPU. Seek time is the amount of time that elapses from when the CPU requests a file and when the first byte of the file is delivered to the CPU.

What is a USB port?

The Universal Serial Bus (USB) connectors first appeared on computers in the late 1990s. It has become the most widely used interface to attach peripherals, such as mice, printers and scanners, external storage drives, digital cameras, and other devices to a computer. Unlike older serial ports and parallel ports, USB ports are easy to reach and can easily be plugged in—even while the computer is in use.

What is a pixel?

A pixel (from the words "pix," for picture, and "element") is the smallest element on a video display screen. A screen contains thousands of pixels, each of which can be made up of one or more dots or a cluster of dots. On a simple, monochrome screen, a pixel is one dot; the two colors of image and background are created when the pixel is switched either on or off. Some monochrome screen pixels can be energized to create different light intensities to allow a range of shades from light to dark. On color screens, three dot colors are included in each pixel—red, green, and blue. The simplest screens have just one dot of each color, but more elaborate screens have pixels with clusters of each color. These more elaborate displays can show a large number of colors and intensities. On color screens, black is created by leaving all three colors off; white by all three colors on; and a range of grays by equal intensities of all the colors.

Who invented the computer mouse?

Douglas C. Engelbart

A computer mouse is a handheld input device that, when rolled across a flat surface, causes a cursor to move in a corresponding way on a display screen. A prototype mouse was part of an input console demonstrated by Douglas C. Engelbart (1925–2013) in 1968 at the Fall Joint Computer Conference in San Francisco. Popularized in 1984 by the Macintosh from Apple Computer Company, the mouse was the result of fifteen years devoted to exploring ways to make communicating with computers simpler and more flexible. The physical appearance of the small box with the dangling, tail-like wire suggested the name of "mouse."

In recent years, the mouse has evolved into other shapes and forms. One type is the wireless (or "tail-less") mouse, which does not have a cord to connect to the computer. Wireless mice use radio signals or infrared to connect to the computer.

The resolution of a computer monitor is expressed as the number of pixels on the horizontal axis and the number of pixels on the vertical axis; for example, a monitor described as 800×600 has 800 pixels on the horizontal axis and 600 pixels on the vertical axis. The higher the numbers, i.e., 1600×1200, the better the resolution.

What are the three types of sensors used in touch screens?

The three types of sensors for touch screens are 1) resistive screens; 2) capacitive screens; and 3) surface acoustic wave screens. Resistive sensors are the most common in touch screens. The basic design is two thin sheets—a conductive and a resistive metallic layer—separated by a grid of plastic dots or spacers on a normal glass panel. Each sheet conducts electricity. Four wires (hence the common name four-wire resistive screen) measure the currents on the screen. The sheets contact each other at the spot where the user touches the screen. Five-wire screens add a sheet. The additional sheet increases the durability of the screen since the user does not touch the screen that carries the current. An eight-wire screen has even greater durability since it has an extra set of wires.

A capacitive touch screen consists of a single, thin sheet on the glass panel. Touching the screen transfers the charge to the user, decreasing the charge on the capacitive layer. The decrease is measured, and the computer or device can calculate the exact location of the physical contact. Capacitive sensors have a clearer screen than screens with resistive sensors. A disadvantage of capacitive sensors is that they cannot sense physical contact with nonconductive objects, such as gloved fingers.

Touch screens that use wave-interruption system sensors have two transducers, one for receiving and one for sending, that detect where the interference occurs. They have no metallic layers, and they offer the clearest resolution for displaying detailed graphics.

How do touch screens work?

Touch screen technology relies on the physical touch of the screen by the user using a finger or stylus for input. Instead of using a computer mouse to activate the cursor, merely touching the screen identifies the location and allows the user to modify the information on the screen. The basic underlying principle is that an electrical current is running through a sensor on the screen. Touching the screen causes a voltage change indicating the location of the physical contact. Specialized hardware converts the voltage changes on the sensors into signals that the computer can receive. Finally, software relates to the computer or other device, for example, smartphone, what is happening on the sensor. The computer or device reacts to the inputted information accordingly.

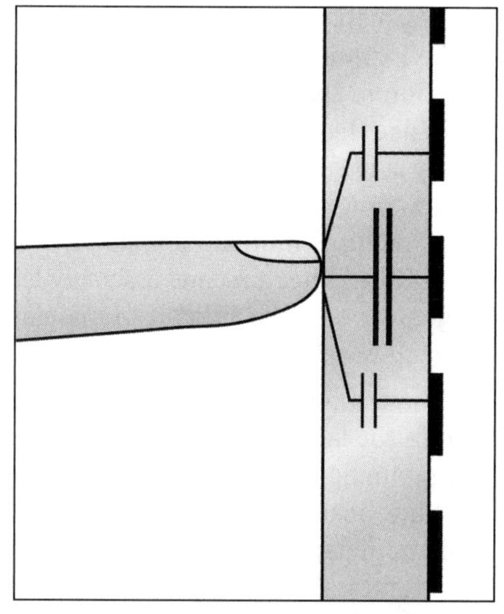

A schematic of a capacitive screen, which works by transferring the electric charge from screen to user, which allows the computer to identify where on the screen a person is touching.

Who was the first programmer?

According to historical accounts, Lord Byron's (1788–1824) daughter, Augusta Ada Byron (1815–1852), the Countess of Lovelace, was the first person to write a computer program for Charles Babbage's "analytical engine." This machine was to work by means of punched cards that could store partial answers that could later be retrieved for additional operations and would then print the results. Her work with Babbage and the essays she wrote about the possibilities of the "engine" established her as a kind of founding parent of the art and science of programming. The programming language called "Ada" was named in her honor by the U.S. Department of Defense. In modern times, Commodore Grace Murray Hopper (1906–1992) of the U.S. Navy is acknowledged as one of the first programmers of the Mark I computer in 1944.

Is assembly language the same as machine language?

While the two terms are often used interchangeably, assembly language is a more "user-friendly" translation of machine language. Machine language is the collection of patterns of bits recognized by a central processing unit (CPU) as instructions. Each particular CPU design has its own machine language. The machine language of the CPU of a microcomputer generally includes about seventy-five instructions; the machine language of the CPU of a large mainframe computer may include hundreds of instructions. Each of these instructions is a pattern of 1s and 0s that tells the CPU to perform a specific operation.

Assembly language is a collection of symbolic, mnemonic names for each instruction in the machine language of its CPU. Like the machine language, the assembly language is tied to a particular CPU design. Programming in assembly language requires intimate familiarity with the CPU's architecture, and assembly language programs are difficult to maintain and require extensive documentation.

The winner of the 1984 Turing Award, Niklaus Wirth created several important programming languages, including PASCAL.

The computer language C first developed in the late 1980s and is a high-level programming language that can be compiled into machine languages for almost all computers, from microcomputers to mainframes, because of its functional structure. It was the first series of programs that allowed a computer to use higher-level language programs and is the most widely used programming language for personal computer software development. C++ was first released in 1985 and is still widely used today.

Who invented the COBOL computer language?

COBOL (common business-oriented language) is a prominent computer language designed specifically for commercial uses, created in 1960 by a team drawn from several computer makers and the Pentagon. The best-known individual associated with COBOL was then-Lieutenant Grace Murray Hopper, who made fundamental contributions to the U.S. Navy standardization of COBOL. COBOL excels at the most common kinds of data processing for business—simple arithmetic operations performed on huge files of data. The language endures because its syntax is very much like English and because a program written in COBOL for one kind of computer can run on many others without alteration.

Who invented the PASCAL computer language?

Niklaus Wirth (1934–), a Swiss computer programmer, created the PASCAL computer language in 1970. PASCAL introduced the concept of "structured programming." Structured programming requires dividing the program into general steps, then refining each step into smaller modules that are methodically and carefully nested within each other. Structured programming leads to simpler programs. It was used in the academic setting as a teaching tool for computer scientists and programmers. Although Wirth did not intend to use PASCAL for commercial applications, it became a standard methodology for computer language development and has widespread use.

Which was the first widely used high-level programming language?

FORTRAN (FORmula TRANslator) was developed by IBM in the late 1950s. John Backus (1924–2007) was the head of the team that developed FORTRAN. Designed for scientific work containing mathematical formulas, FORTRAN allowed programmers to use algebraic expressions rather than cryptic assembly code. The FORTRAN compiler translated the algebraic expressions into machine-level code. By the late 1960s, FORTRAN was available on almost every computer, especially IBM machines, and utilized by many users.

When was Java developed?

Java was released by Sun Microsystems in 1995. A team of developers headed by James Gosling (1955–) began working on a refinement of C++ that ultimately led to Java. Unlike other computer languages, which are either compiled or interpreted, Java compiles the source code into a format called bytecode. The bytecode is then executed by an interpreter. Java was adapted to the emerging World Wide Web and formed the basis of the Netscape Internet browser.

Why is Python so popular?

Python was developed in 1989 by Guido van Rossum (1956–) and first released in 1991. It is a high-level, object-oriented, general-purpose programming language that may be used for many different projects. It has a clear syntax that makes it a good language for first-time programmers. Python is an open-source language.

What are the most popular computer programming languages?

According to the Tiobe Index produced by Tiobe, a software quality company, the most popular programming languages at the end of 2017 were:

- Java
- C++
- C#
- C
- Python

The data are compiled by calculating the number of hits for the programming language of the most popular Internet search engines. Other lists of popular programming languages include:

JavaScript
SQL
PHP
Ruby

Are coding and programming the same thing or two different things?

Many people use the terms "coding" and "programming" interchangeably. Both terms refer to the writing of instructions to tell the computer what to do. All instructions for computers are in machine code, which is a binary language. Everything is translated as either "0" or "1" for computers to perform the multitude of functions and tasks that they do. Programmers use languages such as Java, Python, C, C#, or C++ as the first step in creating source code. These languages are then run through a program compiled to translate the source code into binary (machine) code. Many refer to the act of writing source code as coding. Since it is part of the computer programming procedure, it may also be called programming. Creating software includes writing the source code, running the compiler, and fixing errors. Since computer languages were first introduced, they have become more simplified, allowing more individuals to write code.

What is the idea behind open-source software?

Open-source software is computer software where the code (the rules governing its operation) is available for users to modify. This is in contrast to proprietary code, where the software vendor veils the code so users cannot view and, hence, manipulate (or steal) it. The software termed open-source is not necessarily free, that is, without charge; authors can charge for its use, usually only nominal fees. According to the Free Software Foundation, "free" software is a matter of liberty, not price. To understand the concept, you should think of "free" as in "free speech," not as in "free food." Free software is a matter of the users' freedom to run, copy, distribute, study, change, and improve the software. Despite this statement, most of it is available without charge. Licensing agreements are available in order to use open-source software. It involves a culture of sharing, cooperation, and mutual innovation.

What are the tasks of an operating system?

An operating system is found in all computers and other electronic devices, such as cell phones. The operating system manages all the hardware and software resources of the computer. Operating systems manage data and devices, such as printers, in the computer. Operating systems today have the ability to multitask, allowing the user to keep several different applications open at the same time. Popular operating systems for computers are Windows (Microsoft), OS X (Macintosh), and Linux. The most common operating systems for mobile devices, such as smartphones and tablets, are iOS (for Apple products) and Android (for non-Apple products).

Android is a popular operating system commonly used on non-Apple smartphones.

How are new releases of the Android operating system named?

Android is an open-source, Linux-based software operating system. It was developed by a startup company in 2003, with backing from Google, and bought by Google in 2005. Each new major release of a version of Android is named after a dessert since, according to Google, the devices that use Android "make our lives so sweet." The names of releases of Android have been:

- Android 1.5 Cupcake
- Android 1.6 Donut
- Android 2.0–2.1 Eclair
- Android 2.2 Froyo
- Android 2.3 Gingerbread
- Android 3.0–3.2 Honeycomb
- Android 4.0 Ice Cream Sandwich
- Android 4.1–4.3 Jelly Bean
- Android 4.4 KitKat
- Android 5.0–5.1 Lollipop
- Android 6.0 Marshmallow
- Android 7.0–7.1 Nougat
- Android 8.0–8.1 Oreo
- Android 9.0 Pie

How did the Linux operating system get its name?

The name Linux is a combination of the first name of its principal programmer, Finland's Linus Torvalds (1970–), and the UNIX operating system. Linux (pronounced with a short "i") is an open-source computer operating system that is comparable to more powerful, expansive, and usually costly UNIX systems, of which it resembles in form and function. Linux allows users to run an amalgam of reliable and hearty open-source software tools and interfaces, including powerful web utilities, such as the popular Apache server, on their home computers. Anyone can download Linux for free or can obtain it

Where did the term "bug" originate?

The slang term "bug" is used to describe problems and errors occurring in computer programs. The term may have originated during the early 1940s at Harvard University, when computer pioneer Grace Murray Hopper discovered that a dead moth had caused the breakdown of a machine on which she was working. When asked what she was doing while removing the corpse with tweezers, she replied, "I'm debugging the machine." The moth's carcass, taped to a page of notes, is preserved with the trouble log notebook at the Naval Surface Warfare Center Computer Museum in Virginia.

on disk for only a marginal fee. Torvalds created the kernel—or heart of the system—"just for fun" and released it freely to the world, where other programmers helped further its development. The world, in turn, has embraced Linux and made Torvalds into a computer folk hero.

What is a computer virus, and how is it spread?

Taken from the obvious analogy with biological viruses, a computer "virus" is a name for a type of computer program that searches out uninfected computers, "infects" them by causing them to execute the virus, and then attempts to spread to other computers. A virus does two things: execute code on a computer and spread to other computers.

Most viruses are spread throughout the Internet through infected file attachments or e-mails or by visiting disreputable websites. The executed code can accomplish anything that a regular computer program can do: it can delete files, send e-mails, install programs, cause servers to crash, and copy information from one place to another. These actions can happen immediately or after some set delay. It is often not noticed that a virus has infected a computer because it will mimic the actual actions of the infected computer. By the time it is recognized that the computer is infected, much damage may have occurred. Malware (short for malicious software programs) is installed on a computer without the knowledge of the owner, causing data loss and other damage. It can cause entire networks to crash.

As the name indicates, a 3-D printer is a remarkable device that can make three-dimensional copies of items that are scanned and then printed layer by layer.

What is 3-D printing?

Charles Hull (1939–) invented stereolithography, commonly referred to as 3-D printing, in 1984 as a way to build models in plastic layer by layer. The method uses UV light to cure and bond one polymer layer on top of the next layer. Although 3-D printing techniques were first used in research and development labs, it quickly became a tool to create models and prototypes in industry and manufacturing. It is now used to create automobile and aircraft components, artificial limbs, artwork, musical instruments, and endless other products. It is one of the products that has spurned the maker movement. Hull was awarded U.S. Patent #4,575,330 in 1986 for his invention and inducted into the National Inventors Hall of Fame in 2014.

What is one of the earliest applications of 3-D printing for fertility?

A research team successfully implanted 3-D-printed ovaries into mice. The organs were printed by overlapping pieces of biocompatible gelatin. The researchers then inserted follicles into each ovary. The structures were able to produce hormones and contain eggs. Two ovaries were implanted into seven sterile mice. The mice mated with male mice. After a normal gestation period of about three weeks, three of the female mice gave birth to healthy litters. A possible outgrowth of the experiment will focus on bioprosthetics for humans.

COMMUNICATIONS

How are radio waves important for communications?

Radio waves are at the lowest range of the electromagnetic spectrum since they have the longest wavelength. They are important for voice, data, navigation systems, and entertainment such as television. A radio wave is generated by a transmitter and then detected by a receiver. Individual transmitters and receivers are typically designed to operate over a limited range of frequencies. The radio spectrum is generally divided into nine bands according to frequency range and wavelength range.

Band	Frequency Range	Wavelength Range	Uses
Extremely Low Frequency (ELF)	< 3 kHz	> 62 miles (100 kilometers	Very long range, but frequency is lower than the range of audible sound. Used for digital data at a very slow rate
Very Low Frequency (VLF)	3 to 30 kHz	6–62 miles (10–100 kilometers)	

Band	Frequency Range	Wavelength Range	Uses
Low Frequency (LF)	30 to 300 kHz	3.2 feet to 6 miles (1 meter to 10 kilometers)	Marine and aviation radio, commercial AM radio
Medium Frequency (MF)	300 kHz to 3MHz	328 feet to 0.6 miles (10 meters to 1 kilometer)	Marine and aviation radio, commercial AM radio
High Frequency (HF)	3 to 30 MHz	32–328 feet (10–100 meters)	FM radio, broadcast television sound, public service radio, cell phones, and GPS
Very High Frequency (VHF)	30 to 300 MHz	3.2–32 feet (1–10 meters)	FM radio, broadcast television sound, public service radio, cell phones, and GPS
Ultra High Frequency (UHF)	300 MHz to 3 GHz	4 inches to 1 foot (10 centimeters to 1 meter)	FM radio, broadcast television sound, public service radio, cell phones, and GPS
Super High Frequency (SHF)	3 to 30 GHz	4 to 0.4 inches (10 to 1 centimeter)	Wi-Fi, Bluetooth, wireless USB, space and satellite microwave communication
Extremely High Frequency (EHF)	30 to 300 GHz	0.04–0.4 inches (1 millimeter to 1 centimeter)	Highest frequency in the radio band, but since air tends to absorb these frequencies, the applications are limited to radio astronomy

How do submerged submarines communicate?

Using frequencies from Very High (VHF) to Extremely Low (ELF), submarines can communicate by radio when submerged if certain conditions are met and depending on whether or not avoiding detection is important. Submarines seldom transmit on long-range high radio frequencies if detection is important, as in war. However, Super (SHF), Ultra (UHF), or Very High (VHF) frequency two-way links with cooperating aircraft, surface ships, via satellite, or with the shore are fairly safe with high data rates, though they all require that the boat show an antenna above the water or send a buoy to the surface.

Why do FM radio stations have a limited broadcast range?

Usually, radio waves higher in frequency than approximately 50–60 megahertz are not reflected by Earth's ionosphere and are lost in space. Television, FM radio, and high-frequency communications systems are therefore limited to approximately that of line-of-sight ranges. The line-of-sight distance depends on the terrain and antenna height but is usually limited to 50–100 miles (80–161 kilometers). FM (frequency-modulation) radio uses a wider band than AM (amplitude-modulation) radio to give broadcasts high fidelity, especially noticeable in music—crystal clarity to high frequencies and rich resonance to base notes, all with a minimum of static and distortion. Invented by Edwin Howard Armstrong (1891–1954) in 1933, FM receivers became available in 1939. Armstrong was inducted into the National Inventors Hall of Fame in 1980 for his invention of FM radio.

Why do AM stations have a wider broadcast range at night?

The variation of broadcast range is caused by the nature of the ionosphere of Earth. The ionosphere consists of several different layers of rarefied gases in the upper atmosphere that have become conductive through the bombardment of the atoms of the atmosphere by solar radiation, by electrons and protons emitted by the Sun, and by cosmic rays. One of these layers reflects AM radio signals, enabling AM broadcasts to be received by radios that are great distances from the transmitting antenna. With the coming of night, the ionosphere layers partially dissipate and become an excellent reflector of the short-waveband AM radio waves. This causes distant AM stations to be heard more clearly at night.

What technology is being developed that could potentially replace UPC codes and/or magnetic strips to identify and track information?

Radio frequency identification systems (RFID) are gaining popularity to identify and track information. An RFID system has two parts: a tag or a card that can store and modify information and a transmitter with an antenna to communicate the information. RFID tags can be either passive or active. Passive tags do not have a power supply. The reading signal induces the power to transmit the response. They tend to be small and lightweight but can only be read from distances of a few inches to a few yards. Passive tags are often attached to merchandise to reduce store loss. They are often used in smart cards for transit systems. Active RFID tags have their own battery to supply power so they can initiate communication with the reader. The signal is much stronger and may be read over greater distances. Active RFID tags have been used to track cattle and on shipping containers.

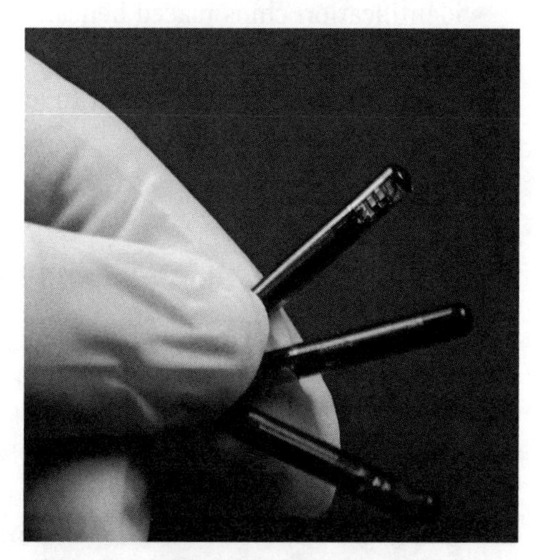

Radio frequency identification systems (RFID) are compact devices used to track anything to which they are attached.

Who invented RFID technology?

The earliest development of RFID technology dates back to World War II. Once radar was invented in 1935, German, Japanese, British, and American planes were detectable while they were still miles away. However, it was impossible to determine whether a plane was an enemy aircraft or friendly aircraft. The Germans discovered that if pilots rolled their planes on their approach to the airfield, it would change the radio signal reflected back, thus alerting the radar crew on the ground that they were German planes. This is the first passive RFID system. The British devised the first active RFID system by installing a transmitter on each plane. When the transmitter received signals from radio stations on the ground, friendly aircraft would broadcast a signal back, identifying it as friendly aircraft.

Uses of radio frequency communications continued to grow after World War II as a way to identify objects remotely. It was developed as a technique to detect shoplifters in retail locations. The first patent (U.S. Patent #3,713,148A, entitled "Transponder Apparatus and System") for an active RFID tag with rewritable memory was awarded to Mario W. Caradullo (1935–) on January 23, 1973. Charles Walton (1921–2011), often referred to as the father of RFID technology, received U.S. Patent #3,752,960 in August 1973 for a passive transponder or "key card" used to unlock a door without using a traditional key.

What are some uses of RFID tags?

RFID tags are used in a wide variety of applications, including:

- automatic fare-payment systems for transit systems
- automatic toll payments for highways and bridges
- student ID cards
- U.S. passports
- tracking cattle
- identification chips placed beneath the skin of pets to help identify lost pets and return them to their owners
- tracking goods from shipment to inventory
- smart cards for locks

How did Bluetooth become the name for the wireless technology used over short distances?

Bluetooth technology was named for King Harald "Bluetooth" Gormsson (c. 958–c. 986), a Danish king who united Denmark and Norway in 958. He was nicknamed Bluetooth since he had a dead tooth that was dark blue/gray in color. When the leaders of Intel, Ericsson, and Nokia were planning the short-range radio technology, Bluetooth was the temporary code name for the project which they hoped would unite the PC and cellular industries with a short-range wireless link.

What are the two methods that microwaves can be transmitted?

Microwaves fall in the Super High Frequency (SHF) end of the electromagnetic spectrum. The first way they can transmit is the line-of-sight approach. A microwave transmitter is pointed at a microwave receiver that is no more than 30 kilometers away. The second method of transmission is to send the signal up to a satellite that receives it and retransmits it back down to the receiver.

Bluetooth technology allows wireless communication over short distances between devices, which makes it convenient for such things as using a headset with a smartphone.

How does Bluetooth technology use radio waves?

Bluetooth sends and receives radio waves in a band of frequencies in the 2.4 GHz range. These frequencies are part of the ISM (Industrial Scientific Medical) bands and are separate from the frequencies for radio, television, and cell phones. It is designed to communicate over short distances, usually less than 30 feet (10 meters). Bluetooth is a wireless communication often used to download photos from a camera to a PC, hook up a wireless mouse to a laptop, or link a hands-free headset to a cell phone. A big advantage is no wires are necessary. Devices with Bluetooth capability have built-in radio antennas with transmitters and receivers so they can simultaneously send and receive wireless signals to other Bluetooth-enabled devices.

How do cell phones work?

Cell phones use the UHF region of the electromagnetic spectrum of 300 MHz to 3 GHz. The service area for a cell phone is divided up into hexagonally shaped cells, each one served by base stations with antenna towers at three corners of the cell. The stations can both receive and transmit information to cell phones. The stations are connected to a network that uses fiber optic cables. When a cell phone is turned on, it searches for available services according to a list stored in the phone. It selects the correct frequencies to transmit and receive data, then sends its serial and phone numbers to the system, registering itself in that cell. The network makes sure that the phone number is part of its system and that money is in the account to pay for a call. After registration is complete and a call is made to that phone, the network can direct the call to the correct cell. The cell phone always searches for the strongest signal from a tower. If the phone moves during the conversation, then the signal strengths will change and the phone will "hand off" the cell to a different base station.

Cell phones digitize the voice signals. Special circuits, called digital signal processors, then compress the voice signals and insert codes that can detect errors in trans-

Where is the "cloud"?

Cloud computing refers to storing and accessing data and programs via the Internet instead of on a local computer's hard drive. Any web-based program for storage or applications is considered cloud computing. Many cloud programs and applications sync with local stored files and programs. Examples of cloud computing are Google Drive and Apple iCloud.

mission. Compression is achieved by sending only the changes in the digital signals, not what stays the same. Cell phone systems also send many different conversations simultaneously. One method, called TDMA (time division multiple access), splits up three compressed calls and sends them together. Another method, CDMA (code division multiple access), uses TDMA to pack three calls together and then puts six more calls at two other frequencies. Each of the nine (or more) calls is assigned a unique code so that it can be directed to the correct recipient. Spread CDMA systems use a wide band of frequencies that permit more simultaneous calls.

How does a smartphone differ from a cell phone?

No precise definition of a smartphone exists, but generally, it is a cell phone with built-in applications and Internet access. The earliest smartphones essentially combined features of a cell phone with those of a personal digital assistant (PDA). IBM designed the first smartphone in 1992 called the Simon personal communicator. Today, smartphones are generally equipped with an operating system that allow them to send e-mails (often syncing with a computer), browse the Internet, view spreadsheets and documents, send and receive text messages, operate as an MP3 player, listen to music, watch movies, read electronic books, and play games as well as having a camera for taking digital pictures or videos. More than cell phones, smartphones are comparable to private, miniature computers.

What is Wi-Fi?

Wi-Fi is a wireless local area network. The Wi-Fi Alliance certifies that network devices comply with the appropriate standard, IEEE 802.11. A Wi-Fi hotspot is the geographic boundary covered by a Wi-Fi access point. It is estimated that by 2021, 542 million public hotspots will be available worldwide.

ENERGY

Why is energy important in a modern, technological society?

Scientists have learned how to change energy from one form to another form and use it to do work. We use energy to heat, cool, and light our homes and offices;

power our appliances, machines, and gadgets; provide power to our factories and manufacturing industries; and provide fuel for our transportation, including automobiles, buses, airplanes, and ships. New methods allow us to generate and use energy more efficiently.

How do primary energy sources differ from secondary energy sources?

Primary energy sources may come from renewable or nonrenewable sources. Nonrenewable sources are sources that will be depleted with use over time. Renewable sources may be replenished and cannot be depleted. Some examples of nonrenewable sources of energy are coal, natural gas, petroleum, and uranium (nuclear). Some examples of renewable energy sources are solar, wind, water (hydropower), geothermal, and biomass. Both renewable and nonrenewable sources of primary energy may be converted to secondary sources, such as electricity and hydrogen.

How is energy measured?

The basic unit used to measure energy in the British system, used in the United States, is the British Thermal Unit, or BTU. The basic unit used to measure energy in the metric system is the joule (J). The joule is named for James Prescott Joule (1818–1889), who discovered that heat is a type of energy. According to the International System of Units (SI), 1 BTU = 1055.06 J.

Different types of energy sources are measured in different physical units.

Energy Source	Unit of Measure
Coal	Short tons
Liquid fuels	Barrels or gallons
Natural gas	Cubic feet
Electricity	Kilowatts or kilowatthours

How much primary energy is consumed in the United States?

During the year 2017, the total U.S. primary energy consumption was equal to about 97.7 quadrillion (97,728,000,000,000,000) Btu.

Year	Total Primary Energy Consumption in U.S. (Btu)
2016	97.4 quadrillion Btu
2015	97.7 quadrillion Btu
2014	98.3 quadrillion Btu
2013	97.5 quadrillion Btu
2012	95.0 quadrillion Btu

Annually, the electric power sector consumes approximately 38 percent of the primary energy in the United States in order to generate electricity.

Are all nonrenewable sources of energy fossil fuels?

Coal, oil, and natural gas are all considered fossil fuels since they were formed from the buried remains of plants and animals that lived millions of years ago. All fossil fuels are nonrenewable sources of energy since their supply will diminish eventually to the point of being too expensive or too environmentally damaging to retrieve for use. Uranium, essential for the production of nuclear energy, is also considered a nonrenewable source of energy since a limited supply of uranium exists. However, it is not a fossil fuel.

What is the relationship between ancient plants and coal formation?

Coal, formed from ancient plant material, is organic. Most of the coal mined today was formed from prehistoric remains of primitive land plants, particularly those of the Carboniferous period, which occurred approximately three hundred million years ago. Five main groups of plants contributed to the formation of coal. The first three groups were all seedless, vascular plants: ferns, club mosses, and horsetails. The last two groups were primitive gymnosperms and the now extinct seed ferns. Forests of these plants were in low-lying, swampy areas that periodically flooded. When these plants died, they decomposed, but as they were covered by water, they did not decompose completely. Over a period of time, the decomposed plant material accumulated and consolidated. Layers of sediment formed over the plant material during each flood cycle. Heat and pressure built up in these accumulated layers and converted the plant material to coal. The various types of coal (lignite, bituminous, and anthracite) were formed as a result of the different temperatures and pressures to which the layers were exposed.

What is shale?

Shale is a fine-grained sedimentary rock that forms from the compaction of silt and clay-sized mineral particles. Black shale contains organic material that can generate oil and natural gas. It also traps the generated oil and natural gas within its pores. It is only in the past decade that technologies of horizontal drilling and hydraulic fracturing have been developed so it is economically feasible to drill and reach the large volumes of oil and natural gas trapped in shale formations.

What does fracking do to rock layers?

Fracking, sometimes referred to as hydro-fracking or hydraulic fracturing, is a well-stimulation technique in which rock is fractured by a pressurized liquid. The process involves the high-pressure injection of "fracking fluid" (usually water containing sand or other materials suspended with the aid of thickening agents) into a wellbore to create cracks in the deep rock formations, through which natural gas, petroleum, and brine will flow more freely. When the hydraulic pressure is removed from the well, small grains of hydraulic fracturing material (either sand or aluminum oxide) hold the fractures open. Hydraulic fracturing began as an experiment in 1947, and the first commercially successful application followed in 1950. Fracking is necessary to achieve adequate flow rates in shale gas, tight gas, tight oil, and coal seam gas wells. Hydraulic

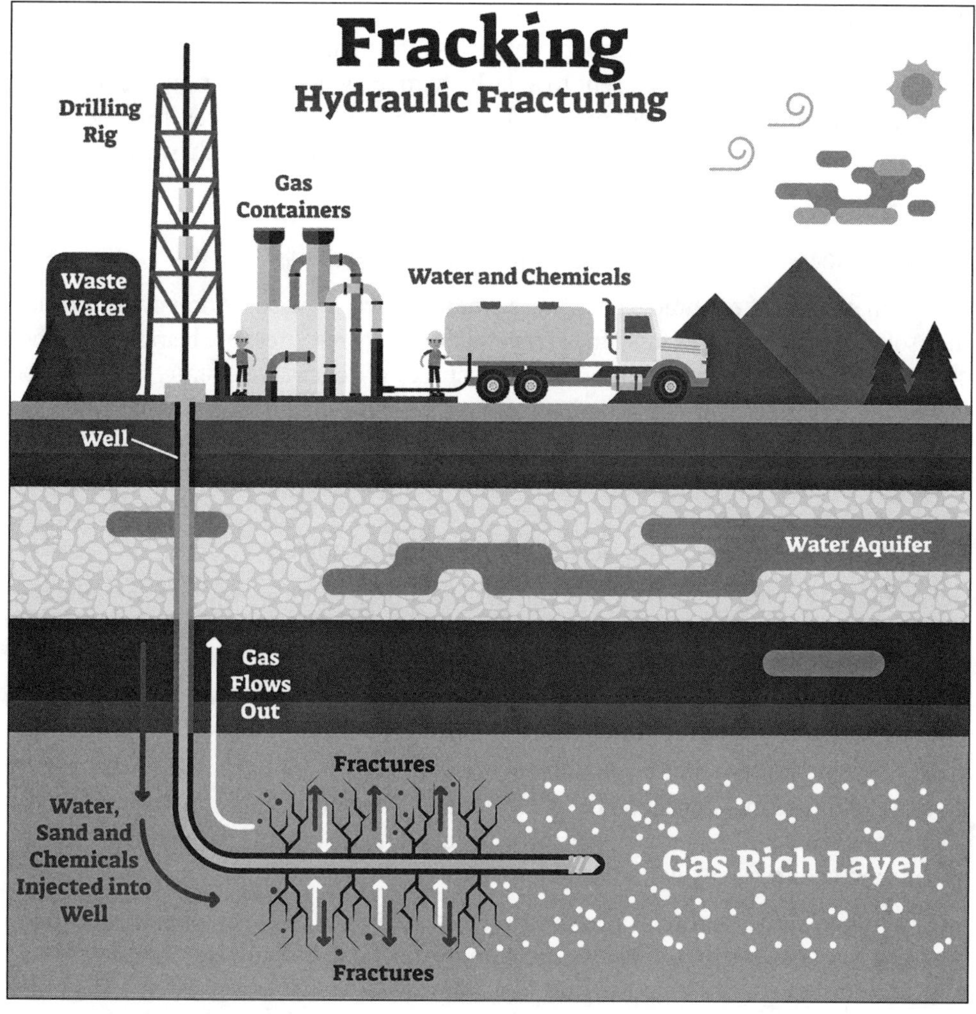

Fracking (hydraulic fracturing) is a method of extracting petroleum and natural gas from deep-rock formations by breaking them up with the high-pressure injection of fluids.

fracturing is controversial in many countries. Those who advocate fracking cite the economic benefits of more accessible hydrocarbons. Opponents argue that these are outweighed by the potential environmental impacts, including risks of ground and surface water contamination, air and noise pollution, and the triggering of earthquakes, along with the consequential hazards to public health and the environment.

What is the process known as hydrocarbon cracking?

Cracking is a process that uses heat to decompose complex substances. Hydrocarbon cracking is the decomposition by heat, with or without catalysts, of petroleum or heavy petroleum fractions (groupings) to give materials lower boiling points. Thermal cracking, developed by William Burton (1865–1954) in 1913, uses heat and pressure to break some of the large, heavy hydrocarbon molecules into smaller, gasoline-grade ones. The cracked hydrocarbons are then sent to a flash chamber, where the various fractions are separated. Thermal cracking not only doubles the gasoline yield but has improved gasoline quality, producing gasoline components with good antiknock characteristics (no premature fuel ignition).

How do renewable energy resources differ from fossil fuels?

The main sources of renewable energy are biofuels, such as wood, hydropower, geothermal, solar, and wind. Unlike fossil fuels, renewable energy resources are being replenished continuously and will never run out. Some sources of renewable energy, such as solar power and wind power, are perpetual, meaning the wind will blow and the Sun will shine no matter how much energy is used. Sources of renewable energy that rely on agriculture, such as woods, are renewable as long as they are not depleted and exploited too rapidly.

What are the advantages and disadvantages of solar power?

Solar energy is a clean, abundant, and safe energy source. More energy falls from the Sun on Earth in 1 hour than is used by everyone in the world in one year. Over a two-week period, Earth gets as much energy from the Sun as is stored in all known reserves of coal, oil, and natural gas. Solar energy can be used to heat water and spaces for homes and businesses or can be converted into electricity. The major advantages of solar power are that it does not produce air pollutants or emit greenhouse gases. Also, solar energy systems have minimal impact on the environment. Solar energy accounts for only about 3 percent of the total renewable energy resources, and less than 1 percent of electricity generation comes from solar power.

The disadvantages of solar energy are that sunlight is not constant. The amount of sunlight reaching Earth varies with location, time of day, season, and weather. Also, a large amount of energy from the Sun is not delivered to any one place at one time. A large surface area is required to collect the energy at a useful rate.

How is solar energy converted into electricity?

Solar energy is converted into electricity using photovoltaic (PV) cells or concentrating solar power plants. Photovoltaic cells convert sunlight directly into electricity. Individual PV cells are combined in modules of about forty cells to form a solar panel. Ten to twenty solar panels are used to power a typical home. The panels are usually mounted on the home facing south or mounted onto a tracking device that follows the Sun for the maximum exposure to sunlight. Power plants and other industrial locations combine more solar panels to generate electricity.

Concentrating solar power plants collect the heat (energy) from the Sun to heat a fluid that produces steam that drives a generator to produce electricity. The three main types of concentrating solar power systems are parabolic trough, solar dish, and solar power tower, which describe the different types of collectors. Parabolic troughs collectors have a long, rectangular, U-shaped reflector or mirror focused on the Sun with a tube (receiver) along its length. A solar dish looks very much like a large satellite dish that concentrates the sunlight into a thermal receiver, which absorbs and collects the heat and transfers it to the engine generator. The engine produces mechanical power, which is used to run a generator converting mechanical power into electrical power. A solar tower uses a field of flat, sun-tracking mirrors, called heliostats, to collect and concentrate the sunlight onto a tower-mounted heat exchanger (receiver). A fluid is heated in the receiver to generate steam, which is used in a generator to produce electricity.

How does a solar cell generate electricity?

A solar cell, also called a photovoltaic (PV) cell, consists of several layers of silicon-based material. When photons, particles of solar energy from sunlight, strike a photovoltaic cell, they are reflected, pass through, or are absorbed. Absorbed photons provide energy to generate electricity. The top layer, the p-layer, absorbs light energy. This energy frees electrons at the junction layer between the p-layer and the n-layer. The freed electrons collect at the bottom layer, the n-layer. The loss of electrons from the top layer produces "holes" in the layer that are then filled by other electrons. When a connection, or circuit, is completed between the p-layer and the n-layer, the flow of electrons creates an electric current. The photovoltaic effect, including the naming of the p-layer and the n-layer, was discovered by Russell Ohl (1898–1987), a researcher at Bell Labs, in 1940.

When were photovoltaic cells developed?

A group of researchers at Bell Labs, Calvin Fuller (1902–1994), Daryl Chapin (1906–1995), and Gerald Pearson (1905–1987), developed the first practical silicon solar cell

Concentrated solar power plants like this one in California's Mojave Desert generate incredible amounts of power. The brilliance of the towers is so blinding that one should not look directly at them for long. In fact, some airline pilots have complained about them.

in 1954. They filed for a patent in 1954 and were granted U.S. Patent #2,780,765, "Solar Energy Converting Apparatus," in 1957. Fuller, Chapin, and Pearson were inducted into the National Inventors Hall of Fame in 2008 for the invention of the silicon solar cell. The earliest PV cells were used to power U.S. space satellites. The use of PV cells was then expanded to power small items, such as calculators and watches.

TRANSPORTATION

AUTOMOBILES

Who invented the automobile?

The concept of a self-propelled vehicle fascinated inventors since the mid-eighteenth century. Some of the first inventions include:

- Nicolas-Joseph Cugnot (1725–1804) invented a steam-driven contraption. He rode the Paris streets at 2.5 miles (4 kilometers) per hour in 1769.
- Richard Trevithick (1771–1833) also produced a steam-driven vehicle that could carry eight passengers. It first ran on December 24, 1801, in Camborne, England.
- Londoner, Samuel Brown (–1849), built the first practical 4-horsepower gasoline-powered vehicle in 1826.
- The Belgian engineer J. J. Etienne Lenoire (1822–1900) built a vehicle with an internal combustion engine that ran on liquid hydrocarbon fuel in 1862, but he did not test it on the road until September 1863. It traveled 12 miles (19.3 kilometers) in three hours.
- The Austrian inventor Siegfried Marcus (1831–1898) invented a four-wheeled, gasoline-powered handcart in 1864 and a full-size car in 1875; the Viennese police objected to the noise that the car made, and Marcus did not continue its development.
- Edouard Delamare-Deboutteville (1856–1901), of France, invented an 8-horsepower vehicle in 1883, which was not durable enough for road conditions.
- Karl Benz (1844–1929) and Gottlieb Daimler (1834–1900) are both credited with the invention of the gasoline-powered automobile, because they were the first to make their automotive machines commercially practicable. Benz and Daimler worked independently, unaware of each other's endeavors. Both built compact, internal-combustion engines to power their vehicles. Benz built his three-wheeler in 1885; it was steered by a tiller. Daimler's four-wheeled vehicle was produced in 1887. Daimler was inducted into the National Inventors Hall of Fame in 2006 for his design of automobile and motorcycle engines.

Who started the first American automobile company?

Charles Duryea (1861–1938), a cycle manufacturer from Peoria, Illinois, and his brother, Frank (1869–1967), founded America's first auto-manufacturing firm and became the first to build cars for sale in the United States. The Duryea Motor Wagon company, set up in Springfield, Massachusetts, in 1895, built gasoline-powered horseless carriages similar to those built by Benz in Germany.

However, the Duryea brothers did not build the first automobile factory in the United States. Ransom Eli Olds (1864–1950) built the first automobile factory in 1899 in Detroit, Michigan, to manufacture his Oldsmobile. More than 10 vehicles a week were produced there by April 1901, for a total of 433 cars produced in 1901. In 1902 Olds introduced the assembly-line method of production and made over 2,500 vehicles in 1902 and 5,508 in 1904. In 1906, 125 companies made automobiles in the United States.

What were Henry Ford's contributions to the automobile industry?

In 1908, American engineer Henry Ford (1863–1947) improved the automobile assembly-line techniques by adding the conveyor belt system that brought the parts to the workers on the production line; this made automotive manufacture quick and cheap, cutting production time to 93 minutes. His company sold 10,660 vehicles that year. Ford was inducted to the National Inventors Hall of Fame in 1982 for his many contributions to the development of the automobile industry.

How does an electric car work?

An electric car uses an electric motor to convert electric energy stored in batteries into mechanical work. Various combinations of generating mechanisms (solar panels, generative braking, internal combustion engines driving a generator, fuel cells) and storage mechanisms are used in electric vehicles.

How many different types of electric vehicles are available in the United States?

Electric vehicles are classified as either all-electric vehicles (AEVs) or plug-in hybrid electric vehicles (PHEVs). As the name indicates, AEVs run only on electricity with no petroleum-based fuel. Their batteries are mostly charged by being plugged into the electric grid, although they also charge partially by regenerative braking. Regenerative braking generates electricity from some of the energy normally lost during braking. All-electric vehicles have a shorter driving distance range, usually 80–100

Although there were auto pioneers before him, Henry Ford is given a lot of credit for transforming the automobile industry into what it is today.

miles, before they need to be recharged. In contrast, PHEVs run on electricity for short distances (6–40 miles) and then switch to an internal combustion engine when the battery is depleted. The internal combustion engine may be powered by petroleum or an alternative fuel. Some types of PHEVs are called extended-range electric vehicles (EREVs).

Which chemical element is the source of power in fuel cell electric vehicles?

Fuel cell electric vehicles use hydrogen gas as the source of energy. Vehicles are equipped with a fuel cell stack consisting of multiple fuel cells. The most common type of fuel cell in vehicles is the polymer electrolyte membrane (PEM), which is also called the Proton Exchange Membrane fuel cell. An electrolyte membrane is sandwiched between a positive (cathode) and negative (anode) electrode. Hydrogen, which is stored in a tank, is introduced to the anode while oxygen (from air) is introduced to the cathode. An electrochemical reaction occurs splitting the hydrogen ions into positively charged protons and negatively charged electrons. The positively charged protons pass through the membrane to the cathode. The negatively charged electrons travel through an external circuit to the cathode, creating an electrical current. This electrical current provides the power to the electric vehicle. The electrons then recombine with the protons on the cathode side, where the protons, electrons, and oxygen molecules combine to form water. The only emission from the tailpipe of a fuel cell electric vehicle is water.

The advantages of a fuel cell electric vehicle are the clean emissions (only water), they refuel in only a few minutes (similar to a gasoline powered internal combustion engine) and have a driving range of 250 to 300 miles (400 to 480 kilometers, which is also similar to a gasoline powered vehicle). The major challenges for fuel cell electric vehicles are to build a hydrogen fuel infrastructure (for refueling), high manufacturing and maintenance costs and the complexity of the design.

What are the technologies that allow for autonomous vehicles?

Autonomous vehicles, often called self-driving vehicles, rely on three technologies: sensors, connectivity, and software/control algorithms. Sensors are devices that detect and measure changes in the surrounding environment and send a signal to a processing unit. Upon analysis of the change, the processing unit determines the nature of the change and determines what response is required. Based on the feedback, the processing unit then sends a signal to initiate a response to the change. Vehicle sensors exist for brakes, engine, exhaust, transmission, and

The Waymo is a self-driving car created by California-based Google, the company famous for its search-engine technology.

chassis. Some examples of sensors in automotive autonomy are lane-keep assistance, forward-collision warning, and blind-spot monitoring. Autonomous vehicles rely on lidar for many of the sensors. Lidar provides computer-friendly precise measurements. It sends out pulses of laser light millions of times a second, then compiles the results in a point cloud. The point cloud is similar to a 3-D map, with enough detail to spot and identify objects. Once identified, the computer can decide how to react.

Connectivity refers to a vehicle's ability to be aware of the surrounding environment, e.g., traffic conditions, weather and weather-related road conditions, and construction. This information will be processed to determine whether or not an alternate route is required to avoid certain conditions or to anticipate braking.

Finally, the software/control algorithms must reliably and accurately analyze and summarize the data from the sensors and connectivity to make decisions on all aspects of a vehicle's driving—steering, braking, speed, and general route conditions.

AIRPLANES

Which scientific principle allows airplanes to fly?

Bernoulli's principle, named for Swiss mathematician and physicist Daniel Bernoulli (1700–1782), states that an increase in a fluid medium's velocity results in a decrease in pressure. An aircraft's wing is shaped so that the air (a fluid medium) flows faster past the upper surface than the lower surface, thus generating a difference in pressures, causing lift. Centuries later, Bernoulli's principle is still the basis for our understanding of flight and other everyday applications.

What is avionics?

Avionics, a term derived by combining aviation and electronics, describes all of the electronic navigational, communications, and flight-management aids with which airplanes are equipped today. In military aircraft, it also covers electronically controlled weapons, reconnaissance, and detection systems. Until the 1940s, the systems involved in operating aircraft were purely mechanical, electric, or magnetic, with radio apparatus being the most sophisticated instrumentation. The advent of radar and the great advance made in airborne detection during World War II led to the general adoption of electronic distance-measuring and navigational aids. In military aircraft, such devices improve weapon delivery accuracy, and in commercial aircraft, they provide greater safety in operation.

What was the name of the Wright brothers' airplane?

The name of the Wright brothers' plane was the *Wright Flyer*. A wood and fabric biplane, the *Flyer* was originally used by the brothers as a glider and measured 40 feet, 4 inches (12 meters) from wing-tip to wing-tip. For their historic flight, Wilbur (1867–1912) and Orville (1871–1948) Wright outfitted it with a four-cylinder, 12-horsepower gasoline engine and two propellers, all of their own design. On December 17, 1903, at

The *Wright Flyer* is shown here in a photo taken that historic day on December 17, 1903. Orville is piloting and his brother, Wilbur, is running alongside.

Kitty Hawk, North Carolina, Orville Wright made the first engine-powered, heavier-than-air craft flight lying in the middle of the lower wing to pilot the craft, which flew 120 feet (37 meters) in 12 seconds. The brothers made three more flights that day, with Wilbur Wright completing the longest one—852 feet (260 meters) in 59 seconds. The Wright brothers were inducted into the National Inventors Hall of Fame in 1975 for their invention of the airplane and their pioneering work in aviation.

Which aircraft inaugurated commercial passenger air travel?

The Douglas Aircraft Company developed the DC-2 (Douglas Commercial Model Two) in 1933, making its inaugural flight in 1934. The DC-2 was similar to the DC-1 prototype, featuring all-metal construction and retractable landing gear. The fuselage, wings, and engine cowlings were all streamlined. Passenger comfort was considered in the engineering of the aircraft with an effort to lessen the noise from the engines. The DC-2 accommodated fourteen passengers and three crew members. An overnight flight from New York to Los Angeles—with stops in Chicago; Kansas City, Missouri; and Albuquerque, New Mexico—allowed travelers to fly coast to coast without losing a business day. In 1935, the Douglas Aircraft Company received the Collier Trophy for outstanding achievement in the development of a twin-engine transport plane.

Design changes to the DC-2 were started in 1934 resulting in the first flight of the DC-3 in 1935. Regularly scheduled commercial flights on the DC-3 began in 1936 and continued for decades with some still being used today. Many consider the DC-3 one of the greatest airplanes initiating the age of comfortable, reliable, and profitable air travel. One of the improved features of the DC-3 were sleeping berths for overnight flights. The DC-3 accommodated fourteen passengers in sleeping berths or twenty-eight passengers for daytime flights with a crew of three.

Comparison of DC-2 and DC-3

Technical Specification	DC-2	DC-3
First flight	May 11, 1934	December 17, 1935
Wingspan	62 feet (18.9 meters)	95 feet (29 meters)
Length	61 feet 11.75 inches (~19 meters)	64 feet 5.5 inches (19.7 meters)
Height	16 feet 3.75 inches (4.9 meters)	16 feet 3.6 inches (4.9 meters)
Ceiling	22,450 feet (6,843 meters)	20,800 feet (6,340 meters)
Range	1,000 miles (1,609 kilometers)	1,495 miles (2,406 kilometers)
Weight	18,560 pounds (8,419 kilograms)	30,000 pounds (13,608 kilograms)
Engines	Two 875-horsepower Wright Cyclone engines	Two 1,200-horsepower Wright Cyclone radial engines
Speed	200 miles per hour (322 kilometers per hour)	192 miles per hour (309 kilometers per hour)

How does the Boeing Dreamliner aircraft achieve its high fuel efficiency and overall high standard of performance?

The earliest concept and development of the Boeing Dreamliner aircraft began in 2003. Boeing's goal was to develop a super-efficient, mid-sized aircraft. Originally scheduled to be fully in use by 2008, the first 787 Dreamliner was delivered in 2011. The Dreamliner aircraft was designed with advanced aerodynamics, modern engines, composite materials for the structure, and greater concern for passenger comfort. Perhaps the most significant design feature is the use of composite materials, especially in the fuselage and wings. Composite materials are lighter, reducing fuel consumption and increasing aircraft efficiency. They do not fatigue or corrode the same as metallic materials. The Dreamliner is reported to have a twenty percent fuel savings over other passenger jets. In addition, the Dreamliner has 20 to 25 percent fewer carbon dioxide emissions than other aircraft. The passenger experience is enhanced by smart sensors that detect turbulence and adjust to give a smoother ride. Air quality within the cabin, lighting, and noise reduction are controlled to give passengers a more pleasant experience. The 2011 Collier Trophy was awarded to the Boeing Company for the significant advances in the use of materials, technologies, and systems to enhance safety, performance, comfort, and value in a commercial aircraft for the 787 Dreamliner.

Specifications of the Boeing 787 Dreamliner

Specification	787-8	787-9	787–10
Wingspan	197 feet (60 meters)	197 feet (60 meters)	197 feet (60 meters)
Length	186 feet (57 meters)	206 feet (63 meters)	224 feet (68 meters)
Height	56 feet (17 meters)	56 feet (17 meters)	56 feet (17 meters)
Range	7,355 nautical miles (13,620 kilometers)	7,635 nautical miles (14,140 kilometers)	6,430 nautical miles (11,910 kilometers)
Engine	GEnx–1B / Trent 1000	GEnx–1B / Trent 1000	GEnx–1B / Trent 1000
Speed	Mach .85	Mach .85	Mach .85
Passenger capacity	242	290	330

Who made the first supersonic flight?

Supersonic flight is flight at or above the speed of sound. The speed of sound is 760 miles (1,223 kilometers) per hour in warm air at sea level. At a height of about 37,000 feet (11,278 kilometers), its speed is only 660 miles (1,062 kilometers) per hour. The first person credited with reaching the speed of sound (Mach 1) was Major Charles E. (Chuck) Yeager of the U.S. Air Force. In 1947, he attained Mach 1.45 at 60,000 feet (18,288 meters) while flying the Bell X-1 rocket-engine research plane designed by John Stack (1906–1972) and Lawrence Bell (1894–1956). This plane had been carried aloft by a B-29 and released at 30,000 feet (9,144 meters). Based on observational data of how the plane was handling, it is, however, highly likely that the sound barrier was broken on April 9, 1945, by Hans Guido Mutke (1921–2004) in a Messerschmitt Me262, the world's first operational jet aircraft. It is also probable that Chalmers Goodlin (1923–2005) broke the barrier in the Bell X-1 six months prior to Yeager's flight, and the barrier was broken again, by George Welch (1918–1954), in a North American XP-86 Sabre shortly before Yeager's flight. The Me262 and the XP-86 were both jet-powered aircraft. In 1949, the Douglas Skyrocket was credited as the first supersonic jet-powered aircraft to reach Mach 1 when Gene May (1904–1966) flew at Mach 1.03 at 26,000 feet (7,925 meters).

TRAINS

How does MAGLEV technology differ from standard railway design?

MAGLEV (*mag*netic *lev*itation) trains run on a bed of air produced from the repulsion or attraction of powerful magnetic fields (based on the principle that like poles of mag-

> **What is a Mach number?**
>
> A Mach number is the equivalent of the speed of sound. For example, Mach 2 is twice the speed of sound, and Mach 0.5 is half the speed of sound.

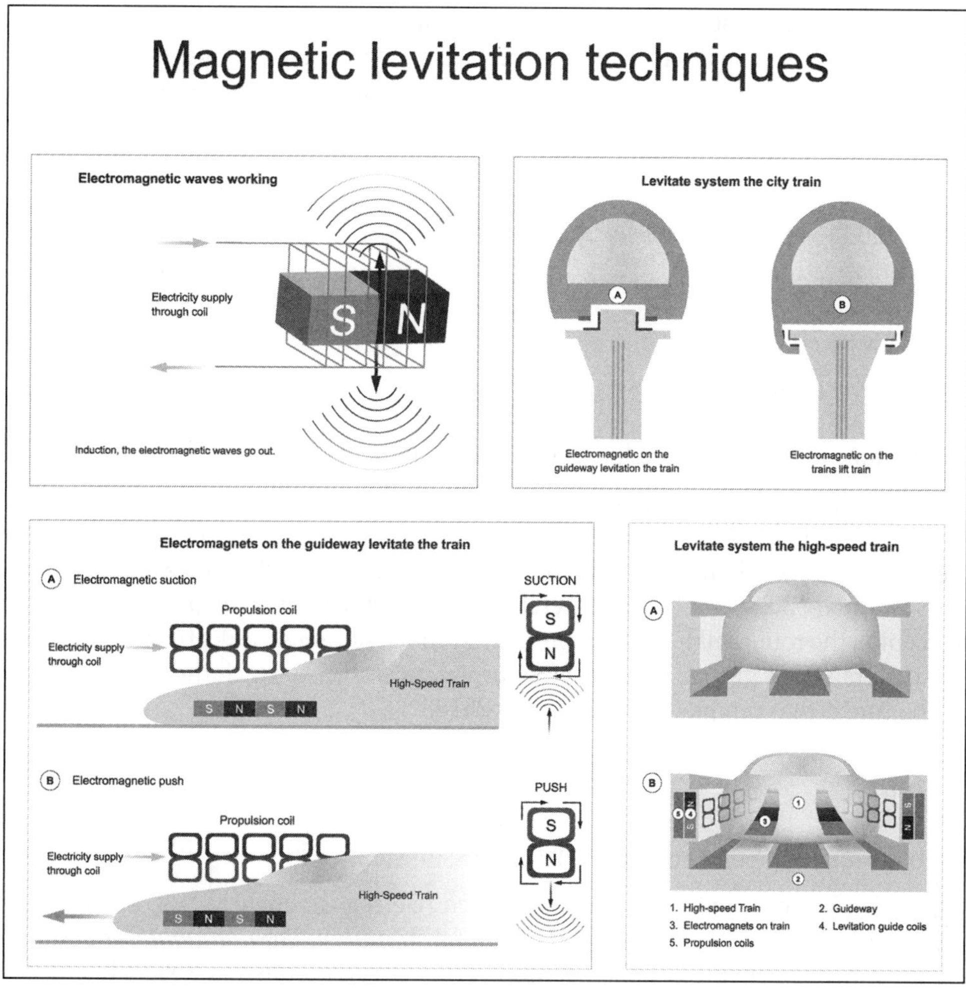

Magnetic levitation trains use magnetic currents to propel the train with great efficiency in large part because the train floats above the track, eliminating friction.

nets repel and unlike poles [north and south] attract). The principle of attraction in magnetism, the employment of winglike flaps extending under the train to fold under a T-shaped guideway, and the use of electromagnets on board (that are attracted to the non-energized magnetic surface) are the guiding components. Interaction between the train's electromagnets and those built on top of the T-shaped track lift the vehicle between 3/8 inch (1 centimeter) and 4 inches (10 centimeters) off the guideway. Another set of magnets along the rail sides provides lateral guidance. The train rides on electromagnetic waves. Alternating current in the magnet sets in the guideway changes their polarity to alternately push and pull the train along. Braking is done by reversing the direction of the magnetic field (caused by reversing the magnetic poles). To increase train speed, the frequency of current is raised.

Systems that use superconducting magnets operate under the same principles of attraction and repulsion as conventional magnets. Superconducting magnets are cooled to less than 450° Fahrenheit below zero (268° Celsius below zero). Once they are cooled, they generate magnetic fields up to ten times stronger than ordinary electromagnets. Since maglev trains do not have any friction against the rails, the ride is very smooth.

When did the first commercially operated, high-speed maglev train begin operation?

Construction of the first commercial maglev railroad began in 2001. It began operations in 2003 between Shanghai International Airport and the outskirts of Shanghai. The 19-mile (30-kilometer) trip takes eight minutes. The highest speed of the train is 267 miles per hour (430 kilometers per hour).

When does Japan anticipate the Linear Chuo Shinkansen line will be operational?

Japan anticipates the Linear Chuo Shinkansen line between Shinagawa Station in Tokyo and Nagoya will begin operating in 2027. This new maglev line will travel the distance of 178 miles (286 kilometers) in 40 minutes. The proposed route will have a minimum curve radius of 3,281 feet (1,000 meters) and a maximum grade of 40 percent. Nearly 86 percent of the route will be underground in a tunnel. At times the depth will be 131 feet (40 meters) below ground. Construction of a vertical shaft to reach the level of the site of the tunnel was completed in November 2018.

SHIPS

Who designed the first commercially successful steamship?

The North River Steamboat of Clermont, often referred to simply as "Clermont", is acknowledged as the first commercially successful steamship. The *Clermont* was designed by the American engineer, Robert Fulton (1765–1815). Fulton and Robert Livingston (1746–1813) demonstrated a 65 foot (20 meter) paddle steamer on the River Seine in Paris in 1803. The steamer had an 8-horsepower engine that was designed in France. The 133-foot (45-meter) *Clermont* had a single-cylinder engine that drove a pair of 15-foot (4.6-meter) paddles, one on each side of the hull. The initial trip from New York to Albany and back, a distance of 240 miles (384 kilometers), took 62 hours, with an average speed of 3.8 miles per hour (6 kilometers per hour). Fulton was inducted into the National Inventors Hall of Fame in 2006 for his invention of the first commercially successful steamboat.

Will autonomous boats and ships become available for transportation?

The development of self-driving, autonomous automobiles has led to the development of automated boats and ships. Maritime design and construction firms are exploring the possibility of ships that are controlled remotely and do not have on-board crews. Potential uses of autonomous boats and ships include ferry boats for canals, tugboats, cargo ships, and remote-controlled container ships to cross the Atlantic and Pacific oceans. Experimental pilot runs have been tested in Boston harbor. Similar to autonomous automobiles, such boats and ships would use sensors to feed information into an artificial intelligence system and output instructions for the boat to maneuver and travel. Safety concerns are already under consideration as this new technology advances.

Fun and Future Travel

Who invented the Segway?

The Segway, the first self-balancing, personal transportation device, was invented by Dean Kamen (1951–). It was first demonstrated to the public in December 2001 but was not available for purchase until November 2002. The Segway is a two-wheeled, motorized scooter that uses a complex array of gyroscopes and computers to mimic the human body's sense of balance. Termed "dynamic stabilization," the scooter combines balancing with maneuverability. To move forward or backward, the rider only has to lean slightly forward or backward. Turning left or right is accomplished by moving the frame to the left or right. The Segway is powered by two rechargeable, phosphate-based, lithium ion batteries. Although it uses energy from the power grid to charge the batteries, it has zero-emissions during operation, making it one of the most environmentally friendly transportation alternatives for short distances. It typically takes eight to ten hours to charge the batteries and 1.04 kilowatt-hour of energy to fully charge the batteries. Depending on the terrain, payload and riding style, the fully charged batteries will permit 16 to 24 miles (26 to 39 kilometers) of travel.

Are there other personal transportation devices besides the Segway?

Since the Segway was introduced in 2001, a variety of other electric personal devices have been introduced as alternatives to automobiles. These include:

- Electric bikes are similar to conventional bicycles but have electric batteries that provide power to assist in pedaling uphill or for easier cruising.
- Electric scooters are devices with two (sometimes three) wheels powered by a rechargeable battery. They can travel up to 20 miles (32 kilometers) on a single charge with speeds of 8 to 25 miles per hour (13 to 40 kilometers per hour). Electric scooters are available in a range of designs from very simple, basic scooters to more sophisticated models with seats.

Introduced in 2001, the Segway is a fun mode of transportation in pedestrian-friendly cities and towns.

- Electric unicycles are one-wheeled devices with pedals on either side of the wheels for the rider to stand on and balance. They use a gyroscope system to help control the ride. The gyroscope system means that as long as the object is spinning—in this case, as long as the wheel is turning—it remains upright. Wheel size, battery size, and overall durability determine the speed and range of distance. Electric unicycles require the rider to use their body to maintain balance.
- Self-balancing scooters (also called the hoverboards) emerged in 2013. They consist of two motorized wheels connected to a pair of pads on which the rider stands. The rider controls the speed by leaning forwards or backwards and the direction by shifting their weight from the left to right foot. The average speed ranges from 5 to 10 miles per hour (8 to 16 kilometers per hour). Most models have a range of 12 miles (19 kilometers) on a full battery charge.

When will laypeople be able to travel to space?

Dennis Tito (1940–), an American businessman, was the first space tourist. He paid an estimated $20 million to travel to space with a Russian crew to spend time on the International Space Station (ISS). He departed aboard the Russian *Soyuz* spacecraft on April 28, 2001, and returned to Earth on May 6, 2001. Commercial companies hope that space tourism becomes a reality within the next several years. Since government funding for space exploration has dwindled, competition from companies in the private sector has sparked innovation and enthusiasm for future space travel. While space tourism may be futuristic, NASA is hoping to contract with private companies to fly astronauts to space. Since the final space shuttle flew in 2011, NASA has relied on Russian spacecraft to fly American astronauts to and from the ISS. Some milestones of commercial space travel are:

- Virgin Galactic had a successful test flight on VSS Unity to space in December 2018. VSS Unity was carried into the air by a larger aircraft from which it was dropped at an altitude of 43,000 feet (13,000 meters). It attained a speed of Mach 2.9 to reach a height of 51.4 miles (82.7 kilometers)—which is space by some definitions—above the surface of Earth. The flight landed fourteen minutes later. Virgin Galactic hopes to carry its first paying space tourists into suborbital space by the end of 2019.
- Blue Origin also hopes to fly passengers to suborbital space aboard its New Shepard rocket system. Both its New Shepard and New Glenn rocket systems are being developed to carry humans and payloads for research purposes.
- SpaceX had a successful six-day, uncrewed test flight of its Crew Dragon spacecraft to the International Space Station in early 2019. The flight carried a dummy astronaut named Ripley to stimulate the effects of the flight on a human astronaut. Crew Dragon was built to carry astronauts for NASA as part of NASA's Commercial Crew Program.
- Boeing is building the CST-100 Starliner spacecraft as part of NASA's Commercial Crew Program. An uncrewed test flight is scheduled for August 2019.

Further Reading

BOOKS

General Science

Bauer, Susan Wise. *The Story of Western Science: From the Writings of Aristotle to the Big Bang Theory*. New York: W.W. Norton & Co., 2015.

Bryson, Bill. *A Short History of Nearly Everything*. New York: Broadway Books, 2003.

Dean, Cornelia. *Making Sense of Science; Separating Substance from Spin*. Cambridge, MA: The Belknap Press of Harvard University Press, 2017.

Fara, Patricia. *Science: A Four Thousand Year History*. Oxford; New York: Oxford University Press, 2009.

Gribben, John R. *The Scientists: A History of Science Told through the Lives of Its Greatest Inventors*. New York: Random House, 2003.

Henry, John. *A Short History of Scientific Thought*. New York: Palgrave Macmillan, 2012.

McGraw-Hill Encyclopedia of Science and Technology. 11th edition. New York: McGraw-Hill, 2012.

Wiggins, Arthur W. *The Five Biggest Unsolved Problems in Science*. Hoboken, NJ: J. Wiley & Sons, 2003.

Wilson, Edward O. *Letters to a Young Scientist*. New York: Liveright Publishing Corporation, 2013.

Mathematics

Barrow, John D. *100 Essential Things You Didn't Know You Didn't Know: Math Explains Your World*. New York: W. W. Norton & Co., Inc., 2009.

Beckmann. Petr. *A History of Pi*. New York: Dorset Press, 1989.

Boyer, Carl B. and Uta Merzbach. *A History of Mathematics*, 3rd ed. Hoboken, NJ: J. Wiley & Sons, 2010.

Cheng, Eugenia. *Beyond Infinity: An Expedition to the Outer Limits of Mathematics*. New York: Basic Books, 2017.

The New York Times Book of Mathematics. Edited by Gina Kolata. New York: Sterling, 2013.

Posamentier, Alfred S. Robert Geretschläger, Charles Li, and Christian Spreitzer. *The Joy of Mathematics: Marvels, Novelties, and Neglected Gems That Are Rarely Taught in Math Class*. Amherst, NY: Prometheus Books, 2017.

Shetterly, Margo Lee. *Hidden Figures: The Story of the Black Women Mathematicians Who Helped Win the Space Race*. New York: William Morrow, 2016.

Smith, Gary. *What the Luck? The Surprising Role of Chance in Our Everyday Lives*. New York: The Overlook Press, 2016.

Stewart, Ian. *Significant Figures: The Lives and Work of Great Mathematicians*. New York: Basic Books, 2017.

Physics

Baggott, Jim. *The Quantum Story: A History in 40 Moments*. Oxford; New York: Oxford University Press, 2011.

Butterworth, Jon. *Atom Land: A Guided Tour through the Strange (and Impossibly Small) World of Particle Physics*. New York: The Experiment, 2018.

Czerski, Helen. *Storm in a Teacup: The Physics of Everyday Life*. New York: W. W. Norton & Co., 2017.

Dyson, Freeman. *Maker of Patterns: An Autobiography through Letters*. New York: Liveright Publishing Corp. 2018.

Einstein, Albert, and Leopold Infeld. *The Evolution of Physics from Early Concepts to Relativity and Quanta*. New York: Simon & Schuster, 1960 (1938).

Feynman, Richard Phillips. *Six Easy Pieces: Essentials of Physics Explained by Its Most Brilliant Teacher*. New York: Basic Books, 2011.

Ford, Kenneth William. *101 Quantum Questions: What You Need to Know about the World You Can't See*. Cambridge, MA: Harvard University Press, 2011.

Gleick, James. *Chaos: Making a New Science*. New York: Viking, 1987.

Hawking, Stephen. *A Brief History of Time*. New York: Bantam Books, 1988.

Hawking, Stephen, and Leonard Mlodinow. *The Grand Design*. New York: Bantam Books, 2010.

Holt, Jim. *When Einstein Walked with Gödel: Excursions to the Edge of Thought*. New York: Farraf, Straus, & Giroux, 2018.

Hossenfelder, Sabine. *Lost in Math: How Beauty Leads Physics Astray*. New York: Basic Books, 2018.

Kakalios, James. *The Physics of Everyday Things: The Extraordinary Science Behind an Ordinary Day*. New York: Crown Publishers, 2017.

Krauss, Lawrence M. *The Greatest Story Ever Told—So Far: Why Are We Here?* New York: Atria Books, 2017.

Lincoln, Don. *The Large Hadron Collider: The Extraordinary Story of the Higgs Boson and Other Stuff That Will Blow Your Mind*. Baltimore, MD: Johns Hopkins University Press, 2014.

Chemistry

Atkins, P. W. *What Is Chemistry?* Oxford and New York: Oxford University Press, 2013.

Brock, William Hodson. *The Chemical Tree: A History of Chemistry*. New York: W. W. Norton & Co., 2000.

Chaline, Eric. *Fifty Minerals That Changed the Course of History*. Buffalo, NY: Firefly Books, 2012.

Challoner, Jack. *The Elements: The New Guide to the Building Blocks of Our Universe*. New York: Sterling, 2012.

Cobb, Cathy, Monty Fetterolf, and Harold Goldwhite. *The Chemistry of Alchemy: From Dragon's Blood to Donkey Dung, How Chemistry Was Forged*. Amherst, NY: Prometheus Books, 2014.

Coffey, Patrick. *Cathedrals of Science: The Personalities and Rivalries That Made Modern Chemistry*. Oxford and New York: Oxford University Press, 2008.

Faraday, Michael. *The Chemical History of a Candle*. Oxford; New York: Oxford University Press, 2011.

Farmer, Steven C. *Strange Chemistry: The Stories Your Chemistry Wouldn't Tell You*. Hoboken, NJ: John Wiley & Sons, Inc., 2017.

Fontani, Marco. *The Lost Elements: The Periodic Table's Shadow Side*. New York: Oxford University Press, 2015.

Garfield, Simon. *Mauve: How One Man Invented a Color That Changed the World*. New York: W. W. Norton & Co., 2001.

Hessen, Dag Olav. *The Many Lives of Carbon*. London, UK: Reaktion Books, Ltd. 2017.

Öhrström, Lars. *The Last Alchemist in Paris & Other Curious Tales from Chemistry*. Oxford: Oxford University Press, 2013.

Astronomy and Space

Baron, David. *American Eclipse: A Nation's Epic Race to Catch the Shadow of the Moon and Win the Glory of the World*. New York: Liveright Publishing Corporation, 2017.

Brown, Mike. *How I Killed Pluto and Why It Had It Coming*. New York: Spiegel & Grau, 2010.

Cruikshank, Dale P., and William Sheehan. *Discovering Pluto: Exploration at the Edge of the Solar System*. Tucson: The University of Arizona Press, 2018.

Forshaw, Jeff. *Universal: A Guide to the Cosmos*. Boston, MA: Da Capo Press, 2017.

Preston, Louisa. *Goldilocks and the Water Bears: The Search for Life in the Universe*. New York: St. Martin's Press, 2016.

Ridpath, Ian. *Stars & Planets: The Complete Guide to the Stars, Constellations, and the Solar System*. Princeton, NJ: Princeton University Press, 2017.

Sobel, Dava. *The Glass Universe: How the Ladies of the Harvard Observatory Took the Measure of the Stars*. New York: Viking, 2016.

Stern, Alan. *Chasing New Horizons: Inside the Epic First Mission to Pluto*. New York: Picador, 2018.

Summers, Michael E., and James Trefil. *Exoplanets: Diamond Worlds, Super Earths, Pulsar Planets, and the New Search for Life Beyond Our Solar System*. Washington, DC: Smithsonian Books, 2017.

Trefil, James. *Space Atlas: Mapping the Universe and Beyond*, 2nd edition. Washington, DC: National Geographic, 2018.

Tyson, Neil DeGrasse. *Astrophysics for People in a Hurry*. New York: W. W. Norton & Co., Inc., 2017.

Geology and Earth Science

Alvarez, Walter. *A Most Improbable Journey: A Big History of Our Planet and Ourselves*. New York: W. W. Norton & Co., 2017.

Bonewitz, Ronald. *Rocks and Minerals*. New York: DK Publishing, 2012.

Brusatte, Stephen. *The Rise and Fall of the Dinosaurs: A New History of a Lost World*. New York: William Morrow, 2018.

Carson, Rachel. *The Sea Around Us*. New York: Oxford University Press, 2003.

Eldredge, Niles. *Extinction and Evolution What Fossils Reveal about the History of Life*, Buffalo, NY: Firefly Books, 2014.

Fortey, Richard A. *Earth: An Intimate History*. New York: Alfred A. Knopf, 2004.

Goodell, Jeff. *The Water Will Come: Rising Seas, Sinking Cities, and the Remaking of the Civilized World*. New York: Little Brown & Co., 2017.

Harlow, George E., and Anna S. Sofianides. *Gems & Crystals: From One of the World's Greatest Collections*. Revised edition. New York: Sterling, 2015.

Luhr, James F., and Jeffrey E. Post, eds. *Earth: The Definitive Visual Guide*. Revised and updated edition. New York: DK Publishing, 2013.

Miles, Kathryn. *Quakeland: On the Road to America's Next Devastating Earthquake*. New York: Dutton, 2017.

O'Hara, Kieran D. *A Brief History of Geology*. Cambridge, UK: Cambridge University Press, 2018.

Oppenheimer, Clive. *Eruptions That Shook the World*. New York: Cambridge University Press, 2011.

Paul, Gregory S. *The Princeton Field Guide to Dinosaurs*, 2nd ed. Princeton, NJ: Princeton University Press, 2016.

Pim, Keiron. *Dinosaurs the Grand Tour: Everything Worth Knowing about Dinosaurs from Aardonyx to Zuniceratops*. New York: The Experiment, 2014.

Prothero, Donald R. *The Story of Life in 25 Fossils: Tales of Intrepid Fossil Hunters and the Wonders of Evolution*. New York: Columbia University Press, 2015.

———. *The Story of the Earth in 25 Rocks: Tales of Important Geological Puzzles and the People Who Solved Them*. New York: Columbia University Press, 2018.

Roberts, Callum. *The Ocean of Life: The Fate of Man and the Sea*. New York: Viking, 2012.

Rohling, Eelco. *The Oceans: A Deep History*. Princeton, NJ: Princeton University Press, 2017.

Stager, Curt. *Still Waters: The Secret World of Lakes*. New York: W. W. Norton & Co., 2018.

Taylor, Paul D., and Aaron O'Dea. *A History of Life in 100 Fossils*. Washington, DC: Smithsonian Books, 2014.

Winchester, Simon. *Krakatoa: The Day the World Exploded, August 27, 1883*. New York: HarperCollins Publishers, 2003.

Meteorology and Climatology

Barnett, Cynthia. *Rain: A Natural and Cultural History*. New York: Crown Publishers, 2015.

Buckley, Bruce, Edward J. Hopkins, and Richard Whitaker. *Weather: A Visual Guide*. Buffalo, NY: Firefly Books, 2004.

Cullen, Heidi. *The Weather of the Future: Heat Waves, Extreme Storms, and Other Scenes from a Climate-changed Planet*. New York: HarperCollins Publishers, 2010.

Global Weirdness: Severe Storms, Deadly Heat Waves, Relentless Drought, Rising Seas, and the Weather of the Future. New York: Pantheon Books, 2012.

Logan, William Bryant. *Air: The Restless Shaper of the World*. New York: W. W. Norton & Co., 2012.

Marshall, George. *Don't Even Think about It: Why Our Brains Are Wired to Ignore Climate Change*. New York: Bloomsbury Publishing, 2014.

McGuire, Bill. *Waking the Giant: How a Changing Climate Triggers Earthquakes, Tsunamis, and Volcanoes*. Oxford; New York: Oxford University Press, 2012.

Miles, Kathryn. *Superstorm: Nine Days Inside Hurricane Sandy*. New York: Dutton, 2014.

Nuccitelli, Dana. *Climatology Versus Pseudoscience: Exposing the Failed Predictions of Global Warming Skeptics*. Santa Barbara, CA: Praeger, 2015.

Revkin, Andrew. *Weather: An Illustrated History: From Clouds Atlases to Climate Change*. New York: Sterling, 2018.

Weart, Spencer R. *The Discovery of Global Warming*. Revised and expanded ed. Cambridge, MA: Harvard University Press, 2008.

Williams, Jack. *The AMS Weather: The Ultimate Guide to America's Weather*. Chicago: University of Chicago Press, 2009.

Biology

Anton, Ted. *Planet of Microbes the Perils and Potential of Earth's Essential Life Forms*. Chicago: The University of Chicago Press, 2017.

Bone, Eugenia. *Microbia: A Journey into the Unseen World around You*. New York: Rodale, 2018.

Fuller, Randall. *The Book that Changed America: How Darwin's Theory of Evolution Ignited a Nation*. New York: Viking, 2017.

Gerald, Michael C. *The Biology Book: From the Origin of Life to Epigenetics, 250 Milestones in the History of Biology*. New York: Sterling Publishing Co., 2015.

Harris, Henry. *The Birth of the Cell*. New Haven, CT: Yale University Press, 1999.

Hudler, George W. *Magical Mushrooms, Mischievous Molds*. Princeton, NJ: Princeton University Press, 1998.

Margulis, Lynn. *Five Kingdoms: An Illustrated Guide to the Phyla of Life on Earth,* 3rd edition. New York: W. H. Freeman, 1998.

———. *What Is Life?* Berkeley, CA: University of California Press, 2000.

Roossinck, Marilyn J. *Virus: An Illustrated Guide to 101 Incredible Microbes*. Princeton, NJ: Princeton University Press, 2016.

Smoller, Jordan. *The Other Side of Normal: How Biology Is Providing the Clues to Unlock the Secrets of Normal and Abnormal Behavior*. New York: HarperCollins, 2012.

Thomas, Lewis. *The Lives of a Cell; Notes of a Biology Watcher*. New York: Viking Press, 1974.

Toomey, David. *Weird Life: The Search for Life That Is Very, Very Different from Our Own*. New York: W. W. Norton & Co., 2013.

Wilson, Edward O. *The Social Conquest of Earth*. New York: W. W. Norton & Co., 2012.

Young, Ed. *I Contain Multitudes: The Microbes within Us and a Grander View of Life*. New York: Ecco, 2016.

Genetics

Browne, Janet. *Darwin's Origin of Species: A Biography*. New York: Atlantic Monthly Press, 2006.

Cobb, Matthew. *Life's Greatest Secret: The Race to Crack the Genetic Code*. New York: Basic Books, 2015.

Cockell, Charles. *The Equations of Life: How Physics Shapes Evolution*. New York: Basic Books, 2018.

Costa, James T. *Darwin's Backyard: How Small Experiments Led to a Big Theory*. New York: W. W. Norton & Co., 2017.

Coyne, Jerry A. *Why Evolution Is True*. New York: Viking, 2009.

Darwin, Charles. *The Origin of Species by Means of Natural Selection; or, The Preservation of Favored Races in the Struggle for Life*. New York: Modern Library, 1993.

Dawkins, Richard. *The Selfish Gene*. Oxford; New York: Oxford University Press, 2006.

Doudna, Jennifer A., and Samuel H. Sternberg. *A Crack in Creation Gene Editing and the Unthinkable Power to Control Evolution*. Boston: Houghton Mifflin Harcourt, 2017.

Grasset, Léo. *How the Zebra Got Its Stripes: Darwinian Stories Told through Evolutionary Biology*. New York: Pegasus Books, 2017.

Heine, Steven J. *DNA Is Not Destiny: The Remarkable, Completely Misunderstood Relationship between You and Your Genes*. New York: W. W. Norton & Co., 2017.

Henig, Robin Marantz. *The Monk in the Garden: The Lost and Found Genius of Gregor Mendel, Father of Genetics*. Boston: Houghton Mifflin, 2000.

Kean, Sam. *The Violinist's Thumb: And Other Lost Tales of Love, War, and Genius as Written by Our Genetic Code*. New York: Little, Brown & Co., 2012.

Laron, Edward J. *Evolution: The Remarkable History of a Scientific Theory*. New York: Modern Library, 2004.

Lister, Adrian. *Darwin's Fossils: The Collection that Shaped the Theory of Evolution*. Washington, DC: Smithsonian Books, 2018.

Parrington, John. *Redesigning Life: How Genome Editing Will Transform the World*. Oxford, UK: Oxford University Press, 2016.

Rheinberger, Hans-Jörg. *The Gene: From Genetics to Postgenomics*. Chicago: University of Chicago Press, 2017.

Sulston, John, and Georgina Ferry. *The Common Thread: A Story of Science, Politics, Ethics, and the Human Genome*. Washington, DC: Joseph Henry Press, 2002.

Watson, James D. *DNA: The Story of the Genetic Revolution*. Revised and updated ed. New York: Alfred A. Knopf, 2017.

———. *The Double Helix: A Personal Account of the Discovery of the Structure of DNA*. New York: Atheneum, 1968.

Wilson, David Sloan. *Evolution for Everyone: How Darwin's Theory Can Change the Way We Think about Our Lives*. New York: Delacorte Press, 2007.

Zimmer, Carl. *She Has Her Mother's Laugh: The Powers, Perversion, and Potential of Heredity*. New York: Dutton, 2018.

Botany

Capon, Brian. *Botany for Gardeners*. Portland, OR: Timber Press, 2010.

Bateman, Helen, and others, eds. *Edible: An Illustrated Guide to the World's Food Plants*. Washington, DC: National Geographic, 2008.

Bayton, Ross, and Simon Maughan. *Plant Families: A Guide for Gardeners and Botanists*. Chicago: The University of Chicago Press, 2017.

Flora: Inside the Secret World of Plants. New York: DK Publishing, 2018.

Frey, Kate. *The Bee-friendly Garden: Design and Abundant, Flower-filled Yard That Nurtures Bees and Supports Biodiversity*. Berkeley, CA: Ten Speed Press, 2016.

Hanson, Thor. *The Triumph of Seeds: How Grains, Nuts, Kernels, Pulses, and Pips, Conquered the Plant Kingdom and Shaped Human History*. New York: Basic Books, 2015.

Howell, Catherine Herbert. *Flora Mirabilis: How Plants Have Shaped World Knowledge, Health, Wealth, and Beauty*. Washington, DC: National Geographic, 2009.

Kassinger, Ruth. *A Garden of Marvels: How We Discovered That Flowers Have Sex, Leaves Eat Air, and Other Secrets of Plants*. New York: William Morrow, 2014.

Kingsbury, Noël. *Garden Flora: The Natural and Cultural History of the Plants in Your Garden*. Portland, OR: Timber Press, 2016.

———. *The Glory of the Tree: An Illustrated History*. Buffalo, NY: Firefly Books, 2014.

Largo, Michael. *The Big, Bad Book of Botany*. New York: William Morrow, 2014.

Laws, Bill. *Fifty Plants That Changed the Course of History*. Buffalo, NY: Firefly Books, 2010.

Lincoff, Gary. *The Complete Mushroom Hunter: An Illustrated Guide to Foraging, Harvesting, and Enjoying Wild Mushrooms: Including New Sections on Growing Your Own Edibles and Off-Season Collecting*. Beverly, MA: Quarry Books, 2017.

Mabey, Richard. *The Cabaret of Plants: Forty Thousand Years of Plant Life and the Human Imagination*. New York: W. W. Norton & Co., 2016.

Stewart, Amy. *The Drunken Botanist: The Plants That Create the World's Great Drinks*. Chapel Hill, NC: Algonquin Books of Chapel Hill, 2013.

Wohlleben, Peter. *The Hidden Life of Trees: What They Feel, How They Communicate: Discoveries from a Secret World*. Vancouver, BC: David Suzuki Institute/Greystone Books, 2016.

ZOOLOGY

Ackerman, Jenifer. *The Genius of Birds*. New York: Penguin Press, 2016.

Attenborough, David. *Life in the Undergrowth*. Princeton, NJ: Princeton University Press, 2006.

———. *Life in Cold Blood*. Princeton, NJ: Princeton University Press, 2008.

Balcombe, Jonathan P. *What a Fish Knows: The Inner Lives of Our Underwater Cousins*. New York: Farrar, Straus, & Giroux, 2016.

Berwald, Juli. *Spineless: The Science of Jellyfish and the Art of Growing a Backbone*. New York: Riverhead Books, 2017.

Birkhead, Tim. *Bird Sense: What It's Like to Be a Bird*. New York: Walker & Co., 2012.

Blakeslee, Nate. *American Wolf: A True Story of Survival and Obsession in the West*. New York: Crown Publishers, 2017.

Chadwick, Fergus, Steve Alton, Emma Sarah Tennant, Bill Fitzmaurice, and Judy Earl. *The Bee Book*. New York: DK Publishing, 2016.

Chaline, Eric. *Fifty Animals That Changed the Course of History*. Richmond Hill, ON: Firefly Books, 2011.

Eisner, Thomas, Maria Eisner, and Melody Siegler. *Secret Weapons: Defenses of Insects, Spiders, Scorpions, and Other Many-legged Creatures*. Cambridge, MA: Belknap Press of Harvard University Press, 2005.

Fossey, Dian. *Gorillas in the Mist*. Boston, MA: Houghton Mifflin, 1983.

Goodall, Jane. *In the Shadow of Man*. Boston: Houghton Mifflin, 1971.

Hanson, Thor. *Buzz: The Nature and Necessity of Bees*. New York: Basic Books, 2018.

Illustrated Atlas of Wildlife. Berkeley, CA: University of California Press, 2009.

Jones, Richard. *House Guests, House Pests: A Natural History of Animals in the Home*. New York: St. Martin's Press, 2015.

Lederer, Roger J. *Beaks, Bones, and Bird Songs: How the Struggle for Survival Has Shaped Birds and Their Behavior*. Portland, OR: Timber Press, 2016.

MacNeal, David. *Bugged: The Insects Who Rule the World and the People Obsessed with Them*. New York: St. Martin's Press, 2017.

Moore, Robin. *In Search of Lost Frogs: The Campaign to Discover the World's Rarest Amphibians*. Buffalo, NY: Firefly Books, 2014.

Strycker, Noah K. *Birding without Borders: An Obsession, a Quest, and the Biggest Year in the World*. Boston: Houghton Mifflin Harcourt, 2017.

Waal, F. B. M. de. *Are We Smart Enough to Know How Smart Animals Are?* New York: W. W. Norton & Co., Inc., 2016.

Anatomy and Physiology

Berns, Gregory. *What It's Like to Be a Dog: And Other Adventures in Animal Neuroscience*. New York: Basic Books, 2017.

Cronin, Thomas W., Sönke Johnsen, N. Justin Marshall, and Eric J. Warrant. *Visual Ecology*. Princeton, NJ: Princeton University Press, 2014.

Denny, Mark. *Engineering Animals: How Life Works*. Cambridge, MA: Belknap Press of Harvard University, 2011.

Dowling, John E. *Understanding the Brain: From Cells to Behavior to Cognition*. New York: W. W. Norton & Co., 2018.

Dowling, John E., and Joseph L. Dowling, Jr. *Vision: How It Works and What Can Go Wrong*. Cambridge, MA: The MIT Press, 2016.

Richardson, Ruth. *The Making of Mr. Gray's Anatomy*. Oxford; New York: Oxford University Press, 2008.

Rinzler, Carol Ann. *Leonardo's Foot: How 10 Toes, 52 Bones, and 66 Muscles Shaped the Human World*. New York: Bellevue Literary Press, 2013.

Wilkinson, Matt. *Restless Creatures: The Story of Life in Ten Movements*. New York: Basic Books, 2015.

Ecology

Anthony, Leslie. *The Aliens among Us: How Invasive Species are Transforming the Planet—and Ourselves*. New Haven, CT: Yale University Press, 2017.

Bedford, Daniel, and John Cook. *Climate Change: Examining the Facts*. Santa Barbara, CA: ABC-CLIO, 2016.

Burdick, Alan. *Out of Eden: An Odyssey of Ecological Invasion*. New York: Farrar, Straus, & Giroux, 2005.

Carson, Rachel. *Silent Spring*. Boston: Houghton Mifflin, 2002.

Charman, Isobel. *The Zoo: The Wild and Wonderful Tale of the Founding of London Zoo, 1826–1851*. New York: Pegasus Books, 2017.

Egan, Dan. *The Death and Life of the Great Lakes*. New York: W. W. Norton & Co., 2017.

Hamilton, Garry. *Super Species: The Creatures That Will Dominate the Planet*. Buffalo, NY: Firefly Books, 2010.

Lovelock, James. *The Ages of Gaia: A Biography of Our Living Earth*. New York: W. W. Norton & Co., 1988.

Montgomery, David R. *Growing a Revolution: Bringing Our Soil Back to Life*. New York: W. W. Norton & Co., 2017.

Muir, John. *Nature Writings*. New York: The Library of America, 1997.

Nielsen, Larry A. *Nature's Allies: Eight Conservationists Who Changed Our World*. Washington, DC: Island Press, 2017.

Paige, Embry. *Our Native Bees North America's Endangered Pollinators and the Fight to Save Them*. Portland, OR: Timber Press, 2018.

Schilthuizen, Menno. *Darwin Comes to Town: How the Urban Jungle Drives Evolution*. New York: Picador, 2018.

Suzuki, David T., and Ian Hanington. *Just Cool It!: The Climate Crisis and What We Can Do: A Post-Paris Agreement Game Plan*. Vancouver, BC: Greystone Books, 2017.

Vince, Gaia. *Adventures in the Anthropocene: A Journey to the Heart of the Planet We Made*. Minneapolis, MN: Milkweed Editions, 2014.

Wills, Christopher. *Green Equilibrium: The Vital Balance of Humans and Nature*. Oxford and New York: Oxford University Press, 2013.

Woodhouse, Keith Mako. *The Ecocentrists: A History of Radical Environmentalism*. New York: Columbia University Press, 2018.

Applied Science and Technology

Beiser, Vince. *The World in a Grain: The Story of Sand and How It Transformed Civilization*. New York: Riverhead Books, 2018.

Brain, Marshall. *The Engineering Book: From the Catapult to the Curiosity Rover: 250 Milestones in the History of Engineering*. New York: Sterling, 2015.

Burns, Lawrence D. *Autonomy: The Quest to Build the Driverless Car—and How It Will Reshape Our World*. New York: Ecco, 2018.

Day, Mark Stuart. *Bits to Bitcoin: How Our Digital Stuff Works*. Cambridge, MA: MIT Press, 2018.

Dougherty, Martin J. *Drones: An Illustrated Guide to the Unmanned Aircraft That Are Filling Our Skies*. New York: Metro Books, 2015.

Edwards, David A. *Creating Things That Matter*. New York: Henry Holt and Company, 2018.

Fry, Hannah. *Hello World: Being Human in the Age of Algorithms*. New York: W. W. Norton & Co., 2018.

Fisher, Adam. *Valley of Genius: The Uncensored History of Silicon Valley, as Told by the Hackers, Founders, and Freaks Who Made It Boom*. New York: Twelve, 2018.

Gilder, George F. *Life after Google: The Fall of Big Data and the Rise of the Blockchain Economy*. Washington, DC: Regnery Gateway, 2018.

Hermann, Andreas, Walter Brenner, and Rupert Stadler. *Autonomous Driving: How the Driverless Revolution Will Change the World*. Bingley, UK: Emerald Publishing, 2018.

Kernighan, Brian W. *Understanding the Digital World: What You Need to Know about Computers, the Internet, Privacy, and Security*. Princeton, NJ: Princeton University Press, 2017.

Panetta, Karen. *Count Girls In: Empowering Girls to Combine Any Interests with STEM to Open Up a World of Opportunity*. Chicago: Chicago Review Press, 2018.

Winchester, Simon. *The Perfectionists: How Precision Engineers Created the Modern World*. New York: Harper, 2018.

PERIODICALS

Air and Space Smithsonian

American Biology Teacher

American Journal of Science

American Scientist

Astronomy
Audubon Magazine
Aviation Week and Space Technology
Biocycle
Bioscience
Daedalus: Proceedings of the American Academy of Arts and Sciences
Discover
Environment
FDA Consumer
Horticulture
Issues in Science and Technology
National Geographic
National Wildlife
Natural History
Nature
New Scientist
Physics Today
Popular Science
Science
Science News
The Science Teacher
Scientific American
Sierra
Sky and Telescope
Smithsonian

WEBSITES

American Association for the Advancement of Science: https://www.aaas.org
American Astronomical Society: https://aas.org
American Chemical Society: https://www.acs.org
American Mathematical Society: https://www.ams.org
American Museum of Natural History: https://www.amhn.org
American Physical Society: https://www.aps.org
Animal Diversity Web: https://animaldiversity.org
Botanical Society of America: https://www.botany.org

Cells Alive!: https://www.cellsalive.com
International Union for Conservation of Nature: https://www.iucn.org
International Union of Pure and Applied Chemistry: https://iupac.org
Lady Bird Johnson Wildflower Center: https://www.wildflower.org
LiveScience: https://www.livescience.com
MadSci Network: www.madsci.org
National Academies of Sciences Engineering Medicine: www.nationalacademies.org
National Aeronautics and Space Administration (NASA): https://nasa.gov
National Audubon Society: https://www.audubon.org
National Center for Biotechnology Information: https://www.ncbi.nlm.nih.gov
National Oceanic and Atmospheric Administration: https://www.noaa.gov
National Science Foundation: https://nsf.gov
National Weather Service: https://www.weather.gov
Plants Database: https://www.plants.usda.gov
Rainforest Action Network: https://www.ran.org
Science—How Stuff Works: https://science.howstuffworks.com
Sierra Club: https://www.sierraclub.org
Smithsonian Institution: https://www.si.edu
U.S. Department of Agriculture: https://www.usda.gov
U.S. Department of Energy (DOE): https://www.energy.gov
U.S. Environmental Protection Agency (EPA): https://www.epa.gov
U.S. Fish and Wildlife Service (FW): https://www.fws.gov
U.S. Geological Survey (USGS): https://usgs.gov
WWF: www.panda.org

Index

Note: (ill.) indicates photos and illustrations.

NUMBERS

A

B

C

D

E

F

G

H

J

K

L

M

N

O

P

Q

R

S

T

U

V

W

X–Y

Z